D1673965

Edited by
Anne Skaja Robinson

Production of Membrane Proteins

Related Titles

Becker, CFW

Chemical Biology of Membrane Proteins

432 pages
Hardcover
ISBN: 978-0-470-87194-2

Carta, G., Jungbauer, A.

Protein Chromatography
Process Development and Scale-Up

364 pages with 178 figures and 45 tables
2010
Hardcover
ISBN: 978-3-527-31819-3

Behme, S.

Manufacturing of Pharmaceutical Proteins
From Technology to Economy

404 pages with 136 figures and 41 tables
2009
Hardcover
ISBN: 978-3-527-32444-6

Spirin, A. S., Swartz, J. R. (eds.)

Cell-free Protein Synthesis
Methods and Protocols

262 pages with 93 figures and 14 tables
2008
Hardcover
ISBN: 978-3-527-31649-6

Walsh, G.

Pharmaceutical Biotechnology
Concepts and Applications

498 pages
2007
Softcover
ISBN: 978-0-470-01245-1

Edited by Anne Skaja Robinson

Production of Membrane Proteins

Strategies for Expression and Isolation

WILEY-VCH Verlag GmbH & Co. KGaA

The Editor

Prof. Anne Skaja Robinson
University of Delaware
Dept. of Biochemical Eng.
150 Academy St.
Newark, DE 19716
USA

We would like to thank Dr. David Salmon and Dr. Krzysztof Palczewski (Polgenix, Inc.) for providing us with the graphic material used in the cover illustration.

Library of Congress Card No.: applied for

British Library Cataloguing-in-Publication Data
A catalogue record for this book is available from the British Library.

Bibliographic information published by the Deutsche Nationalbibliothek
The Deutsche Nationalbibliothek lists this publication in the Deutsche Nationalbibliografie; detailed bibliographic data are available on the Internet at <http://dnb.d-nb.de>.

Cover: Formgeber, Eppelheim
Typesetting: Toppan Best-set Premedia Limited, Hong Kong
Printing and Binding: Fabulous Printers Pte Ltd, Singapore

Printed in Singapore
Printed on acid-free paper

ISBN: 978-3-527-32729-4
ePDF ISBN: 978-3-527-63454-5
oBook ISBN: 978-3-527-63452-1
ePub ISBN: 978-3-527-63453-8
Mobi ISBN: 978-3-527-63455-2

Contents

Preface

With this volume, we have attempted to provide a guide for those interested in structural biology and biochemistry of membrane proteins. It is our hope that this text will be useful both to experts and to those new to the field. The various chapters illustrate the breadth and depth of approaches to, and the challenges associated with, membrane protein production and solubilization. Our goals were to compare and critically analyze methods that have been used successfully for the expression and isolation of membrane proteins, to identify and help disseminate emerging methods, and to provide a practical reference for those working with these proteins.

Interest in membrane proteins has grown substantially in the last 25 years, in part due to the recognition of their importance in major cellular functions, including signaling, transport, and adhesion. As a result of their roles, membrane proteins comprise 30–60% of all drug discovery efforts, but remain a challenge for structure-based discovery efforts because of the paucity of high-resolution structures.

As of the time of writing, about 260 unique high-resolution membrane protein structures have been solved, with an exponential growth of new structures (e.g., 39 unique structures were determined in 2009). Several centers and focused funding efforts, such as the Protein Structure Initiative, are aimed at accelerating the membrane structure rate, although it is not yet clear whether these approaches will increase the pace to equal that of soluble proteins (see http://blanco.biomol.uci.edu/MP_Structure_Progress.html). These initiatives by federal funding agencies and private organizations have yielded new technologies and creative approaches for the production and stabilization of membrane proteins, many of which are described in this volume.

In this collection, the Introduction gives an overview of the challenges and successes of structural biology of membrane proteins. Chapters 1–6 outline best practices for expression in both prokaryotic and eukaryotic hosts. Chapters 7–9 focus on aspects that are specific for individual protein classes and applications. Chapters 10–14 highlight new approaches to producing and working with membrane proteins.

Of course, the credit for this book goes to the outstanding array of chapter authors. On their behalf, I thank Stefanie Volk, the Project Editor at Wiley-VCH, for her support and encouragement, Frank Weinreich at Wiley-VCH for initiating this project, and Rechilda Alba at the University of Delaware for technical assistance.

December 2010 *Anne Skaja Robinson*

List of Contributors

Fatima Alkhalfioui
IREBS Institute
CNRS and University of Strasbourg
Pain and GPCRs Research Group
Boulevard Sébastien Brant
67412 Illkirch
France

Brian J. Bahnson
University of Delaware
Department of Chemistry &
Biochemistry
Newark, DE 19716
USA

Rebecca Batchelor
University of Hull
Department of Chemistry
Cottingham Road
Kingston-upon-Hull HU6 7RX
UK

Michael J. Betenbaugh
Johns Hopkins University
Department of Chemical and
Biomolecular Engineering
221 Maryland Hall
3400 North Charles Street
Baltimore, MD 21218
USA

Olivier Bornert
IREBS Institute
CNRS and University of Strasbourg
Pain and GPCRs Research Group
Boulevard Sébastien Brant
67412 Illkirch
France

Zachary Britton
University of Delaware
Department of Chemical Engineering,
Colburn Laboratory
150 Academy Street
Newark, DE 19716
USA

Özge Can
Acibadem University
School of Medicine
Gulsuyu Mahallesi
Fevzi Cakmak Caddesi
Divan Sokak, No: 1
Maltepe, Istanbul
Turkey

Fabio Casagrande
University of California at San Diego
Department of Chemistry and
Biochemistry
9500 Gilman Drive
La Jolla, CA 92093
USA

Mark Chiu
Department of Structural Biology
Abbott Laboratories
R46Y AP10-LL08
100 Abbott Park Road
Abbott Park, IL 60064
USA

Igor Dodevski
University of Zurich
Department of Biochemistry
Winterthurerstrasse 190
8057 Zurich
Switzerland

David Drew
Imperial College
Membrane Protein Crystallography
Group, Division of Molecular
Biosciences
South Kensington Campus
London SW7 2AZ
UK

Timothy Esbenshade
Abbott Laboratories
Department of Neurosciences
100 Abbott Park Road
Abbott Park, IL 60064
USA

Brian Estvander
Abbott Laboratories
Department of Neurosciences
100 Abbott Park Road
Abbott Park, IL 60064
USA

Dimitra Gialama
Stockholm University
Department of Biochemistry and
Biophysics, Center for Biomembrane
Research
Sv. Arrheniusväg 16c
10691 Stockholm
Sweden

Jan-Willem de Gier
Stockholm University
Department of Biochemistry
and Biophysics
Center for Biomembrane Research
Sv. Arrheniusväg 16c
10691 Stockholm
Sweden

and

Xbrane Bioscience AB
Stureplan 15
111 45 Stockholm
Sweden

Deborah K. Hanson
Argonne National Laboratory
Biosciences Division
9700 South Cass Avenue
Lemnot, IL 60439
USA

Deniz B. Hizal
Johns Hopkins University
Department of Chemical and
Biomolecular Engineering
221 Maryland Hall
3400 North Charles Street
Baltimore, MD 21218
USA

Steve Kakavas
Abbott Laboratories
Advanced Technologies
100 Abbott Park Road
Abbott Park, IL 60064
USA

Hans Kiefer
Biberach University of Applied Sceinces
Karlstrasse 11
88400 Biberach
Germany

Mirjam Klepsch
Stockholm University
Department of Biochemistry
and Biophysics
Center for Biomembrane Research
Sv. Arrheniusväg 16c
10691 Stockholm
Sweden

Kathy Krueger
Abbott Laboratories
Department of Neurosciences
100 Abbott Park Road
Abbott Park, IL 60064
USA

Philip D. Laible
Argonne National Laboratory
Biosciences Division
9700 South Cass Avenue
Lemnot, IL 60439
USA

Marc Lake
Abbott Laboratories
Advanced Technologies
100 Abbott Park Road
Abbott Park, IL 60064
USA

Tzvetana Lazarova
Universitat Autònoma de Barcelona
Unitat de Biofísica
Departament de Bioquímica i de
Biologia Molecular
Facultat de Medicina and
Centre d'Estudis en Biofísica
08193 Bellaterra
Barcelona
Spain

Christel Logez
IREBS Institute
CNRS and University of Strasbourg
Pain and GPCRs Research Group
Boulevard Sébastien Brant
67412 Illkirch
France

Patrick Loll
Drexell University College of Medicine
Department of Biochemistry and
Molecular Biology
Room 10-102 New College Building
245 N. 15th Street, Mailstop 497
Philadelphia, PA 19102-1192
USA

Mark Lorch
University of Hull
Department of Chemistry
Cottingham Road
Kingston-upon-Hull HU6 7RX
UK

Víctor Lórenz-Fonfría
Universitat Autònoma de Barcelona
Unitat de Biofísica
Departament de Bioquímica i de
Biologia Molecular
Facultat de Medicina and
Centre d'Estudis en Biofísica
08193 Bellaterra
Barcelona
Spain

Sunny Mai
Johns Hopkins University
Department of Chemical and
Biomolecular Engineering
221 Maryland Hall
3400 North Charles Street
Baltimore, MD 21218
USA

Klaus Maier
Membrane Receptor Technologies
9381 Judicial Drive
Suite 140
San Diego, CA 92121
USA

Patrick McNeely
University of Delaware
Department of Chemical Engineering,
Colburn Laboratory
150 Academy Street
Newark, DE 19716
USA

Donna L. Mielke
Argonne National Laboratory
Biosciences Division
9700 South Cass Avenue
Lemnot, IL 60439
USA

Andrea Naranjo
University of Delaware
Department of Chemical Engineering
Colburn Laboratory
150 Academy Street
Newark, DE 19716
USA

Erika Ohsfeldt
Johns Hopkins University
Department of Chemical and
Biomolecular Engineering
221 Maryland Hall
3400 North Charles Street
Baltimore, MD 21218
USA

Stanley J. Opella
University of California at San Diego
Department of Chemistry and
Biochemistry
9500 Gilman Drive
La Jolla, CA 92093
USA

Esteve Padrós
Universitat Autònoma de Barcelona
Unitat de Biofísica
Departament de Bioquímica i de
Biologia Molecular
Facultat de Medicina and
Centre d'Estudis en Biofísica
08193 Bellaterra
Barcelona
Spain

Krzysztof Palczewski
Case Western Reserve University
School of Medicine, Department of
Pharmacology W319
10900 Euclid Avenue
Cleveland, OH 44106-4965
USA

Sang Ho Park
University of California at San Diego
Department of Chemistry and
Biochemistry
9500 Gilman Drive
La Jolla, CA 92093
USA

Alex Perálvarez-Marín
Universitat Autònoma de Barcelona
Unitat de Biofísica
Departament de Bioquímica i de
Biologia Molecular
Facultat de Medicina and
Centre d'Estudis en Biofísica
08193 Bellaterra
Bellaterra
Spain

Ana Pereda-Lopez
Abbott Laboratories
Advanced Technologies
100 Abbott Park Road
Abbott Park, IL 60064
USA

Andreas Plückthun
University of Zurich
Department of Biochemistry
Winterthurerstrasse 190
8057 Zurich
Switzerland

Anne Skaja Robinson
University of Delaware
Department of Chemical Engineering
Colburn Laboratory
150 Academy Street
Newark, DE 19716
USA

David Salom
Polgenix, Inc.
Suite 260
11000 Cedar Avenue
Cleveland, OH 44106
USA

James Samuelson
New England Biolabs
Gene Expression Division
240 County Road
Ipswich, MA 01938
USA

Susan Schlegel
Stockholm University
Department of Biochemistry and Biophysics
Center for Biomembrane Research
Sv. Arrheniusväg 16c
10691 Stockholm
Sweden

Krishna Vukoti
Center for Proteomics and Broinformatics
Case Western Reserve University
BRB 9th Floor
10900 Euclid Avenue
Cleveland, OH 44106
USA

Renaud Wagner
IREBS Institute
CNRS and University of Strasbourg
Pain and GPCRs Research Group
Boulevard Sébastien Brant
67412 Illkirch
France

David Wickström
Stockholm University
Department of Biochemistry and Biophysics
Center for Biomembrane Research
Sv. Arrheniusväg 16c
10691 Stockholm
Sweden

and

Xbrane Bioscience AB
Stureplan 15
111 45 Stockholm
Sweden

Alexei Yeliseev
National Institutes of Health
National Institute on Alcohol Abuse and Alcoholism
5625 Fishers Lane, Room 3N17
Rockville, MD 20852
USA

Carissa Young
University of Delaware
Department of Chemical Engineering,
Colburn Laboratory
150 Academy Street
Newark, DE 19716
USA

Introduction

Anne Skaja Robinson and Patrick J. Loll

Membrane proteins, acting as channels, receptors, and transporters, enable the cell to transport information and materials across the plasma membrane. As such, they are vital for cellular functioning and remain major drug discovery targets. Both to facilitate drug development and to understand the triggers of cellular action and reaction, structural information is required. However, although membrane proteins represent 30% of all proteins in the genome, they represent only around 1% of all high-resolution structures. One reason for this disparity is that biophysical and biochemical studies of membrane proteins require large quantities of purified protein – arguably even more than soluble proteins, because of the optimization of detergent conditions that is required. The first membrane protein structure ever determined – that of the photosynthetic reaction center of *Rhodopseudomonas viridis*, in 1985 – used protein purified from the native source [1]. Additional new structures have continued to appear over the last 25 years, but for many years primarily relied on protein purified from native sources. This is probably one reason why the rate of increase in membrane protein structures has been somewhat slower than the corresponding rate for soluble proteins [2]. Ten years ago, the majority of membrane protein structures were still obtained from proteins available in high natural abundance. More recently, heterologous expression has shown success, but this approach is still challenging, particularly for multispanning proteins (Figure 1) (reviewed by [2–5]).

The path to a high-resolution structure of a membrane protein involves many potential obstacles, including poor expression, limited extraction and purification yields, and challenging structural approaches. Successfully navigating this path requires a combination of scientific insight, creativity, opportunism, and trial-and-error empiricism. In this book, we highlight the state-of-the-art approaches for membrane protein expression, solubilization, and high-resolution structural methods that should provide guidance for those new to the field as well as established membrane protein scientists.

Production of Membrane Proteins: Strategies for Expression and Isolation, First Edition.
Edited by Anne Skaja Robinson.

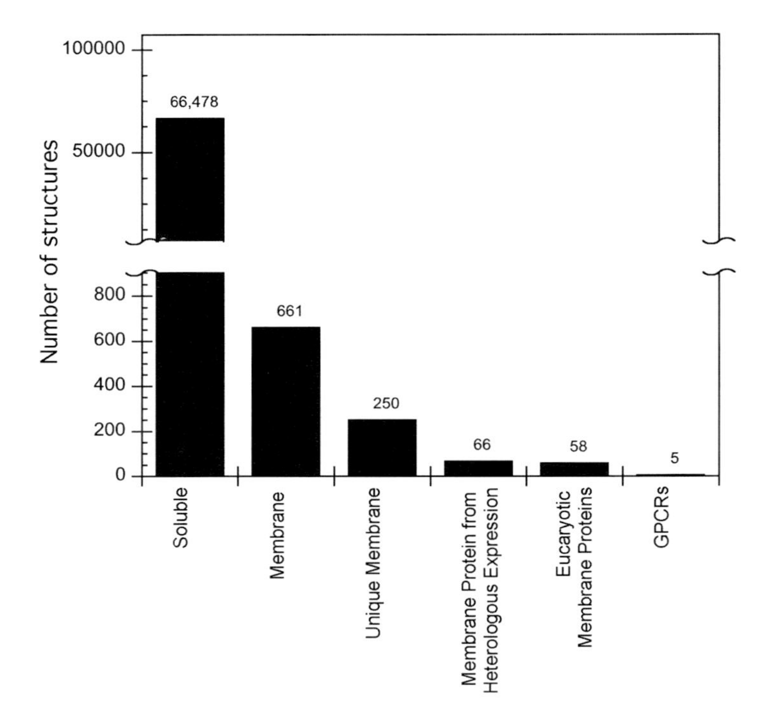

Figure 1 Comparison of numbers of soluble and membrane protein structures determined [6, 7]; proteins from different organisms, as well as mutants of the same structure are included in total membrane structures. Unique membrane structures include membrane protein structures for proteins from different organisms, but exclude protein variants and substrate-bound forms. Data obtained from membrane proteins are further divided into those expressed in non-native hosts (heterologous), eucaryotic proteins, and GPCRs. Note: any errors are attributed to A.S.R. rather than the websites referenced herein.

Expression

Results for many classes of membrane proteins suggest that in unmodified hosts, only a small subset of the desired target proteins will be expressed efficiently. For example, the MePNet consortium screened expression of 103 G-protein-coupled receptors (GPCRs) in Sf9, *Escherichia coli*, or *Pichia* expression systems, and obtained only two that expressed at modest levels in Sf9 cells, eight in *Pichia*, and none in *E. coli* [8]. The Cross and Nakamoto groups found that expression of *Mycobacterium tuberculosis* membrane protein targets in *E. coli* similarly gave low yields of well-expressed, membrane-localized membrane proteins, with no proteins larger than 15 kDa expressing well (Figure 2) [9]. In fact, the report of the Mid-Course Review Panel for the National Institutes of Health (NIH) Structural Biology Working Group states ". . . membrane protein production is still very challenging and much more must be learned to significantly advance the field"

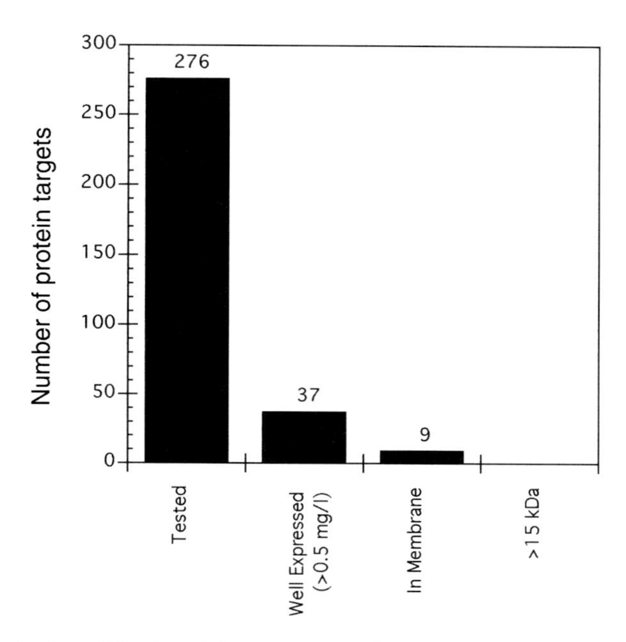

Figure 2 Production of *M. tuberculosis* membrane protein targets in *E. coli* shows that protein production is a major bottleneck to biophysical and structural studies [9]. "In Membrane" designates those proteins determined to inserted into the membrane via detergent solubilization studies.

[10]. Successful approaches to expression of membrane proteins in *E. coli* (Chapter 1), *Saccharomyces cerevisiae* (Chapter 2), *Pichia pastoris* (Chapter 3), insect cells (Chapter 4), mammalian cells (Chapter 5), and photosynthetic bacteria (Chapter 6) are described in Part One of the book. In addition, each chapter contains case studies that highlight successes, and suggested hosts and expression systems to yield the best outcomes for expression.

An interesting class of proteins is the peripheral membrane proteins, which can either exist in an aqueous environment or associated with the membrane. Nonspecific electrostatic interactions, hydrophobic association, covalent lipid anchors, or lipid-binding domains facilitate membrane association of these proteins. Special aspects of expression and purification of these molecules are discussed in Chapter 7.

Perhaps one of the most challenging classes of membrane proteins for expression, solubilization, and crystallization are the multipass membrane proteins, with GPCRs representing the largest family of these. GPCRs are expressed in virtually all human tissues and transmit a wide variety of signals in response to diverse stimuli (including light, hormones, injury, and inflammation). These signals regulate a diverse set of cellular responses via interaction with GTP-binding proteins [11, 12].

The structure of rhodopsin, purified from native tissue, was the first high-resolution GPCR structure determined by X-ray crystallography [13, 14]. In 2007, the human β_2-adrenergic receptor (β_2AR) structure, with either lysozyme or a Fab fragment inserted in cytoplasmic loop 3 (CL3), was determined by Kobilka *et al.*, after many years of tweaking expression and crystallization approaches [15, 16]. A structure for an engineered version of turkey β_1AR, with a truncated C-terminus and a shortened CL3, soon followed [17]. The publication of the high-resolution X-ray structure of the adenosine A_{2A} receptor (A_{2A}R) in late 2008 [18] has generated additional excitement in the GPCR field. To reduce structural flexibility and increase crystallizability of A_{2A}R the C-terminal tail was removed and the CL3 was also substituted with T4 lysozyme, as was done successfully for β_2AR. These structures have revealed that the GPCR family is "extremely adaptable" and can tolerate a wide range of helix distortions or reorientations that impact ligand binding [19].

The protein re-engineering required to facilitate crystallization can alter the native features and function of the receptors, however. The third intracellular loop [20] and the preceding transmembrane domain five are involved in G-protein coupling [21], and it is not surprising that the replacement of the third intracellular loop of A_{2A} with T4 lysozyme inhibited receptor interactions with the G-protein and ligand [18]. Indeed, as impressive as the recent GPCR structural efforts have been, *in silico* ligand design using these structures suggests limited utility due to the inactive state nature of these structures [22]. Continued effort is needed to study this important family and strategies to express GPCRs are described in Chapter 8.

Directed evolution approaches to re-engineering a membrane protein of interest is an alternative to the specialized modifications described above. This method, as described in Chapter 9, relies only on screening and selection for membrane protein function (e.g., through ligand binding activity). One alternative to focusing on re-engineering a particular membrane protein is to screen different membrane proteins in order to identify those that are readily expressed. Most recently, this has been addressed via high-throughput approaches. For example, the University of California at San Francisco Membrane Protein group supported under the NIH Roadmap cites a strategy for "... 'discovery-oriented' selection of *tractable* targets ..." based on expression level, detergent solubilization, and chromatographic behavior; notably, this approach yielded *only* five tractable targets out of an initial 384 candidates drawn from the *S. cerevisiae* membrane proteome [23]. A similar approach using Green Fluorescent Protein (GFP) tagging to screen for "tractable" eukaryotic membrane proteins has yielded notable success (Chapter 10) [24, 25]. One caveat to these approaches is that the screens pick up "soluble" proteins based on GFP fluorescence, but do not test function – a key feature for drug design. For example, membrane insertion and proper localization do not effectively distinguish active GPCRs from inactive ones [26, 27]; similarly, the MePNet consortium found no correlation between expression level and activity in a test of around 50 GPCRs expressed in Sf9 cells and *Pichia* [8]. Further, such screening approaches can only identify those protein targets with the innate ability to be expressed at high levels (low-hanging fruit); they offer no hope for improving poorly expressed targets.

Solubilization and Structural Methods

Once high-level expression is achieved, often the next step is to carry out biophysical characterization or high-resolution structural analysis. Unfortunately, purification and solubilization methods are still primarily trial-and-error approaches, but information in Chapters 11 and 12 (Part Three) should provide a guide for commonly used methods and strategies for any protein of interest.

Structure determination of membrane proteins via nuclear magnetic resonance (NMR) has been ongoing for decades, but has been primarily focused on peptides and small proteins, since detergent micelles, bicelles, or lipids increase the size and heterogeneity of membrane protein mixtures. However, in recent years significant progress has been made in both solution NMR methods for larger proteins and solid-state NMR, such that complete structures have been determined for several membrane proteins [28–30]. In particular, for solution NMR the TROSY (transverse relaxation optimized spectroscopy) method developed by Kurt Wüthrich enabled structure determination of several larger integral membrane proteins (e.g., several β-barrel proteins from *E. coli*). For solid-state NMR (SSNMR), the ability to align structures with bicelles and bilayers enables anisotropic interactions to be resolved [31]. In addition, SSNMR measurements do not limit the sample to solution phase, which enables insoluble or aggregate structure determination. The use of magic angle spinning for SSNMR has also facilitated the resolution of larger structures. Dror Warschawski has developed a database of membrane protein structures determined by NMR (http://www.drorlist.com/nmr/MPNMR.html), which includes 39 unique structures as of 1 June 2010. Although β-barrels dominate the larger structures, there are some larger α-helical structures as well, including that of diacylglycerol kinase, a trimer of 43 kDa. In addition to providing high-resolution structural information, NMR can determine protein conformational changes and flexibility under near-physiological conditions, which is a key feature to understanding function for many integral membrane proteins. This book provides information to assist in developing metabolically labeled proteins for NMR characterization (Chapter 13); readers interested in reviews of NMR methods applied to membrane proteins should examine several good references in the literature [29–32].

Protein structure determination by X-ray crystallography predates NMR methods by many years. However, just as obstacles are faced with NMR approaches, so too the production of suitable crystals has proven a significant stumbling block for crystallography. In fact, not so long ago crystallization of membrane proteins was widely held to be impossible. Thankfully, during the past three decades membrane protein crystallization stopped being considered impossible, moved through a stage during which it was thought of as somewhat heroic, and has now entered a phase in which it is generally accepted to be a practical undertaking.

The first crystals of OmpF porin, a bacterial outer membrane channel, were reported in 1980 [33]. Porin crystallization raised several interesting points, many of which remain relevant today. First, the protein was purified from an overproducing strain of *E. coli* [34] and was not made by the recombinant methodologies that now dominate the structural biology of soluble proteins. This reflected the difficulties associated with heterologous overexpression of membrane proteins; such difficulties continue to dog membrane protein work today and are discussed at length in this book. Second, porin binds tightly to bacterial lipopolysaccharide, and the key to producing good crystals turned out to be the patient and careful removal of as much of this lipid as possible [34]. While this precise formula has by no means proven universal – indeed, there are instances where overzealous removal of lipids compromises crystallization – it presaged the emergence of protein–lipid interactions as important contributors to both protein stability and crystallization behavior. Finally, workers studying the porin system were the first to identify the detergent cloud point as a region of the phase diagram that is fraught with unusual interest for the crystallization of detergent-solubilized membrane proteins [35, 36]; investigators probing the physical mechanisms underlying membrane protein crystallization have returned to this observation time and again in intervening years.

While porin is credited as the first membrane protein to yield well-ordered crystals, the first reported structure belongs to the photosynthetic reaction center from the bacterium *R. viridis* [1]. Crystals of this protein were obtained after development of a rigorous screening procedure aimed at fostering the packing of protein–detergent complexes into a crystalline lattice (so-called "Type II" crystals [37]). One key concept to emerge from this work is the notion that small amphiphiles can help fine-tune the structures and/or interactions of protein–detergent complexes. While such molecules have not proven to be magic bullets that facilitate crystallization for all membrane proteins, they have proven useful in many cases, and are extremely important insofar as they highlight the idea that micelle structure and phase behavior can be altered in rational ways. Approaches to obtaining protein for high-resolution crystallography are described in Chapter 14.

A particularly interesting advance made during the last 15 years was the development of lipidic mesophase crystallization methods [38]. Lipidic mesophases such as the cubic phase form gel-like materials in which bilayers assemble into complex three-dimensional structures. Proteins embedded in these phases bilayers encounter a native-like environment (since they are in a bilayer), while the spatial ordering of the lipid enables them to diffuse through three dimensions; presumably this facilitates crystal contacts not only within the plane of the bilayer, but also orthogonal to it [39]. This methodology has produced some notable successes, most prominently the formation of well-ordered crystals of a variety of seven-transmembrane helical proteins, including both GPCRs and their prokaryotic homologs. While it seems unlikely that "*in cubo*" or "*in meso*" methods, as they are known, will entirely supplant the direct crystallization of protein-detergent complexes – the majority of structures currently being published still rely on the latter method – the stunning successes provided by mesophase approaches ensure them a lasting place in the crystallizer's tool kit.

Membrane protein structural biology is now well established, and in recent years the number of new crystal and NMR structures has grown exponentially [2, 6, 29]. Nonetheless, challenges remain. Efforts aimed at finding "low-hanging fruit" for high-resolution studies have generated much new structural information, but have not provided the insights necessary to develop rational approaches for more challenging targets. Eukaryotic membrane proteins are still under-represented in the structural databases, which most likely reflects difficulties with protein production, low stability, and/or inherent flexibility on the part of many of these molecules. Hence, in the coming decade proponents of membrane protein structural biology will not lack for suitable challenges.

Abbreviations

β_2AR	β_2-adrenergic receptor
$A_{2A}R$	adenosine A_{2A} receptor
CL3	cytoplasmic loop 3
GFP	Green Fluorescent Protein
GPCR	G-protein-coupled receptor
NIH	National Institutes of Health
NMR	nuclear magnetic resonance
SSNMR	solid-state NMR
TROSY	transverse relaxation optimized spectroscopy

References

1 Deisenhofer, J., Epp, O., Miki, K., Huber, R., and Michel, H. (1985) Structure of the protein subunits in the photosynthetic reaction centre of *Rhodopseudomonas viridis* at 3 Å resolution. *Nature*, **318**, 618–624.

2 White, S.H. (2004) The progress of membrane protein structure determination. *Protein Sci.*, **13**, 1948–1949.

3 Tate, C.G. (2001) Overexpression of mammalian integral membrane proteins for structural studies. *FEBS Lett.*, **504**, 94–98.

4 Chiu, M.L., Tsang, C., Grihalde, N., and MacWilliams, M.P. (2008) Over-expression, solubilization, and purification of G protein-coupled receptors for structural biology. *Comb. Chem. High Throughput Screen.*, **11**, 439–462.

5 White, S.H. (2009) Membrane Proteins of Known 3D Structure, http://blanco.biomol.uci.edu/Membrane_Proteins_xtal.html (accessed 10 January 2009).

6 White, S.H. (2010) Progress in Membrane Protein Structure Determination, http://blanco.biomol.uci.edu/MP_Structure_Progress.html (accessed 13 July 2010).

7 Berman, H.M., Westbrook, J., Feng, Z., Gilliland, G., Bhat, T.N., Weissig, H., Shindyalov, I.N., and Bourne, P.E. (2000) The protein data bank. *Nucleic Acids Res.*, **28**, 235–242

8 Lundstrom, K., Wagner, R., Reinhart, C., Desmyter, A., Cherouati, N., Magnin, T., Zeder-Lutz, G., Courtot, M., Prual, C., Andre, N., Hassaine, G., Michel, H., Cambillau, C., and Pattus, F. (2006) Structural genomics on membrane proteins: comparison of more than 100 GPCRs in 3 expression systems. *J. Struct. Funct. Genomics*, **7**, 77–91.

9 Korepanova, A., Gao, F.P., Hua, Y., Qin, H., Nakamoto, R.K., and Cross, T.A. (2005) Cloning and expression of multiple integral membrane proteins

from *Mycobacterium tuberculosis* in *Escherichia coli*. *Protein Sci.*, **14**, 148–158.
10 Buchanan, S.K., Laughlin, M.R., Ramm, L., Rees, D.C., Stenkamp, R.E., Tamm, L., and White, S.H. (2009) Mid-Course Review – Report on the Structural Biology Working Group Membrane Proteins, http://nihroadmap.nih.gov/structuralbiology/midcoursereview/summary2008.asp (accessed 8 June 2009).
11 Lee, N.H. and Kerlavage, A.R. (1993) Molecular biology of G-protein-coupled receptors. *Trends Biomed. Res.*, **6**, 488–497.
12 Strader, C.D., Fong, T.M., Tota, M.R., Underwood, D., and Dixon, R.A. (1994) Structure and function of G protein-coupled receptors. *Annu. Rev. Biochem.*, **63**, 101–132.
13 Bourne, H.R. and Meng, E.C. (2000) Structure. Rhodopsin sees the light. *Science*, **289**, 733–734.
14 Palczewski, K., Kumasaka, T., Hori, T., Behnke, C.A., Motoshima, H., Fox, B.A., Le Trong, I., Teller, D.C., Okada, T., Stenkamp, R.E., Yamamoto, M., and Miyano, M. (2000) Crystal structure of rhodopsin: a G protein-coupled receptor. *Science*, **289**, 739–745.
15 Cherezov, V., Rosenbaum, D.M., Hanson, M.A., Rasmussen, S.G., Thian, F.S., Kobilka, T.S., Choi, H.J., Kuhn, P., Weis, W.I., Kobilka, B.K., and Stevens, R.C. (2007) High-resolution crystal structure of an engineered human beta2-adrenergic G protein-coupled receptor. *Science*, **318**, 1258–1265.
16 Rasmussen, S.G., Choi, H.J., Rosenbaum, D.M., Kobilka, T.S., Thian, F.S., Edwards, P.C., Burghammer, M., Ratnala, V.R., Sanishvili, R., Fischetti, R.F., Schertler, G.F., Weis, W.I., and Kobilka, B.K. (2007) Crystal structure of the human beta2 adrenergic G-protein-coupled receptor. *Nature*, **450**, 383–387.
17 Warne, T., Serrano-Vega, M.J., Baker, J.G., Moukhametzianov, R., Edwards, P.C., Henderson, R., Leslie, A.G., Tate, C.G., and Schertler, G.F. (2008) Structure of a beta1-adrenergic G-protein-coupled receptor. *Nature*, **454**, 486–491.
18 Jaakola, V.P., Griffith, M.T., Hanson, M.A., Cherezov, V., Chien, E.Y., Lane, J.R., Ijzerman, A.P., and Stevens, R.C. (2008) The 2.6 angstrom crystal structure of a human A_{2A} adenosine receptor bound to an antagonist. *Science*, **322**, 1211–1217.
19 White, S.H. (2009) Biophysical dissection of membrane proteins. *Nature*, **459**, 344–346.
20 Audet, M and Bouvier, M. (2008) Insights into signaling from the beta2-adrenergic receptor structure. *Nat. Chem. Biol.*, **4**, 397–403.
21 Khafizov, K., Lattanzi, G., and Carloni, P. (2009) G protein inactive and active forms investigated by simulation methods. *Proteins*, **75**, 919–930.
22 Kolb, P., Rosenbaum, D.M., Irwin, J.J., Fung, J.J., Kobilka, B.K., and Shoichet, B.K. (2009) Structure-based discovery of beta2-adrenergic receptor ligands. *Proc. Natl. Acad. Sci. USA*, **106**, 6843–6848.
23 Li, M., Hays, F.A., Roe-Zurz, Z., Vuong, L., Kelly, L., Ho, C.M., Robbins, R.M., Pieper, U., O'Connell, J.D., 3rd, Miercke, L.J., Giacomini, K.M., Sali, A., and Stroud, R.M. (2009) Selecting optimum eukaryotic integral membrane proteins for structure determination by rapid expression and solubilization screening. *J. Mol. Biol.*, **385**, 820–830.
24 Drew, D., Slotboom, D.J., Friso, G., Reda, T., Genevaux, P., Rapp, M., Meindl-Beinker, N.M., Lambert, W., Lerch, M., Daley, D.O., Van Wijk, K.J., Hirst, J., Kunji, E., and De Gier, J.W. (2005) A scalable, GFP-based pipeline for membrane protein overexpression screening and purification. *Protein Sci.*, **14**, 2011–2017.
25 Newstead, S., Kim, H., von Heijne, G., Iwata, S., and Drew, D. (2007) High-throughput fluorescent-based optimization of eukaryotic membrane protein overexpression and purification in *Saccharomyces cerevisiae*. *Proc. Natl. Acad. Sci. USA*, **104**, 13936–13941.
26 Butz, J.A., Niebauer, R.T., and Robinson, A.S. (2003) Co-expression of molecular chaperones does not improve the heterologous expression of mammalian G-protein coupled receptor expression in yeast. *Biotechnol. Bioeng.*, **84**, 292–304.
27 O'Malley, M.A., Mancini, J.D., Young, C.L., McCusker, E.C., Raden, D., and

Robinson, A.S. (2009) Progress toward heterologous expression of active G-protein-coupled receptors in *Saccharomyces cerevisiae*: linking cellular stress response with translocation and trafficking. *Protein Sci.*, **18**, 2356–2370.

28 Doreleijers, J.F., Mading, S., Maziuk, D., Sojourner, K., Yin, L., Zhu, J., Markley, J.L., and Ulrich, E.L. (2003) BioMagResBank database with sets of experimental NMR constraints corresponding to the structures of over 1400 biomolecules deposited in the Protein Data Bank. *J. Biomol. NMR*, **26**, 139–146.

29 Kim, H.J., Howell, S.C., Van Horn, W.D., Jeon, Y.H., and Sanders, C.R. (2009) Recent advances in the application of solution NMR spectroscopy to multi-span integral membrane proteins. *Prog. Nucl. Magn. Reson. Spectrosc.*, **55**, 335–360.

30 McDermott, A. (2009) Structure and dynamics of membrane proteins by magic angle spinning solid-state NMR. *Annu. Rev. Biophys.*, **38**, 385–403.

31 Opella, S.J. and Marassi, F.M. (2004) Structure determination of membrane proteins by NMR spectroscopy. *Chem. Rev.*, **104**, 3587–3606.

32 Opella, S.J., Nevzorov, A., Mesleb, M.F., and Marassi, F.M. (2002) Structure determination of membrane proteins by NMR spectroscopy. *Biochem. Cell Biol.*, **80**, 597–604.

33 Garavito, R.M. and Rosenbusch, J.P. (1980) Three-dimensional crystals of an integral membrane protein: an initial X-ray analysis. *J. Cell Biol.*, **86**, 327–329.

34 Misra, R. and Reeves, P.R. (1987) Role of *micF* in the *tolC*-mediated regulation of OmpF, a major outer membrane protein of *Escherichia coli* K-12. *J. Bacteriol.*, **169**, 4722–4730.

35 Rosenbusch, J.P. (1990) The critical role of detergents in the crystallization of membrane proteins. *J. Struct. Biol.*, **104**, 134–138.

36 Zulauf, M. (1991) Detergent phenomena in membrane protein crystallization, in *Crystallization of Membrane Proteins* (ed. H. Michel), CRC Press, Boca Raton, FL, pp. 53–72.

37 Michel, H. (1983) Crystallization of membrane proteins. *Trends Biochem. Sci.*, **8**, 56–59.

38 Landau, E.M. and Rosenbusch, J.P. (1996) Lipidic cubic phases: a novel concept for the crystallization of membrane proteins. *Proc. Natl. Acad. Sci. USA*, **93**, 14532–14535.

39 Nollert, P., Qiu, H., Caffrey, M., Rosenbusch, J.P., and Landau, E.M. (2001) Molecular mechanism for the crystallization of bacteriorhodopsin in lipidic cubic phases. *FEBS Lett.*, **504**, 179–186.

Part One
Expression Systems

1
Bacterial Systems

James Samuelson

1.1
Introduction

The study of membrane protein structure and function is limited by various challenges. In native cells, membrane protein copy number is often very low, so the study of individual proteins is often not feasible. Alternatively, overexpression of these hydrophobic molecules in heterologous hosts is not a routine endeavor as it is for many water-soluble proteins. Most modern bacterial expression systems have been engineered for maximal output of recombinant protein. This characteristic is ideal for well-behaved soluble proteins, but less desirable when the target protein normally resides within a lipid environment. A compounding problem in the study of membrane proteins is that the isolated target protein may exhibit polydispersity, meaning that diverse oligomeric complexes can spontaneously accumulate. This latter concern may be influenced by the expression method, but primarily depends on the detergent/lipid and buffer used for solubilization. This chapter highlights preferred strategies for membrane protein expression in bacteria that will increase the likelihood of isolating adequate amounts of homogenous target protein. Many sections will also detail the features of expression strains that are relevant to the yield and quality of expressed protein.

In this chapter, the term membrane protein will generally be used to represent α-helical membrane proteins that reside within a phospholipid bilayer environment of either eukaryotic or prokaryotic cells. Such integral membrane proteins are the most difficult to manipulate since each contains hydrophobic transmembrane (TM) regions as well as hydrophilic extramembrane regions or domains. In the case of single-spanning membrane proteins, often the catalytic domain is a water-soluble entity that may be studied by expression of a ΔTM variant. However, multispanning membrane proteins such as ion channels must be expressed without gross deletions of hydrophobic residues.

Membrane proteins with β-barrel structure such as those found in the Gram-negative bacterial outer membrane or the mitochondrial outer membrane are typically expressed at high levels as inclusion bodies within the *Escherichia coli* cytoplasm. Isolation and washing of these inclusion bodies often leads to a

Production of Membrane Proteins: Strategies for Expression and Isolation, First Edition.
Edited by Anne Skaja Robinson.

relatively pure sample of recombinant protein and the literature contains many examples of refolding of β-barrel proteins, such as Omp proteins from *E. coli* [1]. In contrast, refolding of α-helical integral membrane protein is quite a difficult challenge, although some successes have been reported [2–4]. The default method of expressing α-helical membrane proteins should be to direct them to the membrane fraction of the host cell and to perform purification procedures beginning with isolation of the cellular membrane fraction.

1.2 Understanding the Problem

Each recombinant membrane protein clone should be assumed to be "toxic" to the host cell. This is particularly true when bacterial hosts are employed. It is well established that uncontrolled expression of most membrane proteins in *E. coli* will lead to induction of cellular stress responses and occasionally cell death. In some cases, the plasmid transformation step may fail because the transformed cell cannot recover due to the uncontrolled expression of membrane protein. Therefore, the first bit of advice in designing expression clones is to use a vector that propagates at 40 copies or less per cell (pMB1+*rop*, oriV, p15A, pSC101 replication origins). Accordingly, a vector with a pUC-derived origin should be avoided. Secondly, the promoter driving protein expression should be controllable (inducible). Much of this chapter is allocated to describing appropriate host/vector/promoter combinations (see Table 1.1 for a summary).

In bacteria, passage through the inner membrane Sec translocase [5] is recognized as the primary bottleneck during the overexpression of recombinant membrane protein. Yet, many other factors may contribute to a limited expression yield. There are reports of Sec-independent membrane translocation, but true host protein-independent membrane assembly by a heterologous protein has not been clearly substantiated in the literature. For example, membrane assembly of Mistic fusion proteins [6] may be initiated by the affinity of the Mistic protein for the cytoplasmic face of the *E. coli* inner membrane; however, proper membrane assembly of the fused protein of interest must still require assistance from the Sec translocase when large extracellular hydrophilic domains need to be translocated across the inner membrane.

Our lab has investigated several possible modes of Sec-independent membrane assembly without arriving at any evidence that a heterologous integral membrane protein can bypass the Sec translocase (unpublished data). Furthermore, we have attempted to increase the efficiency of membrane integration by overexpressing the endogenous YidC protein that is thought to aid the Sec translocase or act independently as a membrane insertase [7]. We specifically chose to study the effect of YidC on the membrane integration of phage M13 p8 fusion proteins, as p8 protein by itself requires YidC for inner membrane assembly [8]. To our surprise, a 10-fold increased level of YidC had no effect on the membrane translocation of p8-derived fusion proteins containing a C-terminal PhoA domain as a reporter. One conclusion of this experiment is that the activity of SecA ATPase may be the limiting factor for the translocation of the large hydrophilic PhoA domain. Recently we determined that the p8 fusion partner (p8CBDek described in Luo *et al.* 2009 [9]) utilizes the cotranslational signal recognition particle (SRP) pathway [10–12], the route traveled by most endogenous membrane proteins. During cotranslational membrane protein assembly, there is less opportunity for hydrophobic amino acid segments to aggregate after emerging from the ribosome tunnel. Perhaps the limiting factor in p8 fusion protein expression and the over-expression of most membrane proteins is simply the rate of protein translation (or efficiency of translation initiation) at the ribosome. With this thought in mind, we tested various ribosomal binding sites (RBSs) and found a distinct difference in the efficiency of p8CBDek-mediated polytopic membrane protein assembly. Strikingly, the clone containing the much weaker RBS (AGGACGGCCGGatg) produced a greater level of protein per cell after a 20-h expression period at 20 °C. In contrast, the stronger RBS provided more protein per cell in the first stage of expression, but also resulted in jamming of the translocation pathway and cessation of culture growth. Thus, the take-home message from our recent work is to express recombinant membrane proteins "in moderation." This advice may seem obvious, but many expression systems do not allow for careful control of expression. The solution of genetically engineering the appropriate RBS for the protein of interest may not be a preferred method of optimization. Instead, a much simpler solution for expression optimization is to employ a promoter that allows fine control of the level of mRNA encoding the membrane protein of interest.

1.3 Vector/Promoter Types

The most-studied bacterial promoters are those controlling operons for sugar metabolism (*lacZYA, araBAD, rhaBAD*). Many variants of the *lac* promoter have been isolated but all suffer to some degree from the inability to completely shut off expression with the LacI repressor protein. The wild-type *lac* promoter is a good choice for membrane protein expression due to its moderate strength. However, very few expression vectors encode the unmodified *lac* promoter. Vectors pUC18/pUC19 carry a simple *lac* promoter, but again pUC derivatives are not good choices due to high copy number and overproduction of β-lactamase (AmpR) that enables the growth of cells lacking plasmid. Vectors utilizing modified *lac* promoters are highlighted in Table 1.2. The *lacUV5* promoter has two mutations within the −10 region of the *lac* promoter. In addition, a mutation is present at −66 within the catabolite gene activator protein (CAP) binding site. These mutations increase the promoter strength relative to the wild-type *lac* promoter and expression from *lacUV5* is less subject to catabolite repression [13]. The *tac* promoter was first described by deBoer *et al.* [14–15]. This strong promoter is a hybrid of the −10 region of the *lacUV5* promoter and the −35 region of the *trp* promoter. Amann *et al.* reported that the *tac* promoter is at least 5 times more efficient than the

Table 1.1 Recommended *E. coli* strains for membrane protein expression.

Strain	Source	Distinguishing features	Growth/expression guidelines	Compatible expression vectors
BL21(DE3)	NEB; Novagen; Invitrogen; Stratagene; Genlantis; Lucigen	chromosomal DE3 prophage expresses T7 RNA polymerase under control of the *lacUV5* promoter; BL21 derivatives lack Lon and OmpT proteases, which may stabilize expression of some recombinant protein	exhibits significant basal T7 expression, thus addition of 1% glucose to the growth medium is recommended for toxic clones	pET or T7 vectors with *lacI* gene and *T7–lac* promoter
BL21(DE3) pLysS	Novagen; Invitrogen; Stratagene; Genlantis; Lucigen	same as above, plus the pLysS plasmid produces wild-type T7 lysozyme to reduce basal T7 expression of the gene of interest; pLysS is compatible with plasmids containing the ColE1 or pMB1 origin (most pET vectors)	Cam 34 μg/ml to maintain pLysS	pET or T7 vectors with *lacI* gene and *T7–lac* promoter
Lemo21(DE3)	NEB; Xbrane Bioscience	pLemo plasmid produces amidase-negative T7 lysozyme (*lysY*) from a tunable promoter (*Prha*); pLemo is compatible with plasmids containing the ColE1 or pMB1 origin (most pET vectors)	Cam 34 μg/ml to maintain pLemo; expression trials typically benefit from rhamnose addition	pET or T7 vectors with *lacI* gene and *T7–lac* promoter
T7 Express LysY/I^q	NEB	T7 RNA polymerase gene encoded within the *lac* operon; combination of *lysY* and $lacI^q$ control offers strict control of basal T7 expression	incompatible with CamR expression vectors	pET or other T7 vector, presence of *lacI* gene is less important
BL21-AI	Invitrogen	T7 RNA polymerase gene is controlled by the *araBAD* promoter; if using a pET vector, IPTG is also required for induction to titrate LacI repressor away from the *T7–lac* promoter on the vector	T7 gene 1 is induced by L-arabinose; 0.2% glucose will repress	recommended for use with Invitrogen pDEST vectors
BL21	Various companies	protein expression from non-T7 promoters metabolizes arabinose, but still suitable for membrane expression from *ParaBAD*	none	nearly all non-T7 vectors
NEB Express I^q	NEB	BL21 derivative, same features as above except additional control of IPTG-inducible promoters suitable for membrane expression from *ParaBAD*	incompatible with CamR expression vectors	non-T7 vectors

TOP10; LMG194	Invitrogen	K-12 strains that do not metabolize L-arabinose may provide slight improvement when expressing membrane protein directly from *ParaBAD*	0.2% glucose will repress expression of toxic proteins	pBAD vectors
EXP strains	UCLA, J. Bowie	these are TOP10 mutants selected for improved expression of recombinant membrane protein; DE3 prophage has been added to original isolates	none	nearly all vectors with DE3 strains
Single Step KRX	Promega	T7 RNA polymerase gene is controlled by *PrhaBAD*; induce with 0.1% L-rhamnose; if using pET, pF1A or pF1K vectors, IPTG is also required for induction to titrate LacI repressor away from the *T7–lac* promoter on the vector	0.4% glucose in LB plates or starter culture to stabilize toxic clones	pET vectors T7 Flexi vectors (Promega)
C41(DE3); C43(DE3)	Lucigen	BL21(DE3) derivatives with much lower levels of T7 RNA polymerase under noninducing and inducing conditions	glucose addition is not necessary as with BL21(DE3)	pET or other T7 vector, presence of *lacI* gene is less important
Tuner; Tuner(DE3)	Novagen	*lacZY* derivatives of BL21; the *lac* permease mutation (*lacY1*) allows uniform entry of IPTG into all cells in the population, which produces a concentration-dependent level of induction		pET or T7 vectors with *lacI* gene and *T7–lac* promoter
Tuner (DE3) pLysS	Novagen	expresses T7 lysozyme to control T7 expression in addition to the *lac* permease mutation		pET or T7 vectors with *lacI* gene and *T7–lac* promoter
Rosetta; Rosetta2	Novagen	Rosetta host strains are BL21 *lacZY* (Tuner) derivatives designed to enhance the expression of proteins that contain codons rarely used in *E. coli*; these strains express tRNAs for rare codons on a compatible CamR plasmid	Cam 34 μg/ml	pET or T7 vectors with *lacI* gene and *T7–lac* promoter
Rosetta pLysS; Rosetta2 pLysS	Novagen	in the pLysS strains, the rare tRNA genes and T7 lysozyme gene are carried by the same plasmid	Cam 34 μg/ml	pET or T7 vectors with *lacI* gene and *T7–lac* promoter
BL21 CodonPlus; (DE3)-RIPL	Stratagene	RIPL indicates that the cells carry extra copies of the *argU*, *ileW*, *leuY*, and *proL* tRNA genes for increased recognition of the AGA/AGG (Arg), AUA (Ile), CUA (Leu), and CCC (Pro) codons	Cam 50 μg/ml	pET or other T7 vectors

Table 1.2 Common vectors/promoters/types of regulation (for more options, a comprehensive vector database is maintained by the EMBL Protein Expression and Purification Core Facility: http://www.pepcore.embl.de/strains_vectors/vectors/bacterial_expression.html).

Promoter	RNA polymerase	Inducer/relative strength	Distinguishing features	Vectors	Source
lac	*E. coli*	IPTG/+	repressed by glucose; controlled by LacI	pUC; NR	NEB
lacUV5	*E. coli*	IPTG/++	repressed by glucose; controlled by LacI		
tac	*E. coli*	IPTG/+++	CAP site absent, not affected by glucose well controlled by LacI	pMAL; p8CBDek; pKK223-3	NEB
trc	*E. coli*	IPTG/+++	CAP site absent, not affected by glucose well controlled by LacI	pTrc99a; pPROEX; pKK233-2	
T5	*E. coli*	IPTG/+++	with pQE, need to supply expression of LacI repressor (multicopy pREP4 or a strain carrying *lacI*^{q} gene)	pQE	Qiagen
araBAD	*E. coli*	L-arabinose/++	repressed by 0.2% glucose; tunable from 0.001–0.2% L-arabinose	pBAD	Invitrogen
rhaBAD	*E. coli*	L-rhamnose/++	repressed by 0.2% glucose; very tunable when vector carries *rhaRS* genes; inducer range = 10–2000 μM L-rhamnose	pRHA-67	Xbrane Bioscience
tetA	*E. coli*	aTc/++	pASK75 carries *tetR*; DH5αZ1 strain or other *tetR* strain is recommended	pASK75; pZ series	A. Skerra; H. Bujard/Expressys
T7 plain	T7	IPTG/+++	very high basal expression; controlled by T7 lysozyme	pET3, 9, 14, 17, 20, 23	Novagen
T7–lac	T7	IPTG/+++	high basal expression; controlled by T7 lysozyme	pET21, etc.; Various	Novagen

+, ++, +++ relative strength of induction.
NR, not recommended for membrane protein expression.

lacUV5 promoter [16]. The *trc* promoter is equivalent to the *tac* promoter since the 1-bp difference in spacing between the −35 and −10 consensus sequences does not affect promoter strength [17]. Note that the *tac* and *trc* promoters are not subject to catabolite repression as the CAP binding site is missing. *Ptac* and *Ptrc* systems are generally well controlled by LacI repression. When employing any type of modified *lac* promoter, LacI should be overexpressed from a *lacI* or *lacI*^{q} gene carried by the expression vector. Also, isopropyl-β-D-thiogalactopyranoside (IPTG) induction should be tested in the low range (e.g., 0, 10, 100 versus 400 μM). The *lacI*^{q} mutant was reported by Calos in 1978 and this mutation is simply an "up" promoter mutation resulting in a 10-fold enhancement of LacI repressor expression [18].

The pQE vectors from Qiagen utilize the phage T5 promoter that is controlled by two *lac* operator sequences. The T5 promoter is recognized by the *E. coli* RNA polymerase and induction is accomplished by IPTG addition to release the Lac repressor from the dual operator sequence. Since pQE vectors do not carry the *lacI* gene, the host strain must supply an excess of Lac repressor. Two options exist for LacI supplementation: copropagation of multicopy pREP4 (QIAexpress manual) or use of a strain that carries the *lacI*^{q} gene. Many K-12 strains (e.g., JM109) carry the *lacI*^{q} gene, but few B strains offer LacI overexpression. One recommendation is NEB Express I^{q}, which is a BL21 derivative that carries a miniF-*lacI*^{q} which does not require antibiotic selection (Table 1.1).

Guzman *et al.* characterized the *araBAD* promoter in exquisite detail in 1995 [19], and the resulting the pBAD vector series offers many options for gene cloning and expression using L-arabinose induction. Note that some pBAD vectors do not encode RBS sites, so the gene insert must contain an appropriate translation initiation sequence. When glucose is added to the outgrowth media, expression from *araBAD* is essentially shut off (Table 1 in Guzman *et al.* [19]). For many years, the *araBAD* system was a first choice for tightly regulated expression, as protein output appears to correlate very well with the amount of inducer (Figure 4 in Guzman *et al.* [19]) However, careful studies of the *araBAD* promoter by Siegele *et al.* [20] and Giacalone *et al.* [21] both agreed that at subsaturating levels of L-arabinose, protein expression cultures contain a mixed population with only some of the cells expressing protein. In addition, the potential for protein overexpression is generally lower when a pBAD vector is compared to T7-mediated expression from a pET construct.

A more recently characterized sugar promoter is derived from the rhamnose operon. The *rhaBAD* promoter is induced by L-rhamnose. When protein is expressed directly from *PrhaBAD*, the expression level within each cell falls within a range that correlates very well with the amount of inducer added to the culture [21]. In fact, Giacalone *et al.* presents convincing data that the pRHA-67 vector is more tunable and is capable of higher output than a high-copy vector containing the *araBAD* promoter. The pRHA-67 vector is commercially available from Xbrane Bioscience. Data presented by Haldimann *et al.* [22] indicates that expression from the *rhaBAD* promoter is very tightly regulated, yet this system also offers the potential for 5800-fold induction when glycerol is used as the primary carbon source.

The tetracycline inducible system is also very tightly regulated. Although we do not have experience with this system, Skerra *et al.* [23] reports that the pASK75 vector utilizing the *tetA* promoter/operator and encoding the cognate repressor gene (*tetR*) displays tightly regulated and high-level expression of heterologous protein in several *E. coli* K-12 and B strains. Induction is accomplished with low concentrations of anhydrotetracycline (aTc) and the induction potential is comparable to that of the *lacUV5* promoter. Lutz and Bujard [24] described additional aTc-inducible vectors that make use of the engineered $P_{LtetO\text{-}1}$ promoter, which is also controlled by TetR repression. The pZ vectors offer low, medium or high level expression from $P_{LtetO\text{-}1}$ corresponding to the copy number dictated by the pSC101, p15A, or ColE1 origins of replication, respectively. The aTc-inducible pZ vectors require expression in strains overexpressing TetR (e.g., DH5αZ1). The pSC101 version offers the most strictly regulated expression with an induction/repression ratio of 5000.

1.4 T7 Expression System

Over the last 20 years, the most common vector series for bacterial protein expression is the pET series (*p*lasmid for *e*xpression by *T7* RNA polymerase). The T7 expression system was developed primarily by F. William Studier and colleagues at Brookhaven National Laboratory [25]. The T7 system is best recognized for the capacity to generate a high level of recombinant protein as the phage T7 RNA polymerase is very active and also very selective for phage T7 promoters (e.g., ϕ10). Therefore, T7 transcription within a bacterial cell can be specifically directed at a single promoter within the pET vector carrying the gene of interest. In most T7 expression strains, the chromosomal DE3 prophage carries the T7 RNA polymerase gene (T7 gene 1), which is expressed from the *lacUV5* promoter. Since this promoter is not completely shut off by LacI, some molecules of T7 RNA polymerase are continuously expressed and are able to make considerable amounts of target mRNA in the absence of IPTG. With respect to membrane protein expression, this is an unacceptable situation. An early partial solution to this problem was to include the *lacI* repressor gene on the multicopy pET vector. Thus, LacI repressor protein is produced in large excess relative to its operator binding site present in the *lacUV5* promoter driving T7 gene 1. Another partial solution to leaky T7 expression was the introduction of the *T7–lac* hybrid promoter to the pET vector series. In vectors beginning with pET-10, the *lac* operator sequence overlaps the T7 promoter so that excess LacI is able to inhibit T7-mediated transcription of the target gene. However, even with this improvement uninduced expression is observed in many experiments employing BL21(DE3). Uninduced expression of even mildly toxic gene products may be lethal to BL21(DE3) at the transformation step.

A very effective means to control T7 expression is to coexpress T7 lysozyme, the natural inhibitor of T7 RNA polymerase. Until recently, three types of lysozyme strains were available and all were designed to produce lysozyme at a relatively constant level: pLysS and pLysE express wild-type T7 lysozyme from a low-copy plasmid, and in NEB *lysY* strains, an amidase-negative variant of T7 lysozyme (K128Y) is expressed from a single-copy miniF plasmid. The K128Y variant does not degrade the peptidoglycan layer of the *E. coli* cell wall [26] and, accordingly, *lysY* results in greater overall culture stability when membrane proteins are targeted to the cell envelope. In constitutive lysozyme systems, the level of lysozyme is sufficient to sequester the basal level of T7 RNA polymerase by a 1 : 1 protein interaction. When IPTG is added, the level of T7 RNA polymerase is present in large excess and target protein expression proceeds. If a membrane protein expression plasmid does not yield transformants when using BL21(DE3) or other basic T7 expression strains, the first response should be to test transformation into a lysozyme strain. LysY or pLysS strains may yield normal colonies and express the protein of interest at moderate to high levels. Finally, it should be noted that the choice of lysY or pLysS should take into account downstream processing of cells. Strains expressing active lysozyme often lyse spontaneously upon one freeze–thaw cycle and the resulting cell pellets may be difficult to process.

1.5 Tunable T7 Expression Systems

A recent development in T7 expression is the ability to tune the level of expression. Tunable expression provides a means for optimizing the traffic flow into the membrane translocation pathway. Four commercial strains promote this feature: Tuner™ from Novagen, BL21-AI from Invitrogen, the KRX strain from Promega, and the Lemo21(DE3) strain from New England Biolabs.

- The Tuner strain does not express lac permease (*lacY*) and this allows more uniform uptake of IPTG. However, T7 expression in Tuner strains may still be too robust for membrane protein expression unless the plasmid has a *T7–lac* promoter and lysozyme is coexpressed.
- BL21-AI offers greater potential for expressing toxic gene products as the *araBAD* promoter controls the expression of the T7 RNA polymerase. The associated pDEST expression vectors contain a plain T7 promoter (no *lac* operator site).
- In the Single Step KRX strain, T7 gene 1 expression is controlled by the *rhaBAD* promoter, so greater potential for toxic protein expression is expected. This K-12 strain has been designed for cloning and protein expression.
- The Lemo21(DE3) strain [27] is a tunable T7 expression strain derived from BL21(DE3). Lemo means “less is more” as often less expression results in more protein produced in the desired form. The Lemo strain is distinct from other T7 host strains since the *fraction of functionally active* T7 RNA polymerase is regulated by varying the level of T7 lysozyme (lysY). Fine-tuning is possible

since the LysY inhibitor protein is expressed from the L-rhamnose inducible promoter. The wide-ranging expression potential of Lemo21(DE3) is sampled to find the appropriate level for each target membrane protein. When using Lemo21(DE3), expression media should lack glucose since this carbon source affects lysozyme expression from *PrhaBAD*.

1.6
Other Useful Membrane Protein Expression Strains

C41(DE3) and C43(DE3) have been employed as membrane protein expression strains since their isolation from parent strain BL21(DE3) in 1996 [28]. Recently, Wagner *et al.* [26] reported that these two strains carry mutations within the promoter driving expression of the T7 RNA polymerase. Therefore, the characteristic robust T7 expression of DE3 strains is attenuated in C41(DE3) and C43(DE3), and this accounts for the advantage observed in the expression of some toxic proteins.

More recently, the TOP10 strain was subjected to a genetic selection procedure that produced several mutant strains exhibiting improved expression of heterologous membrane protein. This work was completed by Elizabeth Massey-Gendel *et al.* under the direction of James Bowie at the University of California at Los Angeles (UCLA). Target membrane proteins were expressed with a C-terminal cytoplasmic fusion to mouse dihydrofolate reductase (DHFR) (providing resistance to trimethoprim) or to a kanamycin resistance protein. A positive hit in the selection was obtained when a mutant strain was capable of expressing both fusion proteins at a level sufficient to provide resistance to both drugs. Five of the selected strains have been characterized in some detail [29] and the genomes of two such strains have been sequenced. At the January 2010 Peptalk meeting in San Diego, Professor Bowie reported that his lab is currently investigating the relevance of the mutations identified in the TOP10 derivatives designated as EXP-Rv1337-1 and EXP-Rv1337-5. The results of this investigation are widely anticipated. The DE3 prophage has been added to the EXP strains so that T7 expression is possible.

The Single Protein Production System (SPP System™) was developed by Masayori Inouye [30] and is marketed by Takara Bio. This is a two-vector system suitable for use in most *E. coli* strains. The target protein is expressed from a vector with the cold-inducible *E. coli cspA* promoter, which is of course consistent with membrane protein expression. The unique, enabling feature of the SPP System is the inducible expression of a site-specific mRNA interferase (MazF) from a second plasmid, which degrades endogenous mRNA by acting at ACA sites. Accordingly, the gene of interest must be synthesized to lack ACA sequences. The net result is that the target mRNA persists and becomes a preferential substrate for the translation machinery. The elimination of most host-derived mRNA is reported to create a quasidormant cell where expression of the target membrane protein is sustained. If Sec translocase function is also sustained, then this system may offer an advantage, as the target protein should encounter less competition from endogenous proteins on the membrane translocation pathway.

1.7 Clone Stability

When expressing membrane proteins, clone stability should always be a concern. The first indication of clone toxicity is often realized during the initial cloning/transformation step. Poor transformation results may indicate that mutant genes are being selected during the cloning step, so sequence verification is always advised and is absolutely critical if the gene has been amplified by polymerase chain reaction. If a clone is suspected to be toxic, certain precautions should be followed. First, lower growth temperatures are often stabilizing. Also, it is beneficial to include 0.1% glucose in selection plates in many situations. Glucose will repress basal expression from *Plac, PlacUV5, ParaBAD,* and *Prha.* (Note: *Ptac* and *Ptrc* are not subject to glucose repression as the CAP binding site is absent from these promoters). Glucose containing plates are also advantageous when transforming clones into T7 Express and DE3 expression strains, as the T7 RNA polymerase gene is controlled by *Plac* and *PlacUV5* in these strains, respectively. One exception is transformation into Lemo21(DE3) where glucose repression is not stabilizing. When transforming extremely toxic clones into Lemo21(DE3), 500 μM rhamnose addition to selection plates and starter cultures will reduce the basal expression to an undetectable level (Figure 1.1).

During the outgrowth stage for protein expression, plasmid maintenance should be examined. This is especially critical when propagating AmpR vectors, as the resistance protein (β-lactamase) is secreted and ampicillin may be completely degraded. Plasmid maintenance is easily checked by plating cells at the point of induction onto drug containing plates versus nondrug plates. If a significantly lower number of colonies are counted on the drug plates (below 80% the number counted on nondrug plates), then modifications to the protocol or the clone may be necessary. If plasmid maintenance is an issue with AmpR constructs, increasing the level of ampicillin to 200 μg/ml is recommended. Alternatively, initiate growth with 100 μg/ml ampicillin and then spike in another dose (100 μg/ml) at

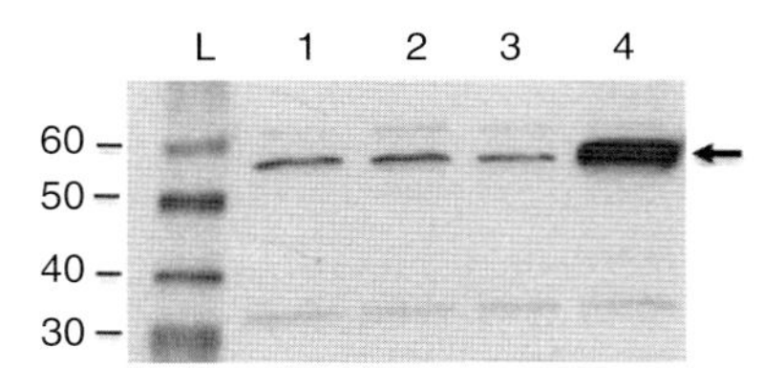

Figure 1.1 T7 expression is tightly regulated in Lemo21(DE3) cells. Whole-cell lysates were subjected to SDS–PAGE, and target protein was detected using anti-YidC serum that recognizes both endogenous wild-type YidC and recombinant 6His-YidC membrane protein expressed from pET28c. (1) No vector control indicating endogenous YidC level; (2) cells containing pET28-6hisyidC, no IPTG, no rhamnose; (3) cells containing pET28-6hisyidC, 500 μM rhamnose, no IPTG; (4) cells containing pET28-6hisyidC, 500 μM rhamnose, 400 μM IPTG. Arrow indicates YidC target.

mid-log stage. Carbenicillin (at 50–200 μg/ml) may be used in place of ampicillin. According to the Novagen pET system manual, pET (AmpR) clones may be stabilized by using high concentrations of carbenicillin and by changing the medium twice prior to induction. Carbenicillin is more stable than ampicillin in low pH conditions, which may be encountered after extended fermentation periods.

Vectors expressing KanR or CamR are preferred for creating membrane protein clones. One versatile KanR vector is pET28, which allows for simple construction of genes tagged at either end with the polyhistidine coding sequence. pBAD33 (CamR) is also a good choice as expression is tightly regulated. (Note: when cloning into the pBAD33 polylinker, a translation initiation signal (RBS site) must be included with the gene insert). In extreme cases plasmid maintenance systems can be incorporated. For example, the *hok/sok* system has been utilized by groups expressing G-protein-coupled receptors (GPCRs) in *E. coli* [31–32].

1.8
Media Types

The type of media is also an important consideration. Although the use of LB is commonly cited, we generally observe a greater level of membrane protein expression in Terrific Broth (TB). This conclusion was made after multiple expression trials using a *tac* promoter, which is insensitive to glucose repression. TB is a rich broth buffered by potassium phosphate and containing glycerol as a carbon source [33]. When the target protein is expressed from *Plac, PlacUV5, ParaBAD*, or *Prha*, a rich media containing a low-level of glucose may be more appropriate. With respect to controlling expression in BL21(DE3), Pan and Malcolm [34] found that 1% glucose addition to either TB or M9 starter cultures minimized basal expression to a level equal to pLysS-containing strains. These researchers further demonstrated that glucose addition is less important in a strain expressing lysozyme to control basal T7 expression. When target protein expression is driven directly from a sugar promoter, then glucose repression is advised. For example, pBAD constructs may be stabilized by growth in media containing 0.1% glucose, which should be metabolized by the point of induction with arabinose. Such a protocol leads to a discussion of “autoinduction” media [35] marketed as the Overnight Express™ autoinduction system for simplified T7 expression. The advantages of this system are: (i) manual IPTG induction is not required and (ii) expression trials are more reproducible as growth is carried out in a defined media containing a mix of carbon sources (generally glucose, lactose, and glycerol). When glucose is depleted, lactose serves to induce expression of the T7 RNA polymerase from the *lacUV5* promoter in DE3 strains. The actual inducer molecule is allolactose that is produced by β-galactosidase (the *lacZ* gene product). Thus, Studier points out that autoinduction should be performed in strains encoding an intact *lac* operon. Note: The T7 Express line of strains (NEB) are not suitable for autoinduction protocols as the T7 RNA polymerase gene disrupts the *lacZ* open reading frame (ORF).

1.9 Fusion Partners/Membrane Targeting Peptides

A first step in cloning or characterizing heterologous membrane protein ORFs is the analysis of membrane topology using more than one algorithm. Four common predictors are: SPOCTOPUS [36] (octopus.cbr.su.se), TopPred (http://mobyle.pasteur.fr/cgi-bin/portal.py?form=toppred), Phobius [37] (phobius.sbc.su.se), and TargetP 1.1 [38] (http://www.cbs.dtu.dk/services/TargetP/). Nielsen *et al.* [39] showed more than 10 years ago that eukaryotic secretory proteins (e.g., membrane receptors) are expressed with N-terminal signals that resemble cleavable signal peptides found in bacteria. *E. coli* proteins exported to the periplasm or outer membrane are expressed with a signal peptide that is cleaved by signal peptidase. In contrast, most endogenous *E. coli* membrane proteins are expressed with a more hydrophobic N-terminal signal that remains uncleaved (signal anchor). A more hydrophobic N-terminal signal increases the probability of *E. coli* SRP recognition and targeting of a protein to the cotranslational membrane insertion pathway [40]. Thus, when designing constructs for *E. coli* expression, the hydrophobicity of the N-terminal residues should be evaluated. If necessary, the N-terminal signal of the protein of interest may be replaced by a relatively hydrophobic signal from a different protein. For example, Chang *et al.* [41] tested eight different membrane-targeting peptides to maximize the *E. coli* expression of a plant derived P450 enzyme 8-cadinene hydroxylase (CAH). The results varied widely and surprisingly the signal from a bovine CAH performed the best. One note of caution regarding heterologous signal peptides is that rare codons may significantly impact expression results. Although using strains that correct for rare codons is an option (Table 1.1) the selection of an appropriate signal sequence is an empirical process because fully optimized translation at the 5′-end of the message is not necessarily advantageous. Other options are to replace the native signal sequence with an N-terminal fusion partner that travels the SRP pathway. For example, the *E. coli* GlpF protein has been used as a fusion partner [42] or the rationally designed P8CBDek fusion partner will also facilitate membrane targeting/expression of foreign membrane proteins in *E. coli* [9].

Maltose binding protein (MBP) from *E. coli* is a tried-and-true N-terminal fusion partner for enhancing the expression/solubility of heterologous protein. Native MBP is a periplasmic protein that is exported via the Sec pathway. Therefore, proteins fused to MBP (containing its native signal peptide) are targeted to the Sec translocase and, as such, have the opportunity to be integrated into the inner membrane upon completion of MBP export. This method has been applied to facilitate the expression of several eukaryotic membrane proteins. For example, the human cannabinoid receptor CB2 [32], human serotonin 5-HT_{1A} [43], rat neurotensin receptor [31], and the prokaryotic Glvi proton-gated ion channel [44] have been expressed in *E. coli* in functional form as fusions to MBP. Several studies have established that a large fraction of MBP molecules are delivered to the membrane and SecA in a post-translational manner after recognition by the cytoplasmic chaperone SecB. Accordingly, a reasonable assumption is that MBP–membrane protein fusion expression might be improved by the overexpression of chaperones such as SecB, DnaK/DnaJ, and GroEL/GroES in order to protect hydrophobic segments from aggregation within the cytoplasm before they engage the Sec translocase.

1.10
Chaperone Overexpression

Many studies have demonstrated that cytoplasmic chaperone overexpression improves the expression of heterologous soluble proteins in *E. coli*. This approach has also been tested to aid the overexpression of CorA, an atypical membrane protein from *E. coli*. CorA lacks an N-terminal hydrophobic signal and is believed to integrate into the inner membrane by a post-translational process. Chen *et al.* [45] carried out a comprehensive study to determine the optimal conditions for CorA overexpression and to determine the effect of overexpressing several different *E. coli* factors relevant to protein biogenesis. A conclusion of this study was that increasing DnaK 8-fold resulted in a 4-fold increase of membrane integrated CorA when expression was carried out at 37 °C. However, the same net result (13–15 mg CorA) was obtained by simply lowering the expression temperature to the range of 15–30 °C. Thus, the underlying conclusion of this work was that the yield of membrane protein per cell may be increased by reducing the synthesis rate of the target protein. In fact, most studies indicate that membrane protein expression is optimal at 20–30 °C, with 20 °C being more favorable for more difficult polytopic membrane proteins. Chen *et al.* calculated the translation rate of CorA at the extremes: in wild-type *E. coli*, the average rate of CorA synthesis was estimated to be 600 molecules/cell/min at 15 °C and 5500 molecules/cell/min at 37 °C, whereas export of proOmpA through the Sec translocase was estimated to be 450–900 molecules/cell/min at 37 °C [46]. Since protein synthesis rate can easily exceed the Sec translocase capacity, this clearly points out the fact that the Sec translocase is a bottleneck for membrane protein expression in *E. coli*. Engineering a strain with greater translocase activity has not been achievable so far, so the most practical option is to express recombinant membrane proteins “in moderation.” Moderate expression at lower temperatures helps to ensure that the chaperone pool is not exceeded and also that the translation rate does not far exceed the capacity of the Sec translocase.

Link *et al.* [47] recently examined the overexpression of the type 1 cannabinoid receptor CB1 and found a positive effect from several different helper proteins. In this study, DnaK/DnaJ coexpression from a plasmid again showed promise in increasing membrane protein yield. Furthermore, overexpression of Ffh (SRP protein component) and Trigger Factor (cotranslational chaperone) provided some benefit (2- to 3-fold enhancement). Most remarkably, the overexpression of FtsH (an inner membrane protease) resulted in up to 8-fold enhancement of GPCR expression and nearly a 2-fold higher cell density after 36 h at 12 °C. Link *et al.* also state in their Discussion that “FtsH overexpression also increases the bacterial production of native *E. coli* membrane proteins such as YidC." The rationale behind the FtsH effect will be discussed in the next section.

1.11 Cautionary Notes Related to Chaperone Overexpression

DnaK overexpression may result in a situation where a fraction of the target protein is isolated as a complex with DnaK. This is a common occurrence during overexpression and Ni-NTA isolation of soluble protein and this outcome was also reported in the CorA–DnaK study. DnaK contamination after one or more chromatography steps most certainly indicates that a fraction of the target protein is not folded properly. Furthermore, attempts to remove DnaK contamination may be futile (personal experience with overexpression of soluble protein). DnaJ chaperone expression in wild-type strains apparently acts to inhibit the expression some membrane proteins according to Skretas and Georgiou [48]. This conclusion was made after a DnaJ null strain displayed a large increase in CB1–Green Fluorescent Protein (GFP) fluorescence and in the production of membrane-integrated CB1. When designing constructs for helper protein (e.g., DnaK) expression, do not employ high-copy vectors with inducible promoters. A simple and effective method is to clone the helper protein gene with its native promoter onto a low-copy vector such as pACYC184. Then the helper protein will be moderately overexpressed and its expression may be naturally regulated upon the induction of cellular stress responses.

1.12 Emerging Role of Quality Control Proteases

Although bacterial cells are extremely efficient factories for protein production, some recombinant proteins fall off or become stalled on the pathway to their final folded destination. This is especially reasonable to imagine for heterologous, hydrophobic proteins. So what happens to such proteins? One outcome is that such proteins serve as aggregation targets and the continuous supply of induced recombinant protein accumulates as inclusion bodies within the cytoplasm. During this process, the cell responds by upregulating chaperones *and* proteases to take care of the state of disarray [49]. Thus, if endogenous membrane protein assembly fails, protease degradation is a natural response. One bit of direct evidence for this comes from the work of van Bloois *et al.* [50] where FtsH protease was able to be crosslinked to the YidC, a protein with a loosely defined membrane protein chaperone activity [51–52]. FtsH is capable of processive, ATP-dependent degradation of *E. coli* membrane substrates such as YccA and SecY [53]. In cases of recombinant membrane protein overexpression, this activity apparently clears away those molecules that become stalled during the post- or cotranslational integration process. Other proteases must certainly play a role in clearing the membrane translocation pathway. In retrospect, at least two studies suggested that the cytoplasmic Lon protease acts as a quality control factor during membrane protein biogenesis as well as soluble protein biogenesis. The first clue came from work in Tom Silhavy's lab published in 1992, where William Snyder was studying the *prlF1* host mutation, which suppresses Sec pathway jamming by the LamB–LacZ fusion protein [54]. Nearly 20 years ago, it was difficult to explain why the PrlF1 phenotype was only observed in *prlF1/lon*$^+$ strains. However, now it is certain that Lon functions to clear away fully translated proteins that are misfolded or partially translated proteins that are stalled at the ribosome [55]. The relevance of the *prlF* gene product is still a bit ambiguous, but recently Schmidt *et al.* [56] showed that PrlF is an antitoxin that counteracts the bacteriostatic effect of its toxin partner YhaV. The *prlF1* 7-bp insertion has two effects: (i) the downstream *yhaV* toxin gene is expressed to a much lesser extent and (ii) the PrlF1 antitoxin is stabilized by amino acids changes at its C-terminus. The overall net result appears to be that deactivation of the toxin system allows cells to better recover from jamming of the Sec pathway. We have tested the hypothesis that Lon is a key factor in rescuing ribosomes that stall during the cotranslational membrane insertion process. Our preliminary data indicates that Lon complementation improves the growth of cells overexpressing a polytopic membrane protein. However, so far an improved yield of membrane protein per cell has not been demonstrated. A role for Lon in membrane protein biogenesis is also suggested by Harris Bernstein's characterization of the SRP pathway. In a study published in 2001, Bernstein and Hyndman [57] reported that proteases Lon and ClpQ become essential when the SRP level is reduced, suggesting that SRP-deficient cells require an increased capacity to degrade mislocalized inner membrane proteins.

Other cell envelope proteases may be essential for membrane protein biogenesis. For example, Wang *et al.* [58] studied YidC-depleted cells and found that several cell envelope proteases were elevated. One such enzyme is HtpX, an inner membrane protease regulated by the Cpx envelope stress response [59].

Contrary to the immediate discussion, deletion of some host proteases may be beneficial. For example, OmpT acts at dibasic sequences and it may be detrimental during the protein isolation stage. Thus, preferred protein expression strains (e.g., BL21 derivatives) are OmpT minus. Note that most K-12 strains express OmpT. However, the K-12 KRX strain has been engineered to lack OmpT protease. Also, uncharacterized differences in protease expression between B and K-12 strains may affect the quality of expressed protein. One specific example is demonstrated by Luo *et al.* [9] where the same protein was resistant to proteolysis when expressed in NEB Express (a BL21 derivative), but was less stable during expression in MC1061, a robust K-12 strain.

1.13
Tag Selection

The selection of affinity tags/detection epitopes is an important consideration when constructing a recombinant membrane protein clone. As expression levels are characteristically low, Western analysis is a routine procedure. However, extremely hydrophobic proteins may run anomalously on standard gel systems or may transfer poorly to Western detection membranes. Before performing sodium dodecyl sulfate–polyacrylamide gel electrophoresis (SDS–PAGE) analysis, it may be necessary to heat samples at 37 °C rather than 95 °C to avoid aggregation of expressed membrane protein. One alternative to Western analysis is to monitor the expression/purification by expressing the protein of interest as a C-terminal fusion to GFP [60].

Metal-affinity chromatography is undoubtedly the preferred method for isolating recombinant membrane proteins from *E. coli*. Many researchers make use of eight histidines to improve the yield from low expressing clones. This may also reduce the background of *E. coli* metal binding proteins, as the protein of interest should be eluted at a higher imidazole concentration. If the isolation procedure includes an ultracentrifugation step to pellet the membrane fraction, then soluble *E. coli* metal binding proteins are less of a concern. The position of a polyhistidine tag may affect expression levels, behavior in solution and the propensity for membrane protein crystal formation [61]. Furthermore, if the protein N-terminus may be subject to signal peptidase cleavage, then of course the affinity tag needs to be located elsewhere. A polyhistidine tag may even be placed within a cytoplasmic loop of the target protein [62]. Histidine-rich sequences may affect membrane translocation, so a cytoplasmic location is recommended. Other effective purification tags include the Strep tag, the FLAG tag, or other immunoaffinity sequences [63]. For a more complete discussion of suitable fusion tags, consult the recent reviews by Xie *et al* [64]. The FLAG tag is rich with charged residues DYKDDDDK yet this tag is efficiently translocated across the inner membrane in the context of the P8CBD fusion partner [9]. The FLAG tag conveniently encodes the enterokinase protease site DDDDK and this protease works well in detergent-containing buffer. Thrombin is also recommended for removal of tags from membrane proteins as this protease shows reliable activity in many detergents. In contrast, tobacco etch virus (TEV) protease suffers from poor activity in several common detergents such as octyl-glucopyranoside [65].

1.14 Potential Expression Yield

Most importantly, have modest expectations. Always remember that quality is more important than quantity when attempting to overexpress membrane proteins in bacteria. Regarding prokaryotic proteins: any yield of membrane-integrated protein above 3 mg/l of culture is a good result. According to the NEB catalog 2007–2008, this level corresponds to approximately 2% of total cellular protein. Levels in the 10–20 mg/l range might be obtained for native *E. coli* membrane proteins (e.g., CorA study). In contrast, proteins from higher organisms will be expressed at much lower levels in most cases. An expression yield (in the membrane fraction) approaching 1 mg/l is an outstanding achievement for eukaryotic polytopic membrane proteins. This discrepancy even after gene/codon optimization is not completely understood, but the following factors may be responsible:

i) The eukaryotic protein translation rate is generally lower than the rate in bacteria even when grown at low temperatures. Thus, eukaryotic membrane proteins may have evolved to require different chaperone requirements or folding timescales.

ii) The molecular composition of eukaryotic membranes varies considerably from the inner membrane composition of *E. coli*. For example, the bacterial plasma (inner) membrane is less rigid than the plasma membrane of mammalian cells due to the lack of cholesterol. Accordingly, membrane protein stability may be influenced.

iii) Wild-type *E. coli* cells do not offer the possibility of post-translational modifications (e.g., glycosylation) that may be necessary to stabilize some eukaryotic proteins.

iv) Bacteria express a different repertoire of membrane proteases that may act on some sequences/structures presented in heterologous proteins.

1.15
Strategies to Overcome Protein Instability

Protein instability may be the result of many factors: unproductive membrane insertion may lead to degradation by cellular proteases. In contrast, proteins failing to assemble properly in the membrane may aggregate *in vivo* or, postexpression, the protein of interest may aggregate as a result of nonoptimal buffer conditions during membrane solubilization or affinity chromatography. Expression and isolation of “stable” membrane protein is still a very empirical process. If a detergent/buffer screen fails to give a protein sample amenable to characterization, then it may be wise to screen homologs, truncation mutants, or point mutants. Screening proteins from multiple organisms is a common practice in order to find one member of a family that behaves well enough for characterization. Proteins from thermophilic organisms may be better behaved during expression and/or crystallization but this is not a general rule. Screening truncation mutants is a common practice with soluble proteins, but this type of systematic approach is not well tested with membrane proteins. Finally, point mutations may provide stabilization but the study of a random variant protein is less desirable. The radical approach of engineering a stabilization domain (T4 lysozyme) into a cytoplasmic loop of the human β_2-adrenergic receptor aided in expression within Sf9 insect cells and influenced the formation of quality protein crystals [66]. Another proven method for stabilizing the same GPCR is cocrystallization of a monoclonal antibody that binds to an inherently flexible region of the receptor [67]. Perhaps similar approaches will be fruitful for bacterial expression and/or crystallization of unstable polytopic membrane proteins.

Acknowledgments

Lab members Jianying Luo, Carine Robichon, and Julie Choulet are recognized for many contributions including: sharing of unpublished data, critical reading of the manuscript, and assistance with preparation of figures/tables. Additionally, Bill Jack and Elisabeth Raleigh of New England Biolabs are thanked for critical reading of the manuscript.

Abbreviations

aTc	anhydrotetracycline
CAH	8-cadinene hydroxylase
CAP	catabolite gene activator protein
DHFR	dihydrofolate reductase
GFP	Green Fluorescent Protein
GPCR	G-protein-coupled receptor
IPTG	isopropyl-β-D-thiogalactopyranoside
MBP	Maltose binding protein
RBS	ribosomal binding site
SDS	sodium dodecyl sulfate
SRP	signal recognition particle
PAGE	polyacrylamide gel electrophoresis
TB	Terrific Broth
TEV	tobacco etch virus
TM	transmembrane
UCLA	University of California at Los Angeles

References

1 Buchanan, S.K. (1999) Beta-barrel proteins from bacterial outer membranes: structure, function and refolding. *Curr. Opin. Struct. Biol.*, **9**, 455–461.

2 Kiefer, H. (2003) *In vitro* folding of alpha-helical membrane proteins. *Biochim. Biophys. Acta*, **1610**, 57–62.

3 Rogl, H., Kosemund, K., Kuhlbrandt, W., and Collinson, I. (1998) Refolding of *Escherichia coli* produced membrane protein inclusion bodies immobilized by nickel chelating chromatography. *FEBS Lett.*, **432**, 21–26.

4 Sarramegna, V., Muller, I., Milon, A., and Talmont, F. (2006) Recombinant G protein-coupled receptors from expression to renaturation: a challenge towards structure. *Cell. Mol. Life Sci.*, **60**, 1–16.

5 Xie, K. and Dalbey, R.E. (2008) Inserting proteins into the bacterial cytoplasmic membrane using the Sec and YidC translocases. *Nat. Rev. Microbiol.*, **6**, 234–244.

6 Roosild, T.P., Greenwald, J., Vega, M., Castronovo, S., Riek, R., and Choe, S. (2005) NMR structure of Mistic, a membrane-integrating protein for membrane protein expression. *Science*, **307**, 1317–1321.

7 Serek, J., Bauer-Manz, G., Struhalla, G., van den Berg, L., Kiefer, D., Dalbey, R., and Kuhn, A. (2004) *Escherichia coli* YidC

is a membrane insertase for Sec-independent proteins. *EMBO J.*, **23**, 294–301.

8 Samuelson, J.C., Chen, M., Jiang, F., Moller, I., Wiedmann, M., Kuhn, A., Phillips, G.J., and Dalbey, R.E. (2000) YidC mediates membrane protein insertion in bacteria. *Nature*, **406**, 637–641.

9 Luo, J., Choulet, J., and Samuelson, J.C. (2009) Rational design of a fusion partner for membrane protein expression in *E. coli*. *Protein Sci.*, **18**, 1735–1744.

10 Luirink, J., von Heijne, G., Houben, E., and de Gier, J.-W. (2005) Biogenesis of inner membrane proteins in *Escherichia coli*. *Annu. Rev. Microbiol.*, **59**, 329–355.

11 Bahari, L., Parlitz, R., Eitan, A., Stjepanovic, G., Bochkareva, E.S., Sinning, I., and Bibi, E. (2007) Membrane targeting of ribosomes and their release require distinct and separable functions of FtsY. *J. Biol. Chem.*, **282**, 32168–32175.

12 Weiche, B., Bürk, J., Angelini, S., Schiltz, E., Thumfart, J.O., and Koch, H.G. (2008) A cleavable N-terminal membrane anchor is involved in membrane binding of the *Escherichia coli* SRP receptor. *J. Mol. Biol.*, **377**, 761–773.

13 Hirschel, B.J., Shen, V., and Schlessinger, D. (1980) Lactose operon transcription from wild-type and L8-UV5 *lac* promoters in *Escherichia coli* treated with chloramphenicol. *J. Bacteriol.*, **143**, 1534–1537.

14 deBoer, H.A., Comstock, L.J., Yansura, D.G., and Heynecker, H.L. (1982) Construction of a tandem *trp–lac* promoter and a hybrid *trp–lac* promoter for efficient and controlled expression of the human growth hormone gene in *Escherichia coli*, in *Promoters: Structure and Function* (eds R.L. Rodriguez and M.J. Chamberlin), Praeger, New York, pp. 462–481.

15 deBoer, H.A., Comstock, L.J., and Vasser, M. (1983) The *tac* promoter: a functional hybrid derived from *trp* and *lac* promoters. *Proc. Natl. Acad. Sci. USA*, **80**, 21–25.

16 Amann, E., Brosius, J., and Ptashne, M. (1983) Vectors bearing a hybrid *trp–lac* promoter useful for regulated expression of cloned genes in *Escherichia coli*. *Gene*, **25**, 167–178.

17 Mulligan, M.E., Brosius, J., and Clure, W.R. (1985) Characterization *in vitro* of the effect of spacer length on the activity of *Escherichia coli* RNA polymerase at the *tac* promoter. *J. Biol. Chem.*, **260**, 3529–3538.

18 Calos, M.P. (1978) DNA Sequence for a low-level promoter of the *lac* repressor gene and an “up” promoter mutation. *Nature*, **274**, 762–765.

19 Guzman, L.M., Belin, D., Carson, M.J., and Beckwith, J. (1995) Tight regulation, modulation, and high-level expression by vectors containing the arabinose PBAD promoter. *J. Bacteriol.*, **177**, 4121–4130.

20 Giacalone, M.J., Gentile, A.M., Lovitt, B.T., Berkley, N.L., Gunderson, C.W., and Surber, M.W. (2006) Toxic protein expression in *Escherichia coli* using a rhamnose-based tightly regulated and tunable promoter system. *Biotechniques*, **40**, 355–364.

21 Siegele, D.A. and Hu, J.C. (1997) Gene expression from plasmids containing the araBAD promoter at subsaturating inducer concentrations represents mixed populations. *Proc. Natl. Acad. Sci. USA*, **94**, 8168–8172.

22 Haldimann, A., Daniels, L.L., and Wanner, B.L. (1998) Use of new methods for construction of tightly regulated arabinose and rhamnose promoter fusions in studies of the *Escherichia coli* phosphate regulon. *J. Bacteriol.*, **180**, 1277–1286.

23 Skerra, A. (1994) Use of the tetracycline promoter for the tightly regulated production of a murine antibody fragment in *Escherichia coli*. *Gene*, **151**, 131–135.

24 Lutz, R. and Bujard, H. (1997) Independent and tight regulation of transcriptional units in *Escherichia coli* via the *lac*R/O, the TetR/O and AraC/I1-I2 regulatory elements. *Nucleic Acids Res.*, **25**, 1203–1210.

25 Studier, F.W., Rosenberg, A.H., Dunn, J.J., and Dubendorff, J.W. (1990) Use of T7 RNA polymerase to direct expression of cloned genes. *Methods Enzymol.*, **185**, 60–89.

26 Cheng, X., Zhang, X., Pflugrath, J.W., and Studier, F.W. (1994) The structure of

bacteriophage T7 lysozyme, a zinc amidase and an inhibitor of T7 RNA polymerase. *Proc. Natl. Acad. Sci. USA*, **91**, 4034–4038.

27 Wagner, S., *et al.* (2008) Tuning *Escherichia coli* for membrane protein overexpression. *Proc. Natl. Acad. Sci. USA*, **105**, 14371–14376.

28 Miroux, B. and Walker, J.E. (1996) Over-production of proteins in *Escherichia coli*: mutant hosts that allow synthesis of some membrane proteins and globular proteins at high levels. *J. Mol. Biol.*, **260**, 289–298.

29 Massey-Gendel, E., Zhao, A., Boulting, G., Kim, H.-Y., Balamotis, M.A., Seligman, L.M., Nakamoto, R.K., and Bowie, J.U. (2009) Genetic selection system for improving recombinant membrane protein expression in *E. coli*. *Protein Sci.*, **18**, 372–383.

30 Suzuki, M., Mao, L., and Inouye, M. (2007) Single protein production (SPP) system in *Escherichia coli*. *Nat. Protoc.*, **2**, 1802–1810.

31 Tucker, J. and Grisshammer, R. (1996) Purification of a rat neurotensin receptor expressed in *Escherichia coli*. *Biochem. J.*, **317**, 891–899.

32 Yeliseev, A.A., Wong, K.K., Soubias, O., and Gawrisch, K. (2005) Expression of human peripheral cannabinoid receptor for structural studies. *Protein Sci.*, **14**, 2638–2653.

33 Tartof, K.D. and Hobbs, C.A. (1988) New cloning vectors and techniques for easy and rapid restriction mapping. *Gene*, **30**, 169–182.

34 Pan, S.H. and Malcolm, B.A. (2000) Reduced background expression and improved plasmid stability with pET vectors in BL21(DE3). *Biotechniques*, **29**, 1234–1237.

35 Studier, F.W. (2005) Protein production by auto-induction in high-density shaking cultures. *Protein Expr. Purif.*, **41**, 207–234.

36 Viklund, H., Bernsel, A., Skwark, M., and Elofsson, A. (2008) SPOCTOPUS: a combined predictor of signal peptides and membrane protein topology. *Bioinformatics*, **24**, 2928–2929.

37 Käll, L., Krogh, A., and Sonnhammer, E.L. (2007) Advantages of combined transmembrane topology and signal peptide prediction – the Phobius web server. *Nucleic Acids Res.*, **35**, W429–W432.

38 Emanuelsson, O., Brunak, S., von Heijne, G., and Nielsen, H. (2006) Locating proteins in the cell using TargetP, SignalP and related tools. *Nat. Protoc.*, **2**, 95371.

39 Nielsen, H., Engelbrecht, J., Brunak, S., and von Heijne, G. (1997) Identification of prokaryotic and eukaryotic signal peptides and prediction of their cleavage sites. *Protein Eng.*, **10**, 1–6.

40 Lee, H.C. and Bernstein, H.D. (2001) The targeting pathway of *Escherichia coli* presecretory and integral membrane proteins is specified by the hydrophobicity of the targeting signal. *Proc. Natl. Acad. Sci. USA*, **98**, 3471–3476.

41 Chang, M.C.Y., Eachus, R.A., Trieu, W., Ro, D.-K., and Keasling, J.D. (2007) Engineering *Escherichia coli* for production of functionalized terpenoids using plant P450s. *Nat. Chem. Biol.*, **3**, 274–277.

42 Neophytou, I., Harvey, R., Lawrence, J., Marsh, P., Panaretou, B., and Barlow, D. (2007) Eukaryotic integral membrane protein expression utilizing the *Escherichia coli* glycerol-conducting channel protein (GlpF). *Appl. Microbiol. Biotechnol.*, **77**, 375–381.

43 Bertin, B., Freissmuth, M., Breyer, R.M., Schutz, W., Strosberg, A.D., and Marullo, S. (1992) Functional expression of the human serotonin 5-HT1A receptor in *Escherichia coli*. *J. Biol. Chem.*, **267**, 8200–8206.

44 Bocquet, N., Prado de Carvalho, L., Cartaud, J., Neyton, J., Le Poupon, C., Taly, A., Grutter, T., Changeux, J.P., and Corringer, P.J. (2007) A prokaryotic proton-gated ion channel from the nicotinic acetylcholine receptor family. *Nature*, **445**, 116–119.

45 Chen, Y., Song, J., Sui, S.-f., and Wang, D.-N. (2003) DnaK and DnaJ facilitated the folding process and reduced inclusion body formation of magnesium transporter CorA overexpressed in *Escherichia coli*. *Protein Exp. Purif.*, **32**, 221–231.

46 De Keyzer, J., van der Does, C., and Driessen, A.J. (2002) Kinetic analysis of the translocation of fluorescent precursor

proteins into *Escherichia coli* membrane vesicles. *J. Biol. Chem.*, **277**, 46059–46065.
47 Link, A.J., Skretas, G., Strauch, E.-M., Chari, N.S., and Georgiou, G. (2008) Efficient production of membrane-integrated and detergent-soluble G protein-coupled receptors in *Escherichia coli*. *Protein Sci.*, **17**, 1857–1863.
48 Skretas, G. and Georgiou, G. (2009) Genetic analysis of G protein-coupled receptor expression in *Escherichia coli*: inhibitory role of DnaJ on the membrane integration of the human central cannabinoid receptor. *Biotechnol. Bioeng.*, **102**, 357–367.
49 Wagner, S., Baars, L., Ytterberg, A.J., Klussmeier, A., Wagner, C.S., Nord, O., Nygren, P.A., van Wijk, K.J., and de Gier, J.W. (2007) Consequences of membrane protein overexpression in *Escherichia coli*. *Mol. Cell Proteomics*, **6**, 1527–1550.
50 van Bloois, E., Dekker, H.L., Fröderberg, L., Houben, E.N., Urbanus, M.L., de Koster, C.G., de Gier, J.W., and Luirink, J. (2008) Detection of cross-links between FtsH, YidC, HflK/C suggests a linked role for these proteins in quality control upon insertion of bacterial inner membrane proteins. *FEBS Lett.*, **582**, 1419–1424.
51 Nagamori, S., Smirnova, I.N., and Kaback, H.R. (2004) Role of YidC in folding of polytopic membrane proteins. *J. Cell Biol.*, **165**, 53–62.
52 Kol, S., Nouwen, N., and Driessen, A.J. (2008) Mechanisms of YidC-mediated insertion and assembly of multimeric membrane protein complexes. *J. Biol. Chem.*, **283**, 31269–31273.
53 Kihara, A., Akiyama, Y., and Ito, K. (1999) Dislocation of membrane proteins in FtsH-mediated proteolysis. *EMBO J.*, **18**, 2970–2981.
54 Snyder, W.B. and Silhavy, T.J. (1992) Enhanced export of beta-galactosidase fusion proteins in prlF mutants is Lon-dependent. *J. Bacteriol.*, **174**, 5661–5668.
55 Choy, J.S., Aung, L.L., and Karzai, A.W. (2007) Lon protease degrades transfer-messenger RNA-tagged proteins. *J. Bacteriol.*, **189**, 6564–6571.
56 Schmidt, O., Schuenemann, V.J., Hand, N.J., Silhavy, T.J., Martin, J., Lupas, A.N., and Djuranovic, S. (2007) *prlF* and *yhaV* encode a new toxin–antitoxin system in *Escherichia coli*. *J. Mol. Biol.*, **372**, 894–905.
57 Bernstein, H.D. and Hyndman, J.B. (2001) Physiological basis for conservation of the signal recognition particle targeting pathway in *Escherichia coli*. *J. Bacteriol.*, **183**, 2187–2197.
58 Wang, P., Kuhn, A., and Dalbey, R.E. (2010) Global change of gene expression and cell physiology in YidC-depleted *E. coli*. *J. Bacteriol.*, **192**, 2193–2209.
59 Shimohata, N., Chiba, S., Saikawa, N., Ito, K., and Akiyama, Y. (2002) The Cpx stress response system of *Escherichia coli* senses plasma membrane proteins and controls HtpX, a membrane protease with a cytosolic active site. *Genes Cells*, **7**, 653–662.
60 Drew, D., Lerch, M., Kunji, E., Slotboom, D.-J., and de Gier, J.-W. (2006) Optimization of membrane protein overexpression and purification using GFP fusions. *Nat. Methods*, **3**, 303–313.
61 Mohanty, A.K. and Wiener, M.C. (2004) Membrane protein expression and production: effects of polyhistidine tag length and position. *Protein Expr. Purif.*, **33**, 311–325.
62 Paetzel, M., Strynadka, N.C., Tschantz, W.R., Casareno, R., Bullinger, P.R., and Dalbey, R.E. (1997) Use of site-directed chemical modification to study an essential lysine in *Escherichia coli* leader peptidase. *J. Biol. Chem.*, **272**, 9994–10003.
63 Takayama, H., Chelikani, P., Reeves, P.J., Zhang, S., and Khorana, H.G. (2008) High-level expression, single-step immunoaffinity purification and characterization of human tetraspanin membrane protein CD81. *PLoS ONE*, **3**, e2314.
64 Xie, H., Guo, X.-M., and Chen, H. (2009) Making the most of fusion tags technology in structural characterization of membrane proteins. *Mol. Biotechnol.*, **42**, 135–145.
65 Mohanty, A.K., Simmons, C., and Wiener, M.C. (2003) Inhibition of tobacco

etch virus protease activity by detergents. *Protein Expr. Purif.*, **27**, 109–114.

66 Cherezov, V., *et al.* (2007) High-resolution crystal structure of an engineered human beta2-adrenergic G protein-coupled receptor. *Science*, **318**, 1258–1265.

67 Rasmussen, S.G.F., *et al.* (2007) Crystal structure of the human beta2-adrenergic G protein-coupled receptor. *Nature*, **450**, 383–387.

2
Membrane Protein Expression in *Saccharomyces cerevisiae*

Zachary Britton, Carissa Young, Özge Can, Patrick McNeely, Andrea Naranjo, and Anne Skaja Robinson

2.1
Introduction

Saccharomyces cerevisiae contains around 6000 genes compared to around 25 000 human genes, yet *S. cerevisiae* possesses similar mechanisms for protein synthesis, maturation, and secretory pathway trafficking. Due to this simplicity and the ability of *S. cerevisiae* to perform homologous recombination – a process of DNA repair by homologous end-joining – genetic manipulation of *S. cerevisiae* is straightforward [1]. For example, several methods have been developed that permit the specific introduction or removal of DNA into the yeast chromosome [2–4], which enables the creation of unique protein expression strains and facilitates stable expression of foreign proteins.

Potential concerns with expressing heterologous membrane proteins in *S. cerevisiae* may not be easily overcome, especially with regard to lipid and membrane differences. For example, ergosterol in *S. cerevisiae* plays an analogous role to cholesterol in higher eukaryotes. Cholesterol analogs, which are lacking in prokaryotic expression systems, have been shown in some cases to directly interact with membrane proteins (e.g., G-protein-coupled receptors (GPCRs)). In addition to sterols, other membrane components, including lipids and their derivatives, are also present in similar – although not identical – levels in *S. cerevisiae*. Specific interactions with membrane components, and more generally effects imparted by specific membrane properties, play an important role in the folding and trafficking of membrane proteins.

Despite these differences, successful heterologous membrane protein expression in *S. cerevisiae* has resulted in high-resolution crystal structures. Of the over 50 000 protein structures in the Protein Data Bank, only 39 are eukaryotic membrane proteins and five of these (two species of monoamine oxidase A, mitochondrial ATP synthase, H^+-ATPase 2, and Ca^{2+}-ATPase SERCA1a) were expressed in *S. cerevisiae* [5, 6]. Jidenko *et al.* compared the purified Ca^{2+}-ATPase SERCA1a protein with the native protein from rabbit, and showed that calcium-dependent ATPase activity and calcium transport are intact after purification and reconstitution in proteoliposomes. Through crystal structure determination, they also

Production of Membrane Proteins: Strategies for Expression and Isolation, First Edition.
Edited by Anne Skaja Robinson.

showed that ligands contact their previously identified binding sites, which demonstrates in this example that there are no structural differences between the protein expressed in yeast and the protein expressed in its native system [7].

Given the tools available and the number of successful membrane protein expression studies in the literature, there are a few considerations to take into account while designing experiments to explore the behavior of overexpressed endogenous and heterologous proteins in *S. cerevisiae*. This chapter will highlight some of the fundamentals of protein expression in *S. cerevisiae* (e.g., promoter systems, host strains, plasmids, etc.) and will demonstrate the potential for membrane protein expression. We emphasize aspects of cotranslational machinery, post-translational modifications, and the lipid environment to which nascent proteins are exposed, as well as dynamic cellular responses to protein expression. Furthermore, case studies have been chosen which demonstrate useful *S. cerevisiae*-specific methods for structure–function studies and illuminate potential pitfalls in experimental designs.

2.2 Getting Started

2.2.1 Promoter Systems

2.2.1.1 Constitutive Promoters

Constitutive promoters for cytochrome *c* oxidase (P_{CYC1}), phosphatidyl glycerol kinase 1 (P_{PGK}), translational elongation factor 1α (P_{TEF}), alcohol dehydrogenase 1 (P_{ADH}), and glyceraldehyde-3-phosphate dehydrogenase (P_{GPD}) have all been used extensively for protein expression in *S. cerevisiae* Table 2.1 [8]. However, constitutive promoters may not be ideal for membrane protein expression due to the potential toxicity of expressed proteins. Therefore, inducible promoter systems have been exploited for this purpose.

2.2.1.2 Inducible Promoters

Inducible promoter systems permit the control of timing and gene expression levels. Examples of yeast inducible promoters include the *MET25* gene, negatively regulated by methionine [9]; the *PHO5* gene, negatively regulated by inorganic phosphate [10]; the metallothionein-encoding *CUP1* gene, activated by Cu(II); and the *GAL1*, *GAL 4*, *GAL7*, and *GAL10* genes, activated by galactose and repressed by glucose [11]. For the purpose of this chapter, discussion of yeast inducible promoters will be limited to the most common promoters for applied gene expression, promoters of *GAL* and *CUP1* genes [12]. For further insights into other inducible promoters and more in-depth rationale for regulation, see the review by Maya *et al.* [13].

The *GAL* gene family is a set of structural and regulatory genes that encode for enzymes required for galactose utilization [14]. *GAL* gene products are proteins

Table 2.1 *S. cerevisiae* promoters.

Promoter	Gene	Regulation	Reference
Constitutive			
P_{ADH}	alcohol dehydrogenase	ethanol	[21]
P_{CYC1}	cytochrome *c* oxidase	glucose (–)	[22]
P_{PGK}	phosphatidyl glycerol kinase-1	glucose	[23]
P_{TEF}	translational elongation factor-1α	glucose	[24, 25]
P_{GPD}	glyceraldehyde-3-phosphate dehydrogenase	glucose	[26, 27]
Inducible			
P_{CUP1}	metallothionein	Cu(II) (+)	[18]
P_{GAL1}	galactokinase	glucose (–)/galactose (+)	[28]
$P_{GAL1\text{-}10}$	–	glucose (–)/galactose (+)	
P_{MET25}	methionine and cysteine synthase	methionine (–)	[9]
P_{PHO5}	repressible acid phosphatase	inorganic phosphate (–)	[10]

that transport galactose into cells, convert intracellular galactose to glucose-1-phosphate, and demonstrate galactosidase activity [15]. Galactose-inducible promoters (P_{GAL1} and $P_{GAL1\text{-}10}$) have been used extensively for the production of proteins due to precise and efficient regulatory mechanisms. For example, *GAL* genes can be found in three carbon source-dependent states: inactive, repressed (glucose); inactive, nonrepressed (glycerol); and active, induced to high-level expression (galactose) [15], where expression levels are typically tens to hundreds of milligrams per liter. This strict regulation of expression may be particularly useful for control of membrane proteins that may be toxic to *S. cerevisiae.*

In *S. cerevisiae,* the *CUP1* promoter (P_{CUP1}) modulates expression levels of metallothionein in response to potentially lethal levels of copper in the growth medium, where metallothionein is a cysteine-rich protein that binds and sequesters Cu(II) and provides the yeast with resistance to Cu(II) [16]. P_{CUP1} has been applied to both endogenous [17, 18] and heterologous [19] protein expression, where the strength of gene transcription increases monotonically with Cu(II) concentration to a maximum equal to transcription controlled by P_{PGK} [20].

2.2.2 Host Strains, Selection Strategies, and Plasmids

2.2.2.1 Host Strains

Numerous *S. cerevisiae* strains have been developed for use with common auxotrophic and antibiotic selection strategies; however, each cell strain has been developed for a unique characteristic. For example, strains that are protease-deficient may improve the observed expression levels of susceptible membrane proteins [29]. Another potential factor in membrane protein expression is the lipid composition of membranes. For example, Duport *et al.* showed that the sterol content and capacity for glycosylation of the host strain impacted human β_2- and β_3-adrenergic receptor expression levels in *S. cerevisiae* [30].

2.2.2.2 Selection Strategies

Auxotrophic Selection Markers The original and most commonly used strategies for transformation and selection have been auxotrophic markers: *TRP1* [31], *HIS3* [32], *LEU2, URA3* [33], and more recently *MET15* and *ADE2* [34]. Expression plasmids are paired with widely available *S. cerevisiae* strains that are auxotrophic for tryptophan, histidine, leucine, uracil, methionine, and adenosine, respectively, by carrying full or functional knock-outs of these auxotrophic genes (for a comprehensive list refer to http://www.yeastgenome.org/alleltable.shtml). *URA3* and *LYS2* are particularly useful because there are methods for counterselection, where *ura3*⁻ and *lys2*⁻ cells can be selected for resistance to 5-fluoro-orotic acid [35] and α-aminoadipic acid [36–38], respectively. Auxotrophic selection, however, requires continued selection by growth in minimal media lacking the relevant nutrient [39].

Antibiotic Selection Markers Antibiotic selection offers an attractive alternative because antibiotics provide a selection strategy independent of auxotrophic strain limitations and can be used for selection in rich growth medium. Additionally, multiple antibiotic selections may be used within the same strain because these antibiotics differ in function and mechanism. Common antibiotic selection strategies use aminoglycoside phosphotransferase (*APT*), hygromycin B phosphotransferase (*HPH*) and the zeocin™ resistance (*Sh ble*) genes that supply resistances to G418 [40], hygromycin B (hphB) [41], and zeocin [42], respectively. G418 interferes with 80S ribosome function, hygromycin B inhibits protein synthesis by disrupting translocation and promoting mistranslation at the 80S ribosome [43, 44], and zeocin binds to and cleaves DNA [45]. It is important to note that because these antibiotics affect ribosome function, expression studies should be performed with rich liquid media lacking antibiotics.

2.2.2.3 Plasmids and Homologous Recombination

Yeast plasmids can be divided into three general categories: (i) low-copy, (ii) high-copy, and (iii) integrating plasmids [8, 46, 47]. Low-copy plasmids contain yeast centromere (CEN) and autonomous replicating sequences (ARSs) that make these plasmids mitotically stable in yeast. *S. cerevisiae* transformed with a CEN/ARS plasmid typically contain one or two copies per cell [48]. High-copy plasmids contain the 2μ sequence [46, 49, 50], which is a 6.3-kb plasmid naturally present in most *S. cerevisiae* strains at approximately 100 copies per haploid genome [51–53]. *S. cerevisiae* transformed with 2μ plasmids typically contain 20 copies per cell [42]. Despite their widespread application, both low-copy and high-copy plasmids are lost in 1.5–5% of progeny per doubling time during nonselective growth [42]; therefore, plasmid-based expression may be unpredictable due to plasmid and copy number instability [51, 54]. Overcoming these obstacles has been achieved through use of integrating plasmids: plasmids that contain a bacterial origin of replication, selection markers, and yeast chromosomal DNA for targeted integration by homologous recombination [47, 55]. Following the initial transformation and selection of a genomic integration, cells are genetically stable and no longer require selection. See Table 2.2.

Table 2.2 *S. cerevisiae* plasmids.

Plasmid	Type	Selection	Reference
pRS401	yIP, integrating	*MET15*	[34]
pRS402	yIP, integrating	*ADE2*	[34]
pRS303	yIP, integrating	*HIS3*	[47]
pRS403	yIP, integrating	*HIS3*	[47]
pRS304	yIP, integrating	*TRP1*	[47]
pRS404	yIP, integrating	*TRP1*	[47]
pRS305	yIP, integrating	*LEU2*	[47]
pRS405	yIP, integrating	*LEU2*	[47]
pRS306	yIP, integrating	*URA3*	[47]
pRS406	yIP, integrating	*URA3*	[47]
pRS411	yCP, centromeric	*MET15*	[34]
pRS412	yCP, centromeric	*ADE2*	[34]
pRS313	yCP, centromeric	*HIS3*	[47]
pRS413	yCP, centromeric	*HIS3*	[47]
pRS314	yCP, centromeric	*TRP1*	[47]
pRS414	yCP, centromeric	*TRP1*	[47]
pRS315	yCP, centromeric	*LEU2*	[47]
pRS415	yCP, centromeric	*LEU2*	[47]
pRS316	yCP, centromeric	*URA3*	[47]
pRS416	yCP, centromeric	*URA3*	[47]
pRS317	yCP, centromeric	*LYS2*	[84]
pRS421	yEP, 2μ-based	*MET15*	[34]
pRS422	yEP, 2μ-based	*ADE2*	[34]
pRS423	yEP, 2μ-based	*HIS3*	[42]
pRS424	yEP, 2μ-based	*TRP1*	[42]
pRS425	yEP, 2μ-based	*LEU2*	[42]
pRS426	yEP, 2μ-based	*URA3*	[42]
pITy	yIP, integrating	G418	[58]

MET15, S. cerevisiae auxotrophy marker, encodes O-acetyl homoserine-O-acetyl serine sulfhydrylase.
ADE2, S. cerevisiae auxotrophy marker, encodes phosphoribosylaminoimidazole carboxylase.
HIS3, S. cerevisiae auxotrophy marker, encodes imidazoleglycerol-phosphate dehydratase.
TRP1, S. cerevisiae auxotrophy marker, encodes phosphoribosylanthranilate isomerase.
LEU2, S. cerevisiae auxotrophy marker, encodes β-isopropylmalate dehydrogenase.
URA3, S. cerevisiae auxotrophy marker, encodes orotidine-5′-phosphate decarboxylase.
LYS2, S. cerevisiae auxotrophy marker, encodes α-aminoadipate reductase.
G418, *S. cerevisiae* antibiotic marker.

Polymerase chain reaction (PCR)-mediated techniques have been used to amplify DNA for genomic integration [56]. Both integrating plasmids and, in particular, PCR-based methods for generating and integrating expression cassettes are ideal because they can be used to target integration at a single site or multiple sites based on the promiscuity of the yeast chromosomal DNA target [57, 58]. In addition to applications of homologous recombination to directed integration, the emergence of *in vivo* methods for homologous recombination has permitted ligation of genes into expression plasmids [59–61] and construction of protein variants, as demonstrated for the mouse TRPM5 [62].

2.2.3 Expression Conditions

There are a number of experimental parameters that affect production and function of membrane proteins expressed in *S. cerevisiae*, including pH, temperature, induction strategy, and expression duration. Although *S. cerevisiae* may respond differently to each membrane protein expressed, the purpose of this section is to outline common strategies used to optimize membrane protein production.

Bonander *et al.* have described a set of experiments to examine the reasons for successes and failures in membrane protein production by using the yeast glycerol facilitator, Fps1p, as a model membrane protein [63]. Under the control of the constitutive triose phosphate isomerase promoter (P_{TPI}), the study systematically quantified the effects of pH, temperature and expression duration on Fps1p-expressing cultures in high-performance bioreactors. Their results suggest that the optimal conditions for functional Fps1p production do not correlate with conditions that promote rapid cell growth. For example, the highest Fsp1p expression levels were observed at 35 °C and pH 5, but the most membrane-associated Fsp1p, which is a more suitable indicator for functional production of membrane proteins, was observed at 20 °C and pH 5. Lowering the growth temperature reduces the cellular growth rate, but may have other advantages, such as promoting protein folding and membrane insertion, reducing rates of proteolysis, and upregulating cold-shock chaperones that aid in protein folding. Bonander *et al.* also note the importance of expression duration on membrane protein yields, where they observe a distinct maximum in Fsp1p levels prior to glucose exhaustion despite constant transcript levels. Differences in Fsp1p expression levels correlate with changes in gene expression for genes involved in membrane protein secretion and yeast cellular physiology [63], but may also result from translational control or increased proteolysis.

Although examples in the literature may serve as a guide for successful membrane protein expression, ultimately the conditions must be optimized for each target protein. For example, a thorough set of experiments to investigate the effects of temperature, pH, cell induction concentration, and galactose concentration has been described for the optimization of human adenosine A_{2A} receptor expression in *S. cerevisiae* [64] and may serve as a useful guide.

2.3
Special Considerations

This section describes protein-specific considerations that may be relevant for membrane protein expression in *S. cerevisiae*, including post-translational modifications, lipid requirements, signal sequences, protein topology, and cellular responses to protein expression.

2.3.1
Post-Translational Modifications

2.3.1.1 Glycosylation

Glycosylation is a highly specific and complex form of post-translational modification [65]. Structure–function studies have shown that the specific carbohydrate presented on a glycoprotein endows important biological character, including immunogenicity, solubility, protein and cellular recognition, resistance to proteolysis, protein folding, and biological function [66, 67].

In *S. cerevisiae*, glycosylation may occur in either the endoplasmic reticulum (ER) or Golgi apparatus. Glycosylation of protein inner core residues occurs in the ER, whereas glycosylation of protein outer core residues may occur within either the ER or Golgi apparatus. *N*-linked glycosylation may occur at sites including asparagine–X–serine (N–X–S) or asparagine–X–threonine (N–X–T), where X is any amino acid except proline (P) [68]. These sites are required for *N*-linked glycosylation but do not ensure glycosylation will occur. Preliminary stages of *N*-linked glycosylation occur in the ER and include the attachment of two *N*-acetylglucosamine molecules, nine mannose sugars, and three glucose sugars to the asparagine residue of glycosylation sites. The initial complex carbohydrate is processed prior to ER exit, where three glucose and one mannose sugars are removed. This initial process in *S. cerevisiae* is similar to higher eukaryotes, whereas further glycosylation processes differ in later stages.

In higher eukaryotes, mannosidases within the Golgi apparatus remove mannose residues, and additional sugars may be added, ultimately resulting in carbohydrate chains terminated by sugars such as galactose and sialic acid [11]. *S. cerevisiae* lack Golgi mannosidases, which limit carbohydrate chain length in higher eukaryotes. Furthermore, the absence of Golgi mannosidases permits hyperglycosylation or heterogeneous addition of 40 or more mannose residues [67, 69]. Modifications to the carbohydrate chain made in the Golgi apparatus can be different for each glycoprotein and may affect protein homogeneity and function.

Strategies to control glycosylation state have been developed. For example, site-directed mutagenesis of glycosylation sites has been achieved but may result in altered glycoprotein function. Additionally, *S. cerevisiae* mutants have been identified that exhibit decreased glycosylation processing; mutants in mannan biosynthesis (*mnn*) that reduce outer chain addition events have been isolated [70]. However, advantages provided by limiting *N*-linked glycosylation are reduced by the effects that mannan mutations have on cell growth [71]. If hyperglycosylation is a problem, switching yeast species may be worthwhile. Other yeast species like *Pichia pastoris* and *Hansenula polymorpha* are less prone to hyperglycosylation of heterologous proteins, where *P. pastoris*- and *H. polymorpha*-derived invertase had average mannose chain lengths of 8–14, whereas *S. cerevisiae*-derived invertase contained 50–100 [67, 69]. The extent of glycosylation may be protein-specific (e.g., many GPCRs expressed in *S. cerevisiae* appear nonglycosylated [72]).

2.3.1.2 **Disulfide Bond Formation**

Disulfide bonds covalently cross-link cysteine residues and stabilize favored conformations. Formation of disulfide bonds may be required for proper folding and function of membrane proteins. Therefore, it is important to maintain proper disulfide bonds during expression. For example, overexpression of protein disulfide isomerase, a protein that plays a key role in disulfide bond formation, improved secretion of human platelet-derived growth factor B homodimer, *Schizosaccharomyces pombe* acid phosphatase, and single-chain antibody fragments [73, 74]. However, expression improvements may be protein-dependent as coexpression of protein disulfide isomerase did not improve adenosine A_{2A} or substance P receptor localization or yields [75].

2.3.2 Lipid Requirements

Biological membranes are composed of diverse lipids, where variations in the lipid head-group and acyl chain give rise to unique lipid properties. In addition, differences in the lipid composition of membranes give rise to unique membrane properties important for membrane protein structure and function. For example, lipid composition affects membrane protein translocation efficiency, topology, stability, complex assembly, transport within the secretory pathway, and function. Therefore, careful consideration of the phospholipid, glycolipid, and sterol content of the expression host may permit the optimization of membrane protein yield and activity. In this section, components of *S. cerevisiae* membranes will be defined and specific examples of their importance to authentic protein production and function reviewed.

It should be noted that both lipid biosynthesis and membrane protein biogenesis occur within the membrane of the ER, where both the lipid and protein components of the membrane are balanced [76]. In addition, the specific lipid composition of *S. cerevisiae* membranes is constantly changing. For example, ratios of specific phospholipids vary among strains, but also depend on the carbon source and culture conditions [77]. This complex regulation of membrane components can present a challenge for membrane protein expression. For example, surplus membranes were synthesized in *S. cerevisiae* overexpressing an integral membrane protein, 3-hydroxy-3-methylglutaryl coenzyme A [78].

2.3.2.1 Glycerophospholipids

Glycerophospholipids are a major structural component of biological membranes. Each glycerophospholipid consists of a polar head group and two long hydrophobic chains, which define inherent properties. Specific characteristics of glycerophospholipids and their roles in *S. cerevisiae* are described below.

Anionic glycerophospholipids such as phosphatidylserine (PS), phosphotidylinositol (PI), and its phosphorylated derivatives, and cardiolipin (CL) are required for proper topology of membrane proteins, where topology of the protein generally follows the positive-inside rule to position cationic residues flanking the transmembrane domain in proximity to anionic phospholipids on the cytosolic side of the membrane [79–81].

PS is a minor component of organelle membranes in *S. cerevisiae* except for the plasma membrane, where it represents around 34% of total phospholipids (Table 2.3) [82]. PS is selectively arranged on the inner leaflet of the endoplasmic reticulum and plasma membranes by phospholipid flippases [83].

PI and its phosphorylated derivatives, generally called phosphoinositides (PIPs), provide structural and signaling roles in the cytoplasmic face of cellular membranes. PI contains a unique head-group among lipids in that three of the hydroxyl groups of the inositol ring may be phosphorylated. Phosphorylation of PI results in the recruitment of specific proteins with phosphoinositide binding domains. For further review of the synthesis and function of PI and PIPs in *S. cerevisiae*, see the review by Strahl [85].

CL (bis-phosphatidylglycerol) is a dianionic phospholipid that is composed of two phosphatidyl residues connected by a glycerol moiety [86]. CL interacts with a variety of mitochondrial proteins through hydrophobic and electrostatic interactions, where they have been shown to stabilize proteins in the mitochondrial respiratory chain [87–89]. A more extensive assessment of the cellular functions of CL in yeast has been reviewed elsewhere [90].

Phosphatidylcholine (PC) is the most abundant glycerophospholipid in *S. cerevisiae* [75], where it is a major structural component of organelle membranes and

Table 2.3 Percentages of total phospholipid content in each organelle membrane (adapted with permission from Zinser *et al.* [82]).

	Percentage of total phospholipid in each organelle membrane				
	Mitochondria	**Microsomes**	**Plasma membrane**	**Vacuole**	**Peroxisomes**
PC	40.2	51.3	16.8	46.5	48.2
PE	26.5	33.4	20.3	19.4	22.9
PS	3.0	6.6	33.6	4.4	4.5
PI	14.6	7.5	17.7	18.3	15.8
CL	13.3	0.4	0.2	1.6	7.0
Ergosterol/phospholipid ratio	0.2	0.3	3.3	0.2	0.4

serves as a reservoir for signaling molecules [91, 92]. In addition, PC biosynthesis is involved in the regulation of vesicle trafficking in yeast [93].

Phosphatidylethanolamine (PE) is a neutral or zwitterionic glycerophospholipid with a small head-group that forms nonbilayer structures. PE localizes in close proximity to and often copurifies with membrane proteins, where PE improves protein stability and relieves disturbances in the membrane bilayer caused by polytopic membrane protein insertion.

PE depletion in *S. cerevisiae* decreased the activities of the arginine (Can1p), proline (Put4p), and general amino acid (Gap1p) permeases [94] as well as uracil (Fur4p) and maltose (Mal6p) proton motive force-driven transporters, whereas activities of Pma1p (H^+-ATPase) and a hexose transporter (Hxtp1p) were largely unaffected [95]. Although both Can1p and Pma1p associate with lipid rafts in wild-type cells, Can1p and Pma1p occupy distinct nonoverlapping membrane domains at the cell surface [96]. PE depletion hindered progress of Can1p but not Pma1p through the secretory pathway leading to Can1p accumulation in the Golgi instead of the plasma membrane.

2.3.2.2 Sphingolipids

Sphingolipids are a major structural component of *S. cerevisiae* membranes, but also act as second messengers in signal transduction pathways. Sphingolipids have a ceramide backbone, which is made up of a long-chain base phytosphingosine that has been modified with a hydroxyl C_{26} fatty acid. There are only three types of sphingolipids in *S. cerevisiae* and differ only in their inositol moiety: inositolphosphate ceramide, mannosyl-inositolphosphate-ceramide, and mannosyl-diinositolphosphate-ceramide [97]. Sphingolipids have roles in protein trafficking/exocytosis, lipid microdomains, calcium homeostasis, cellular aging, and nutrient uptake. For an excellent review of sphingolipid function and metabolism in yeast, see [98].

2.3.2.3 Sterols

Sterols are rigid hydrophobic molecules with a polar hydroxyl group. Sterols affect the fluidity of the plasma membrane, where membrane fluidity, in turn, affects the lateral movement and activity of membrane proteins [97]. In addition, local sterol-rich domains may also form membrane protein insertion sites. The plasma membrane of *S. cerevisiae* and other yeasts contains mostly ergosterol and low levels of zymosterol, whereas the most abundant sterol in higher eukaryotes is cholesterol [82]. This substitution may or may not affect membrane protein structure and function; however, it has been noted that the slow diffusion of proteins and lipids in yeast membranes may be the result of the higher ergosterol content of yeast membranes [97].

Ergosterol in *S. cerevisiae* plays an analogous role to cholesterol in higher eukaryotes. In the case of heterologous expression of membrane proteins, this difference provides a similar – although not identical – molecular environment in the cell membranes of yeast as compared to that of the native system. For example, ergosterol may replace cholesterol for sterol-dependent mammalian membrane proteins or specific sterols (sito-, stigma-, and campesterols) for plant membrane proteins. However, this sterol substitution may not result in fully functional heterologous proteins. For example, human MDR1 expressed in *S. cerevisiae* showed impaired drug binding, where it was found that ergosterol inhibits azidopine binding [99]. In addition, the human μ-opioid receptor [100] and dopamine D_{2S} receptor [101] ligand binding were affected by the lipid environment of *S. cerevisiae* compared to their natural environment. Interestingly, ligand-binding affinities for the human D_{2S} receptor were different when the receptor was expressed in *S. pombe* and *S. cerevisiae*. In contrast to these receptors, the ligand binding activity of human adenosine A_{2A} receptor [75] and human β_2-adrenergic receptor [102] in *S. cerevisiae* appears to be unaffected.

2.3.3 Signal Sequences

Although not required for all membrane proteins (some membrane proteins contain inherent or cleavable N-terminal signal sequences), utilization of host-specific signal sequences has become a common strategy to improve and facilitate membrane insertion of proteins destined for the secretory pathway. Common secretory pathway targeting sequences used in *S. cerevisiae* are those of the Ste2 receptor [101–105], the pre-propeptide of α-mating factor [106, 107], and the signal peptide of acid phosphatase [108, 109].

Despite potential complications to membrane protein expression, signal sequences have been routinely used in the expression of GPCRs, proteins oriented with their N-terminus facing the ER lumen in the ER and the extracellular environment in the plasma membrane. One recent study investigated the role of the signal peptide in the expression and processing of glucagon-like peptide-1 receptor in human embryonic kidney HEK-293 cells [110]. Glucagon-like peptide-1 receptor is a human GPCR and a target for the treatment of type II diabetes. This study demonstrated that even though signal sequence cleavage is not necessary for the expression of the GPCR, it was required for glycosylation and trafficking to the plasma membrane.

It has also been shown that the cleavable, putative signal sequence of the human endothelin B receptor is necessary for translocation through the ER of COS.M6 cells [111]. Receptor mutants lacking the putative signal sequence were retained in the ER in a nonfunctional form. Although the mechanism is still unknown, the adjacent N-terminal sequence of the endothelin B receptor immediately followed by a signal peptide was also found to be important for efficient translocation [112]. Taken together, these results signify the importance of signal sequences for efficient ER translocation.

2.3.4 Topology

The topology of a membrane protein – specifically the identification of membrane spanning regions and in/out orientations relative to the membrane – is a basic yet extremely powerful feature to guide experimental studies. Multiple experimental approaches have been used to generate topology information including *in situ* proteolytic digestion, chemical modifications, selective placement of epitope tags combined with antibody binding, various gene fusions, and design of hybrid, chimeric proteins. Classical topological reporter constructs include an internal 53-amino-acid peptide of invertase (Suc2p) containing three glycosylation sites that are efficiently glycosylated when translocated across the ER membrane [113]. A SUC2–HIS4C dual-topology reporter, where HIS4C harbors the gene encoding histidinol dehydrogenase that acts on its substrate, histidinol, if located in the cytosol [114], enabled global topology maps of the *S. cerevisiae* membrane proteome [115, 116]. Shorter reporter sequences have been useful for this topology mapping, including factor Xa protease cleavage sites in hydrophilic loops [117] and ubiquitin (Ub)-specific cleavage for an ER translocation assay [118, 119]. Green Fluorescent Protein (GFP) tags have facilitated high-throughput fluorescent-based optimization of endogenous and heterologous membrane protein expression and purification in *S. cerevisiae* [59, 61]. In all cases, the choice of tag and expression strategy must always include appropriate controls. For example, to verify the stability and function of Ost4p, a 36-amino-acid subunit of oligosaccharyltransferase in *S. cerevisiae*, a HA/Suc2/His4C fusion construct was appropriately compared to a single HA tag [115]. To ensure that a fusion does not compromise cellular function, complementation with mutant or deletion strains may be completed, growth assays conducted (e.g., His plates supplemented with histidinol, exemplified by HIS4C fusions), or localization of a tagged heterologous protein can be compared to an endogenous organelle marker by confocal microscopy using live cell imaging or immunofluorescence techniques.

In the absence of high-resolution structural data, topology models are required for structure–function studies of membrane proteins. Many algorithms exist which attempt to predict membrane protein topology, e.g., PRODIV-TMHMM [120], TMHMM [121], HMMTOP [122], and others, and until recently have lacked any estimate of prediction reliability [123]. TOPCONS (topcons.net) combines an arbitrary number of topology predictions into one consensus output and quantifies the reliability of the prediction [124]. Most algorithms do not distinguish between an N-terminal transmembrane region and signal peptide since both contain hydrophobic residues. However, SPOCTOPUS (octopus.cbr.su.se) is an algorithm that maximizes discrimination between signal peptide and transmembrane domain predictions [125]. To facilitate the identification and characterization of favorable transmembrane regions, MPEx (http://blanco.biomol.uci.edu/mpex) allows for physical-scale hydropathy analysis, translocon-scale hydropathy analysis, β-barrel analysis, hydropathy plots for comparison of sequence mutagenesis, and design of membrane peptides [126]. Information regarding polytopic membrane N_{out}–C_{in} or N_{in}–C_{out} orientation is detailed elsewhere [127, 128], and protein insertion into the ER membrane is reviewed by White and von Heijne [129].

2.3.5
Cellular Responses to Membrane Protein Expression

S. cerevisiae has developed mechanisms to cope with defects in protein synthesis and maturation. Two of these, the unfolded protein response (UPR) and the heat-shock response (HSR), are relevant to membrane protein expression in *S. cerevisiae*. The continued induction of cellular responses leads to reduced expression levels [130].

2.3.5.1 UPR

An abundance of improperly folded protein within the ER leads to activation of the UPR, a universal eukaryotic quality control mechanism that is responsible for maintaining cell homeostasis and limiting ER stress (see reviews of eukaryotic UPR in Chapman *et al.* [131] and Kaufman [132]). In *S. cerevisiae*, the UPR is triggered by accumulation of misfolded protein within the ER [133], which leads to Hac1-mediated upregulation of chaperones, foldases, and many components of the secretory pathway in order to promote proper folding and accelerate protein transport from the ER. Proteins that fail to fold correctly are targeted to the vacuole by the ER-associated degradation (ERAD) proteolytic pathway [134, 135].

In *S. cerevisiae*, overexpression of membrane proteins has been reported to activate the UPR. For example, heterologous overexpression of human GPCRs [72] and overexpression of P2 H^+/adenosine transporter [136] induced the UPR, where strong activation of the UPR correlated with decreased functional production of the P2 transporter. Therefore, tuning synthesis rates to match the ER's folding capacity and avoid UPR activation may result in improved functional yields.

2.3.5.2 HSR

Another *S. cerevisiae* quality control mechanism that may be relevant to membrane protein expression is the HSR, which is responsible for limiting cytosolic stress. The HSR is triggered by accumulation of misfolded protein within the cytosol [137, 138]. Analogous to the UPR, the HSR mediates transcription of molecular chaperones and components of proteolytic pathways [139]. The HSR is regulated by transcription factor Hsf1 [140], which when activated binds heat shock elements in the promoters of HSR genes [141, 142]. Not all HSR target genes, however, are unique from UPR target genes and there is some evidence that the HSR may relieve ER stress [143], perhaps, through common transcriptional targets.

2.4
Case Studies

This section will use Ste2p, Pma1p, and cystic fibrosis (cystic fibrosis CF) transmembrane conductance regulator (CFTR) as examples to illustrate approaches to overcome specific obstacles to membrane protein expression and structure–function studies in *S. cerevisiae*.

2.4.1
Ste2p

GPCRs are a major family of membrane proteins, which contain seven-transmembrane helices and are responsible for mediating signaling across the cellular membrane through interactions with the heterotrimeric G-protein. *S. cerevisiae* contain three GPCRs; the most well-studied of these is the α-factor receptor, Ste2p, which plays a role in cellular conjugation. Ste2p has used extensively as a model system for structure–function studies of ligand–GPCR interactions, mechanisms of GPCR activation, and regulation of heterotrimeric G-protein signaling [144].

Successful overexpression and purification of Ste2p has been achieved in *S. cerevisiae*. For example, Ste2p was expressed from a 2μ-based plasmid under the control of P_{GPD} in a vacuolar protease-deficient strain, BJ2168 (*MATa prc1-407 prb1-1122 pep4-3 leu2 trp1 ura3-52*), at 26 °C [145]. Using these conditions, up to 1 mg Ste2p (200 l of culture) was purified to 95% homogeneity from membrane preparations using metal-affinity chromatography. Ligand binding of purified Ste2p was drastically reduced after reconstitution in lipid vesicles but was restored by supplementing the vesicles with yeast membranes deficient in the receptor [145]. Thus, the importance of membrane components on membrane activity is illustrated as Ste2p activity has some as yet unidentified lipid dependence.

A glycosylation- and cysteine-deficient Ste2p was generated by site-directed mutagenesis to improve homogeneity of the expressed receptor [146]. In this case, the Ste2p variant maintained full wild-type biological activity [146, 147], and motivated large-scale overexpression and purification. To accomplish this, the Ste2p variant was expressed using a 2μ-plasmid under the control of P_{GPD} in a vacuolar protease deficient strain, BJ2168 (*MATa prc1-407 prb1-1122 pep4-3 leu2 trp1 ura3-52 ste2::-Kan*^{R}). Using these conditions 120 μg mutant Ste2p was purified per liter of culture by affinity chromatography using a monoclonal antibody to the rhodopsin (1D4) epitope [146].

The Ste2p example not only demonstrates the successful overexpression and purification of a membrane protein in *S. cerevisiae*, but also illustrates effects of lipids and post-translational modifications on membrane protein activity. This case also demonstrates that functional GPCRs may be successfully expressed in *S. cerevisiae*, and many groups are applying *S. cerevisiae* to generate and characterize heterologous GPCRs [148–150] (Table 2.4).

2.4.2
Pma1p

The H^+-ATPase of *S. cerevisiae*, encoded by the *PMA1* gene, belongs to a widespread family of cation transporters referred to as the P_2-type ATPases [191]. Accounting for up to 20% of total protein content of the plasma membrane, Pma1p generates the electrochemical proton gradient essential for nutrient uptake by secondary active transport [192]. Structurally, Pma1p is an oligomeric molecule with asymmetric membrane topology composed of a large central cytoplasmic domain and an extracytoplasmic domain that consists of small stretches connecting 10 transmembrane domains [193, 194]. Newly synthesized Pma1p is assembled as an oligomer in the ER but its association with lipid rafts in the secretory pathway is critical for plasma membrane localization and activity [192, 195, 196].

Pma1p has emerged as a prototype for studies of plasma membrane biogenesis (see review [194]) and mechanisms for protein sorting and trafficking to the cell surface. Complementary approaches to improve expression have included temperature-sensitive strains, which are used to block successive steps of the secretory pathway and enable Pma1p trafficking to the cell surface; point mutations that have generated insights into the structural requirements for protein folding and trafficking; determination of cellular machinery required for membrane protein degradation; and enhancers of defective Pma1p variants that have revealed specialized roles for secretory pathway components. Using the Pma1p example as a guide may enable one to develop and apply appropriate strategies required for successful membrane protein expression.

Isolation of temperature-sensitive mutants exhibiting defective H^+-ATPase trafficking is possible due to Pma1p stability at the plasma membrane (half-life greater than 12 h) and lack of recycling [197, 198]. By using a genetic screen for mutations that blocked growth when combined with a mutation of a known coat protein gene, *SEC13* and *LST1* were discovered. Due to Lst1p's 23% sequence homology to Sec24p, the coat component of COPII [199] and its deletion mutant, which inhibits ER transport, Pma1p has been used to infer individual roles of vesicle trafficking components [200]. Alternative experimental approaches have used Pma1p variants to determine novel components of the vacuolar degradation pathway. For example, screening with a high-copy genomic library [8] and subsequent studies that screened an insertion library resulted in the identification of suppressors and enhancers that allowed Pma1p variants to bypass ERAD and traffic appropriately to the plasma membrane [200, 201]. Of the identified suppressors, vacuolar protein sorting (VPS) genes were identified that control biogenesis of newly synthesized vacuolar proteins involved in Golgi-to-endosome or endosome-to-vacuole trafficking, which suggests a Golgi-based quality control mechanism for misfolded Pma1p variant.

Comprehensive mutational studies of Pma1p have uncovered numerous structure–function relationships and allowed the isolation of sorting-defective alleles [195, 197, 202]. In fact, nearly 300 site-directed mutations have been introduced throughout this ATPase, while substitutions at 45 positions have led to defects in biogenesis (see review [202]). Pma1p variants include the least severe, G381A, which displays an intermediate folding defect (i.e., allows slow growth of cells co-expressing wild-type H^+-ATPase) that is transiently arrested in the ER but escapes to the plasma membrane, where it is recycled to the vacuole for degradation; to D378N where the Asp residue involved in phosphorylation has been replaced and exhibits a severe biogenetic defect and results in a dominant lethal phenotype [203]. Furthermore, Pma1p-D378N is improperly folded in the ER and is retrotranslocated to the cytosol for degradation by the proteasome [204]. Pma1p

Table 2.4 Production of functional heterologous GPCRs in *S. cerevisiae*.

Receptor Family	Species	GPCR subtype	Strain	Plasmid	Plasmid type	Promoter	Signal/fusion peptide
Acetylcholine (muscarinic)	human	M_1	DJ213-6-3-a	pVTI02-U	NR	P_{ADH}	–
	human	M_1	MPY578fc	p416GPD	centromeric	P_{GPD}	–
	human	M_1	MPY578fc	p416GPD	centromeric	P_{GPD}	–
	rat	M_3	MPY578fc	p416GPD	centromeric	P_{GPD}	–
	rat	M_3	MPY578fc	p416GPD	centromeric	P_{GPD}	–
	rat	M_3	MPY578q5	p416GPD	centromeric	P_{GPD}	–
	rat	M_3	MPY578q5	p416ADH	centromeric	P_{ADH}	–
	rat	M_3	MPY578q5	p416TEF	centromeric	P_{TEF}	–
	rat	M_3	MPY578q5	p416GPD	centromeric	P_{GPD}	–
	rat	M_3	MPY578q5	p416GPD	centromeric	P_{GPD}	–
	rat	M_3	MPY578q5	p416GPD	centromeric	P_{GPD}	–
	rat	M_3	MPY578q5	p416GPD	centromeric	P_{GPD}	–
	human	M_5	MPY578fc	p416GPD	centromeric	P_{GPD}	–
	human	M_5	MPY578fc	p416GPD	centromeric	P_{GPD}	–
	rat	M_5	20B12	pYAS1	2μ-based	$P_{MF\alpha}$	MFα

N-terminal tag(s)	C-terminal tag(s)	Notable alterations (fusions, mutations, truncations)	Ligand binding	G-protein coupling	Yield	References
–	–	–	+	ND	0.02 pmol/mg	[151]
–	–	–	+	+	ND	[152]
–	–	Δi3 (Pro231–Gly345)	+	+	ND	[152]
HA	–	–	+	+	ND	[152]
HA	–	Δi3 (Ala274–Lys469)	+	+	ND	[152]
HA	–	Δi3 (Ala274–Lys469), Q490L, random mutagenesis (TM 5–7)	+	+	ND	[153]
HA	–	Δi3 (Ala274–Lys469), Q490L, random mutagenesis (TM 5–7)	+	+	ND	[153]
HA	–	Δi3 (Ala274–Lys469), Q490L, random mutagenesis (TM 5–7)	+	+	ND	[153]
HA	–	Δi3 (Ala274–Lys469), D113N, random mutagenesis	+	+	ND	[154]
HA	EGFP	Δi3 (Ala274–Lys469) and random mutagenesis	+	+	ND	[155]
HA	EGFP	Δi3 (Ala274–Lys469) and random mutagenesis (e2)	+	+	ND	[156]
HA	FLAG	Δi3 (Ala274–Lys469) and random mutagenesis	+	+	ND	[157]
–	–	Δi3 (Thr237–Pro413)	+	+	ND	[152]
–	–	–	+	+	ND	[152]
–	–	23-amino-acid N-terminal truncation	+	–	0.13 pmol/mg	[107]

(*Continued*)

Table 2.4 (*Continued*)

Receptor Family	Species	GPCR subtype	Strain	Plasmid	Plasmid type	Promoter	Signal/fusion peptide
Adenosine	human	A_1	CY12660	NR	2μ-based	P_{PGK}	MFα
	human	A_1	MMY11	p426GPD	2μ-based	P_{GPD}	–
	human	A_{2A}	BJ5464	pITy	integrated	$P_{GAL1\text{-}10}$	pre-pro
	human	A_{2A}	BJ5464	pITy	integrated	$P_{GAL1\text{-}10}$	pre-pro
	human	A_{2A}	NFV220	pITy	integrated	$P_{GAL1\text{-}10}$	pre-pro
	human	A_{2A}	BJ5464	pITy	integrated	$P_{GAL1\text{-}10}$	pre-pro
	human	A_{2A}	BJ5464	pITy	integrated	$P_{GAL1\text{-}10}$	pre-pro
	human	A_{2A}	BJ5464	pITy	integrated	$P_{GAL1\text{-}10}$	pre-pro
	human	A_{2A}	BJ5464	pITy	integrated	$P_{GAL1\text{-}10}$	pre-pro
	human	A_{2A}	BJ5464	pITy	integrated	$P_{GAL1\text{-}10}$	pre-pro
	human	A_{2B}	MMY9	pPGK	2μ-based	P_{PGK}	–
	human	A_{2B}	MMY11	pPGK	2μ-based	P_{PGK}	–
	human	A_{2B}	MMY11	pDT-PGK	2μ-based	P_{PGK}	–
	human	A_{2B}	MRS1706	pDT-PGK	2μ-based	P_{PGK}	–
	human	A_{2B}	BJ5464	pITy	integrated	$P_{GAL1\text{-}10}$	pre-pro
Adrenergic	human	α_2c2	SC261	pYEDS14	2μ-based	P_{GAL1}	–
	human	α_2c2	SC261	pYEDS14	2μ-based	P_{GAL1}	–
	human	α_2c2	SC261	pYEDS14	2μ-based	P_{GAL1}	–
	human	β_2	NNY19	yEP24	2μ-based	P_{GAL1}	Ste2
	human	β_2	SC261	yEP24	2μ-based	P_{GAL1}	Ste2
	human	β_2	FY1679-28C	pYeDP60	2μ-based	$P_{GAL10\text{-}CYC1}$	–
	human	β_2	W303-1A	pYeDP61	2μ-based	$P_{GAL10\text{-}CYC1}$	–
	human	β_2	TGY47-1	pYeDP62	2μ-based	$P_{GAL10\text{-}CYC1}$	–
	human	β_3	FY1679-28C	pYeDP63	2μ-based	$P_{GAL10\text{-}CYC1}$	-
	human	β_3	W303-1A	pYeDP64	2μ-based	$P_{GAL10\text{-}CYC1}$	-
	human	β_3	TGY47-1	pYeDP65	2μ-based	$P_{GAL10\text{-}CYC1}$	-
Calcitonin	human	CRLR	CY16463	CP1289	2μ-based	P_{PGK}	MFα(1–89)
	human	CRLR	CY16463	CP1290	2μ-based	P_{PGK}	MFα(1–89)
Cannabinoid	human	CB_1	NR	NR	NR	NR	NR

N-terminal tag(s)	C-terminal tag(s)	Notable alterations (fusions, mutations, truncations)	Ligand binding	G-protein coupling	Yield	References
–	–	–	+	+	ND	[158]
–	–	–	+	+	ND	[159]
–	c-myc	–	+	ND	ND	[75]
–	EGFP	–	+	ND	ND	[75]
–	c-*myc*	–	+	ND	ND	[75]
–	EGFP	–	+	ND	ND	[160]
–	EGFP	–	+	ND	~4 mg/L	[161]
–	EGFP	–	+	ND	~4 mg/L	[64]
–	His_{10}	–	+	ND	6 mg/L	[162]
–	EGFP-His10	–	+	ND	7.4 mg/L	[162]
–	–	–	+	+	ND	[163]
–	–	–	+	+	ND	[163]
–	–	random mutagenesis	+	+	ND	[164]
–	–	random mutagenesis	+	+	ND	[165]
–	EGFP	–	+	ND	ND	[72]
His5	–	–	+	ND	7–70 pmol/mg	[103]
His5	–	–	+	ND	ND	[166]
His5	–	–	+	ND	ND	[167]
–	–	–	+	+	115 pmol/mg	[102]
–	–	–	+	+	36 pmol/mg	[103]
–	–	–	+	+	ND	[30]
–	–	–	+	+	ND	[30]
–	–	–	+	+	ND	[30]
–	–	–	+	+	ND	[30]
–	–	–	+	+	ND	[30]
–	–	–	+	+	ND	[30]
–	–	–	+	+	ND	[168]
–	FLAG	–	+	+	ND	[168]
NR	NR	NR	+	+	ND	[169]

(Continued)

Table 2.4 (*Continued*)

Receptor Family	Species	GPCR subtype	Strain	Plasmid	Plasmid type	Promoter	Signal/fusion peptide
Chemoattractant	human	C5a	BY1142	p1303	NR	P_{PGK}	–
	human	C5a	BY1142	p1303	NR	P_{PGK}	–
	human	C5a	BY1142	p1303	NR	P_{PGK}	–
Chemokine	human	CXCR4	CY12946	CP4258	2μ-based	P_{PGK}	MFα(1–89)
	human	CXCR4	CY12946	CP4181		P_{PGK}	–
	human	CXCR4	CY12946	CP4258	2μ-based	P_{PGK}	MFα(1–89)
	human	CXCR4	CY12946	CP4258	2μ-based	P_{PGK}	MFα(1–89)
	human	CXCR4	CY12946	CP4258	2μ-based	P_{PGK}	MFα(1–89)
Dopamine	human	D_{1A}	DK102	pGEM-T	2μ-based	P_{GPD}	–
	human	D_{1A}	BJ2168	pGEM-T	2μ-based	P_{GPD}	–
	human	D_{1A}	DK102	pGEM-T	2μ-based	P_{GPD}	Ste2
	human	D_{1A}	BJ2168	pGEM-T	2μ-based	P_{GPD}	Ste2
	human	D_{2S}	cI3-ABYS-86	pRS421	2μ-based	P_{PMA1}	–
	human	D_{2S}	cI3-ABYS-86	pRS421	2μ-based	P_{PMA1}	Ste2
	human	D_{2S}	WCG	YEp51	2μ-based	$P_{GAL1\text{-}10}$	–
	human	D_{2S}	YMTA	YEp51	2μ-based	$P_{GAL1\text{-}10}$	–
	human	D_{2S}	YMTAB	YEp51	2μ-based	$P_{GAL1\text{-}10}$	–
	human	D_{2S}	WCG	YEp51	2μ-based	$P_{GAL1\text{-}10}$	Ste2
	human	D_{2S}	YMTA	YEp51	2μ-based	$P_{GAL1\text{-}10}$	Ste2
	human	D_{2S}	YMTAB	YEp51	2μ-based	$P_{GAL1\text{-}10}$	Ste2
Formyl peptide	human	FPRL-1	CY1141	Cp1651	2μ-based	P_{PGK}	–
Growth hormone-releasing hormone	human	GHRH	LY252	pMP3	2μ-based	$P_{GAL1\text{-}10}$	–
Histamine	human	H_2	NR	NR	NR	NR	NR
Lysophosphatidic acid/S1P	human	Edg2	SY2069	pYEUra3	centromeric	P_{GAL1}	–
Melatonin	human	ML_{1A}	SDY102	pDT-PGK	2μ-based	P_{PGK}	–
	human	ML_{1B}	MMY9	pPGK	2μ-based	P_{PGK}	–
	human	ML_{1B}	MMY11	pPGK	2μ-based	P_{PGK}	–
Neurotensin	human	NT_1	BJ2168	pEMR1675	2μ-based	P_{GRAP1}	–
	human	NT_1	BJ2168	pEMR1231	2μ-based	P_{GRAP1}	–
Opioid	human	μ-opioid	LY296	pEMR516	2μ-based	P_{GRAP1}	Ste2

N-terminal tag(s)	C-terminal tag(s)	Notable alterations (fusions, mutations, truncations)	Ligand binding	G-protein coupling	Yield	References
–	–	random mutagenesis (TM3, 5–7)	+	+	ND	[170]
–	–	random mutagenesis (TM1, 2, 4)	+	+	ND	[171]
–	–	random mutagenesis (EC2)	+	+	ND	[172]
–	–	random mutagenesis	+	+	ND	[173]
–	–	–	+	+	ND	[174]
c-*myc*	His_6	random mutagenesis	+	+	ND	[175]
–	–	–	+	+	ND	[176]
–	–	N119S	+	+	ND	[176]
–	FLAG-His_6	six C-terminal amino acids from Ste2	+	ND	0.13 pmol/mg	[105]
–	FLAG-His_6	six C-terminal amino acids from Ste2	+	ND	0.13 pmol/mg	[105]
–	FLAG-His_6	six C-terminal amino acids from Ste2	+	ND	0.13 pmol/mg	[105]
–	FLAG-His_6	six C-terminal amino acids from Ste2	+	ND	0.13 pmol/mg	[105]
–	–	–	+	ND	1–2 pmol/mg	[177]
–	–	–	+	ND	1–2 pmol/mg	[177]
–	–	–	+	ND	1–2 pmol/mg	[104]
–	–	–	+	ND	1–2 pmol/mg	[104]
–	–	–	+	ND	1–2 pmol/mg	[104]
–	–	–	+	ND	1–2 pmol/mg	[104]
–	–	–	+	ND	1–2 pmol/mg	[104]
–	–	–	+	ND	1–2 pmol/mg	[104]
–	–	–	+	+	ND	[178].
–	–	–	+	+	ND	[179]
NR	NR	NR	+	+	ND	[180]
–	–	–	+	+	ND	[181]
–	–	site-directed mutagenesis	+	+	ND	[182]
–	–	–	+	+	ND	[163]
–	–	–	+	+	ND	[163]
–	–	–	+	ND	ND	[183]
–	c-*myc*	–	+	ND	ND	[183]
–	–	–	+	–	0.4 pmol/mg	[100]

(*Continued*)

Table 2.4 (*Continued*)

Receptor Family	Species	GPCR subtype	Strain	Plasmid	Plasmid type	Promoter	Signal/fusion peptide
Purinergic	human	$P2Y_1$	MPY578q5	p426GPD	2μ-based	P_{GPD}	–
	human	$P2Y_1$	MMY9	pPGK	2μ-based	P_{PGK}	–
	human	$P2Y_1$	MMY11	pPGK	2μ-based	P_{PGK}	–
	human	$P2Y_2$	MMY9	pPGK	2μ-based	P_{PGK}	–
	human	$P2Y_2$	MMY11	pPGK	2μ-based	P_{PGK}	–
	human	$P2Y_{12}$	MPY578fc	p426GPD	2μ-based	P_{GPD}	–
	mouse	$P2Y_{12}$	MPY578fc	p426GPD	2μ-based	P_{GPD}	–
Rhodopsin	bovine	opsin	CKY96	yEpRF1	2μ-based	$P_{GAL1\text{-}10}$	–
Serotonin	human	$5HT_{1A}$	MMY9	pPGK	2μ-based	P_{PGK}	–
	human	$5HT_{1A}$	MMY11	pPGK	2μ-based	P_{PGK}	–
	human	$5HT_{1D}$	MMY9	pPGK	2μ-based	P_{PGK}	–
	human	$5HT_{1D}$	MMY11	pPGK	2μ-based	P_{PGK}	–
	mouse	$5HT_{5A}$	CI3-ABYS-86	PCNNmm	2μ-based	P_{PRB1}	*Bacillus macerans*
	rat	$5HT_{5A}$	NR	NR	NR	NR	–
	rat	$5HT_{5A}$	NR	NR	NR	NR	MFα
Somatostatin	human	SST_2	MMY9	pFL61	2μ-based	P_{PGK}	–
	human	SST_2	MMY11	pFL61	2μ-based	P_{PGK}	–
	rat	SST_2	LY228	pJH2	2μ-based	$P_{GAL1\text{-}10}$	–
	rat	SST_2	LY252	pJH2	2μ-based	$P_{GAL1\text{-}10}$	–
	rat	SST_2	LY754	pJH2	2μ-based	$P_{GAL1\text{-}10}$	–
	rat	SST_2	LY784	pJH2	2μ-based	$P_{GAL1\text{-}10}$	–
	human	SST_5	MMY9	pPGK	2μ-based	P_{PGK}	–
	human	SST_5	MMY11	pPGK	2μ-based	P_{PGK}	–
Vasopressin	human	V2	MPY578fc	p416GPD	centromeric	P_{GPD}	–
	human	V2	MPY578q5	p416GPD	centromeric	P_{GPD}	–
	human	V2	MPY578s5	p416GPD	centromeric	P_{GPD}	–
	human	V2	MPY578fc	p416GPD	centromeric	P_{GPD}	–
	human	V2	MPY578q5	p416GPD	centromeric	P_{GPD}	–
	human	V2	MPY578s5	p416GPD	centromeric	P_{GPD}	–

EGFP = enhanced GFP; MF = mating factor; ND = not determined; NR = not reported.

N-terminal tag(s)	C-terminal tag(s)	Notable alterations (fusions, mutations, truncations)	Ligand binding	G-protein coupling	Yield	References
–	–	–	+	+	ND	[184]
–	–	–	+	+	ND	[163]
–	–	–	+	+	ND	[163]
–	–	–	+	+	ND	[163]
–	–	–	+	+	ND	[163]
–	–	–	+	+	ND	[185]
–	–	–	+	+	ND	[185]
–	–	–	+	–	2 mg/10^{10} cells	[186]
–	–	–	+	+	ND	[163]
–	–	–	+	+	ND	[163]
–	–	–	+	+	ND	[163]
–	–	–	+	+	ND	[163]
c-*myc*	–	–	+	ND	16 pmol/mg	[187]
–	–	–	+	ND	ND	[188]
–	c-*myc*	–	+	ND	ND	[188]
–	–	–	+	+	ND	[163]
–	–	–	+	+	ND	[163]
–	–	–	+	+	0.2 pmol/mg	[189]
–	–	–	+	+	0.2 pmol/mg	[189]
–	–	–	+	+	0.2 pmol/mg	[189]
–	–	–	+	+	0.2 pmol/mg	[189]
–	–	–	+	+	ND	[163]
–	–	–	+	+	ND	[163]
HA	–	–	+	+	ND	[190]
HA	–	–	+	+	ND	[190]
HA	–	–	+	+	ND	[190]
HA	–	ΔM145	+	+	ND	[190]
HA	–	ΔM145	+	+	ND	[190]
HA	–	ΔM145	+	+	ND	[190]

variants have been studied as a paradigm of misfolded membrane proteins that commence ERAD, ubiquitination, and proteasomal degradation [199, 205, 206]. Similarly, cellular quality control processes, such as the UPR, HSR, and autophagy have been examined in wild-type Pma1p and its variants; their affiliations with chaperones of the ER and cytosol have yielded an improved understanding of vacuolar targeting and cytoplasm-to-vacuole targeting/autophagy pathways [206–208].

Association with lipid rafts has been shown to play a role in biosynthetic delivery to the plasma membrane in yeast and has been demonstrated in the delivery of Pma1p to the plasma membrane [196, 209, 210]. Disruption of rafts has led to mistargeting of wild-type Pma1p to the vacuole while Pma1-7, the ATPase variant that is mistargeted to the vacuole, was shown to exhibit impaired raft association. Additionally, one of the previously identified suppressors, multicopy AST1, not only restored surface delivery but also raft association of Pma1-7. Furthermore, it has been shown that the surface transport of Pma1p is independent of newly synthesized sterols but that sphingolipids with C26 very-long-chain fatty acids are crucial for raft association and proper localization, while more recent experimental studies using yeast strain SLC1-1 in which the essential function of sphingolipids is substituted for glycerophospholipids containing C26 very long chain fatty acids have suggested that sphingolipids are dispensable for raft association and surface delivery of Pma1p, but the C26 fatty acid is crucial [211, 212].

2.4.3 CFTR

The ATP-binding cassette (ABC) transporters mediate ATP-driven transport of diverse substrates across cellular and organellar membranes [213]. The structural core of ABC transporters is composed of two membrane-spanning domains, which typically contain multiple transmembrane segments, and a nucleotide-binding domain (NBD) that couples ATP hydrolysis and transport. The NBD is characterized by three conserved motifs, Walker A and Walker B, as well as the signature consensus motif "LSGGQ" (referred to as the C motif).

The yeast genome consists of 30 ABC transporters. The existence of functionally redundant transporters have enabled gene deletion studies that have improved our understanding of specific transporter function (reviewed in [214]). In recent years, significant contributions of yeast ABC transporters include the elucidation of Ub as a trafficking signal for endocytosis (e.g., Ste6, a lipopeptide mating pheromone a-factor Ste6p), analysis of the ER quality control machinery and the identification of ER-associated compartments, mutagenesis-mediated structure–function analysis, and the development of integrated membrane yeast two-hybrid (iMYTH) technology to identify membrane protein interactors, Ycf1 and Yor1 [215–220].

Loss-of-function mutations in ABC transporters have been implicated in several inherited diseases, including the lung disease CF. CFTR is a 1480-residue membrane glycoprotein that functions as a cAMP-regulated chloride and bicarbonate channel in the apical membrane of epithelial cells [221]. Deletion of phenylalanine at position 508 (ΔF508-CFTR) is associated with CF and is one of the most common autosomal recessive disorders in individuals of European descent. Approximately 80% of wild-type CFTR is targeted for ERAD and essentially all of the ΔF508 is degraded [222]. Interestingly, expression of CFTR and ΔF508-CFTR fail to elicit a UPR response in yeast, and the growth rate is similar to cells that lack a CFTR expression vector [223]. ΔF508-CFTR is not transported to the plasma membrane yet does function as a chloride channel in its native host, where decreased temperature, high glycerol content, or addition of sodium 4-phenylbutyrate has been shown to improve the trafficking (and restore function) of ΔF508-CFTR [224–227]. Expression of CFTR in *S. cerevisiae* has been used to delineate degradation components including cytosolic and ER luminal chaperones (i.e., Hsp70, Hsp90, Hsp40, sHsp), E3 Ub ligases, and lectins; to demonstrate that COPII machinery sorts ERAD substrates to a degradation ER subcompartment in yeast; and evaluate CFTR degradation by autophagy [227–232]. Furthermore, the use of chimeras, specifically endogenous yeast Ste6 fused to ΔF508-CFTR (STE6/ΔF508-CFTR), has led to the isolation of two novel revertant mutations (I539T and G550E) that increased chloride channel activity [233].

2.5 Conclusions

S. cerevisiae has been established as a valuable host for membrane protein expression due to advantages of genetic flexibility, low cost of microbial growth, and production of biologically relevant proteins. This chapter has served as an introduction to membrane protein expression in *S. cerevisiae*, identified potential obstacles that affect functional membrane protein production, and illustrated methods for successful membrane protein expression, which used Ste2p, Pma1p, and CFTR as examples. In several cases, this has led to successful production of membrane protein crystals (Table 2.5).

Table 2.5 Crystal structures obtained from expression in *S. cerevisiae*.

Species	Membrane protein	Structural resolution (Å)	Vector	Strain	Protein description	Promoter	Tag	Reference
Human	monoamine oxidase A	2.2	YEp51	BJ2168	mitochondrial outer membrane protein	P_{GAL}	His_6	[234]
Rat	monoamine oxidase A	3.2	YEp51	BJ2168	mitochondrial outer membrane protein	P_{GAL}	His_6	[235, 236]
Yeast	ATP synthase	3.9	–	–	mitochondrial ATP synthase	–	–	[237]
Yeast	F1 ATPase	2.8	–	DMY301	ATP unit	–	His_6	[238, 239]
Yeast	cytochrome bc_1 complex	2.3	–	–	multisubunit mitochondrial transmembrane lipoprotein	–	–	[240]

Abbreviations

ABC	ATP-binding cassette
ARS	autonomous replicating sequence
CEN	centromere
CF	cystic fibrosis
CFTR	cystic fibrosis transmembrane conductance regulator
CL	cardiolipin
ER	endoplasmic reticulum
ERAD	ER-associated degradation
GFP	Green Fluorescent Protein
GPCR	G-protein-coupled receptor
HSR	heat-shock response
iMYTH	integrated membrane yeast two-hybrid
NBD	nucleotide-binding domain
PC	Phosphatidylcholine
PCR	Polymerase chain reaction
PE	Phosphatidylethanolamine
PI	phosphotidylinositol
PIP	phosphoinositide
PS	phosphatidylserine
Ub	ubiquitin
UPR	unfolded protein response
VPS	vacuolar protein sorting

References

1 Wach, A. (1996) PCR-synthesis of marker cassettes with long flanking homology regions for gene disruptions in *S. cerevisiae*. *Yeast*, **12**, 259–265.

2 Longtine, M.S., *et al.* (1998) Additional modules for versatile and economical PCR-based gene deletion and modification in *Saccharomyces cerevisiae*. *Yeast*, **14**, 953–961.

3 Guldener, U., *et al.* (1996) A new efficient gene disruption cassette for repeated use in budding yeast. *Nucleic Acids Res.*, **24**, 2519–2524.

4 Storici, F. and Resnick, M.A. (2006) The delitto perfetto approach to *in vivo* site-directed mutagenesis and chromosome rearrangements with synthetic oligonucleotides in yeast. *Methods Enzymol.*, **409**, 329–345.

5 Carpenter, E.P., *et al.* (2008) Overcoming the challenges of membrane protein crystallography. *Curr. Opin. Struct. Biol.*, **18**, 581–586.

6 White, S.H. (2010) Membrane Proteins of Known 3D Structure. Available from: http://blanco.biomol.uci.edu/Membrane_Proteins_xtal.html (accessed 10 January 2009).

7 Jidenko, M., *et al.* (2005) Crystallization of a mammalian membrane protein overexpressed in *Saccharomyces cerevisiae*. *Proc. Natl. Acad. Sci. USA*, **102**, 11687–11691.

8 Mumberg, D., Muller, R., and Funk, M. (1995) Yeast vectors for the controlled expression of heterologous proteins in different genetic backgrounds. *Gene*, **156**, 119–122.

9 Mumberg, D., Muller, R., and Funk, M. (1994) Regulatable promoters of *Saccharomyces cerevisiae*: comparison of transcriptional activity and their use for heterologous expression. *Nucleic Acids Res.*, **22**, 5767–5768.
10 Hinnen, A., Meyhack, B., and Heim, J. (1989) Heterologous gene expression in yeast. *Biotechnology*, **13**, 193–213.
11 Romanos, M.A., Scorer, C.A., and Clare, J.J. (1992) Foreign gene expression in yeast: a review. *Yeast*, **8**, 423–488.
12 Wang, Z. (2006) Controlled expression of recombinant genes and preparation of cell-free extracts in yeast. *Methods Mol. Biol.*, **313**, 317–331.
13 Maya, D., *et al.* (2008) Systems for applied gene control in *Saccharomyces cerevisiae*. *Biotechnol. Lett.*, **30**, 979–987.
14 Johnston, M. (1987) A model fungal gene regulatory mechanism: the GAL genes of *Saccharomyces cerevisiae*. *Microbiol. Rev.*, **51**, 458–476.
15 Lohr, D., Venkov, P., and Zlatanova, J. (1995) Transcriptional regulation in the yeast GAL gene family: a complex genetic network. *FASEB J.*, **9**, 777–787.
16 Butt, T.R., *et al.* (1984) Cloning and expression of a yeast copper metallothionein gene. *Gene*, **27**, 23–33.
17 Robinson, A.S., *et al.* (1996) Reduction of BiP levels decreases heterologous protein secretion in *Saccharomyces cerevisiae*. *J. Biol. Chem.*, **271**, 10017–10022.
18 Etcheverry, T. (1990) Induced expression using yeast copper metallothionein promoter. *Methods Enzymol.*, **185**, 319–329.
19 Holz, C. and Lang, C. (2004) High-throughput expression in microplate format in *Saccharomyces cerevisiae*. *Methods Mol. Biol.*, **267**, 267–276.
20 Butt, T.R. and Ecker, D.J. (1987) Yeast metallothionein and applications in biotechnology. *Microbiol. Rev.*, **51**, 351–364.
21 Price, V.L., *et al.* (1990) Expression of heterologous proteins in *Saccharomyces cerevisiae* using the ADH2 promoter. *Methods Enzymol.*, **185**, 308–318.
22 Guarente, L., *et al.* (1984) Distinctly regulated tandem upstream activation sites mediate catabolite repression of the CYC1 gene of *S. cerevisiae*. *Cell*, **36**, 503–511.
23 Tuite, M.F., *et al.* (1982) Regulated high efficiency expression of human interferon-alpha in *Saccharomyces cerevisiae*. *EMBO J.*, **1**, 603–608.
24 Schirmaier, F. and Philippsen, P. (1984) Identification of two genes coding for the translation elongation factor EF-1 alpha of *S. cerevisiae*. *EMBO J.*, **3**, 3311–3315.
25 Nagashima, K., *et al.* (1986) Structure of the two genes coding for polypeptide chain elongation factor 1 alpha (EF-1 alpha) from *Saccharomyces cerevisiae*. *Gene*, **45**, 265–273.
26 Musti, A.M., *et al.* (1983) Transcriptional mapping of two yeast genes coding for glyceraldehyde 3-phosphate dehydrogenase isolated by sequence homology with the chicken gene. *Gene*, **25**, 133–143.
27 Bitter, G.A. and Egan, K.M. (1984) Expression of heterologous genes in *Saccharomyces cerevisiae* from vectors utilizing the glyceraldehyde-3-phosphate dehydrogenase gene promoter. *Gene*, **32**, 263–274.
28 Johnston, S.A., Salmeron, J.M., and Dincher, S.S. (1987) Interaction of positive and negative regulatory proteins in the galactose regulon of yeast. *Cell*, **50**, 143–146.
29 Jones, E.W. (1991) Tackling the protease problem in *Saccharomyces cerevisiae*. *Methods Enzymol.*, **194**, 428–453.
30 Duport, C., Loeper, J., and Strosberg, A.D. (2003) Comparative expression of the human beta$_2$ and beta$_3$ adrenergic receptors in *Saccharomyces cerevisiae*. *Biochim. Biophys. Acta*, **1629**, 34–43.
31 Tschumper, G. and Carbon, J. (1980) Sequence of a yeast DNA fragment containing a chromosomal replicator and the TRP1 gene. *Gene*, **10**, 157–166.
32 Struhl, K. and Davis, R.W. (1980) A physical, genetic and transcriptional map of the cloned his3 gene region of *Saccharomyces cerevisiae*. *J. Mol. Biol.*, **136**, 309–332.
33 Rose, M., Grisafi, P., and Botstein, D. (1984) Structure and function of the

yeast URA3 gene: expression in *Escherichia coli*. *Gene*, **29**, 113–124.

34 Brachmann, C.B., *et al.* (1998) Designer deletion strains derived from *Saccharomyces cerevisiae* S288C: a useful set of strains and plasmids for PCR-mediated gene disruption and other applications. *Yeast*, **14**, 115–132.

35 Boeke, J.D., LaCroute, F., and Fink, G.R. (1984) A positive selection for mutants lacking orotidine-5′-phosphate decarboxylase activity in yeast: 5-fluoro-orotic acid resistance. *Mol. Gen. Genet.*, **197**, 345–346.

36 Barnes, D.A. and Thorner, J. (1986) Genetic manipulation of *Saccharomyces cerevisiae* by use of the LYS2 gene. *Mol. Cell Biol.*, **6**, 2828–2838.

37 Chattoo, B.B., *et al.* (1979) Selection of *lys2* mutants of the yeast *Saccharomyces cerevisiae* by the utilization of alpha-aminoadipate. *Genetics*, **93**, 51–65.

38 Fleig, U.N., Pridmore, R.D., and Philippsen, P. (1986) Construction of LYS2 cartridges for use in genetic manipulations of *Saccharomyces cerevisiae*. *Gene*, **46**, 237–245.

39 Guthrie, C. (ed.) (1991) Guide to yeast genetics and molecular biology. *Methods Enzymol.*, **194**, 1–863.

40 Hadfield, C., *et al.* (1990) G418-resistance as a dominant marker and reporter for gene expression in *Saccharomyces cerevisiae*. *Curr. Genet.*, **18**, 303–313.

41 Gritz, L. and Davies, J. (1983) Plasmid-encoded hygromycin B resistance: the sequence of hygromycin B phosphotransferase gene and its expression in *Escherichia coli* and *Saccharomyces cerevisiae*. *Gene*, **25**, 179–188.

42 Christianson, T.W., *et al.* (1992) Multifunctional yeast high-copy-number shuttle vectors. *Gene*, **110**, 119–122.

43 Davies, J. and Davis, B.D. (1968) Misreading of ribonucleic acid code words induced by aminoglycoside antibiotics. The effect of drug concentration. *J. Biol. Chem.*, **243**, 3312–3316.

44 Cabanas, M.J., Vazquez, D., and Modolell, J. (1978) Dual interference of hygromycin B with ribosomal translocation and with aminoacyl-tRNA recognition. *Eur. J. Biochem.*, **87**, 21–27.

45 Gatignol, A., Durand, H., and Tiraby, G. (1988) Bleomycin resistance conferred by a drug-binding protein. *FEBS Lett.*, **230**, 171–175.

46 Parent, S.A., Fenimore, C.M., and Bostian, K.A. (1985) Vector systems for the expression, analysis and cloning of DNA sequences in *S. cerevisiae*. *Yeast*, **1**, 83–138.

47 Sikorski, R.S. and Hieter, P. (1989) A system of shuttle vectors and yeast host strains designed for efficient manipulation of DNA in *Saccharomyces cerevisiae*. *Genetics*, **122**, 19–27.

48 Clarke, L. and Carbon, J. (1980) Isolation of a yeast centromere and construction of functional small circular chromosomes. *Nature*, **287**, 504–509.

49 Armstrong, K.A., *et al.* (1989) Propagation and expression of genes in yeast using 2-micron circle vectors. *Biotechnology*, **13**, 165–192.

50 Broach, J.R. (1983) Construction of high copy yeast vectors using 2-microns circle sequences. *Methods Enzymol.*, **101**, 307–325.

51 Futcher, A.B. (1988) The 2 micron circle plasmid of *Saccharomyces cerevisiae*. *Yeast*, **4**, 27–40.

52 Murray, J.A. (1987) Bending the rules: the 2-mu plasmid of yeast. *Mol. Microbiol.*, **1**, 1–4.

53 Volkert, F.C., Wilson, D.W., and Broach, J.R. (1989) Deoxyribonucleic acid plasmids in yeasts. *Microbiol. Rev.*, **53**, 299–317.

54 Mead, D.J., Gardner, D.C., and Oliver, S.G. (1986) The yeast 2 micron plasmid: strategies for the survival of a selfish DNA. *Mol. Gen. Genet.*, **205**, 417–421.

55 Orr-Weaver, T.L., Szostak, J.W., and Rothstein, R.J. (1983) Genetic applications of yeast transformation with linear and gapped plasmids. *Methods Enzymol.*, **101**, 228–245.

56 Lorenz, M.C., *et al.* (1995) Gene disruption with PCR products in *Saccharomyces cerevisiae*. *Gene*, **158**, 113–117.

57 Lee, F.W. and Da Silva, N.A. (1997) Sequential delta-integration for the regulated insertion of cloned genes in

Saccharomyces cerevisiae. *Biotechnol. Prog.*, **13**, 368–373.

58 Parekh, R.N., Shaw, M.R., and Wittrup, K.D. (1996) An integrating vector for tunable, high copy, stable integration into the dispersed Ty delta sites of *Saccharomyces cerevisiae*. *Biotechnol. Prog.*, **12**, 16–21.

59 Newstead, S., *et al.* (2007) High-throughput fluorescent-based optimization of eukaryotic membrane protein overexpression and purification in *Saccharomyces cerevisiae*. *Proc. Natl. Acad. Sci. USA*, **104**, 13936–13941.

60 Li, M., *et al.* (2009) Selecting optimum eukaryotic integral membrane proteins for structure determination by rapid expression and solubilization screening. *J. Mol. Biol.*, **385**, 820–830.

61 Drew, D., *et al.* (2008) GFP-based optimization scheme for the overexpression and purification of eukaryotic membrane proteins in *Saccharomyces cerevisiae*. *Nat. Protoc.*, **3**, 784–798.

62 Ito, K., *et al.* (2008) Advanced method for high-throughput expression of mutated eukaryotic membrane proteins in *Saccharomyces cerevisiae*. *Biochem. Biophys. Res. Commun.*, **371**, 841–845.

63 Bonander, N., *et al.* (2005) Design of improved membrane protein production experiments: quantitation of the host response. *Protein Sci.*, **14**, 1729–1740.

64 Wedekind, A., *et al.* (2006) Optimization of the human adenosine A2a receptor yields in *Saccharomyces cerevisiae*. *Biotechnol. Prog.*, **22**, 1249–1255.

65 Meynial-Salles, I. and Combes, D. (1996) *In vitro* glycosylation of proteins: an enzymatic approach. *J. Biotechnol.*, **46**, 1–14.

66 Olden, K., *et al.* (1982) Function of the carbohydrate moieties of glycoproteins. *J. Cell. Biochem.*, **18**, 313–335.

67 Wang, C., *et al.* (1996) Influence of the carbohydrate moiety on the stability of glycoproteins. *Biochemistry*, **35**, 7299–7307.

68 Aubert, J.P., Biserte, G., and Loucheux-Lefebvre, M.H. (1976) Carbohydrate–peptide linkage in glycoproteins. *Arch. Biochem. Biophys.*, **175**, 410–418.

69 Romanos, M.A. (1995) Advances in the use of *Pichia pastoris* for high-level gene-expression. *Curr. Opin. Biotechnol.*, **5**, 527–533.

70 Tsai, P.K., Frevert, J., and Ballou, C.E. (1984) Carbohydrate structure of *Saccharomyces cerevisiae* mnn9 mannoprotein. *J. Biol. Chem.*, **259**, 3805–3811.

71 Eckart, M.R. and Bussineau, C.M. (1996) Quality and authenticity of heterologous proteins synthesized in yeast. *Curr. Opin. Biotechnol.*, **7**, 525–530.

72 O'Malley, M.A., *et al.* (2009) Progress toward heterologous expression of active G-protein-coupled receptors in *Saccharomyces cerevisiae*: linking cellular stress response with translocation and trafficking. *Protein Sci.*, **18**, 2356–2370.

73 Robinson, A.S., Hines, V., and Wittrup, K.D. (1994) Protein disulfide isomerase overexpression increases secretion of foreign proteins in *Saccharomyces cerevisiae*. *Biotechnology*, **12**, 381–384.

74 Shusta, E.V., *et al.* (1998) Increasing the secretory capacity of *Saccharomyces cerevisiae* for production of single-chain antibody fragments. *Nat. Biotechnol.*, **16**, 773–777.

75 Butz, J.A., Niebauer, R.T., and Robinson, A.S. (2003) Co-expression of molecular chaperones does not improve the heterologous expression of mammalian G-protein coupled receptor expression in yeast. *Biotechnol. Bioeng.*, **84**, 292–304.

76 Schneiter, R. and Toulmay, A. (2007) The role of lipids in the biogenesis of integral membrane proteins. *Appl. Microbiol. Biotechnol.*, **73**, 1224–1232.

77 Daum, G., *et al.* (1999) Systematic analysis of yeast strains with possible defects in lipid metabolism. *Yeast*, **15**, 601–614.

78 Wright, R., *et al.* (1988) Increased amounts of HMG-CoA reductase induce "karmellae": a proliferation of stacked membrane pairs surrounding the yeast nucleus. *J. Cell Biol.*, **107**, 101–114.

79 van Klompenburg, W., *et al.* (1997) Anionic phospholipids are determinants of membrane protein topology. *EMBO J.*, **16**, 4261–4266.

80 van Dalen, A. and de Kruijff, B. (2004) The role of lipids in membrane insertion and translocation of bacterial proteins. *Biochim. Biophys. Acta*, **1694**, 97–109.
81 Osborne, A.R., Rapoport, T.A., and van den Berg, B. (2005) Protein translocation by the Sec61/SecY channel. *Annu. Rev. Cell Dev. Biol.*, **21**, 529–550.
82 Zinser, E., *et al.* (1991) Phospholipid synthesis and lipid composition of subcellular membranes in the unicellular eukaryote *Saccharomyces cerevisiae*. *J. Bacteriol.*, **173**, 2026–2034.
83 Bretscher, M.S. (1972) Asymmetrical lipid bilayer structure for biological membranes. *Nat. New Biol.*, **236**, 11–12.
84 Sikorski, R.S. and Boeke, J.D. (1991) *In vitro* mutagenesis and plasmid shuffling: from cloned gene to mutant yeast. *Methods Enzymol.*, **194**, 302–318.
85 Strahl, T. and Thorner, J. (2007) Synthesis and function of membrane phosphoinositides in budding yeast, *Saccharomyces cerevisiae*. *Biochim. Biophys. Acta*, **1771**, 353–404.
86 Lecocq, J. and Ballou, C.E. (1964) On the structure of cardiolipin. *Biochemistry*, **3**, 976–980.
87 Schlame, M., Rua, D., and Greenberg, M.L. (2000) The biosynthesis and functional role of cardiolipin. *Prog. Lipid Res.*, **39**, 257–288.
88 Hoch, F.L. (1992) Cardiolipins and biomembrane function. *Biochim. Biophys. Acta*, **1113**, 71–133.
89 Fry, M. and Green, D.E. (1981) Cardiolipin requirement for electron transfer in complex I and III of the mitochondrial respiratory chain. *J. Biol. Chem.*, **256**, 1874–1880.
90 Joshi, A.S., *et al.* (2009) Cellular functions of cardiolipin in yeast. *Biochim. Biophys. Acta*, **1793**, 212–218.
91 Exton, J.H. (1994) Phosphatidylcholine breakdown and signal transduction. *Biochim. Biophys. Acta*, **1212**, 26–42.
92 Kent, C. and Carman, G.M. (1999) Interactions among pathways for phosphatidylcholine metabolism, CTP synthesis and secretion through the Golgi apparatus. *Trends Biochem. Sci.*, **24**, 146–150.
93 Howe, A.G. and McMaster, C.R. (2001) Regulation of vesicle trafficking, transcription, and meiosis: lessons learned from yeast regarding the disparate biologies of phosphatidylcholine. *Biochim. Biophys. Acta*, **1534**, 65–77.
94 Robl, I., *et al.* (2001) Construction of phosphatidylethanolamine-less strain of *Saccharomyces cerevisiae*. Effect on amino acid transport. *Yeast*, **18**, 251–260.
95 Opekarova, M., Robl, I., and Tanner, W. (2002) Phosphatidyl ethanolamine is essential for targeting the arginine transporter Can1p to the plasma membrane of yeast. *Biochim. Biophys. Acta*, **1564**, 9–13.
96 Malinska, K., *et al.* (2003) Visualization of protein compartmentation within the plasma membrane of living yeast cells. *Mol. Biol. Cell*, **14**, 4427–4436.
97 van der Rest, M.E., *et al.* (1995) The plasma membrane of *Saccharomyces cerevisiae*: structure, function, and biogenesis. *Microbiol. Rev.*, **59**, 304–322.
98 Dickson, R.C., Sumanasekera, C., and Lester, R.L. (2006) Functions and metabolism of sphingolipids in *Saccharomyces cerevisiae*. *Prog. Lipid Res.*, **45**, 447–465.
99 Saeki, T., *et al.* (1991) Expression of human P-glycoprotein in yeast cells – effects of membrane component sterols on the activity of P-glycoprotein. *Agric. Biol. Chem.*, **55**, 1859–1865.
100 Lagane, B., *et al.* (2000) Role of sterols in modulating the human mu-opioid receptor function in *Saccharomyces cerevisiae*. *J. Biol. Chem.*, **275**, 33197–33200.
101 Sander, P., *et al.* (1994) Expression of the human D2S dopamine receptor in the yeasts *Saccharomyces cerevisiae* and *Schizosaccharomyces pombe*: a comparative study. *FEBS Lett.*, **344**, 41–46.
102 King, K., *et al.* (1990) Control of yeast mating signal transduction by a mammalian beta 2-adrenergic receptor and Gs alpha subunit. *Science*, **250**, 121–123.
103 Sizmann, D., *et al.* (1996) Production of adrenergic receptors in yeast. *Receptors Channels*, **4**, 197–203.

104 Sander, P., *et al.* (1994) Heterologous expression of the human D2S dopamine receptor in protease-deficient *Saccharomyces cerevisiae* strains. *Eur. J. Biochem.*, **226**, 697–705.

105 Andersen, B. and Stevens, R.C. (1998) The human D_{1A} dopamine receptor: heterologous expression in *Saccharomyces cerevisiae* and purification of the functional receptor. *Protein Expr. Purif.*, **13**, 111–119.

106 Brake, A.J., *et al.* (1984) Alpha-factor-directed synthesis and secretion of mature foreign proteins in *Saccharomyces cerevisiae. Proc. Natl. Acad. Sci. USA*, **81**, 4642–4646.

107 Huang, H.J., *et al.* (1992) Functional expression of rat M5 muscarinic acetylcholine receptor in yeast. *Biochem. Biophys. Res. Commun.*, **182**, 1180–1186.

108 Phongdara, A., *et al.* (1998) Cloning and characterization of the gene encoding a repressible acid phosphatase (PHO1) from the methylotrophic yeast *Hansenula polymorpha. Appl. Microbiol. Biotechnol.*, **50**, 77–84.

109 Kaur, P., *et al.* (2007) APHO1 from the yeast *Arxula adeninivorans* encodes an acid phosphatase of broad substrate specificity. *Antonie Van Leeuwenhoek*, **91**, 45–55.

110 Huang, Y., Wilkinson, G.F., and Willars, G.B. (2010) Role of the signal peptide in the synthesis and processing of the glucagon-like peptide-1 receptor. *Br. J. Pharmacol.*, **159**, 237–251.

111 Kochl, R., *et al.* (2002) The signal peptide of the G protein-coupled human endothelin B receptor is necessary for translocation of the N-terminal tail across the endoplasmic reticulum membrane. *J. Biol. Chem.*, **277**, 16131–16138.

112 Alken, M., *et al.* (2009) The sequence after the signal peptide of the G protein-coupled endothelin B receptor is required for efficient translocon gating at the endoplasmic reticulum membrane. *Mol. Pharmacol.*, **75**, 801–811.

113 Green, G.N., Hansen, W., and Walter, P. (1989) The use of gene-fusions to determine membrane protein topology in *Saccharomyces cerevisiae. J. Cell Sci. Suppl.*, **11**, 109–113.

114 Sengstag, C. (2000) Using SUC2–HIS4C reporter domain to study topology of membrane proteins in *Saccharomyces cerevisiae. Methods. Enzymol.*, **327**, 175–190.

115 Kim, H., Melen, K., and von Heijne, G. (2003) Topology models for 37 *Saccharomyces cerevisiae* membrane proteins based on C-terminal reporter fusions and predictions. *J. Biol. Chem.*, **278**, 10208–10213.

116 Kim, H., *et al.* (2006) A global topology map of the *Saccharomyces cerevisiae* membrane proteome. *Proc. Natl. Acad. Sci. USA*, **103**, 11142–11147.

117 Wilkinson, B.M., Critchley, A.J., and Stirling, C.J. (1996) Determination of the transmembrane topology of yeast Sec61p, an essential component of the endoplasmic reticulum translocation complex. *J. Biol. Chem.*, **271**, 25590–25597.

118 Cheng, Z. and Gilmore, R. (2006) Slow translocon gating causes cytosolic exposure of transmembrane and lumenal domains during membrane protein integration. *Nat. Struct. Mol. Biol.*, **13**, 930–936.

119 Jiang, Y., *et al.* (2008) An interaction between the SRP receptor and the translocon is critical during cotranslational protein translocation. *J. Cell Biol.*, **180**, 1149–1161.

120 Viklund, H. and Elofsson, A. (2004) Best alpha-helical transmembrane protein topology predictions are achieved using hidden Markov models and evolutionary information. *Protein Sci.*, **13**, 1908–1917.

121 Krogh, A., *et al.* (2001) Predicting transmembrane protein topology with a hidden Markov model: application to complete genomes. *J. Mol. Biol.*, **305**, 567–580.

122 Tusnady, G.E. and Simon, I. (1998) Principles governing amino acid composition of integral membrane proteins: application to topology prediction. *J. Mol. Biol.*, **283**, 489–506.

123 Melen, K., Krogh, A., and von Heijne, G. (2003) Reliability measures for membrane protein topology prediction algorithms. *J. Mol. Biol.*, **327**, 735–744.

124 Bernsel, A., *et al.* (2009) TOPCONS: consensus prediction of membrane

protein topology. *Nucleic Acids Res.*, **37**, W465–W468.

125 Viklund, H., *et al.* (2008) SPOCTOPUS: a combined predictor of signal peptides and membrane protein topology. *Bioinformatics*, **24**, 2928–2929.

126 Snider, C., *et al.* (2009) MPEx: a tool for exploring membrane proteins. *Protein Sci.*, **18**, 2624–2628.

127 Hessa, T., *et al.* (2007) Molecular code for transmembrane-helix recognition by the Sec61 translocon. *Nature*, **450**, 1026–1030.

128 Lundin, C., *et al.* (2008) Molecular code for protein insertion in the endoplasmic reticulum membrane is similar for N_{in}–C_{out} and N_{out}–C_{in} transmembrane helices. *Proc. Natl. Acad. Sci. USA*, **105**, 15702–15707.

129 White, S.H. and von Heijne, G. (2008) How translocons select transmembrane helices. *Annu. Rev. Biophys.*, **37**, 23–42.

130 Kauffman, K.J., *et al.* (2002) Decreased protein expression and intermittent recoveries in BiP levels result from cellular stress during heterologous protein expression in *Saccharomyces cerevisiae. Biotechnol. Prog.*, **18**, 942–950.

131 Chapman, R., Sidrauski, C., and Walter, P. (1998) Intracellular signaling from the endoplasmic reticulum to the nucleus. *Annu. Rev. Cell. Dev. Biol.*, **14**, 459–485.

132 Kaufman, R.J. (1999) Stress signaling from the lumen of the endoplasmic reticulum: coordination of gene transcriptional and translational controls. *Genes Dev.*, **13**, 1211–1233.

133 Kimata, Y., *et al.* (2004) A role for BiP as an adjustor for the endoplasmic reticulum stress-sensing protein Ire1. *J. Cell Biol.*, **167**, 445–456.

134 Kaufman, R.J., *et al.* (2002) The unfolded protein response in nutrient sensing and differentiation. *Nat. Rev. Mol. Cell Biol.*, **3**, 411–421.

135 Spear, E. and Ng, D.T. (2001) The unfolded protein response: no longer just a special teams player. *Traffic*, **2**, 515–523.

136 Griffith, D.A., *et al.* (2003) A novel yeast expression system for the overproduction of quality-controlled membrane proteins. *FEBS Lett.*, **553**, 45–50.

137 Nicolet, C.M. and Craig, E.A. (1991) Inducing and assaying heat-shock response in *Saccharomyces cerevisiae. Methods Enzymol.*, **194**, 710–717.

138 Boorstein, W.R. and Craig, E.A. (1990) Transcriptional regulation of SSA3, an HSP70 gene from *Saccharomyces cerevisiae. Mol. Cell. Biol.*, **10**, 3262–3267.

139 Mager, W.H. and Ferreira, P.M. (1993) Stress response of yeast. *Biochem. J.*, **290**, 1–13.

140 Sorger, P.K., Lewis, M.J., and Pelham, H.R. (1987) Heat shock factor is regulated differently in yeast and HeLa cells. *Nature*, **329**, 81–84.

141 Hahn, J.S., *et al.* (2004) Genome-wide analysis of the biology of stress responses through heat shock transcription factor. *Mol. Cell. Biol.*, **24**, 5249–5256.

142 Eastmond, D.L. and Nelson, H.C. (2006) Genome-wide analysis reveals new roles for the activation domains of the *Saccharomyces cerevisiae* heat shock transcription factor (Hsf1) during the transient heat shock response. *J. Biol. Chem.*, **281**, 32909–32921.

143 Liu, Y. and Chang, A. (2008) Heat shock response relieves ER stress. *EMBO J.*, **27**, 1049–1059.

144 Ladds, G., *et al.* (2005) A constitutively active GPCR retains its G protein specificity and the ability to form dimers. *Mol. Microbiol.*, **55**, 482–497.

145 David, N.E., *et al.* (1997) Expression and purification of the *Saccharomyces cerevisiae* alpha-factor receptor (Ste2p), a 7-transmembrane-segment G protein-coupled receptor. *J. Biol. Chem.*, **272**, 15553–15561.

146 Lee, B.K., *et al.* (2007) Affinity purification and characterization of a G-protein coupled receptor, *Saccharomyces cerevisiae* Ste2p. *Protein Expr. Purif.*, **56**, 62–71.

147 Akal-Strader, A., *et al.* (2002) Residues in the first extracellular loop of a G protein-coupled receptor play a role in signal transduction. *J. Biol. Chem.*, **277**, 30581–30590.

148 Sarramegna, V., *et al.* (2003) Heterologous expression of G-protein-

coupled receptors: comparison of expression systems fron the standpoint of large-scale production and purification. *Cell. Mol. Life Sci.*, **60**, 1529–1546.
149 Sarramegna, V., *et al.* (2006) Recombinant G protein-coupled receptors from expression to renaturation: a challenge towards structure. *Cell. Mol. Life Sci.*, **63**, 1149–1164.
150 Chiu, M.L., *et al.* (2008) Over-expression, solubilization, and purification of G protein-coupled receptors for structural biology. *Comb. Chem. High Throughput Screen.*, **11**, 439–462.
151 Payette, P., *et al.* (1990) Expression and pharmacological characterization of the human M1 muscarinic receptor in *Saccharomyces cerevisiae. FEBS Lett.*, **266**, 21–25.
152 Erlenbach, I., *et al.* (2001) Functional expression of M_1, M_3 and M_5 muscarinic acetylcholine receptors in yeast. *J. Neurochem.*, **77**, 1327–1337.
153 Schmidt, C., *et al.* (2003) Random mutagenesis of the M3 muscarinic acetylcholine receptor expressed in yeast. Identification of point mutations that "silence" a constitutively active mutant M3 receptor and greatly impair receptor/G protein coupling. *J. Biol. Chem.*, **278**, 30248–30260.
154 Li, B., *et al.* (2005) Random mutagenesis of the M3 muscarinic acetylcholine receptor expressed in yeast: identification of second-site mutations that restore function to a coupling-deficient mutant M3 receptor. *J. Biol. Chem.*, **280**, 5664–5675.
155 Li, B., *et al.* (2007) Rapid identification of functionally critical amino acids in a G protein-coupled receptor. *Nat. Methods*, **4**, 169–174.
156 Scarselli, M., *et al.* (2007) Multiple residues in the second extracellular loop are critical for M3 muscarinic acetylcholine receptor activation. *J. Biol. Chem.*, **282**, 7385–7396.
157 Thor, D., *et al.* (2008) Generation of an agonistic binding site for blockers of the M(3) muscarinic acetylcholine receptor. *Biochem. J.*, **412**, 103–112.
158 Campbell, R.M., *et al.* (1999) Selective A1-adenosine receptor antagonists identified using yeast *Saccharomyces cerevisiae* functional assays. *Bioorg. Med. Chem. Lett.*, **9**, 2413–2418.
159 Stewart, G.D., *et al.* (2009) Determination of adenosine A1 receptor agonist and antagonist pharmacology using *Saccharomyces cerevisiae*: implications for ligand screening and functional selectivity. *J. Pharmacol. Exp. Ther.*, **331**, 277–286.
160 Niebauer, R.T., Wedekind, A., and Robinson, A.S. (2004) Decreases in yeast expression yields of the human adenosine A2a receptor are a result of translational or post-translational events. *Protein Expr. Purif.*, **37**, 134–143.
161 Niebauer, R.T. and Robinson, A.S. (2006) Exceptional total and functional yields of the human adenosine (A2a) receptor expressed in the yeast *Saccharomyces cerevisiae. Protein Expr. Purif.*, **46**, 204–211.
162 O'Malley, M.A., *et al.* (2007) High-level expression in *Saccharomyces cerevisiae* enables isolation and spectroscopic characterization of functional human adenosine A2a receptor. *J. Struct. Biol.*, **159**, 166–178.
163 Brown, A.J., *et al.* (2000) Functional coupling of mammalian receptors to the yeast mating pathway using novel yeast/mammalian G protein alpha-subunit chimeras. *Yeast*, **16**, 11–22.
164 Beukers, M.W., *et al.* (2004) Random mutagenesis of the human adenosine A2B receptor followed by growth selection in yeast. Identification of constitutively active and gain of function mutations. *Mol. Pharmacol.*, **65**, 702–710.
165 Li, Q., *et al.* (2007) ZM241385, DPCPX, MRS1706 are inverse agonists with different relative intrinsic efficacies on constitutively active mutants of the human adenosine A2B receptor. *J. Pharmacol. Exp. Ther.*, **320**, 637–645.
166 Kapat, A., *et al.* (2000) Production and purification of recombinant human alpha 2C2 adrenergic receptor using *Saccharomyces cerevisiae. Bioseparation*, **9**, 167–172.
167 Liitti, S., *et al.* (2001) Immunoaffinity purification and reconstitution of human alpha(2)-adrenergic receptor

subtype C2 into phospholipid vesicles. *Protein Expr. Purif.*, **22**, 1–10.
168 Miret, J.J., *et al.* (2002) Functional expression of heteromeric calcitonin gene-related peptide and adrenomedullin receptors in yeast. *J. Biol. Chem.*, **277**, 6881–6887.
169 Horswill, J.G., *et al.* (2007) PSNCBAM-1, a novel allosteric antagonist at cannabinoid CB1 receptors with hypophagic effects in rats. *Br. J. Pharmacol.*, **152**, 805–814.
170 Baranski, T.J., *et al.* (1999) C5a receptor activation. Genetic identification of critical residues in four transmembrane helices. *J. Biol. Chem.*, **274**, 15757–15765.
171 Geva, A., *et al.* (2000) Genetic mapping of the human C5a receptor. Identification of transmembrane amino acids critical for receptor function. *J. Biol. Chem.*, **275**, 35393–35401.
172 Klco, J.M., *et al.* (2005) Essential role for the second extracellular loop in C5a receptor activation. *Nat. Struct. Mol. Biol.*, **12**, 320–326.
173 Zhang, W.B., *et al.* (2002) A point mutation that confers constitutive activity to CXCR4 reveals that T140 is an inverse agonist and that AMD3100 and ALX40-4C are weak partial agonists. *J. Biol. Chem.*, **277**, 24515–24521.
174 Sachpatzidis, A., *et al.* (2003) Identification of allosteric peptide agonists of CXCR4. *J. Biol. Chem.*, **278**, 896–907.
175 Zhang, W.B., *et al.* (2004) Functional expression of CXCR4 in *S. cerevisiae*: development of tools for mechanistic and pharmacologic studies. *Ernst Schering Res. Found. Workshop*, 125–152.
176 Evans, B.J., *et al.* (2009) Expression of CXCR4, a G-protein-coupled receptor for CXCL12 in yeast identification of new-generation inverse agonists. *Methods Enzymol.*, **460**, 399–412.
177 Sander, P., *et al.* (1994) Constitutive expression of the human D2S-dopamine receptor in the unicellular yeast *Saccharomyces cerevisiae. Biochim. Biophys. Acta*, **1193**, 255–262.
178 Klein, C., *et al.* (1998) Identification of surrogate agonists for the human FPRL-1 receptor by autocrine selection in yeast. *Nat. Biotechnol.*, **16**, 1334–1337.
179 Kajkowski, E.M., *et al.* (1997) Investigation of growth hormone releasing hormone receptor structure and activity using yeast expression technologies. *J. Recept. Signal Transduct. Res.*, **17**, 293–303.
180 Gatehouse, D., *et al.* (1988) Investigations into the genotoxic potential of loxtidine, a long-acting H_2-receptor antagonist. *Mutagenesis*, **3**, 57–68.
181 Erickson, J.R., *et al.* (1998) Edg-2/Vzg-1 couples to the yeast pheromone response pathway selectively in response to lysophosphatidic acid. *J. Biol. Chem.*, **273**, 1506–1510.
182 Kokkola, T., *et al.* (1998) Mutagenesis of human Mel1a melatonin receptor expressed in yeast reveals domains important for receptor function. *Biochem. Biophys. Res. Commun.*, **249**, 531–536.
183 Leplatois, P., *et al.* (2001) Neurotensin induces mating in *Saccharomyces cerevisiae* cells that express human neurotensin receptor type 1 in place of the endogenous pheromone receptor. *Eur. J. Biochem.*, **268**, 4860–4867.
184 Niebauer, R.T., *et al.* (2005) Signaling of the human $P2Y_1$ receptor measured by a yeast growth assay with comparisons to assays of phospholipase C and calcium mobilization in 1321N1 human astrocytoma cells. *Purinergic Signal.*, **1**, 241–247.
185 Pausch, M.H., *et al.* (2004) Functional expression of human and mouse P2Y12 receptors in *Saccharomyces cerevisiae. Biochem. Biophys. Res. Commun.*, **324**, 171–177.
186 Mollaaghababa, R., *et al.* (1996) Structure and function in rhodopsin: expression of functional mammalian opsin in *Saccharomyces cerevisiae. Proc. Natl. Acad. Sci. USA*, **93**, 11482–11486.
187 Bach, M., *et al.* (1996) Pharmacological and biochemical characterization of the mouse 5HT5A serotonin receptor heterologously produced in the yeast *Saccharomyces cerevisiae. Receptors Channels*, **4**, 129–139.
188 Weiss, H.M., Haase, W., and Reilander, H. (1998) Expression of an integral

membrane protein, the $5HT_{5A}$ receptor. *Methods Mol. Biol.*, **103**, 227–239.

189 Price, L.A., *et al.* (1995) Functional coupling of a mammalian somatostatin receptor to the yeast pheromone response pathway. *Mol. Cell Biol.*, **15**, 6188–6195.

190 Erlenbach, I., *et al.* (2001) Single amino acid substitutions and deletions that alter the G protein coupling properties of the V2 vasopressin receptor identified in yeast by receptor random mutagenesis. *J. Biol. Chem.*, **276**, 29382–29392.

191 Catty, P., de Kerchove d'Exaerde, A., and Goffeau, A. (1997) The complete inventory of the yeast *Saccharomyces cerevisiae* P-type transport ATPases. *FEBS Lett.*, **409**, 325–332.

192 Serrano, R., Kielland-Brandt, M.C., and Fink, G.R. (1986) Yeast plasma membrane ATPase is essential for growth and has homology with (Na^+ + K^+), K^+- and Ca^{2+}-ATPases. *Nature*, **319**, 689–693.

193 Kuhlbrandt, W. (2004) Biology, structure and mechanism of P-type ATPases. *Nat. Rev. Mol. Cell Biol.*, **5**, 282–295.

194 Ferreira, T., Mason, A.B., and Slayman, C.W. (2001) The yeast Pma1 proton pump: a model for understanding the biogenesis of plasma membrane proteins. *J. Biol. Chem.*, **276**, 29613–29616.

195 Ambesi, A., *et al.* (2000) Biogenesis and function of the yeast plasma-membrane H^+-ATPase. *J. Exp. Biol.*, **203**, 155–160.

196 Bagnat, M., Chang, A., and Simons, K. (2001) Plasma membrane proton ATPase Pma1p requires raft association for surface delivery in yeast. *Mol. Biol. Cell*, **12**, 4129–4138.

197 Chang, A. and Fink, G.R. (1995) Targeting of the yeast plasma membrane [H^+]ATPase: a novel gene AST1 prevents mislocalization of mutant ATPase to the vacuole. *J. Cell Biol.*, **128**, 39–49.

198 Benito, B., Moreno, E., and Lagunas, R. (1991) Half-life of the plasma membrane ATPase and its activating system in resting yeast cells. *Biochim. Biophys. Acta*, **1063**, 265–268.

199 Harris, S.L., *et al.* (1994) Dominant lethal mutations in the plasma membrane H^+-ATPase gene of *Saccharomyces cerevisiae*. *Proc. Natl. Acad. Sci. USA*, **91**, 10531–10535.

200 Shimoni, Y., *et al.* (2000) Lst1p and Sec24p cooperate in sorting of the plasma membrane ATPase into COPII vesicles in *Saccharomyces cerevisiae*. *J. Cell Biol.*, **151**, 973–984.

201 Luo, W. and Chang, A. (1997) Novel genes involved in endosomal traffic in yeast revealed by suppression of a targeting-defective plasma membrane ATPase mutant. *J. Cell Biol.*, **138**, 731–746.

202 Morsomme, P., Slayman, C.W., and Goffeau, A. (2000) Mutagenic study of the structure, function and biogenesis of the yeast plasma membrane H^+-ATPase. *Biochim. Biophys. Acta*, **1469**, 133–157.

203 DeWitt, N.D., *et al.* (1998) Phosphorylation region of the yeast plasma-membrane H+-ATPase. Role in protein folding and biogenesis. *J. Biol. Chem.*, **273**, 21744–21751.

204 Luo, W. and Chang, A. (2000) An endosome-to-plasma membrane pathway involved in trafficking of a mutant plasma membrane ATPase in yeast. *Mol. Biol. Cell*, **11**, 579–592.

205 Nakamoto, R.K., *et al.* (1998) Substitutions of aspartate 378 in the phosphorylation domain of the yeast PMA1 H^+-ATPase disrupt protein folding and biogenesis. *J. Biol. Chem.*, **273**, 7338–7344.

206 Han, S., Liu, Y., and Chang, A. (2007) Cytoplasmic Hsp70 promotes ubiquitination for endoplasmic reticulum-associated degradation of a misfolded mutant of the yeast plasma membrane ATPase, PMA1. *J. Biol. Chem.*, **282**, 26140–26149.

207 Wang, Q. and Chang, A. (1999) Eps1, a novel PDI-related protein involved in ER quality control in yeast. *EMBO J.*, **18**, 5972–5982.

208 Mazon, M.J., Eraso, P., and Portillo, F. (2007) Efficient degradation of misfolded mutant Pma1 by endoplasmic reticulum-associated degradation requires Atg19 and the Cvt/autophagy pathway. *Mol. Microbiol.*, **63**, 1069–1077.

209 Toulmay, A. and Schneiter, R. (2007) Lipid-dependent surface transport of the proton pumping ATPase: a model to study plasma membrane biogenesis in yeast. *Biochimie*, **89**, 249–254.

210 Bagnat, M., *et al.* (2000) Lipid rafts function in biosynthetic delivery of proteins to the cell surface in yeast. *Proc. Natl. Acad. Sci. USA*, **97**, 3254–3259.

211 Gaigg, B., *et al.* (2005) Synthesis of sphingolipids with very long chain fatty acids but not ergosterol is required for routing of newly synthesized plasma membrane ATPase to the cell surface of yeast. *J. Biol. Chem.*, **280**, 22515–22522.

212 Gaigg, B., Toulmay, A., and Schneiter, R. (2006) Very long-chain fatty acid-containing lipids rather than sphingolipids per se are required for raft association and stable surface transport of newly synthesized plasma membrane ATPase in yeast. *J. Biol. Chem.*, **281**, 34135–34145.

213 Higgins, C.F. (1992) ABC transporters: from microorganisms to man. *Annu. Rev. Cell Biol.*, **8**, 67–113.

214 Paumi, C.M., *et al.* (2009) ABC transporters in *Saccharomyces cerevisiae* and their interactors: new technology advances the biology of the ABCC (MRP) subfamily. *Microbiol. Mol. Biol. Rev.*, **73**, 577–593.

215 Wemmie, J.A. and Moye-Rowley, W.S. (1997) Mutational analysis of the *Saccharomyces cerevisiae* ATP-binding cassette transporter protein Ycf1p. *Mol. Microbiol.*, **25**, 683–694.

216 Huyer, G., *et al.* (2004) A striking quality control subcompartment in *Saccharomyces cerevisiae*: the endoplasmic reticulum-associated compartment. *Mol. Biol. Cell*, **15**, 908–921.

217 Falcon-Perez, J.M., *et al.* (1999) Functional domain analysis of the yeast ABC transporter Ycf1p by site-directed mutagenesis. *J. Biol. Chem.*, **274**, 23584–23590.

218 Mason, D.L., *et al.* (2003) A region within a lumenal loop of *Saccharomyces cerevisiae* Ycf1p directs proteolytic processing and substrate specificity. *Eukaryot. Cell*, **2**, 588–598.

219 Mason, D.L. and Michaelis, S. (2002) Requirement of the N-terminal extension for vacuolar trafficking and transport activity of yeast Ycf1p, an ATP-binding cassette transporter. *Mol. Biol. Cell*, **13**, 4443–4455.

220 Paumi, C.M., *et al.* (2007) Mapping protein-protein interactions for the yeast ABC transporter Ycf1p by integrated split-ubiquitin membrane yeast two-hybrid analysis. *Mol. Cell*, **26**, 15–25.

221 Pilewski, J.M. and Frizzell, R.A. (1999) Role of CFTR in airway disease. *Physiol. Rev.*, **79** (Suppl.), S215–S255.

222 Cheng, S.H., *et al.* (1990) Defective intracellular transport and processing of CFTR is the molecular basis of most cystic fibrosis. *Cell*, **63**, 827–834.

223 Zhang, Y., *et al.* (2001) Hsp70 molecular chaperone facilitates endoplasmic reticulum-associated protein degradation of cystic fibrosis transmembrane conductance regulator in yeast. *Mol. Biol. Cell*, **12**, 1303–1314.

224 Dalemans, W., *et al.* (1991) Altered chloride ion channel kinetics associated with the delta F508 cystic fibrosis mutation. *Nature*, **354**, 526–528.

225 Denning, G.M., *et al.* (1992) Processing of mutant cystic fibrosis transmembrane conductance regulator is temperature-sensitive. *Nature*, **358**, 761–764.

226 Sato, S., *et al.* (1996) Glycerol reverses the misfolding phenotype of the most common cystic fibrosis mutation. *J. Biol. Chem.*, **271**, 635–638.

227 Rubenstein, R.C. and Zeitlin, P.L. (2000) Sodium 4-phenylbutyrate downregulates Hsc70: implications for intracellular trafficking of deltaF508-CFTR. *Am. J. Physiol. Cell Physiol.*, **278**, C259–C267.

228 Youker, R.T., *et al.* (2004) Distinct roles for the Hsp40 and Hsp90 molecular chaperones during cystic fibrosis transmembrane conductance regulator degradation in yeast. *Mol. Biol. Cell*, **15**, 4787–4797.

229 Ahner, A., *et al.* (2007) Small heat-shock proteins select deltaF508-CFTR for endoplasmic reticulum-associated degradation. *Mol. Biol. Cell*, **18**, 806–814.

230 Gnann, A., Riordan, J.R., and Wolf, D.H. (2004) Cystic fibrosis transmembrane conductance regulator

degradation depends on the lectins Htm1p/EDEM and the Cdc48 protein complex in yeast. *Mol. Biol. Cell*, **15**, 4125–4135.

231 Fu, L. and Sztul, E. (2003) Traffic-independent function of the Sar1p/COPII machinery in proteasomal sorting of the cystic fibrosis transmembrane conductance regulator. *J. Cell Biol.*, **160**, 157–163.

232 Fu, L. and Sztul, E. (2009) ER-associated complexes (ERACs) containing aggregated cystic fibrosis transmembrane conductance regulator (CFTR) are degraded by autophagy. *Eur. J. Cell Biol.*, **88**, 215–226.

233 DeCarvalho, A.C., Gansheroff, L.J., and Teem, J.L. (2002) Mutations in the nucleotide binding domain 1 signature motif region rescue processing and functional defects of cystic fibrosis transmembrane conductance regulator delta f508. *J. Biol. Chem.*, **277**, 35896–35905.

234 Son, S.Y., *et al.* (2008) Structure of human monoamine oxidase A at 2.2-A resolution: the control of opening the entry for substrates/inhibitors. *Proc. Natl. Acad. Sci. USA*, **105**, 5739–5744.

235 Ma, J., *et al.* (2004) Structure of rat monoamine oxidase A and its specific recognitions for substrates and inhibitors. *J. Mol. Biol.*, **338**, 103–114.

236 Ma, J. and Ito, A. (2002) Tyrosine residues near the FAD binding site are critical for FAD binding and for the maintenance of the stable and active conformation of rat monoamine oxidase A. *J. Biochem.*, **131**, 107–111.

237 Stock, D., Leslie, A.G., and Walker, J.E. (1999) Molecular architecture of the rotary motor in ATP synthase. *Science*, **286**, 1700–1705.

238 Kabaleeswaran, V., *et al.* (2006) Novel features of the rotary catalytic mechanism revealed in the structure of yeast F1 ATPase. *EMBO J.*, **25**, 5433–5442.

239 Mueller, D.M., *et al.* (2004) Ni-chelate-affinity purification and crystallization of the yeast mitochondrial F1-ATPase. *Protein Expr. Purif.*, **37**, 479–485.

240 Hunte, C., *et al.* (2000) Structure at 2.3 A resolution of the cytochrome bc_1 complex from the yeast *Saccharomyces cerevisiae* co-crystallized with an antibody Fv fragment. *Structure*, **8**, 669–684.

3
Expression Systems: *Pichia pastoris*

Fatima Alkhalfioui, Christel Logez, Olivier Bornert, and Renaud Wagner

3.1
Introduction

Among the most widespread, popular, effective, and inexpensive microorganisms developed for heterologous expression, *Pichia pastoris* has become a system of choice not only for the production of cytosoluble and industrially relevant proteins, but also for a growing panel of eukaryotic membrane proteins expressed at levels compatible with structural studies. Up to now indeed, more than 150 different representative membrane proteins have been expressed in *P. pastoris* and this has led to the acquisition of high-resolution structures for a dozen of them, making *P. pastoris* one of the most performant heterologous expression system for the structural studies of eukaryotic membrane proteins. This chapter gives a global overview on (i) how the *P. pastoris* system basically operates, (ii) how it performs for the recombinant expression of membrane proteins, and (iii) the different strategies and tips that can be applied to improve the system.

3.2
A (Brief) Summary on the (Long) History of *P. pastoris*

P. pastoris is an ascosporous yeast that is naturally present in tree fluxes from European and north American forests [1]. It was first isolated in 1919 in France from the exudate of a chestnut tree [2] and half a century later was described for its ability to use methanol as a sole carbon source [3]. From this time point, the potential of *P. pastoris* for biotechnological applications has been continuously explored both in industry and academia. While its use as a potential source of single-cell protein for animal feed did not meet the expected economic viability [4], this yeast was rapidly recognized as a remarkable production platform for a wide class and number of heterologous proteins, and up to now more than 500 candidates have been successfully recombinantly expressed in this system (reviewed in [5, 6] among others), several of them being biopharmaceuticals already on the market. Recently, phylogenetic analyses based on rRNA sequence

Production of Membrane Proteins: Strategies for Expression and Isolation, First Edition.
Edited by Anne Skaja Robinson.

comparisons led to the transfer of *Pichia pastoris* into the *Komagataella* genus [7], and the commonly used biotechnological strains are now classified into two dinstinct species, *K. pastoris* and *K. phaffii* [8]. The saga of this special yeast, that we will continue to call *P. pastoris* for the sake of simplicity, will be certainly boosted in the coming years after the very recent release of its genome sequence [9]. This new wealth of information will indeed open new possibilities for the engineering of enhanced biotechnological strains.

3.3 Introducing *P. pastoris* as a Biotechnological Tool: Its (Extended) Strengths and (Limited) Weaknesses

The success of *P. pastoris* as an efficient protein factory is attributable to a series of advantages related to both its yeast nature and its particular methylotrophic metabolism. This organism indeed presents a short generation time (2 h), grows on very simple and inexpensive media, and is very easy to handle. A comprehensive panel of plasmids and strategies is available for the expression of recombinant genes and genetic manipulation is nearly as straightforward as for *Saccharomyces cerevisiae*. As a methylotroph, *P. pastoris* possesses a peculiar methanol utilization pathway relying on some of the strongest and most tightly regulated known promoters that can be used for very-high-level expression of recombinant genes. In addition, *P. pastoris* can reach very high cell densities (up to 130 g/l dry cell weight [5]), and various fermentation processes and formats have been developed in the industry so that up-scaling protein production is easily achievable [10, 11]. Contrary to *Escherichia coli*, *P. pastoris* is a eukaryotic microorganism capable of complex post-translational modifications including disulfide isomerization, sulfation, phosphorylation, N-terminal acetylation, C-terminal methylation, myristoylation, farnesylation, and glycosylation (reviewed in [12]), which are often very essential for the proper targeting, biological activity, and stability of the expressed recombinant proteins. Regarding glycosylation, which is central for many membrane proteins [13–15], *P. pastoris* has been notably shown to graft shorter and more authentic oligosaccharide chains to proteins than *S. cerevisiae* does [16], and therefore often appeared as a more appropriate system [11]. Finally, ^{15}N and ^{13}C isotopic labeling of recombinant proteins for nuclear magnetic resonance or spectrometric studies is also achievable with *P. pastoris*, both in a uniform mode using isotopically enriched nitrogen and carbon sources [17] or more selectively using amino acid isotopes and engineered auxotroph strains [18].

This idyllic description has, however, to be tempered with some drawbacks that prevent *P. pastoris* from becoming an ideal expression system. First, a common characteristic for eukaryotic systems that are efficiently overproducing proteins, an overload of the translocation and folding machineries in *P. pastoris* often creates a stress that triggers the activation of sorting and degradation processes, and results in lowered expression levels and heterogeneity of recombinant proteins [18]. Moreover, if glycosylation processes occur in a fashion acceptable for many recombinant proteins, *P. pastoris* is not able to graft the complex carbohydrate motifs that are sometimes critical for the functionality of mammalian proteins. In addition, nonhomogeneous *N*-glycosylation of recombinant proteins is frequently observed, notably in the case of membrane proteins, leading to some degree of heterogeneity [19–22] that can be detrimental in various applications, including structural studies. Recently, several *Pichia* strains have been engineered to generate more complex and more homogeneous *N*-glycosylations (reviewed in [23]). These strains exhibited protein-dependent but promising outcomes, thereby also demonstrating all the potential and possibilities that could be gained from these genetic engineering approaches.

Another characteristic that can have a direct impact on the expression of mammalian membrane proteins is related to the lipidic composition of yeast membranes that varies significantly from that of higher eukaryotes membranes [24]. As membrane proteins do require specific lipids for their proper functions or for their correct folding and stability, these differences may influence both the expression level and functionality of recombinant membrane proteins. This was notably reported in studies where the absence of cholesterol in *Pichia* membranes was shown to profoundly alter the activity and stability of recombinant membrane proteins [25].

Finally, a secondary but not trivial issue is related to the presence of a significant cell wall surrounding *Pichia* cells that cannot only hinder the secretion of certain proteins [26], but also represents an obstacle for the preparation of membrane-embedded proteins as aggressive disruption methods are needed [27]. Engineered strains with weaker cell walls have been recently developed [26, 28], but none has been reported yet for its use in heterologous expression of membrane proteins and their benefit for a facilitated cell disruption has still to be assayed.

3.4 Basics of the *P. pastoris* Expression System

This section is intended at give a global overview on how the system functions for the heterologous expression of proteins in general, before giving specific details on how it performs in particular for membrane proteins (Section 3.5). Further details on the system as well as additional information concerning secreted and/or cytosoluble proteins can be found in several excellent and comprehensive reviews that have been published on the topic [5, 6, 11, 29].

3.4.1 Methanol Utilization Pathway

Together with a small set of methylotrophic yeasts from the *Pichia*, *Komagataella*, *Candida* and *Ogatae* genera, *P. pastoris* has developed a specific metabolism for the utilization of methanol as sole carbon source. Briefly, methanol enters specialized microbodies, the peroxisomes, where it is oxidized by specific oxidases that are encoded by the two genes *AOX1* and *AOX2* to generate formaldehyde and hydrogen peroxide. While the latter compound is decomposed to water and molecular oxygen by a peroxisomal catalase, formaldehyde leaves the peroxisome to enter both the cytosolic dissimilatory pathway to yield energy and the assimilatory pathway for generation of biomass [30]. The genes encoding the specific enzymes related to this peculiar metabolism are repressed when cells are grown on nonmethanol carbon sources (glucose, glycerol, ethanol, etc.) and are dramatically induced in presence of methanol; alcohol oxidases representing as high as 30% of the total soluble protein content. These enzymes are thus very tightly regulated and their promoters represent ideal components to be used for recombinant expression – the basis for the development of the *P. pastoris* expression system.

3.4.2
Host Strains and Plasmids

The principal strains used for recombinant expression derived either from the NRRL Y-11430 (Northern Regional Research Laboratories, Peoria, IL) or the NRRL Y-48124 (Invitrogen expression kit) strains, both being from the *K. phaffii* type [8]. This limited number of strains is listed in Table 3.1. They mainly differ in their auxotrophic behavior, principally relying on a histidinol dehydrogenase deficiency (*his4*), allowing, upon transformation, for the positive selection of recombinant expression vectors. Some of them bear additional deficiencies in endogenous proteases (SMD series); others were recently engineered for their capacity in performing "human-like" *N*-glycosylations [23].

Expression vectors (Table 3.2) are built on a classical *E. coli*/yeast shuttle model

Table 3.1 Most commonly used strains of *P. pastoris.*

Strain	Genotype	Phenotype
NRRL[a] Y-11430	wild-type	Mut^+
X-33	wild-type	Mut^+
GS115	*his4*	Mut^+, His^-
KM71	*his4, arg4, aox1::ARG4*	Mut^S, His^-, Arg^+
SMD1163	*his4, pep4, prb1*	Mut^+, His^-, $Prot^-$ (A^-, B^-, $CarbY^-$)
SMD1165	*his4, prb1*	Mut^+, His^-, $Prot^-$ (B^-)
SMD1168	*his4, ura3, pep4::URA3*	Mut^+, His^-, $Prot^-$ (A^-, B^S, $CarbY^-$)
PichiaPink® Strain 1	*ade2*	Mut^+, Ade^-
PichiaPink® Strain 2	*ade2, pep4*	Mut^+, Ade^-, $Prot^-$ (A^-, B^S, $CarbY^-$)
PichiaPink® Strain 3	*ade2, prb1*	Mut^+, Ade^-, $Prot^-$ (B^-)
PichiaPink® Strain 4	*ade2, pep4, prb1*	Mut^+, Ade^-, $Prot^-$ (A^-, B^-, $CarbY^-$)

a) NRRL, Northern Regional Research Laboratories, Peoria, IL.
See text for the explanation of the different elements.

Table 3.2 *P. pastoris* expression vectors.

Name	Selection markers	Phenotype of transformants	Promoter	Secretion sequence	Added tags
pAO815	*HIS4*	His^+	P_{AOX1}	none	none
pPIC3.5K	*HIS4, Kan*	His^+, $G418^R$	P_{AOX1}	none	none
pPIC9K	*HIS4, Kan*	His^+, $G418^R$	P_{AOX1}	α factor	none
pPICZ A, B, C	*Ble*	Zeo^R	P_{AOX1}	none	c-Myc/His_6
pPICZα A, B, C	*Ble*	Zeo^R	P_{AOX1}	α factor	c-Myc/His_6
pPIC6 A, B, C	*Bsd*	Bla^R	P_{AOX1}	none	c-Myc/His_6
pHIL-D2	*HIS4*	His^+	P_{AOX1}	none	none
pHIL-S2	*HIS4*	His^+	P_{AOX1}	PHO1	none
pFLD	*Ble*	Zeo^R	P_{FLD1}	none	V5 epitope/His_6
pFLDα	*Ble*	Zeo^R	P_{FLD1}	α factor	V5 epitope/His_6
pGAPZ A, B, C	*Ble*	Zeo^R	P_{GAP}	none	c-Myc/His_6
pGAPZα A, B, C	*Ble*	Zeo^R	P_{GAP}	α factor	c-Myc/His_6
pPink-HC	*ADE2*	Ade^+	P_{AOX1}	none	none
pPink-LC	*ADE2*	Ade^+	P_{AOX1}	none	none
pPinkα-HC	*ADE2*	Ade^+	P_{AOX1}	α factor	none

HIS4, P. pastoris auxotrophy marker, encodes a histidinol dehydrogenase; *Kan* gene, confers resistance to kanamycine (Kan^R, *E. coli*) and G418 ($G418^R$, *P. pastoris*); *Ble, Streptoalloteichus hindustanus ble* gene, confers resistance to zeocin (Zeo^R); *Bsd* gene, confers resistance to blasticidin (Bla^R); *ADE2, P. pastoris* auxotrophy marker, encodes a phosphoribosylaminoimidazole carboxylase; P_{AOX1}, promoter sequence of the alcohol oxidase-encoding *AOX1* gene from *P. pastoris*; P_{FLD1}, promoter sequence of the formaldehyde dehydrogenase-encoding *FLD1* gene from *P. pastoris*; P_{GAP}, promoter sequence of the glyceraldehyde-3-phosphate dehydrogenase-encoding *GAP* gene from *P. pastoris*; α factor, encodes the native *S. cerevisiae* α factor secretion signal; PHO1, encodes the native *P. pastoris* acid phosphatase secretion signal; V5 epitope, GKPIPNPLLGLDST peptide; c-Myc, C-terminal *myc* epitope, EQKLISEEDL.

with components required for *E. coli* amplification (classically one origin of replication and one antibiotic selection marker) and specific elements for heterologous gene expression in *P. pastoris*. These typically include selectable auxotrophy markers (*HIS4, ADE2*) and/or antibiotic resistance bacterial genes (*kan, bsd,* and *ble*), a range of promoter and terminator sequences, a multiple cloning cassette, and supplementary signal sequences and other fusion sequences that can be added to improve the secretion and detection of the expressed proteins.

Among the panel of constitutive and inducible promoters that have been introduced in expression vectors (listed in [31]), P_{AOX1} is by far the most widely used as it is the most strongly induced in the presence of methanol. Moreover, an original P_{AOX1} synthetic promoter library was developed in a recent study that demonstrated enhanced P_{AOX1} variants could reach higher expression levels of a tested recombinant Green Fluorescent Protein (GFP) [31].

A comprehensive set of vectors and strains is commercially available from Invitrogen, each of them being accessible either individually or included in expression kits.

3.4.3
Transformation and Clone Selection Strategies

As for many other yeasts, transformation of *P. pastoris* is rather straightforward. Several robust methods are available, either based on chemically competent (spheroplasts, PEG1000, LiCl) or electrocompetent cells, thus being accessible to a large majority of operators in standard labs. Moreover, these protocols are well described and can be easily found on numerous websites (convenient *Pichia* manuals can be downloaded from www.invitrogen.com).

Except for a limited set of autoreplicative plasmids that are not yet frequently employed [32–35], most of the transforming expression vectors are designed to be maintained as integrative elements in the genome of *P. pastoris*. This is generally achieved through recombination events between linearized sequences borne by the plasmids (typically *HIS4* or P_{AOX1}) and their homologous sequence counterparts present on the genome, leading to the targeted insertion of the expression vectors. Moreover, such plasmid insertions frequently occur in tandem in yeasts and thus lead to the multiple integration of the genes of interest with a correlated impact on their subsequent expression levels.

Alternatively, integration can be obtained by a gene replacement strategy. In this case, a double recombination event must be realized between the *AOX1* promoter and terminator sequences present on the transforming DNA (containing the gene of interest and a selection marker) and the corresponding homologous sequences present on *P. pastoris* genome. This double recombination event ends up with the replacement of the *AOX1* gene by the construct of interest.

The phenotype of the resulting transformants then depends not only on the selection marker present on the chosen vector (auxotrophy and/or antibiotic resistance), but also on the selected integration strategy (plasmid insertion versus gene replacement) that dictates their methanol utilization behavior. Indeed, while a plasmid insertion does not affect the methanol utilization ability of the transformed strain (Mut^+, methanol utilization plus phenotype), the gene replacement of *AOX1* leads to a Mut^S (methanol utilization slow) phenotype. In several cases, these differences in methanol utilization have been reported as an important parameter to consider for enhancing the performance of recombinant protein expression [36].

3.4.4
Expression Conditions and Culturing Formats

Once transformants have been obtained, the next step usually consists in screening for the clones and conditions exhibiting the best expression levels of the recombinant protein. For expression strategies based on P_{AOX1}-dependent vectors, this is practically achieved by growing the cells in repressive media to an appropriate cell density and growth phase, before starting the production phase by transferring the cells to a methanol-containing induction media. For clones and expression condition screenings, small-scale culturing procedures most often rely on shaken baffled-flasks or on tubes of smaller volumes provided an appropriate aeration is maintained. Several parameters are then usually adjusted for an optimal expression, such as the duration and the temperature of the induction phase, as well as the media formulation including the methanol concentration and the use of additive compounds.

For the production of large amounts of heterologous proteins, shake-flask culture is usually not recommended due to the limitations of volume, oxygen transfer, substrate addition, and an inability to monitor these factors efficiently. The use of bioreactors is therefore preferred, since all of these parameters can be monitored and controlled simultaneously, allowing more efficient production of the desired heterologous protein. Accordingly, a number of robust fermentation processes including fed-batch techniques and continuous culturing procedures have been developed and are routinely employed. Detailed descriptions of these methods with their benefits and limitations can be found in several excellent and comprehensive reviews that we recommend [6, 10, 11, 37].

3.5 Successful Large-Scale Expression of Membrane Proteins Using *P. pastoris*

3.5.1 *P. pastoris* for Membrane Protein Expression

The first use of *P. pastoris* as a host for the expression of an integral membrane protein was reported in 1995 when Helmut Reilander and his colleagues successfully expressed a member of the G-protein-coupled receptor (GPCR) family – the mouse serotonin receptor 5-HT_{5A} [38]. Few additional membrane proteins were then assayed in the following years before the system became more and more popular in the 2000s: a thorough survey we conducted on the last decade of published results revealed 100 references encompassing more than 150 different membrane proteins expressed in *P. pastoris* (Table 3.3). In this list where only integral membrane proteins were considered (soluble domains of membrane proteins as well as membrane-anchored proteins were excluded), all classes of eukaryotic membrane proteins are equally represented. This includes monotopic receptors and enzymes, several aquaporins and ion channels, many members of the GPCR family, as well as large polytopic transporters bearing up to 17 putative transmembrane domains. This survey also highlights the great potential of *P. pastoris* for coexpression approaches, including studies on two membrane subunits of multimeric protein complexes, two interacting membrane protein partners, as well as a membrane protein and a cytosolic partner, in strategies where the coexpressed genes are either borne on a same vector or on two distinct vectors. Successful coexpression was actually recorded for the α and β subunits of Na/K-ATPase [39–42], and for α/β and phospholemman (a membrane modulator of the enzyme) [43]. Similarly, coexpression strategies were also reported for $K_v1.2$, a membrane subunit, and $K_v\beta2$, a cytosoluble partner, of the rat voltage-dependent K^+ channel [44–46].

Table 3.3 Recombinant membrane proteins produced using the *P. pastoris* expression system.

Protein name	Organism	kDa	Transmembrane domains	Strains
Transporters				
P-glycoprotein MDR3	mouse	140	12	GS115
P-glycoproteins MDR3 (S430T, S1073T)	mouse	140	12	GS115
Multidrug resistance protein MRP1	human	165	17	GS115, KM71
Phosphate transporter *Mt*PT1	*Medicago truncatula*	45	ND	GS115
Intestinal peptide transporter hPEPT1	human	71	ND	GS115
P-glycoproteins MDR1, MDR3 (unglycosylated)	human, mouse	140	12	GS115
P-glycoprotein MDR1 (Cys-less)	human	140	12	GS115
Antimalarial drug resistance protein Pfcrt (codon-optimized)	*Plasmodium falciparum*	57	10	KM71, GS115
P-glycoprotein MDR3	mouse	140	12	GS115
Serotonin transporter rSERT	rat	50	12	GS115, SMD1168
Breast cancer resistance protein BCRP	human	62	6	KM71
Sodium/glucose cotransporter hSGLT1	human	55	9	GS115
Copper transporter hCTR1	human	23	3	SMD1163
Low-affinity cation transporter LCT1	wheat	ND	ND	GS115
Chloroquine resistance transporter *Pf*CRT (codon-optimized)	*Plasmodium falciparum*	45	10	KM71
P-glycoprotein MDR3 (Cys-less)	mouse	140	12	GS115
Multidrug resistance protein *Pf*MDR1 (codon-optimized)	*Plasmodium falciparum*	161	12	KM71, X-33
16 ABC transporters: ABCC3, ABC A1, A4, B1, C10, C11, C12, G5, G8, B7, B6, D1, E1, F1, G1, G4	human	176	6, 12	KM71
Glucose transporter *Nl*HT1	*Nilaparvata lugens*	40	12	X-33

Vector constructs (plasmid backbone)	Activity	Process	Reference
(pHIL-D2)-**MDR3**-His_6-bio	P: 4.3 μmol/min/mg, 0.35 mg/l	CESP	[47]
(pHIL-D2)-**mutMDR3**-His_6	P: 3 μmol/min/mg, 0.7 mg/l	CESP	[48]
(pHIL-D2)-**MRP1**-HA-His_6	ligand binding assay	CE	[49]
(pPIC3K)-**MtPT1**	functional complementation	CE	[50]
(pGAPZB)-**hPEPT1**-c-Myc-His	transport assay, E: 64 pmol/mg	CE	[51]
(pHIL-D2)-**QQQ-MDR1**-His_{10}, (pHIL-D2)-**QQQ-MDR3**-His_6	ATPase activity, P: 1.2–3.8 U/mg, 0.75–1.25 mg/l	CESP	[52]
(pHIL-D2)-**MDR1**-His_{10}	ATPase activity, P: 0.75 mg/l	CESP	[53]
(pPIC3.5)-**Pfcrt**-bio	ATPase activity	CESP	[54]
(pHIL-D2)-**MDR3**-His_6	ATPase activity	CESP	[55]
(pHIL-D2)-**rSERT**	nonfunctional	CE	[20]
(pHIL)-**BCRP**-His_{10}	ATPase activity, ligand binding assay, E: 80 nmol/min/mg	CE	[56]
(pPICZB)-**hSGLT1**-FLAG-His_6	functional transport, E: 273 nmol/min/mg, P: 3 mg/l	CESPC	[57]
(pPIC3.5K)-HA-**hCTR1N15Q**	functional complementation	CESPCS (6 Å)	[58]
(pPIC3.5K)-**LCT1**	E: 14 pmol/10^6 cells/10 min	CE	[59]
(pPICZA)-**CRT**-His_6	transport activity, P: 487 pmol/mg/min	CESP	[60]
(pHIL-D2)-**MDR3**-His_6	P: 2.7 μmol/min/mg, 60 mg/g cells	CESP	[61]
(pPICZc/pPIC3.5)-**PfMDR1**-His_6-bio	ATPase activity, P: 63 μmol/mg/min	CESP	[62]
(pSGP18)-AF-**ABC**-CBP-His_6	ATPase activity, P: 82 nmol/min/mg, P: 35 mg/g cells	CESP (ABCC3)	[63]
(pPICZB)-**NlHT1**-c-Myc-His_6	transport activity	CE	[64]

(Continued)

Table 3.3 (*Continued*)

Protein name	Organism	kDa	Transmembrane domains	Strains
Vesicular glutamate transporter VGLUT1	rat	61	12	X-33
Glucose transporters GLUT1 and GLUT4	human, rat	42, 46	12	X-33
P-glycoprotein Pgp	mouse	ND	12	GS115
Formate-nitrite transporter *An*NitA	*Aspergillus nidulans*	31	6, 8	GS115
Water channel proteins				
Aquaporin PM28A	spinach	32	6	X-33
Aquaporin *So*PIP2;1	spinach	32	6	ND
Aquaporin *Pv*TIP3;1	plant	25	6	KM71
Aquaporin *So*PIP2.1	spinach	32	6	X-33
Aquaporin AQP6	rat	29	6	X-33, GS115, KM71
Aquaporin hAQP1	human	35	6	X-33
Aquaglyceroporin *Pf*AQP (codon-optimized)	*Plasmodium falciparum*	30	6	X-33
Aquaporin HsAQP5	human	ND	6	ND
Aquaporin *At*PIP2;1	*Arabidopsis thaliana*	55	6	X-33
Aquaporins *Tg*PIP2;1, *Tg*PIP2;2	*Tulipa gesneriana*	31	6	KM71
Aquaporin Aqy1	*Pichia pastoris*	ND	6	GS115 *aqy1*
Aquaporin hAQP4	human	ND	6	X-33
13 Aquaporins (hAQP0 to hAQP12)	human	30	6	X-33
Ion channel proteins				
Voltage-sensitive K^+ channel $K_v1.2/\beta2$ (coexpression)	rat	58/40	6	SMD1163
Voltage-dependent K^+ channel $K_v1.2/\beta2$ (coexpression)	rat	ND	6	SMD1163
Calcium-activated K^+ channel SK2	mammalian	64	6	SMD1163
Chimeric K^+ channel $K_v1.2/K_v2.1/\beta2.1$ (coexpression)	rat	ND	6	SMD1163
Inward-rectifier K^+ channel $K_{ir}2.2$	chicken	ND	6	SMD1163

Vector constructs (plasmid backbone)	Activity	Process	Reference
(pGAPZB)-c-Myc-His-**VGLUT1**	P: 1 mg/l	CESP	[65]
(pPICZB)-**GLUT**-His_8	transport activity, P: 13.1 mg/g cells	CESPC	[66]
(pHIL-D2)-**QQQ-Pgp**-His_6	ATPase activity	CESPCS (3.8 Å)	[67]
(pPICZA)-His_6-**AnNitA**	ND	CESP	[68]
(pPICZB)-**PM28A**-c-Myc-His_6	P: 25 mg/l	CESP	[69]
(ND)-**SoPIP2;1**	water channel activity	CESPCS (5 Å)	[70]
(pPICZ)-**PvTIP3;1**-gly3-His_6	water channel activity	CE	[71]
(pPICZB)-**SoPIP2.1** ± His_6	P: 25 mg/l	CESPCS (2.1, 3.9 Å)	[72]
(pPICHOLi/pPICZ)-**AQP6**-His_6	E: 7 pmol/mg	CE	[73]
(pPICZB)-**hAQP1**-c-Myc-His_6	water channel activity, P: 90 mg/l	CESPC	[74]
(pPICZB)-**PfAQP**-c-Myc-His_6	P: 18 mg/l	CESPC	[75]
(pPICZB)-**HsAQP5**	water channel activity	CESPCS (2 Å)	[76]
(pPICZB)-**WT/mutAtPIP2;1**	water channel activity, P: 65 μg/l	CESP	[77]
(pPICZB)-**TgPIP2**-gly3-His_6	water channel activity	CE	[78]
(pPICZaB)-ΔN36**Aqy1**-His_6	water channel activity	CESPCS (1.5 Å)	[79]
(pPICZ)-His_8-FLAG-**hAQP4**	water channel activity	CESPCS (1.8 Å)	[80]
(pPICZB)-**hAQP**-His_6	water channel activity	CE	[81]
(pPIC3.5K)-His_8-**K_v1.2**; (pPICZC)-strepII-**$K_v\beta 2$**	E: 98 pmol/mg, P: 26 mg, 3.3 nmol/mg	CESPCS (2.1 nm)	[44]
(pPICZC)-His_8-**K_v1.2-β2**	P: 10 mg/ml	CESPCS (2.9 Å)	[45]
(pPIC3.5K)-strepII-His-**SK2**	E: 0.1 pmol/mg	CES	[82]
(pPICZC)-His_{10}-**K_v1.2/K_v2.1/β2.1**	channel activation	CESPCS (2.4 Å)	[46]
(pPICZB)-**K_{ir}2.2**-GFP-1D4	channel activation, P: 8 mg/ml	CESPCS (3.1 Å)	[83]

(Continued)

Table 3.3 (*Continued*)

Protein name	Organism	kDa	Transmembrane domains	Strains
GPCRs				
Endothelin receptor B ETB	human	55	7	SMD1163
Endothelin receptor B ETB	human	55	7	SMD1163
Cannabinoid receptor CB2	human	51	7	X-33
Mu-opioid receptor HuMOR	human	74	7	GS115, SMD1163, SMD1168, X-33
Dopamine receptor D_{2S}	human	40	7	SMD1163
Dopamine receptor D_{2S}	human	51	7	SMD1163
Receptor smoothened hSmo	human	80	7	GS115
Cannabinoid receptor CB1	human	75	7	X-33
Mu-opioid receptor HuMOR	human	45/66	7	SMD1163
20 GPCRs: ADA1B, ADA2B, ACM1, ACM2, HRH2, OPRK, $5HT_{1D}$, $5HT_{1B}$, $5HT_{1A}$, DRD2, NK1R, NK2R, NK3R, NPY1R, $AA_{2A}R$	human, pig, mouse, rat	ND	7	SMD1163
Adenosine A_{2A} receptor $hA_{2A}R$	human	34	7	SMD1163
β_2-Adrenergic receptor β_2AR (codon-optimized)	human	45	7	SMD1168
100 GPCRs	human, pig, mouse, rat, bovine, yeast	36–126	7	SMD1163
EDG-1 receptors	human	69	7	SMD1168
Bradykinin B2 receptor B2R	human	68	7	GS115
Bradykinin B2 receptor B2R	human	55	7	SMD1163

Vector constructs (plasmid backbone)	Activity	Process	Reference
(pPIC9K)-AF-FLAG-**ETB/ΔGPETB**-bio	E: 35–60 pmol/mg	CE	[84]
(pPIC9K)-AF-FLAG-**ΔGPETB**-bio/GFP/His_{10}; (pPIC9K)-AF-FLAG-His_{10}-**ΔGPETB**-bio	E: 7–60 pmol/mg	CESP	[85]
(pPICZa)-AF-**CB2**-c-Myc-His_6	E: 2.6 pmol/mg	CESP	[86]
(pPICZaA)-AF-GFP-**HuMOR**-c-Myc-His_6	E: 1 pmol/mg	CE	[87, 88]
(pPIC9K)-AF-FLAG ± His_6-$\mathbf{D_{2S}}$	P: 10 pmol/mg	CESP	[89]
(pPIC9K)-AF-FLAG ± His_6-$\mathbf{D_{2S}}$, (pPIC9K)-AF-FLAG-$\mathbf{D_{2S}}$-His_{10}/bio	E: 37–80 pmol/mg	CES	[90]
(pAO815)-**hSmo**-CBD-strep-HA-His_6	ND	CESP	[91]
(pPICZa)-AF-FLAG-**CB1**-c-Myc-His_6	E: 3.6 pmol/mg	CESP	[92]
(pPICZ) ± GFP-**HuMOR**-c-Myc-His_6	E: 0.45 pmol/mg	CESP	[93]
(pPIC9K)-AF-FLAG-His_{10}-**GPCR**-bio	E: 0.3–165 pmol/mg	CE	[94]
(pPICZaA)-AF-FLAG-His_{10}-$\mathbf{hA_{2A}R}$	P: 18 nmol/mg	CESP	[95]
(pPIC9K)-AF-His_6-$\mathbf{\beta_2AR}$	P: 11 nmol/mg	CESP	[96]
(pPIC9K)-AF-FLAG-His_{10}-**GPCR**-bio	E: 0.1–180 pmol/mg	CE	[97]
(pPIC9K)-AF-**EDG-1**-GFP	E: 8.2 pmol/mg	CES	[98]
(pPIC9K)-AF-**B2R**-GFP	ND	CES	[99]
(pPIC9K)-AF-FLAG-His_{10}-**B2R**-bio	E: 3.5 pmol/mg	CE	[100]

(*Continued*)

Table 3.3 (Continued)

Protein name	Organism	kDa	Transmembrane domains	Strains
Neuromedin U receptor NmU2R	human	60	7	SMD1163
Cannabinoid receptor CB2	human	42	7	X-33
12 GPCRs: CNR2, NK1R, NK3R, ADA1B, ADA2B, ADA2C, D2DR, OPRK, OPRD, P2RY1, HRH1, PAR1	human, rat	40–50	7	SMD1163
Mu-opioid receptor (ΔN64) ΔN64-HuMOR	human	38	7	SMD1163
β-Adrenergic receptors β_1AR, β_2AR, β_3AR	human	66	7	SMD1163
25 nonglycosylated GPCRs: CNR1, AGTR2, HTR1B, ADORA2A, DRD1, DRD2, DRD4, DRD5, OPRK, CHRM2, PTGER1, PTGER2, PTGER3, PTGER4, TBXRA2, TACR1, TACR2, TACR3, NTSR1, ADRB2, HRH4	human, mouse	ND	7	SMD1163
Leukotriene B_4 receptor BLT1	guinea pig	100	7	GS115
Enzymes				
11β-OH steroid dehydrogenase 11β-HSD1	human	29	ND	GS115
Monoamine oxidase B MAOB	human	60	1	KM71, GS115
Monoamine oxidase B MAOB	human	60	1	KM71, GS115
Monoamine oxidase A MAOA	human	60	1	KM71
11β-OH steroid dehydrogenases 11β-HSD1	human, rat	31	ND	GS115, X-33
Isatin-bound monoamine oxidase B MAOB	human	60	1	KM71, GS115
Na/K-ATPase (α_1, β_1) (coexpression)	pig	112/47	10, 1	SMD1165
Oxidosqualene cyclase hOSC	human	80	1	GS115
Na/K-ATPase (α, β) (coexpression)	pig	ND	10, 1	SMD1165

Vector constructs (plasmid backbone)	Activity	Process	Reference
(pPIC9K)-AF-FLAG-His_{10}-**NmU2R**-bio	E: 6 pmol/mg	CE	[21]
(pPICZa)-AF-FLAG-**CB2**-c-Myc-His_6; (pPICZa)-AF-FLAG-**CB2**-His_6/His_{10}	ligand binding assay	CESP	[101]
(pPIC9K)-AF-FLAG-His_{10}-**GPCR**-bio	ligand binding assay, P: up to 0.9 mg/l	CESP	[102]
(pPICZB) ± AF ± GFP-**ΔN64-HuMOR**-c-Myc-His_6	P: 5 mg/l	CESP	[103]
(pPICZ)-AF-GFP-**βAR**-c-Myc-His_6	ND	CESP	[104]
(pPIC9K)-AF-FLAG-**GPCR**-His_{10}	E: 0–75.4 pmol/mg	CE	[22]
(pPIC3.5K)-AF-FLAG-**BLT1**	E: 50 pmol/mg, P: 0.4 mg/l	CESP	[105]
(pPIC3.5K)-**11β-HSD1**	enzyme activity	CE	[106]
(pPIC3.5K)-**MAOB**	oxidase activity P: 100 mg/l	CESP	[107]
(pPIC3.5K)-**MAOB**	ND	CESPCS (3 Å)	[108]
(pPIC3.5K)-**MAOA**	P: 115 mg/l	CESP	[109]
(pPIC3.5K/pPICZB)-His_6-**11β-HSD1**	dehydrogenase activity, S: 713 pmol/min/mg	CESP	[110]
(pPIC3.5K)-**MAOB**	ND	CESPCS (1.7 Å)	[111]
(pHIL-D2)-α_1/β_1	E: 30–50 pmol/mg	CE	[39]
(pPICZB)-**hOSC**-c-Myc-His_6	IC_{50} values, P: 105 mg/l	CESPC	[112]
(pHIL-D2)-α/His_{10}-β	E: 30 pmol/mg, 1 g/3 l	CESP	[40]

(Continued)

Table 3.3 (*Continued*)

Protein name	Organism	kDa	Transmembrane domains	Strains
Clorgyline-bound monoamine oxidase A MAOA	human	60	1	KM71
Cytochrome P450 2D6 monooxygenase CYP2D6/ NADPH P450 oxidoreductase CPR (coexpression)	human	56/77	1, 1	X-33
Adrenal cytochrome b_{561} Cytb561	bovine	28	5, 6	GS115
Cytochrome P450 *Pc*CYP1f	*Phanerochaete chrysosporium*	60	ND	KM71
Phospholemman (PLM)/ Na/K-ATPase (α_1, β_1) (coexpression)	human PLM, pig (α_1, β_1)	ND	1, 1, 1	SMD1165
Na/K-ATPase (α_1, β_1) (coexpression)	human, pig	ND	10, 1	SMD1165
Cytochrome P450 17α-hydroxylase CYP17	human	54	ND	GS115
Apo and GSH-complexed leukotriene C_4 synthase LTC4S	human	ND	4	KM71H
Na/K-ATPase (α_3, β_1) (coexpression)	pig	110/44	10, 1	GS115, SMD1168
Monoamine oxidase B MAOB	rat	60	1	KM71
Leukotriene C_4 synthase LTC4S	rat	18	4	KM71
Monoamine oxidase A MAOA	rat	60	1	KM71
Monoamine oxidase MAO	zebrafish	60	1	KM71
Other membrane proteins				
Thromboplastin, tissue factor TF	rabbit	31	1	GS115
Immunotoxin Cyt2Aa1 (codon-optimized)	*Bacillus thuringiensis*	60	1?	KM71
Lectin-like oxLDL receptor 1 hLOX-1	human	43	1?	GS115

Vector constructs (plasmid backbone)	Activity	Process	Reference
(pPIC3.5K)-**MAOA**	P: 115 mg/l	CESPCS (3 Å)	[113]
(pPICZA)-**CPR-CYP2D6**	E: 8.8 pmol/min/pmol enzyme	CE	[114]
(pPICZB)-**Cytb561**-His_6	S: 0.7 mg/l, P: 2.7 mg	CESP	[115]
(pPICZA)-**PcCYP1f**	ND	CE	[116]
(pHIL-D2)-α_1/His_{10}-β_1; (pGAPZA)-**PLM**	ATPase activity	CESP	[43]
(pHIL-D2)-α_1/His_{10}-β_1	S: 8–16 μmol/min/mg	CESP	[41]
(pPIC3.5K)-**CYP17**(His)	E: 300 pmol/mg	CES	[117]
(pPICZA)-His_6-**LTC4S**	ND	CESPCS (2, 2.15 Å)	[118]
(pAO815)-α_3/β_1	E: 0.23 mg/l	CESP	[42]
(pPIC3.5K)-**MAOB**	oxidase activity, P: 100 mg/0.5 l	CESP	[119]
(pPICZA)-His_6-**LTC4S**	P: 1 mg/l, 49 μmol/mg/min	CESPC	[120]
(pPIC3.5K)-**MAOA**	E: 700 U/l, P: 200 mg/l	CESP	[121]
(pPIC3.5K)-**zMAO**	P: 200 mg/l, 300 U/l	CESP	[122]
(pIL-D26)-PHO1-**TF**-His_6	P: 0.1 mg/g cells	CESP	[123]
(pPICZB)-CsFvC6.5-**synCyt2Aa1**-c-Myc-His_6	cytotoxic activity, P: 10 mg/l	CESP	[124]
(pPIC9K)-AF-**hLOX-1**-His_6	ND	CESP	[125]

(*Continued*)

Table 3.3 (*Continued*)

Protein name	Organism	kDa	Transmembrane domains	Strains
Metal-dependent hydrolase PAB0107	*Pyrococcus abyssi*	20–46	5	X-33
Chloride channel PAB2010	*P. abyssi*	20–46	10	X-33
Putative membrane protein PAB0965	*P. abyssi*	20–46	3	X-33
Fe^{3+} ABC protein PAB0677	*P. abyssi*	20–46	9	X-33
Carbohydrate transport protein PAB0724	*P. abyssi*	20–46	9	X-33
Tetraspanin hCD81	human	26	4	X-33, GS115
Peroxisomal membrane protein 22 PMP22	rat	22	4	SMD1163
Tetraspanin peripherin/RDS p/RDS	ND	37	4	KM71
Epidermal growth factor receptor 2 HER-2/*neu*	human	200	1	X-33

In the vector constructs column: His_6, His_8, His_{10}, hexa-, octa-, deca-istidine tag; AF, α factor secretion signal; CBP, calmodulin binding peptide; gly3, triglycine tag; strep, streptavidin tag; strepII, StrepII tag (MAWSHPQFEK); 1D4, 1D4 antibody recognition sequence (TETSQVAPA); HA, hemagglutinin A tag; c-Myc, C-terminal *myc* epitope (EQKLISEEDL); FLAG, FLAG tag; bio, biotin acceptor domain.
In the process column CESPCS: C, cloning; E, expression; S, solubilization; P, purification; C, cristallization; S, structure.
Other: ND, not determined.

Moreover, this list not only reports on expression evaluations in *P. pastoris* but also covers solubilization, purification, crystallization, and structural studies of membrane proteins produced with this system. Remarkably, high-resolution structures for a dozen of them were thus obtained (see CEPSCS-labeled references in Table 3.3), which represents about one-third of the recombinantly produced eukaryotic membrane proteins for which a three-dimensional structure is available as of January 2010 (http://blanco.biomol.uci.edu/Membrane_Proteins_xtal.html). Overall, these records highlight *P. pastoris* as one of the most performant heterologous expression system for the structural studies of eukaryotic membrane proteins.

3.5.2
Common Trends for an Efficient Expression of Membrane Proteins in *P. pastoris*

The basic experimental data recorded in Table 3.3 are intended to give some general directions to help the reader in the choice of an adapted procedure to start with for his/her favorite membrane protein to be expressed in *P. pastoris*.

Vector constructs (plasmid backbone)	Activity	Process	Reference
(pPICZB/pPICZa) ± AF-**MP**-c-Myc-His_6	ND	CESP	[126]
(pPICZB/pPICZa) ± AF-**MP**-c-Myc-His_6	ND	CESP	[126]
(pPICZB/pPICZa) ± AF-**MP**-c-Myc-His_6	ND	CESP	[126]
(pPICZB/pPICZa) ± AF-**MP**-c-Myc-His_6	ND	CESP	[126]
(pPICZB/pPICZa) ± AF-**MP**-c-Myc-His_6	ND	CESP	[126]
(pPICZB)-**hCD81**-His_6	P: 1.75 mg/l	CESP	[127]
(pPICZA)-His_{10}-**PMP22**	P: 90 mg/4 l	CESP	[128]
(pPICZA)-**p/RDS**-c-Myc-His_6	P: 0.3 mg/l	CESP	[129]
(pPICZaA)-AF-**HER-2/*neu***-c-Myc-His	ND	CE	[130]

Regarding the plasmidic constructs to select, nearly all kinds of available expression vectors have been assayed, most exclusively based on the P_{AOX1} inducible promoter. To our knowledge, only three noticeable exceptions are reported on the use of vectors bearing the constitutive promoter P_{GAP} for the expression evaluation of membrane proteins, including the human intestinal peptide transporter hPEPT1 [51], the human phospholemman [43], and the rat vesicular glutamate transporter 1 VGLUT1 [65]. In all cases the expression levels were rather significant; however, no comparison was conducted on the benefit of such a constitutive expression over an inducible system.

Overall, no real tendency emerges in the choice of a given vector for a given type of membrane protein, except in the case of water channel proteins where pPICZ constructs were always privileged. Similarly, whereas a secretion sequence is quite systematically added upstream of GPCR open reading frames (ORFs) for an enhanced expression of functional proteins, there is no apparent rule for all the other reported membrane proteins whatever their topology and orientation in the membrane. In the case of the six-transmembrnae domain aquaporins, for instance, where N- and C-termini are intracellularly located, protein expression has been evaluated with or without a fused secretion sequence and both situations proved efficient enough to obtain high-resolution structures of the produced protein ([79] versus [72, 76, 80]).

As for several other expression systems, a large panel of tag sequences are frequently inserted to improve the downstream detection and purification steps, ranging from hexa- or decahistidine (the most widely employed tags), c-Myc, FLAG, HA and StrepII epitopes, to larger peptidic domains such as a biotinylation domain (bio), a calmodulin binding domain (CBD), or the fluorescent protein GFP. In addition, protease cleavage sequences are sometimes included in the constructs such as the Factor Xa or tobacco etch virus (TEV) sequences in order to eliminate the fused tags after or during [102] the purification step.

Nearly all commercially available strains have been used to express membrane proteins. While the criteria used for the choice of a given strain are generally not documented, few studies reported on the membrane protein-dependent differential behavior of strains – a phenomenon that is commonly observed. SMD1163, KM71, and X-33 indeed appeared more performant than GS115 for the expression of a GPCR [38], an ATP-binding cassette (ABC) transporter [49], and a tetraspanin [127], respectively, whereas GS115 performed better than SMD1168 for the recombinant expression of a serotonin transporter [20]. Alternatively, no real variation of the expression level could be observed when a GPCR, the μ-opioid receptor, was evaluated in the X-33, GS115, SMD1163, and SMD1168 strains [87].

In most of the reported studies, functional expression levels of membrane proteins are assessed through both specific immunodetection tests and activity assays. In the situation where these parameters can be compared, as in the particular case of GPCRs, the outcome highlights a very fluctuating performance of the system that not only depends on the expressed membrane protein, but also on the experimental conditions assayed. Therefore, optimization of the expression conditions is often very helpful for the recovery of higher amounts of functional recombinant membrane proteins. This is the issue of the next section that aims at illustrating how expression levels can be enhanced with GPCRs as model membrane proteins.

3.6
Guidelines for Optimizing Membrane Protein Expression in *P. pastoris* Using GPCRs as Models

From the seminal work of Weiss *et al.* [38], more than 30 original articles focusing on GPCR expression in *P. pastoris* have been published so far. This wealth of quantitative and qualitative information relative to hundreds of different receptors from the same membrane protein family represents an ideal source of data to exemplify the different directions that can be undertaken to enhance the expression levels of membrane proteins. Different experimental adjustments conducted in several of these studies indeed proved highly beneficial. They can be divided into two main categories that are detailed below: those allowing us to design and to select for the most performing clones, and those implemented at the level of growth and induction. In addition, some considerations on optimizing yeast cell lysis are also briefly discussed.

3.6.1
Design and Selection of Enhanced Expression Clones

As a general rule in heterologous expression studies, optimizing the coding sequence of the gene to be expressed is often very helpful. For instance, fitting the gene sequence to the codon usage of the host organism has generally shown a beneficial impact on expression levels (reviewed in [131]). In the case of GPCRs expressed in *P. pastoris*, a codon-optimized human β_2-adrenergic receptor (β_2AR) exhibited an activity of 6 pmol/mg in total membrane preparations [96] – a figure to be compared with the 24 pmol/mg functional receptors that were obtained for a nonoptimized human β_2AR expressed in a quite similar context [19]. This codon-optimization engineering of a GPCR thus did not appear really profitable, albeit this single reported approach may probably not be representative and more data are needed. Similarly, with the double goal of improving expression levels and receptor homogeneity, direct mutagenesis of potential *N*-glycosylated residues was evaluated on several GPCRs and revealed a rather average outcome: whereas the receptor homogeneity was generally enhanced, the specific activity of the receptors was lowered in most cases [22, 95], only few of them being improved [22, 84] or remaining unchanged [90]. Larger sequence modifications were also reported, showing important beneficial effects for a C-terminally truncated adenosine A_{2A} receptor [132] and for a 47-amino-acid deletion of an internal loop for an acetylcholine muscarinic receptor [22], but with no real impact on expression levels in the case of a N-terminal deletion of a μ-opioid receptor [103, 133].

Introducing additional fusion sequences may sometimes reveal a fruitful way to increase expression of GPCRs. Several studies notably compared the benefit of secretion signals added upstream the gene of interest, showing a very substantial effect of the signal sequence of the α-factor from *S. cerevisiae* on the expression levels of a serotonergic [38], an opioid [103, 133], and a dopaminergic [90] receptor. This signal sequence is now systematically inserted for GPCR expression.

Addition of short tag sequences including the hexa- or decahistidine, FLAG or the c-Myc regularly proved very useful for the downstream detection and/or purification procedures of GPCRs, but did not result in significant changes in their expression levels. Similarly, GFP fused to several GPCRs either C-terminally [85, 98, 99] or N-terminally [103, 104] did not markedly modify their expression profile. Instead, the GFP fusion appeared a useful multipurpose tool for (i) the selection of overexpressing clones, (ii) the determination of total recombinant protein expression, (iii) the evaluation of solubilization and purification conditions, and (iv) the subcellular localization of the receptors.

Interestingly, a significantly increased production level was observed when the biotinylation domain of the transcarboxylase from *Propionibacterium shermanii* was fused to the C-terminus of several receptors. For instance, for 5-HT_{5A} [19], human ETB endothelin receptor (ETBR) [85], DRD2 [90], and β_2AR (C. Reinhart, personal communication) the number of active receptors per cell was more than doubled. In addition, the absence of this sequence from the GPCR constructs used in the study of Yurugi-Kobayashi *et al.* [22] probably participated to the lower expression levels observed for several of them when compared to the same receptors bearing this biotinylation domain [97]. Addition of this domain likely stabilized the recombinant receptor either by protecting the receptor from direct degradation or maintaining folding fidelity to avoid the unfolded protein response.

Regarding the choice of the cellular host to use, several studies compared the benefit of one strain over the others. For instance, the receptors 5-HT_{5A} and ETA were respectively expressed at higher levels in the protease-deficient strains SMD1163 [38] and SMD1168 [134] than in the GS115 strain (see Table 3.1 for the description of the strains). Similarly, the strain SMD1168 appeared a most appropriate host for the expression of a CB2 receptor when compared to the strain X-33 [101]. In several other cases, however, no significant differences were observed, and strains from the SMD series were mainly retained because of their protease-deficient properties and their inherent lower impact on protein degradation during the downstream preparative steps.

In addition, since multicopy integration events occur with a relatively high frequency in *P. pastoris* transformants, gene dosage is also an important issue that directly impacts expression levels. For instance, a panel of clones resistant to increasing concentrations of zeocin [88] and G418 (geneticine) [19, 90] was selected for its representative content of integrated GPCR gene copy number. These studies and others show that the levels of active receptors increased correlatively with the number of integrated copies up to a plateau after which additional copies had no effect. Most importantly, from our observations (unpublished data), clones bearing the highest antibiotic resistance levels (i.e., the highest copy number of GPCRs genes) were often those presenting the highest amounts of immunodetected receptors, whereas ligand-binding activities were not improved. Such clones are thus generally not desired as they display a large proportion of nonfunctional receptors. Moreover, these observations strongly suggest that the bottleneck for the production of functional receptors lies in folding and/or post-translational processing rather than in the transcription and translation steps. As a consequence, clone selection procedures have to rely both on a representative phenotypic screening followed by an appropriate evaluation of the receptor quantity and activity.

3.6.2
Optimization of the Expression Conditions

Once the most performant clones have been selected, further improvements can be implemented by appropriately adjusting some of the experimental parameters that influence the host cell physiology, and hence its performance for heterologous gene expression, correct protein folding, and proper trafficking. These external factors include culture format and procedures, temperature and time of induction, cell densities, formulation of growth media, or supplementation with stabilizing compounds or chemical chaperones.

As a first step, evaluating the production time course of a GPCR is often very useful as the outcome may vary importantly from one receptor to another. For instance, 10 h was determined as the optimal induction time for different constructs expressing a μ-opioid receptor [87], whereas the highest expression levels were obtained in the range of 18–24 h postinduction for a majority of other receptors and up to 60 h for an engineered ACM2 muscarinic receptor [22]. Similarly, while the induction phase in methanol-containing media is usually performed using cell densities of about 5×10^7 cells/ml (1 OD_{600}/ml), we observed that this parameter was differently affecting the expression level of GPCRs, higher cell densities (up to 10 OD_{600}/ml) being actually more appropriate for several of them (unpublished data).

Formulation of the induction media is also an important issue. Adjustments in the composition of buffered media, pH values, and methanol concentration usually did not bring major benefits in GPCR expression, and a typical induction is generally performed at pH 5–7 in buffered complex media containing 0.5% methanol. Much more substantial improvements, however, can be gained by supplementing these media with some small compounds that are believed to facilitate the folding and processing of the recombinant proteins. Among these molecules, dimethyl sulfoxide (DMSO) added in the induction medium remarkably increased the production yield of 16 out of 20 tested receptors up to 6-fold relative to standard conditions [94]. Similar effects were observed in other studies evaluating GPCR expression not only in the *P. pastoris* system [21, 100], but also using mammalian [21] or insect [135] cells hosts. The precise role of DMSO here is not clear, but it has been shown to dramatically alter the membrane properties of several organisms by increasing their permeability [136], thus possibly influencing the processes of membrane protein translocation. DMSO is also thought to act as a stabilizer of folding intermediates and has been already qualified a chemical chaperone [137]. In a comparable fashion, adding ligands specific to a given GPCR proved highly beneficial for a large majority of tested receptors [19, 90, 94]. In the case of a histaminergic H_2 receptor, the expression level was improved up to 7-fold in the presence of an antagonist compound, cimetidine, in the induction medium [94]. Such small molecules are considered pharmacological chaperones as they have been shown to selectivity promote the proper folding and trafficking of the targeted GPCR [137], therefore limiting the recurrent complications related to misfolding/aggregation and misfolding/degradation pathways.

In addition, it was shown in several studies that lowering temperature during expression to a typical range of 18–24 °C was optimal for various receptors, as measured by ligand binding [22, 87, 90, 94, 95]. Possible explanations for the temperature effect include slowing down protein production and not overloading the translocation machinery, protein processing, or intracellular trafficking. Lowering temperature has also been shown to reduce proteolytic activities and upregulate cold shock proteins such as chaperones.

As illustrated in a work we conducted on a selection of 20 GPCRs [94], adjusting these different parameters at the culturing level always turned out beneficial and every tested clones revealed higher ligand binding values (B_{max}) compared with the standard condition. Strikingly, eight out of these 20 receptors revealed high B_{max} values (above 20 pmol/mg) after optimization. In addition, we and others also found that the amount of functional receptor (in terms of ligand binding) did not scale with the total amount of receptors evaluated either by immunodetection [94] or by fluorescence [88] measurements. Most importantly, the total amount of receptors was not changed after optimization while the total number of binding sites (B_{max}) was increased from 1.3- to more than 8-fold. These data are all in agreement with the concept that GPCRs are expressed in *P. pastoris* under a functional/nonfunctional equilibrium that can be modulated with the expression conditions that are used.

3.6.3
Yeast Cell Lysis

As a common trait of budding yeasts, *P. pastoris* possesses a thick protective cell wall that requires the use of aggressive disruption methods for the recovery and preparation of the membrane protein-containing fractions. The choice of a cell lysis method suited to *P. pastoris* cultured in various volumes and formats is therefore nontrivial, albeit very few studies report on this important issue [27]. We here indicate a selection of techniques and apparatus references that we and others found the most adapted to this key step.

Shearing-based methods involving microbeads are mostly preferred as they are very efficient, compatible with a broad range of sample volumes, and directly accessible to most of the standard-equipped labs. In the simplest and widest used mode, cells are violently shaken with 500-μm diameter beads in cold buffers using a basic vortex apparatus in cycles alternating shaking and ice-cooling phases ([38, 138] among others). In order to achieve more reliable and reproducible results, programmable equipment is recommended, such as the Tissue Lyser from Qiagen [132] or the FastPrep 24 from MP Biomedicals [102] that accommodate various sample volumes and formats (up to 50 ml), or the more sophisticated and expensive grinder series from Dyno Mills [85] that can operate using large volumes both in batch or continuous modes. In addition, pressure-based instruments have also proven efficient for the lysis of *P. pastoris* cells, and besides the well-known French pressure cell press stands a panel of cell disruptors from Constant Systems that can handle from 1 to 20 ml of batch samples and more than 500 ml/min with the continuous flow models. Alternatively, methods involving glucanase enzymes (e.g., helicase from snail digestive juice; zymolyase or lyticase from microbial sources) can be used to remove the cell wall and give rise to spheroplast preparations that can be easily burst. For cost and practical reasons, however, these methods are obviously not recommended for the lysis of large cell volumes.

Noteworthy, we and others commonly observed that the longer the induction phase, the less efficient the lysis step. This is likely to be related to the very dynamic nature of the cell wall that can adapt to various physiological changes in order to maintain the integrity of yeast cells and prevent them from lysis [139]. This issue is notably to be considered when optimizing the induction time, which should ideally strike a balance between expression level and cell lysis efficiency.

3.7
Conclusions and Future Directions

Overall, *P. pastoris* appears a very efficient heterologous system for the production of a large panel of membrane proteins. Importantly, a series of various optimizations including gene and vector sequence engineering, host strains, clone selection, culture format and procedures, temperature and time of induction, cell densities, formulation of growth media, and supplementation with stabilizing compounds or chemical chaperones, always prove very helpful for improving expression levels and protein functionality. As an exemplar illustration of such improvements, the functional expression level of a dopaminergic D_{2S} GPCR was increased from about 1000 receptors per cell for an unmodified receptor up to more than 50 000 receptors per cell for an ultimate engineered αFD2SBio receptor [90].

Even if some general tips emerge, the outcome is, however, often a matter of membrane protein-dependent adjustments and successes still often rely on trial-and-error strategies. An improved success rate will certainly be obtained in the coming years with a more rationalized use of *P. pastoris*, taking advantage of recent breakthroughs. Very recently, the combination of proteomics and genetics brought some mechanistic insights into the biology of recombinant production of membrane proteins both in *E. coli* [140] and *S. cerevisiae* [141], and subsequently allowed the engineering of strains presenting a specific task-adapted physiology and enhanced production properties. Such strain evolution strategies have been already applied to *P. pastoris*, notably for the production of human glycoproteins in glycoengineered strains [23]. They will be probably further exploited with the recent release of the whole-genome sequence of *P. pastoris* [9] that is now fully accessible to proteomic analyses and genetic manipulations.

Acknowledgments

The authors are supported by the CNRS and the University of Strasbourg, and by grants from the French National Research Agency (ANR-06-PCVI-0008 and ANR- 07-PCVI-0024).

Abbreviations

β_2AR	β_2-adrenergic receptor
ABC	ATP-binding cassette
CBD	calmodulin binding domain
DMSO	dimethyl sulfoxide
GFP	Green Fluorescent Protein
GPCR	G-protein-coupled receptor
ORF	open reading frame
TEV	tobacco etch virus

References

1 Phaff, H.J. and Starmer, W.T. (1987) Yeasts associated with plants, insects and soil, in *The Yeasts, Volume 1: Biology of Yeasts*, 2nd edn (eds A.H. Rose and J.S. Harrison), Academic Press, Orlando, FL, pp. 123–180.

2 Guilliermond, C.H. (1919) Seances. *Mem. Soc. Biol.*, **82**, 466–470.

3 Ogata, K., Nishikawa, H., and Ohsugi, M. (1969) A yeast capable of utilizing methanol. *Agric. Biol. Chem.*, **33**, 1519–1520.

4 Wegner, G.H. (1990) Emerging applications of methylotrophic yeast. *FEMS Microbiol. Rev.*, **7**, 279–284.

5 Cereghino, J.L. and Cregg, J.M. (2000) Heterologous protein expression in the methylotrophic yeast *Pichia pastoris*. *FEMS Microbiol. Rev.*, **24**, 45–66.

6 Macauley-Patrick, S., Fazenda, M.L., McNeil, B., and Harvey, L.M. (2005) Heterologous protein production using the *Pichia pastoris* expression system. *Yeast*, **22**, 249–270.

7 Yamada, Y., Matsuda, M., Maeda, K., and Mikata, K. (1995) The phylogenetic relationships of methanol-assimilating yeasts based on the partial sequences of 18S and 26S ribosomal RNAs: the proposal of *Komagataella* gen. nov. (Saccharomycetaceae). *Biosci. Biotechnol. Biochem.*, **59**, 439–444.

8 Kurtzman, C.P. (2009) Biotechnological strains of *Komagataella* (*Pichia*) *pastoris* are *Komagataella phaffii* as determined from multigene sequence analysis. *J. Ind. Microbiol. Biotechnol.*, **36**, 1435–1438.

9 De Schutter, K., Lin, Y.C., Tiels, P., Van Hecke, A., Glinka, S., Weber-Lehmann, J., Rouze, P., Van de Peer, Y., and Callewaert, N. (2009) Genome sequence of the recombinant protein production host *Pichia pastoris*. *Nat. Biotechnol.*, **27**, 561–569.

10 Cereghino, G.P.L., Cereghino, J.L., Ilgen, C., and Cregg, J.M. (2002) Production of recombinant proteins in fermenter cultures of the yeast *Pichia pastoris*. *Curr. Opin. Biotechnol.*, **13**, 329–332.

11 Li, P., Anumanthan, A., Gao, X.G., Ilangovan, K., Suzara, V.V., Düzgünes, N., and Renugopalakrishnan, V. (2007) Expression of recombinant proteins in *Pichia pastoris*. *Appl. Biochem. Biotechnol.*, **142**, 105–124.

12 Eckart, M.R. and Bussineau, C.M. (1996) Quality and authenticity of heterologous proteins synthesized in yeast. *Curr. Opin. Biotechnol.*, **7**, 525–530.

13 Wheatley, M. and Hawtin, S.R. (1999) Glycosylation of G-protein-coupled receptors for hormones central to normal reproductive functioning: its occurence and role. *Hum. Reprod.*, **5**, 356–364.

14 Duvernay, M.T., Filipeanu, C.M., and Wu, G. (2005) The regulatory mechanisms of export trafficking of G protein-coupled receptors. *Cell. Signal.*, **17**, 1457–1465.

15 Cohen, D.M. (2006) Regulation of TRP channels by *N*-linked glycosylation. *Semin. Cell. Dev. Biol.*, **17**, 630–637.

16 Grinna, L.S. and Tschopp, J.F. (1989) Size distribution and general structural features of *N*-linked oligosaccharides from the methylotrophic yeast, *Pichia pastoris*. *Yeast*, **5**, 107–115.

17 Wood, M.J. and Komives, E.A. (1999) Production of large quantities of isotopically labeled protein in *Pichia pastoris* by fermentation. *J. Biomol. NMR*, **13**, 149–159.

18 Whittaker, M.M. and Whittaker, J.W. (2005) Construction and characterization of *Pichia pastoris* strains for labeling aromatic amino acids in recombinant proteins. *Protein Expr. Purif.*, **41**, 266–274.

19 Weiss, H.M., Haase, W., Michel, H., and Reilander, H. (1998) Comparative biochemical and pharmacological characterization of the mouse $5HT_{5A}$ 5-hydroxytryptamine receptor and the human β_2-adrenergic receptor produced in the methylotrophic yeast *Pichia pastoris*. *Biochem. J.*, **330**, 1137–1147.

20 Tate, C.G., Haase, J., Baker, C., Boorsma, M., Magnani, F., Vallis, Y., and Williams, D.C. (2003) Comparison

of seven different heterologous protein expression systems for the production of the serotonin transporter. *Biochim. Biophys. Acta*, **1160**, 141–153.

21 Shukla, A.K., Haase, W., Reinhart, C., and Michel, H. (2007) Heterologous expression and comparative characterization of the human neuromedin U subtype II receptor using the methylotrophic yeast *Pichia pastoris* and mammalian cells. *Int. J. Biochem. Cell Biol.*, **39**, 931–942.

22 Yurugi-Kobayashi, T., Asada, H., Shiroishi, M., Shimamura, T., Funamoto, S., Katsuta, N., Ito, K., Sugawara, T., Tokuda, N., Tsujimoto, H., Murata, T., Nomura, N., Haga, K., Haga, T., Iwata, S., and Kobayashi, T. (2009) Comparison of functional non-glycosylated GPCRs expression in *Pichia pastoris*. *Biochem. Biophys. Res. Commun.*, **380**, 271–276.

23 Hamilton, S.R. and Gerngross, T.U. (2007) Glycosylation engineering in yeast: the advent of fully humanized yeast. *Curr. Opin. Biotechnol.*, **18**, 387–392.

24 Opekarova, M. and Tanner, W. (2003) Specific lipid requirements of membrane proteins – a putative bottleneck in heterologous expression. *Biochim. Biophys. Acta*, **1610**, 11–22.

25 Lifshitz, Y., Petrovich, E., Haviv, H., Goldshleger, R., Tal, D.M., Garty, H., and Karlish, S.J.D. (2007) Purification of the human α2 isoform of Na,K-ATPase expressed in *Pichia pastoris*. Stabilization by lipids and FXYD1. *Biochemistry*, **46**, 14937–14950.

26 Marx, H., Sauer, M., Resina, D., Vai, M., Porro, D., Valero, F., Ferrer, P., and Mattanovich, D. (2006) Cloning, disruption and protein secretory phenotype of the GAS1 homologue of *Pichia pastoris*. *FEMS Microbiol. Lett.*, **264**, 40–47.

27 Hopkins, T.R. (1991) Physical and chemical cell disruption for the recovery of intracellular proteins. *Bioprocess Technol.*, **12**, 57–83.

28 Resina, D., Maurer, M., Cos, O., Arnau, C., Carnicer, M., Marx, H., Gasser, B., Valero, F., Mattanovich, D., and Ferrer, P. (2009) Engineering of bottlenecks in *Rhizopus oryzae* lipase production in *Pichia pastoris* using the nitrogen source-regulated *FLD1* promoter. *Nat. Biotechnol.*, **25**, 396–403.

29 Daly, R. and Hearn, M.T.W. (2005) Expression of heterologous proteins in *Pichia pastoris*: a useful experimental tool in protein engineering and production. *J. Mol. Recognit.*, **18**, 119–138.

30 Gellisen, G. (2000) Heterologous protein production in methylotrophic yeasts. *Appl. Microbiol. Biotechnol.*, **54**, 741–750.

31 Hartner, F.S., Ruth, C., Langenegger, D., Johnson, S.N., Hyka, P., Lin-Cereghino, G.P., Lin-Cereghino, J., Kovar, K., Cregg, J.M., and Glieder, A. (2008) Promoter library designed for fine-tuned gene expression in *Pichia pastoris*. *Nucleic Acids Res.*, **36**, e76.

32 Cregg, J.M., Barringer, K.J., Hessler, A.Y., and Madden, K.R. (1985) *Pichia pastoris* as a host system for transformations. *Mol. Cell. Biol.*, **5**, 3376–3385.

33 Lee, C.C., Williams, T.G., Wong, D.W.S., and Robertson, G.H. (2005) An episomal expression vector for screening mutant gene libraries in *Pichia pastoris*. *Plasmid*, **54**, 80–85.

34 Hong, I.P., Lee, S.J., Kim, Y.S., and Choi, S.G. (2007) Recombinant expression of human cathelicidin (hCAP18/LL-37) in *Pichia pastoris*. *Biotechnol. Lett.*, **29**, 73–78.

35 Sandstrom, A.G., Engstrom, K., Nyhlen, J., Kasrayan, A., and Backvall, J.E. (2009) Directed evolution of *Candida antarctica* lipase A using an episomal replicating yeast plasmid. *Protein Eng. Des. Sel.*, **22**, 413–420.

36 Pla, I.A., Damasceno, L.M., Vannelli, T., Ritter, G., Batt, C.A., and Shuler, M.L. (2006) Evaluation of Mut^+ and Mut^S *Pichia pastoris* phenotypes for high level extracellular scFv expression under feedback control of the methanol concentration. *Biotechnol. Prog.*, **22**, 881–888.

37 Cos, O., Ramón, R., Montesinos, J.L., and Valero, F. (2006) Operational strategies, monitoring and control of heterologous protein production in the methylotrophic yeast *Pichia pastoris*

under different promoters: a review. *Microb. Cell Fact.*, **5**, 17.

38 Weiss, H.M., Haase, W., Michel, H., and Reilander, H. (1995) Expression of functional mouse 5-HT_{5A} serotonin receptor in the methylotrophic yeast *Pichia pastoris*: pharmacological characterization and localization. *FEBS Lett.*, **377**, 451–456.

39 Strugatsky, D., Gottschalk, K.E., Goldshleger, R., Bibi, E., and Karlish, S.J.D. (2003) Expression of Na^+, K^+-ATPase in *Pichia pastoris*: analysis of wild type and D369N mutant proteins by Fe^{2+}-catalyzed oxidative cleavage and molecular modeling. *J. Biol. Chem.*, **278**, 46064–46073.

40 Cohen, E., Goldshleger, R., Shainskaya, A., Tal, D.M., Ebel, C., Le Maire, M., and Karlish, S.J.D. (2005) Purification of Na,K-ATPase expressed in *Pichia pastoris* reveals an essential role of phospholipid–protein interactions. *J. Biol. Chem.*, **280**, 16610–16618.

41 Haviv, H., Cohen, E., Lifshitz, Y., Tal, D.M., Goldshleger, R., and Karlish, S.J.D. (2007) Stabilization of Na^+,K^+-ATPase purified from *Pichia pastoris* membranes by specific interactions with lipids. *Biochemistry*, **46**, 12855–12867.

42 Reina, C., Padoani, G., Carotti, C., Merico, A., Tripodi, G., Ferrari, P., and Popolo, L. (2007) Expression of the $\alpha3/\beta1$ isoform of human Na,K-ATPase in the methylotrophic yeast *Pichia pastoris*. *FEMS Yeast Res.*, **7**, 585–594.

43 Lifshitz, Y., Lindzen, M., Garty, H., and Karlish, S.J.D. (2006) Functional interactions of phospholemman (PLM) (FXYD1) with Na^+, K^+-ATPase: Purification of $\alpha1/\beta1$/PLM complexes expressed in *Pichia pastoris*. *J. Biol. Chem.*, **281**, 15790–15799.

44 Parcej, D.N. and Eckhardt-Strelau, L. (2003) Structural characterization of neuronal voltage-sensitive K^+ channels heterologously expressed in *Pichia pastoris*. *J. Mol. Biol.*, **333**, 103–116.

45 Long, S.B., Campbell, E.B., and MacKinnon, R. (2005) Crystal structure of a mammalian voltage-dependent *Shaker* family K^+ channel. *Science*, **309**, 897–903.

46 Long, S.B., Tao, X., Campbell, E.B., and MacKinnon, R. (2007) Atomic structure of a voltage dependent K^+ channel in a lipid membrane-like environment. *Nature*, **450**, 376–382.

47 Julien, M., Kajiji, S., Kaback, R.H., and Gros, P. (2000) Simple purification of highly active biotinylated P-glycoprotein: enantiomer-specific modulation of drug-stimulated ATPase activity. *Biochemistry*, **39**, 75–85.

48 Urbatsch, I.L., Gimi, K., Wilke-Mounts, S., and Senior, A.E. (2000) Conserved Walker A Ser residues in the catalytic sites of P-glycoprotein are critical for catalysis and involved primarily at the transition state step. *J. Biol. Chem.*, **275**, 25031–25038.

49 Cai, J., Daoud, R., Georges, E., and Gros, P. (2001) Functional expression of multidrug resistance protein 1 in *Pichia pastoris*. *Biochemistry*, **40**, 8307–8316.

50 Chiou, T.J., Liu, H., and Harrison, M.J. (2001) The spatial expression patterns of a phosphate transporter (*Mt*PT1) from *Medicago truncatula* indicate a role in phosphate transport at the root/soil interface. *Plant J.*, **25**, 281–293.

51 Theis, S., Doring, F., and Daniel, H. (2001) Expression of the Myc/His-tagged human peptide transporter hPEPT1 in yeast for protein purification and functional analysis. *Protein Expr. Purif.*, **22**, 436–442.

52 Urbatsch, I.L., Wilke-Mounts, S., Gimi, K., and Senior, A.E. (2001) Purification and characterization of *N*-glycosylation mutant mouse and human P-glycoproteins expressed in *Pichia pastoris* cells. *Arch. Biochem. Biophys.*, **388**, 171–177.

53 Urbatsch, I.L., Gimi, K., Wilke-Mounts, S., Lerner-Marmarosh, N., Rousseau, M.E., Gros, P., and Senior, A.E. (2001) Cysteines 431 and 1074 are responsible for inhibitory disulfide cross-linking between the two nucleotide-binding sites in human P-glycoprotein. *J. Biol. Chem.*, **276**, 26980–26987.

54 Zhang, H., Howard, E.M., and Roepe, P.D. (2002) Analysis of the antimalarial drug resistance protein Pfcrt expressed in yeast. *J. Biol. Chem.*, **277**, 49767–49775.

55 Urbatsch, I.L., Tyndall, G.A., Tombline, G., and Senior, A.E. (2003) P-glycoprotein catalytic mechanism, studies of the ADP-vanadate inhibited state. *J. Biol. Chem.*, **278**, 23171–23179.

56 Mao, Q., Conseil, G., Gupta, A., Cole, S.P.C., and Unadkat, J.D. (2004) Functional expression of the human breast cancer resistance protein in *Pichia pastoris*. *Biochem. Biophys. Res. Commun.*, **320**, 730–737.

57 Tyagi, N.K., Goyal, P., Kumar, A., Pandey, D., Siess, W., and Kinne, R.K.H. (2005) High-yield functional expression of human sodium/D-glucose cotransporter 1 in *Pichia pastoris* and characterization of ligand-induced conformational changes as studied by tryptophan fluorescence. *Biochemistry*, **44**, 15514–15524.

58 Aller, S.G. and Unger, V.M. (2006) Projection structure of the human copper transporter CTR1 at 6-Å resolution reveals a compact trimer with a novel channel-like architecture. *Proc. Natl. Acad. Sci. USA*, **103**, 3627–3632.

59 Diatloff, E., Forde, B.G., and Roberts, S.K. (2006) Expression and transport characterisation of the wheat low-affinity cation transporter (LCT1) in the methylotrophic yeast *Pichia pastoris*. *Biochem. Biophys. Res. Commun.*, **344**, 807–813.

60 Tan, W., Gou, D.M., Tai, E., Zhao, Y.Z., and Chow, L.M.C. (2006) Functional reconstitution of purified chloroquine resistance membrane transporter expressed in yeast. *Arch. Biochem. Biophys.*, **452**, 119–128.

61 Tombline, G., Urbatsch, I.L., Virk, N., Muharemagic, A., Bartholomew White, L., and Senior, A.E. (2006) Expression, purification, and characterization of cysteine-free mouse P-glycoprotein. *Arch. Biochem. Biophys.*, **445**, 124–128.

62 Amoah, L.E., Lekostaj, J.K., and Roepe, P.D. (2007) Heterologous expression and ATPase activity of mutant versus wild type PfMDR1 protein. *Biochemistry*, **46**, 6060–6073.

63 Chloupková, M., Pickert, A., Lee, J.Y., Souza, S., Trinh, Y.T., Connelly, S.M., Dumont, M.E., Dean, M., and Urbatsch, I.L. (2007) Expression of 25 human ABC transporters in the yeast *Pichia pastoris* and characterization of the purified ABCC3 ATPase activity. *Biochemistry*, **46**, 7992–8003.

64 Price, D.R.G., Wilkinson, H.S., and Gatehouse, J.A. (2007) Functional expression and characterisation of a gut facilitative glucose transporter, NlHT1, from the phloem-feeding insect *Nilaparvata lugens* (rice brown planthopper). *Insect Biochem. Mol. Biol.*, **37**, 1138–1148.

65 Cox, H.D., Chao, C.K., Patel, S.A., and Thompson, C.M. (2008) Efficient digestion and mass spectral analysis of vesicular glutamate transporter 1: a recombinant membrane protein expressed in yeast. *J. Proteome Res.*, **7**, 570–578.

66 Alisio, A. and Mueckler, M. (2010) Purification and characterization of mammalian glucose transporters expressed in *Pichia pastoris*. *Protein Expr. Purif.*, **70**, 81–87.

67 Aller, S.G., Yu, J., Ward, A., Weng, Y., Chittaboina, S., Zhuo, R., Harrell, P.M., Trinh, Y.T., Zhang, Q., Urbatsch, I.L., and Chang, G. (2009) Structure of P-glycoprotein reveals a molecular basis for poly-specific drug binding. *Science*, **323**, 1718–1722.

68 Beckham, K.S.H., Potter, J.A., and Unkles, S.E. (2010) Formate-nitrite transporters: optimisation of expression, purification and analysis of prokaryotic and eukaryotic representatives. *Protein Expr. Purif.*, **71**, 184–189.

69 Karlsson, M., Fotiadis, D., Sjovall, S., Johansson, I., Hedfalk, K., Engel, A., and Kjellbom, P. (2003) Reconstitution of water channel function of an aquaporin overexpressed and purified from *Pichia pastoris*. *FEBS Lett.*, **537**, 68–72.

70 Kukulski, W., Schenk, A.D., Johanson, U., Braun, T., de Groot, B.L., Fotiadis, D., Kjellbom, P., and Engel, A. (2005) The 5 Å structure of heterologously expressed plant aquaporin SoPIP2;1. *J. Mol. Biol.*, **350**, 611–616.

71 Daniels, M.J., Wood, M.R., and Yeager, M. (2006) *In vivo* functional assay of a recombinant aquaporin in *Pichia pastoris*. *Appl. Environ. Microbiol.*, **72**, 1507–1514.

72 Tornroth-Horsefield, S., Wang, Y., Hedfalk, K., Johanson, U., Karlsson, M., Tajkhorshid, E., Neutze, R., and Kjellbom, P. (2006) Structural mechanism of plant aquaporin gating. *Nature*, **439**, 688–694.

73 Eifler, N., Duckely, M., Sumanovski, L.T., Egan, T.M., Oksche, A., Konopka, J.B., Luthi, A., Engel, A., and Werten, P.J.L. (2007) Functional expression of mammalian receptors and membrane channels in different cells. *J. Struct. Biol.*, **159**, 179–193.

74 Nyblom, M., Oberg, F., Lindkvist-Petersson, K., Hallgren, K., Findlay, H., Wikstrom, J., Karlsson, A., Hansson, O., Booth, P.J., Bill, R.M., Neutze, R., and Hedfalk, K. (2007) Exceptional overproduction of a functional human membrane protein. *Protein Expr. Purif.*, **56**, 110–120.

75 Hedfalk, K., Pettersson, N., Oberg, F., Hohmann, S., and Gordon, E. (2008) Production, characterization and crystallization of the *Plasmodium falciparum* aquaporin. *Protein Expr. Purif.*, **59**, 69–78.

76 Horsefield, R., Nordén, K., Fellert, M., Backmark, A., Tornroth-Horsefield, S., Terwisscha van Scheltinga, A.C., Kvassman, J., Kjellbom, P., Johanson, U., and Neutze, R. (2008) High-resolution X-ray structure of human aquaporin 5. *Proc. Natl. Acad. Sci. USA*, **105**, 13327–13332.

77 Verdoucq, L., Grondin, A., and Maurel, C. (2008) Structure–function analysis of plant aquaporin *At*PIP2;1 gating by divalent cations and protons. *Biochem. J.*, **15**, 409–416.

78 Azad, A.K., Sawa, Y., Ishikawa, T., and Shibata, H. (2009) Heterologous expression of tulip petal plasma membrane aquaporins in *Pichia pastoris* for water channel analysis. *Appl. Environ. Microbiol.*, **75**, 2792–2797.

79 Fischer, G., Kosinska-Eriksson, U., Aponte-Santamaria, C., Palmgren, M., Geijer, C., Hedfalk, K., Hohmann, S., de Groot, B.L., Neutze, R., and Lindkvist-Petersson, K. (2009) Crystal structure of a yeast aquaporin at 1.15 Å reveals a novel gating mechanism. *PLOS Biol.*, **7**, e1000130.

80 Ho, J.D., Yeh, R., Sandstrom, A., Chorny, I., Harries, W.E.C., Robbins, R.A., Miercke, L.J.W., and Stroud, R.M. (2009) Crystal structure of human aquaporin 4 at 1.8 Å and its mechanism of conductance. *Proc. Natl. Acad. Sci. USA*, **106**, 7437–7442.

81 Öberg, F., Ekvall, M., Nyblom, M., Backmark, A., Neutze, R., and Hedfalk, K. (2009) Insight into factors directing high production of eukaryotic membrane proteins; production of 13 human AQPs in *Pichia pastoris*. *Mol. Membr. Biol.*, **26**, 215–227.

82 Licata, L., Haase, W., Eckhardt-Strelau, L., and Parcej, D.N. (2006) Over-expression of a mammalian small conductance calcium-activated K^+ channel in *Pichia pastoris*: effects of trafficking signals and subunit fusions. *Protein Expr. Purif.*, **47**, 171–178.

83 Tao, X., Avalos, J.L., Chen, J., and MacKinnon, R. (2009) Crystal structure of the eukaryotic strong inward-rectifier K^+ channel Kir2.2 at 3.1 Å resolution. *Science*, **326**, 1668–1674.

84 Schiller, H., Haase, W., Molsberger, E., Janssen, P., Michel, H., and Reilander, H. (2000) The human ETB endothelin receptor heterologously produced in the methylotrophic yeast *Pichia pastoris* shows high-affinity binding and induction of stacked membranes. *Receptors Channels*, **7**, 93–107.

85 Schiller, H., Molsberger, E., Janssen, P., Michel, H., and Reilander, H. (2001) Solubilization and purification of the human ETB endothelin receptor produced by high-level fermentation in *Pichia pastoris*. *Receptors Channels*, **7**, 453–569.

86 Feng, W., Cai, J., Pierce, W.M., Jr, and Song, Z.H. (2002) Expression of CB2 cannabinoid receptor in *Pichia pastoris*. *Protein Expr. Purif.*, **26**, 496–505.

87 Sarramegna, V., Demange, P., Milon, A., and Talmont, F. (2002) Optimizing functional versus total expression of the human μ-opioid receptor in *Pichia pastoris*. *Protein Expr. Purif.*, **24**, 212–220.

88 Sarramegna, V., Talmont, F., Seree de Roch, M., Milon, A., and Demange, P. (2002) Green fluorescent protein as a reporter of human μ-opioid receptor

overexpression and localization in the methylotrophic yeast *Pichia pastoris*. *J. Biotechnol.*, **99**, 23–39.

89 De Jong, L.A.A., Grunewald, S., Franke, J.P., Uges, D.R.A., and Bischoff, R. (2004) Purification and characterization of the recombinant human dopamine D2S receptor from *Pichia pastoris*. *Protein Expr. Purif.*, **33**, 176–184.

90 Grunewald, S., Haase, W., Molsberger, E., Michel, H., and Reilander, H. (2004) Production of the human D2S receptor in the methylotrophic yeast *P. pastoris*. *Receptors Channels*, **10**, 37–50.

91 De Rivoyre, M., Bonino, F., Ruel, L., Bidet, M., Thérond, P., and Mus-Veteau, I. (2005) Human receptor Smoothened, a mediator of Hedgehog signalling, expressed in its native conformation in yeast. *FEBS Lett.*, **579**, 1529–1533.

92 Kim, T.K., Zhang, R., Feng, W., Cai, J., Pierce, W., and Song, Z.H. (2005) Expression and characterization of human CB1 cannabinoid receptor in methylotrophic yeast *Pichia pastoris*. *Protein Expr. Purif.*, **40**, 60–70.

93 Sarramegna, V., Muller, I., Mousseau, G., Froment, C., Monsarrat, B., Milon, A., and Talmont, F. (2005) Solubilization, purification, and mass spectrometry analysis of the human mu-opioid receptor expressed in *Pichia pastoris*. *Protein Expr. Purif.*, **43**, 85–93.

94 André, N., Cherouati, N., Prual, C., Steffan, T., Zeder-Lutz, G., Magnin, T., Pattus, F., Michel, H., Wagner, R., and Reinhart, C. (2006) Enhancing functional production of G protein-coupled receptors in *Pichia pastoris* to levels required for structural studies via a single expression screen. *Protein Sci.*, **15**, 1115–1126.

95 Fraser, N.J. (2006) Expression and functional purification of a glycosylation deficient version of the human adenosine 2a receptor for structural studies. *Protein Expr. Purif.*, **49**, 129–137.

96 Noguchi, S. and Satow, Y. (2006) Purification of human β2-adrenergic receptor expressed in methylotrophic yeast *Pichia pastoris*. *J. Biochem.*, **140**, 799–804.

97 Lundstrom, K., Wagner, R., Reinhart, C., Desmyter, A., Cherouati, N., Magnin, T., Zeder-Lutz, G., Courtot, M., Prual, C., André, N., Hassaine, G., Michel, H., Cambillau, C., and Pattus, F. (2006) Structural genomics on membrane proteins: comparison of more than 100 GPCRs in 3 expression systems. *J. Struct. Funct. Genomics*, **7**, 77–91.

98 Yang, G., Liu, T., Peng, W., Sun, X., Zhang, H., Wu, C., and Shen, D. (2006) Expression and localization of recombinant human EDG-1 receptors in *Pichia pastoris*. *Biotechnol. Lett.*, **28**, 1581–1586.

99 Yang, G.X., Liu, T.L., Zhang, H., Wu, C.Q., and Shen, D.L. (2006) Expression and localization of recombinant human B2 receptors in the methylotrophic yeast *Pichia pastoris*. *Genetika*, **42**, 728–731.

100 Shukla, A.K., Haase, W., Reinhart, C., and Michel, H. (2007) Heterologous expression and characterization of the recombinant bradykinin B2 receptor using the methylotrophic yeast *Pichia pastoris*. *Protein Expr. Purif.*, **55**, 1–8.

101 Zhang, R., Kim, T.K., Qiao, Z.H., Cai, J., Pierce, Jr, W.M., and Song, Z.H. (2007) Biochemical and mass spectrometric characterization of the human CB2 cannabinoid receptor expressed in *Pichia pastoris*–importance of correct processing of the N-terminus. *Protein Expr. Purif.*, **55**, 225–235.

102 Magnin, T., Fiez-Vandal, C., Potier, N., Coquard, A., Leray, I., Steffan, T., Logez, C., Alkhalfioui, F., Pattus, F., and Wagner, R. (2008) A novel, generic and effective method for the rapid purification of G protein-coupled receptors. *Protein Expr. Purif.*, **64**, 1–7.

103 Muller, I., Sarramégna, V., Milon, A., and Talmont, F.J. (2010) The N-terminal end truncated mu-opioid receptor: from expression to circular dichroism analysis. *Appl. Biochem. Biotechnol.*, **160**, 2175–2186.

104 Talmont, F. (2009) Monitoring the human β1, β2, β3 adrenergic receptors expression and purification in *Pichia pastoris* using the fluorescence properties of the enhanced green fluorescent protein. *Biotechnol. Lett.*, **31**, 49–55.

105 Hori, H., Sato, Y., Takahashi, N., Takio, K., Yokomizo, T., Nakamura, M.,

Shimizu, T., and Miyano, M. (2010) Expression, purification and characterization of leukotriene B_4 receptor, BLT1 in *Pichia pastoris. Protein Expr. Purif.*, **72**, 66–74.

106 Blum, A., Martin, H.J., and Maser, E. (2000) Human 11β-hydroxysteroid dehydrogenase 1/carbonylreductase: recombinant expression in the yeast *Pichia pastoris* and *Escherichia coli. Toxicology*, **144**, 113–120.

107 Newton-Vinson, P., Hubalek, F., and Edmondson, D.E. (2000) High-level expression of human liver monoamine oxidase B in *Pichia pastoris. Protein Expr. Purif.*, **20**, 334–345.

108 Binda, C., Newton-Vinson, P., Hubalek, F., Edmondson, D.E., and Mattevi, A. (2002) Structure of human monoamine oxidase B, a drug target for the treatment of neurological disorders. *Nat. Struct. Biol.*, **9**, 22–26.

109 Li, M., Hubalek, F., Newton-Vinson, P., and Edmondson, D.E. (2002) High-level expression of human liver monoamine oxidase A in *Pichia pastoris*: comparison with the enzyme expressed in *Saccharomyces cerevisiae. Protein Expr. Purif.*, **24**, 152–162.

110 Nobel, C.S.I., Dunas, F., and Abrahmsén, L.B. (2002) Purification of full-length recombinant human and rat type 1 11β-hydroxysteroid dehydrogenases with retained oxidoreductase activities. *Protein Expr. Purif.*, **26**, 349–356.

111 Binda, C., Li, M., Hubalek, F., Restelli, N., Edmondson, D.E., and Mattevi, A. (2003) Insights into the mode of inhibition of human mitochondrial monoamine oxidase B from high-resolution crystal structures. *Proc. Natl. Acad. Sci. USA*, **100**, 9750–9755.

112 Ruf, A., Muller, F., D'Arcy, B., Stihle, M., Kusznir, E., Handschin, C., Morand, O.H., and Thoma, R. (2004) The monotopic membrane protein human oxidosqualene cyclase is active as monomer. *Biochem. Biophys. Res. Commun.*, **315**, 247–254.

113 De Colibus, L., Li, M., Binda, C., Lustig, A., Edmondson, D.E., and Mattevi, A. (2005) Three-dimensional structure of human monoamine oxidase A (MAO A): relation to the structures of rat MAO A and human MAO B. *Proc. Natl. Acad. Sci. USA*, **102**, 12684–12689.

114 Dietrich, M., Grundmann, L., Kurr, K., Valinotto, L., Saussele, T., Schmid, R.D., and Lange, S. (2005) Recombinant production of human microsomal cytochrome P450 2D6 in the methylotrophic yeast *Pichia pastoris. ChemBioChem.*, **6**, 2014–2022.

115 Liu, W., Kamensky, Y., Kakkar, R., Foley, E., Kulmacz, R.J., and Palmer, G. (2005) Purification and characterization of bovine adrenal cytochrome b_{561} expressed in insect and yeast cell systems. *Protein Expr. Purif.*, **40**, 429–439.

116 Matsuzaki, F. and Wariishi, H. (2005) Molecular characterization of cytochrome P450 catalyzing hydroxylation of benzoates from the white-rot fungus *Phanerochaete chrysosporium. Biochem. Biophys. Res. Commun.*, **334**, 1184–1190.

117 Kolar, N.W., Swart, A.C., Mason, J.I., and Swart, P. (2007) Functional expression and characterisation of human cytochrome P45017α in *Pichia pastoris. J. Biotechnol.*, **129**, 635–644.

118 Martinez Molina, D., Wetterholm, A., Kohl, A., McCarthy, A.A., Niegowski, D., Ohlson, E., Hammarberg, T., Eshaghi, S., Haeggstrom, J.Z., and Nordlund, P. (2007) Structural basis for synthesis of inflammatory mediators by human leukotriene C_4 synthase. *Nature*, **448**, 613–617.

119 Upadhyay, A.K. and Edmondson, D.E. (2008) Characterization of detergent purified recombinant rat liver monoamine oxidase B expressed in *Pichia pastoris. Protein Expr. Purif.*, **59**, 349–356.

120 Wetterholm, A., Martinez Molina, D., Nordlund, P., Eshaghi, S., and Haeggström, J.Z. (2008) High-level expression, purification, and crystallization of recombinant rat leukotriene C_4 synthase from the yeast *Pichia pastoris. Protein Expr. Purif.*, **60**, 1–6.

121 Wang, J. and Edmondson, D.E. (2010) High-level expression and purification of rat monoamine oxidase A (MAO A) in *Pichia pastoris* : comparison with human MAO A. *Protein Expr. Purif.*, **70**, 211–217.

122 Arslan, B.K. and Edmondson, D.E. (2010) Expression of zebrafish (*Danio rerio*) monoamine oxidase (MAO) in *Pichia pastoris*: purification and comparison with human MAO A and MAO B. *Protein Expr. Purif.*, **70**, 290–297.

123 Brucato, C.L., Birr, C.A., Bruguera, P., Ruiz, J.A., and Sanchez-Martınez, D. (2002) Expression of recombinant rabbit tissue factor in *Pichia pastoris*, and its application in a prothrombin time reagent. *Protein Expr. Purif.*, **26**, 386–393.

124 Gurkan, C. and Ellar, D.J. (2003) Expression in *Pichia pastoris* and purification of a membrane-acting immunotoxin based on a synthetic gene coding for the *Bacillus thuringiensis* Cyt2Aa1 toxin. *Protein Expr. Purif.*, **29**, 103–116.

125 Huang, Z., Zhang, T., Yang, J., Zhu, P., Du, G., and Cheng, K. (2005) Cloning and expression of human lectin-like oxidized low density lipoprotein receptor-1 in *Pichia pastoris*. *Biotechnol. Lett.*, **27**, 49–52.

126 Labarre, C., Van Tilbeurgh, H., and Blondeau, K. (2007) *Pichia pastoris* is a valuable host for the expression of genes encoding membrane proteins from the hyperthermophilic Archeon *Pyrococcus abyssi*. *Extremophiles*, **11**, 403–413.

127 Jamshad, M., Rajesh, S., Stamataki, Z., McKeating, J.A., Dafforn, T., Overduin, M., and Bill, R.M. (2008) Structural characterization of recombinant human CD81 produced in *Pichia pastoris*. *Protein Expr. Purif.*, **57**, 206–216.

128 Egawa, K., Shibata, H., Yamashita, S., Yurimoto, H., Sakai, Y., and Kato, H. (2009) Overexpression and purification of rat peroxisomal membrane protein 22, PMP22, in *Pichia pastoris*. *Protein Expr. Purif.*, **64**, 47–54.

129 Vos, W.L., Vaughan, S., Lall, P.Y., McCaffrey, J.G., Wysocka-Kapcinska, M., and Findlay, J.B.C. (2010) Expression and structural characterization of peripherin/RDS, a membrane protein implicated in photoreceptor outer segment morphology. *Eur. Biophys. J.*, **39**, 679–688.

130 Vlahopoulos, S., Gritzapis, A.D., Perez, S.A., Cacoullos, N., Papamichail, M., and Baxevanis, C.N. (2009) Mannose addition by yeast *Pichia Pastoris* on recombinant HER-2 protein inhibits recognition by the monoclonal antibody herceptin. *Vaccine*, **27**, 4704–4708.

131 Gustafsson, C., Govindarajan, S., and Minshull, J. (2004) Codon bias and heterologous protein expression. *Trends Biotechnol.*, **22**, 346–353.

132 Singh, S., Gras, A., Fiez-Vandal, C., Ruprecht, J., Rana, R., Martinez, M., Strange, P.G., Wagner, R., and Byrne, B. (2008) Large-scale functional expression of WT and truncated human adenosine A_{2A} receptor in *Pichia pastoris* bioreactor cultures. *Microb. Cell Fact.*, **7**, 28.

133 Talmont, F., Sidobre, S., Demange, P., Milon, A., and Emorine, L.J. (1996) Expression and pharmacological characterization of the human mu opioid receptor in the methylotrophic yeast *Pichia pastoris*. *FEBS Lett.*, **394**, 268–272.

134 Cid, G.M., Nugent, P.G., Davenport, A.P., Kuc, R.E., and Wallace, B.A. (2000) Expression and characterization of the human endothelin-A-receptor in *Pichia pastoris*: influence of N-terminal epitope tags. *J. Cardiovasc. Pharmacol.*, **36**, S55–S57.

135 Brillet, K., Perret, B.G., Klein, V., Pattus, F., and Wagner, R. (2008) Using EGFP fusions to monitor the functional expression of GPCRs in the *Drosophila* Schneider 2 cells. *Cytotechnology*, **57**, 101–109.

136 Yu, Z.W. and Quinn, P.J. (1994) Dimethyl sulphoxide: a review of its applications in cell biology. *Biosci. Rep.*, **14**, 259–281.

137 Bernier, V., Lagacé, M., Bichet, D.G., and Bouvier, M. (2004) Pharmacological chaperones: potential treatment for conformational diseases. *Trends Endocrinol. Metab.*, **15**, 222–228.

138 Zeder-Lutz, G., Cherouati, N., Reinhart, C., Pattus, F., and Wagner, R. (2006) Dot-blot immunodetection as a versatile and high-throughput assay to evaluate recombinant GPCRs produced in the yeast *Pichia pastoris*. *Protein Expr. Purif.*, **50**, 118–127.

139 Aguilar-Uscanga, B. and François, J.M. (2003) A study of the yeast cell wall composition and structure in response to growth conditions and mode of cultivation. *Lett. Appl. Microbiol.*, **37**, 268–274.

140 Wagner, S., Klepsch, M.M., Schlegel, S., Appel, A., Draheim, R., Tarry, M., Högbom, M., van Wijk, K.J., Slotboom, D.J., Persson, J.O., and de Gier, J.W. (2008) Tuning *Escherichia coli* for membrane protein overexpression. *Proc. Natl. Acad. Sci. USA*, **105**, 14371–14376.

141 Bonander, N., Darby, R.A., Grgic, L., Bora, N., Wen, J., Brogna, S., Poyner, D.R., O'Neill, M.A., and Bill, R.M. (2009) Altering the ribosomal subunit ratio in yeast maximizes recombinant protein yield. *Microb. Cell. Fact.*, **8**, 10.

4
Heterologous Production of Active Mammalian G-Protein-Coupled Receptors Using Baculovirus-Infected Insect Cells

Mark Chiu, Brian Estvander, Timothy Esbenshade, Steve Kakavas, Kathy Krueger, Marc Lake, and Ana Pereda-Lopez

4.1
Introduction

G-protein-coupled receptors (GPCRs) comprise a family of seven-transmembrane receptors that mediate most of the cell–cell communication in humans via a wide variety of extracellular activators such as hormones, light, neurotransmitters, ions, odorants, and amino acids [1–3]. Owing to their physiological importance in metabolic, endocrine, neuromuscular, and central nervous biology, many pharmaceutical companies have significant efforts to develop therapeutic drugs that act on this family of proteins [4, 5]. The basis of such research efforts follow the success of more than 50% of the marketed drugs that treat diseases by targeting only 20 GPCRs [6]. Bioinformatic analyses of genomic data suggest that there are about 1000 GPCRs in the human genome: 614 annotated entries, with the remainder classified as orphan receptors [7–9]. These as yet untargeted GPCRs may have important physiological roles and unique mechanisms of interaction, which makes GPCRs the class of proteins with the highest drug discovery potential [10].

One of the great challenges in understanding GPCR biology is to elucidate the structure–function relationships that govern the specificity of the binding of agonists, antagonists, and allosteric modulators in the GPCR/G-protein systems as well as GPCR/G-protein–effector interactions [11–14]. This ignorance is due in part to the complex and multifactorial nature of signal transduction regulation [15]. Notwithstanding, biochemical and structural descriptions of GPCR ligand binding and activation could provide early diagnostic tools that could prevent progression of receptor-mediated disease, and be new strategies for more specific and effective therapeutic intervention [16].

In order to elucidate biochemical and structural aspects of GPCR activities, milligram quantities of receptors have to be generated. With the exception of rhodopsins, most mammalian GPCRs are intrinsically produced in low amounts in the cell, sometimes for a short time in the cell cycle. Hence, to obtain milligram

Production of Membrane Proteins: Strategies for Expression and Isolation, First Edition.
Edited by Anne Skaja Robinson.

quantities of GPCRs, high-level overexpression in heterologous systems is required. Production of GPCRs in mammalian cell overexpression systems often provides enough material for the screening of ligand interactions. On the other hand, production of milligram quantities from such expression systems would require fermentation of hundreds of liters, which is cost-prohibitive. In addition, the copurification of related endogenous GPCRs can be problematic for assay interpretation.

Although certain GPCRs can be well expressed in cell-free translation systems, and bacterial and yeast cells, the receptors produced in such host expression systems do not reproduce the native pharmacology found in mammalian cells [17–22]. Presumably this loss of receptor function is due to suboptimal protein folding caused by the alterations of post-translational modifications. By having the machinery to generate post-translational modifications (such as glycosylation, phosphorylation, and palmitoylation found in mammalian cells [23, 24]), recombinant baculovirus infections of insect cell cultures have proved to be useful heterologous expression systems [25–27]. As insect cells produce low levels of receptors homologous to mammalian GPCRs, they can be used for the expression of GPCR families such as class A GPCRs [28–48], class B GPCRs [49], class C GPCRs [49–52], and olfactory receptors [53]. Since the amounts of protein produced are generally present at the cell surface and are made at 25–600 times higher than those obtained in naturally producing mammalian cells [48], insect cell-produced GPCRs have been utilized for assay development and biochemical analyses. Heterologous expression in insect cells has been used to generate the constructs of all high-resolution mammalian GPCR structures [54–58]. Likewise, other structures of eukaryotic ion channels (AMPA ionotropic glutamate receptor [59], acid sensing ion channel [60–62], and ATP-gated P2X4 [63, 64]), and enzymes such as cyclooxygenase-2 [64] have been obtained from heterologous baculovirus expression in insect cells.

Baculoviruses are double-stranded circular supercoiled DNA viruses in rod-shaped capsids with narrow host ranges of insect cells [65]. The most utilized member of the baculoviruses is *Autographa californica* nuclear polyhedrosis virus (*Ac*MNPV) [66], which serves as the backbone for most recombinant baculovirus vectors that are employed for protein expression studies [67]. Extensive reviews have outlined the use of baculovirus-based expression of recombinant proteins [68, 69]. Conveniently, baculovirus infection of insect cells for protein expression can be performed in Biosafety Level I laboratories since the virus host range is restricted to a few insect species and does not propagate in mammalian or plant cells. Helper cell lines and helper viruses are not needed since the baculovirus genome has all of the required genetic information. The main difference between the naturally occurring *in vivo* infection and the recombinant *in vitro* infection is that the naturally occurring polyhedron gene within the baculovirus genome is replaced with a recombinant gene or cDNA. An outline of the baculovirus expression is shown in Figure 4.1. The development of the baculovirus expression system is based on the propagation of two later viral gene products: P10 and polyhedrin. Both genes are under the control of strong promoters that have been used in different expression vectors. Early promoters have also been used for the heterologous production

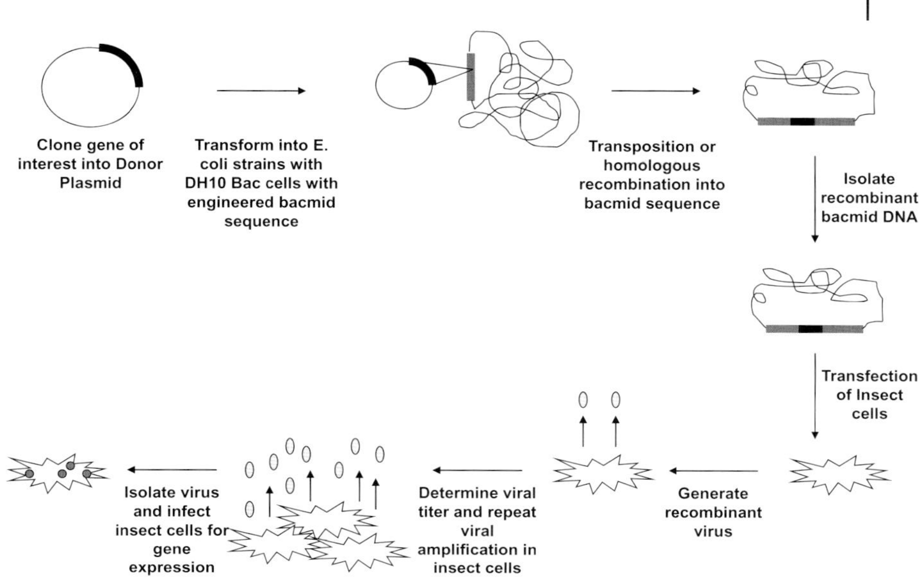

Figure 4.1 Outline of recombinant protein production using baculovirus infection of insect cells. The gene of interest can be cloned into a donor plasmid that is transformed into DH10Bac cells with an engineered bacmid sequence. Transposition using site-specific helpers allows for the generation of recombinant baculovirus. This step can take about 1–3 days. The recombinant virus is then isolated and transfected into insect cells to generate more recombinant virus. This step will take about 3–5 days to evaluate. Repeated rounds of viral amplification and insect cell growth increase the amount of recombinant virus that is isolated and then used for protein production. Each round takes about 1 week.

of proteins, but were found to be less effective [70]. The desired gene of interest is cloned into a region juxtaposed to baculovirus promoters using homologous recombination, ligation of restriction digests, or site-specific transposition. Depending on the method used, this process can take about 1–4 days. The chimeric baculovirus is selected (2–3 days) and then amplified (1 week) for protein expression in insect cells. Each of these steps can be optimized for each particular protein. A list of companies that provide the transfer vectors, transfection reagents, and insect cells for recombinant protein production is shown in Table 4.1.

In this chapter, we review the preparation of GPCRs in insect cells by discussing: (i) the selection of the appropriate expression vector and baculovirus preparation, and (ii) the evaluation of insect cell lines, growth, and infection conditions. We include an example of baculovirus expression of human histamine H_3 receptor and the verification of the GPCR integrity. The rationale and methodology used to generate GPCRs in insect cells can be also applied to other membrane proteins, including transporters, channels, and enzymes. Crude cell membrane preparations prepared from infected insect cells can be a source of active membrane

Table 4.1 Vendors that supply reagents for baculovirus work.

Baculovirus vectors	***Custom baculovirus services***
AB vector	Abgent Protein Expression
BD Biosciences	AB vector
Clontech	AgriVirion
Oxford Expression Technologies	Analytical Biological Services
Invitrogen	Bio X Cell
Novagen	Biologics Process Development
Insect cells	BioSciences Research Associates
BD Biosciences	Blue Sky Biotech
ATCC	Chesapeake PERL
Invitrogen–GIBCO	deltaDOT
Novagen	DIARECT
Transfection reagents	Entopath
EMD Biosciences	Genscript
Invitrogen	IBA Strep-Tag Custom Service
Novagen	Kinnakeet Biotechnology
Cell media	Orbigen
HyClone	Oxford Expression Technologies
Invitrogen–GIBCO	Protein Sciences
Lonza	Q-BioGene
Novagen	Virinova Diagnostics
Sigma-Aldrich	University of Minnesota Biotechnology Resource Center
Good websites about insect cell expression	
baculovirus.com	
http://www.invitrogen.com/site/us/en/home/Products-and-Services/Applications/Cell-Culture/Insect-Cell-Culture.html	
http://www.abvector.com/technology.htm#proeasy	
http://wolfson.huji.ac.il/expression/insect_exp.html	

proteins for ligand binding, drug screening, and other GPCR functional assays involving G-protein and effector protein interactions.

4.2 Experimental

4.2.1 Generation of Recombinant Baculovirus

Insect cell lines allow for the replication of *Ac*MNPV, which will express inserted recombinant genes as part of the viral and replication process. Recombinant baculovirus can be generated by two methods: homologous recombination into the genome of *Ac*MNPV and site-specific transposition into bacmid DNA propagated in *Escherichia coli* [68, 71]. The first reports of the use of *Ac*MNPV virus for protein expression [66, 72] involved the cumbersome selection of recombinant baculoviruses using homologous recombination based on the phenotype of the virus plaques. The gene of interest is cloned into a transfer vector containing a baculovirus promoter flanked by baculovirus DNA derived from a nonessential locus. The recombinant sequence is then inserted into the genome of the parent virus by homologous recombination after transfection into insect cells. The difficult part with this approach is assessing virus-containing recombinant DNA quickly. The typical efficiency of this method is less than 1%, and the screening for recombinant virus, which is performed by plaque assays and hybridizations, is tedious and time-consuming [73]. Typically only 0.1–1% of the virus progeny obtained after transfection is recombinant.

Recombinant baculovirus production can be conducted more efficiently and quickly by the development of recombinant viral DNA with well-defined restriction sites [74] and an *E. coli* cell line that contains modified baculovirus DNA in an autonomously replicating episome known as a bacmid [75]. These technologies have been subsequently commercialized by as the Bac to Bac® Baculovirus Expression System (Invitrogen), BaculoGold™ (BD Pharmingen), BacPAK™ (Clontech), and ProEasy™ and ProFold™ (AB Vector). They are widely applied since there is no need for homologous recombination in insect cells and the use of plaque assays for the selection of cells containing the pure recombinant virus. The gene of interest can be cloned into a transfer vector, which is then transformed into DH10Bac-competent cells that contain the bacmid with an *att:Tn7* target site. Recombination is instead carried out within specialized bacteria, which can be selected for the presence of recombinant baculovirus DNA. Alternatively, recombinant baculoviruses can be made using site-specific transposition with *Tn7* to insert foreign DNA into bacmid DNA propagated in DH10Bac *E. coli*. In the presence of transposition proteins provided by the helper plasmid, recombinant bacmids can be identified by antibiotic selection and blue–white screening, since the transposition results in the disruption of the *lacZα* gene. High-molecular-weight DNA prepared from these selected *E. coli* clones containing the recombinant bacmid is then used to transfect insect cells. In addition, the baculovirus expression vector also serves to induce the production of the transcriptional complex needed to transcribe the foreign gene of interest under the control of a late or very late baculovirus promoter.

Most transfer vectors are designed to have the target gene be inserted into the polyhedrin site in the baculovirus DNA. Generally, expression is controlled by the very strong polyhedrin promoter that is induced by chemical addition (e.g., copper sulface addition) at a time point that is late in the infection cycle, which is when cell lysis is beginning. Gene expression late in the infection cycle can be advantageous since most of the host cell protein production is diminished. The polyhedrin promoter is extremely efficient in protein expression, producing 30% of the total cell protein mass in the later stages of baculovirus infection. Thus, the strong, late promoters are useful for providing maximal quantities of receptors for biochemistry and structural biology.

Application of codon bias to recombinant protein design can also help increase levels of heterologous protein expression in insect cells [76–78]. In addition, base composition around the start codon can have some influence on protein production levels [79]. The introduction of signal sequences that direct proteins into the membrane protein biogenesis machinery [80] has been reported to increase expression levels of membrane proteins such as human β_2-adrenergic receptor [81], human dopamine D_{2S} receptor [40, 82], and TRPV1 ion channel [83]. The heterologous signal peptide sequences such as influenza hemagglutinin [84] and honey bee melittin [85] can work as well as baculovirus-encoded signal sequences such as gp64 [86], egt, and p67 [87]. It is also possible that the N-terminal domains of GPCRs and other membrane proteins can contain cryptic signal sequences that result in functional proteins in virion particles [88]. Green Fluorescent Protein (GFP) can be added to the N-terminal domain to monitor protein levels during purification [32, 33]. By monitoring protein monodispersity by size-exclusion chromatography, many constructs can be screened for protein stability in crude extracts [89]. Chimeric proteins with peptide epitope tags (polyhistidine, FLAG, HA, c-Myc, other monoclonal antibody tags, etc.) and glutathione S-transferase (GST) domains can also be added in the N- and C-terminal domains for purification without loss of activity [54, 55, 58, 90, 91]. Recent GPCR structures have also been derived from N- and C-terminal-tagged protein constructs [54–58, 92]. Nonetheless, careful assessment of protein activity is required to determine whether the tags can affect protein conformation [29, 40].

If maximum protein production is required at an early stage of infection, then early promoters can be utilized, albeit with less cell mass and protein content [93]. Alternatively, stably transformed cells can be used for expression using viral promoters, thereby removing the infection process that leads to cell lysis [94]. Typically the vectors designed for this purpose use weaker promoters and cell lines derived from the fall armyworm *Spodoptera frugiperda* are the most commonly used for its propagation [95].

The preparation of virus stock comes from infecting Sf9 cells grown in mid-log phase in flasks. After 5–10 days, P2 viral stock working solutions are obtained via centrifugation. Viral stocks can be stored at 4 °C for up to 1 year. In the past, obtaining viral titers was important to gage how much virus would be needed to infect insect cells. It is prudent to run small-scale evaluations of protein expression by growing up cells with up to 1-log change in multiplicity of infection (MOI) to cells. Recently a method was developed to bypass viral titer determination methods by implementing titerless infected cell preservation and scale-up (TIPS) methodology for protein expression. In this method, small-scale insect cell cultures are first incubated with a recombinant baculovirus, which replicates in the cells. The baculovirus-infected insect cells (BIICs) are harvested and frozen prior to cell lysis and subsequent escape of the newly replicated virus into the culture supernatant. The thawed BIIC stocks are used for subsequent scale-up. The TIPS method eliminates the need and protracted time required for titering virus supernatants, and provides stable and concentrated storage of recombinant baculovirus in the form of infected cells [96]. The use of the TIPS system allows for uniform starting titers, specific activity, and composition of contaminating proteins that facilitate the development of a reproducible purification process.

4.2.2 Baculovirus Infection of Insect Cells

The general process of overexpression of proteins in insect cells is well described elsewhere [68, 69, 73, 97, 98]. The order Lepidoptera includes the moths and butterflies, which are the hosts for many viruses in the family Baculoviridae, including *Ac*MNPV [99–101]. The insect cells lines that are used predominantly include Sf9, Sf21, MG1, and High Five™ [102, 103]. Sf21 is a line that had been isolated from pupal ovarian tissue of *S. frugiperda* [72, 104–106]. Sf9 is a cell line that is a subclone of IPLB-Sf21-AE. MG1 and High Five (originally named BTI-TN-5B-1-4) are cell lines isolated from adult ovarian tissue of *Trichoplusia ni* (cabbage looper) [103, 107]. Other cell lines have been developed and optimized to produce recombinant proteins with mammalian glycosylation profiles (i.e., Mimic Sf9) and to minimize proteolysis. These cells can be obtained from several different companies, including Invitrogen, Novagen, or the American Type Culture Collection (ATCC) as listed in Table 4.1.

Sf9, Sf21, and High Five cells can grow in either adherent or suspension formats. Thus, it is convenient to perform lab-scale protein production experiments by infecting a few million cells as adherent cultures in plates or flasks or as suspension cultures in flasks. Adherent and suspension cultures of Sf21 or Sf9 cells are also routinely used to plaque-purify and quantify recombinant baculovirus expression vectors. On the other hand, both Sf9 and High Five cells can be scaled-up to varying degrees in spinner flasks, shake flasks, or in airlift, stirred tank [108] or Wave [109] bioreactors to produce larger amounts of recombinant proteins. Suspension culture methods can generate mechanical shear forces that can damage insect cell integrity. Serum concentrations of 5–20% in medium and Pluronic F-68 can be added to serve as shear force protectants.

The conditions and methods used to culture insect cells are quite different from those used to culture mammalian cells. For example, the optimal temperature range for insect cell culture is 25–30 °C, rather than 37 °C. In addition, Sf9, Sf21, and High Five are loosely adherent cell lines, and neither EDTA nor trypsin is required for their subculture. It is not necessary to have a CO_2 incubator to grow insect cells because insect cell culture media are buffered with phosphate, rather than carbonate. Sf9 and High Five cells both can be cultured in either growth media supplemented with serum or in serum-free media. A pH range of 6–6.4 works well for most insect cells lines. The optimal osmolality is 345–380 mOsm/kg. Sf9 cell cultures can be maintained in shake flasks (Bellco) in either serum-containing or serum-free media.

Traditionally, Grace's supplemented medium (also known as TNM-FH medium) has been the medium of choice for insect cell culture. Fetal bovine serum has been the primary growth supplement used in insect culture. Serum has lactalbumin hydrolysate and yeast factors that provide insect cells with growth-promoting factors such as amino acids, peptides, and vitamins, which may not be available in defined basal media formulations. However, other serum and serum-free formulations have evolved since.

Protocols in growing host cells are well established. Typically, maintenance cultures are split each Monday, Wednesday, and Friday of the week, using seeding densities for Sf9 cells of 0.5×10^6 cells/ml. Typical insect cell doubling times should be around 24 h.

The amount of virus added to insect cells for either cell infection or baculovirus amplification is determined by:

$$\text{Virus required (ml)} = \text{desired MOI} \times \text{total number of cells} / [\text{viral titer (PFU)}]$$

Where MOI is the multiplicity of infection and PFU is plaque-forming unit. For amplification of recombinant baculovirus, it is typical to inoculate with a low MOI such as 0.1 as a precautionary measure in case there is nonrecombinant virus present within the viral population. The lower concentration of virus used can decrease the potential for an increase in the nonrecombinant component of the viral population during subsequent amplifications.

For GPCR infection, we typically infect insect cells for protein production using a MOI of 1–10. When high levels of functional receptors are required, optimization of MOI is required since expression patterns of different GPCRs can vary considerably [40]. Ideally, the expression is conducted at least 48 h postinfection. However, extended growth past 72 h can risk the chance of proteolytic degradation of the desired protein. A good measure of protein expression is to measure cell viability using the standard Trypan blue dye exclusion procedure with a hemocytometer [110]. This technique has major shortcomings due to subjective determination of cell counts as well as manual and time-consuming steps. In comparison to the manual techniques, automated cell counters such as the Beckman Vi-Cell system can perform the Trypan blue dye exclusion with video imaging of flow-through cells and give results in minutes. Physiological and morphological changes upon infection by viruses have been observed in insect cells. During the late phase (6–24 h postinfection), infected insect cells stop dividing and increase in diameter [73, 98]. Cell size is increased as a result of viral infection and multiplication

[111–114]. In the very late phase (24–36 h) cells cease to produce budded virus and begin production of recombinant protein. The nucleus and cytoplasm will contain assembled baculovirus, thus resulting in changes in granularity and increases in cell size. There is a direct correlation between the time of maximum modal insect cell diameter and the time of maximum recombinant protein concentration [111, 115]. Typically, we observe that the Sf9 cell diameters change from a range of 13–15 to 18–22 μm for successful infections. Being able to use cell diameter during baculovirus infection of insect cells can decrease the reliance on measuring viral titer and thus save on time. Cell viabilities, as measured by the number of intact cells and cell diameter, should remain high in the cultures.

Other additives can be added into the growth media as the cell viability decreases over time. The lytic infection process can result in degradation of recombinant proteins and limit the use of late promoters for cellular assays. However, the growth media can be supplemented with lipids and protease inhibitors to minimize the degradation of expressed receptors [40, 116–118]. Only a certain amount of added protease inhibitors (below 10 mM) can be tolerated before they affect insect cell viability.

The plasma membrane targeting of GPCRs is a tightly regulated process. Folding of the receptor in the endoplasmic reticulum and its exit to the Golgi represent the limiting steps in maturation and cell surface expression. General molecular chaperones involved in the folding of many membrane proteins have been suggested to play important roles at different steps of this process. In the absence of such mammalian host factors in insect cells, additional ligands can act as pharmacological chaperones that stabilize receptor integrity during protein production [119]. Typically, ligand concentrations of 10–50 times the K_D values are added to the insect cell media during the baculovirus infection step. It is preferable to use ligands that are more water soluble to increase the efficiency of cellular incorporation for receptor stabilization. This strategy has proven to be important in generating large amounts of functional GPCRs for protein crystallization [54–58].

Although insect cells are capable of complex glycosylation, baculovirus infection can result in partially processed high-mannose forms depending on the time period of postinfection. Crude cell membrane extracts from baculovirus-infected cells can contain immature GPCRs, probably representing a fraction of proteins trapped in the membranes of the endoplasmic reticulum and Golgi apparatus as a result of protein biosynthetic pathway saturation. A purification step based on the biological activity of the receptor such as a ligand-based affinity column can be included during purification to eliminate nonfunctional misfolded receptors [49, 120–122].

In summary, several factors are important for insect cell expression: amount of recombinant virus to add; timely addition of lipids, protease inhibitors, and ligands to stabilize the expressed protein; and the time of postinfection growth. Although such optimization has many variables, expression screening of constructs can be more quickly assessed by using cultures of insect cells in 24-deep-well blocks for small-scale optimization of baculovirus-mediated protein expression experiments

[40, 123]. The media we typically use is supplemented with total brain lipids (100 mg/l), histamine (10 μM), and Sigma protease inhibitor cocktail (2 ml/l) at the time of infection.

During harvesting of the cells, rapid collection of the cells and freezing in liquid nitrogen can minimize protein loss during proteolysis from cell breakdown. At 48 h postinfection, the cells are harvested by centrifugation (1560 g) for 7 min at 4 °C, flash frozen in liquid nitrogen, and stored at −80 °C until use. Washing the cells with protein-stability reagents such as proline, arginine, and trehalose can improve receptor stability during the freezing process [124].The washing of the cells can increase ligand-binding activity 2- to 10-fold versus unwashed cells.

4.2.3 Case Study: Histamine H_3 Receptor

Pharmacological characterization of GPCRs has focused largely on receptors in their membrane-bound state in homologous or native cell lines where the receptors are most stable and native G-proteins and other accessory proteins are available. In these cases, minute quantities of the receptors present are sufficient for routine ligand-binding assays, competition studies, and pharmacological assessments of receptor agonists and antagonists. As the study of GPCRs continues to move toward gaining additional biochemical and three-dimensional structural information, particularly as it relates to drug discovery, heterologous expression, solubilization, and purification of milligram quantities of receptor will be required. In addition, new methods for characterizing these now membrane-free receptors in the detergent-soluble state will likely need to be developed as the presence of detergents and the behavior of the solubilized receptor itself may present new challenges to traditional biochemical analyses.

The first step in membrane protein purification – solubilization of membrane proteins from their lipid environment – requires detergent that separates the membrane protein from the membrane lipid bilayer by forming soluble protein–detergent micelles. The solubilized membrane protein is then ready for subsequent purification followed by biochemical or three-dimensional structural analysis. The efficiency with which the detergent can solubilize a membrane protein varies and relies on a number of parameters, including detergent choice and buffer components. A number of GPCRs have been successfully solubilized and subsequent purifications of the solubilized receptors to varying degrees have been reported [50, 125–129]. Usually, the solubilization efficiency is determined qualitatively, most often by Western blot. Less frequent are cases where ligand-binding assays are employed to select for optimal solubilization conditions, even though ligand binding presents a clear demonstration of the structural integrity of the solubilized receptor. However, ligand-binding assays used in drug discovery generally use a radioligand [130, 131], typical for membrane-bound receptors. These assays often fail for solubilized receptors either due to the lack of retention of the solubilized ligand-bound receptor on the filter, the presence of a high background signal from nonspecific binding of radioligand–detergent micelles to the filter, or inactivation of the receptor by the detergent [47, 128]. Attempts to solve this problem have led to the development of some promising new methods, including the use of surface plasmon resonance that utilizes immobilized solubilized receptors (or peptide ligands) for ligand-binding measurements [47, 132–134]. The requirement of specialized equipment, conformationally sensitive antibodies for the receptor of interest, and the current limit of working with peptide substrates, however, may make these methods challenging. In contrast, simple modifications to known ligand-binding methods can greatly reduce or eliminate nonspecific binding problems for membrane-bound receptors [135, 136], leaving the possibility that similar modifications would allow reliable measurement of ligand binding for detergent solubilized receptors.

In the cases where GPCRs have been solubilized and purified, the results of pharmacological comparisons between the isolated and the membrane-bound receptor are mixed. In some cases, the receptor isolated is a fusion protein [125, 127, 137] and the activity of the purified receptor either differs from the native membrane-bound receptor [125] or is simply not reported [127, 137]. While it has been shown that isolated receptors can be reconstituted into the lipid environment and have similar binding affinities compared to membrane-bound receptors [128], there are few examples of this result with soluble, nonreconstituted receptors.

In order to establish the utility of some of the recent progress in the heterologous expression, solubilization, and characterization of recombinant GPCRs, particularly as they relate to targets of pharmaceutical interest, the solubilization and pharmacological characterization of the histamine H_3 receptor is presented. The functional role of the histamine H_3 receptor has made it an attractive target for the pharmaceutical industry for obesity and a number of cognitive disorders [138–140]. Therapeutic interest in this receptor has led to the identification of a number of antagonists and inverse agonists for the receptor [141, 142]. These compounds have been instrumental in enumerating the varying pharmacological profiles not only for the known mammalian receptors, but for the different human isoforms as well [143–145]. The availability of these compounds and the robust ligand-binding assay for the membrane-bound receptor [146] make it an ideal choice for comparing the pharmacological profile of the detergent-soluble and membrane-bound receptor.

In addition to the pharmacological characterization of heterologously expressed human histamine H_3 receptor in insect cells, we also confirm the integrity of the detergent-solubilized GPCRs. The validity of a modified glass fiber filter-binding assay for the soluble histamine H_3 receptor is demonstrated along with data showing that ligand-binding activity is maintained in conditions optimal for solubilization. In addition, it is suggested that the structural integrity of the solubilized receptor is intact based on the results of a pharmacological comparison of the soluble and membrane-bound histamine H_3 receptors.

Ligand-binding activity measurements, Western blot, and mass spectrometry were utilized to confirm the heterologous expression of functional histamine H_3 receptor in membranes derived from cultured insect cells that had been infected with a recombinant baculovirus containing the human histamine H_3 receptor gene. For the initial expression analysis, two constructs differing in their N-terminal signal sequence were expressed in Sf9 and High Five insect cells. Consistent with some of the expression profiles reported by Akermoun *et al.* [29], maximal expression of the histamine H_3 receptor was observed 48 h after baculovirus infection. The presence of a signal sequence has been shown to maximize presentation of heterologously expressed receptors in cell membranes [85]. A representative construct including the honeybee melittin signal sequence is shown in Figure 4.2(a). The sequence of the histamine H_3 construct is shown in Figure 4.2(b). In general, similar expression levels were observed utilizing either construct in either cell type. In High Five cells, saturation binding studies using [^{3}H]-(*N*)-α-methylhistamine (representative saturation curve in Figure 4.3) indicated that the receptor expressed with an apparent B_{max} of 2176 ± 129 fmol/mg and with a K_D value for (*N*)-α-methylhistamine binding of 1.82 ± 0.28 nM. This expression level is greater than that found in HGT1 cells (B_{max} = 54 ± 3 fmol/mg) [148], and in stable clonal lines of both human embryonic kidney (HEK) and C6 rat glioma cells whose B_{max} values are 1573 and 1277 fmol/mg, respectively. Comparable B_{max} values were obtained for the receptor in Sf9 cells; however, the measured K_D values for (*N*)-α-methylhistamine were somewhat elevated (3.41 ± 0.37 nM). For this reason and the larger cell mass obtained using High Five cells, the use of Sf9 cells was discontinued. Since there was no obvious discrimination in B_{max} between the leader sequences in our initial expression trials, the melittin signal sequence construct was used in all subsequent expressions. To assess the G-protein activity of these insect cell receptors, the similarity of GTPγS binding by histamine H_3 agonists and inverse agonists to known mammalian cell constructs was confirmed (Figure 4.4).

The expressed histamine H_3 receptor is found in the membrane fraction of the insect cell lysate (Figure 4.5a). The Western blot identifies the receptor based on recognition of the C-terminal hexahistidine (His_6) tag with the correct molecular weight (Figure 4.5a). Migration on the sodium dodecylsulfate (SDS) gel indicates a molecular weight of approximately 50 kDa, which is close to the theoretical molecular weight of the complete construct of approximately 54 kDa. To confirm the results of the Western blot, gel bands from the SDS gel (bracketing the 50-kDa marker, Figure 4.5b) were excised and processed for in-gel trypsin digest followed by liquid chromatography/tandem mass spectrometry as described above. Protein database search results showed that eight peptide fragments from the histamine H_3 receptor (Table 4.2) were identified in the gel near the 50-kDa molecular weight marker (Figure 4.5). The third cytoplasmic loop, based on secondary structure predictions, consists of 142 amino acids from residue 257 to 398. The peptides found covered 53% of that loop (see Figure 4.2b). The first predicted transmembrane helix starts at residue 79 (see Figure 4.2b), which is 40 amino acids from the N-terminal methionine of the receptor. The fifth peptide described in Table 4.2 represents 67% coverage of the N-terminus of the expressed receptor. The last peptide in Table 4.2 represents a nonnative translated region (a result of cloning) of the gene that links the tobacco etch virus (TEV)

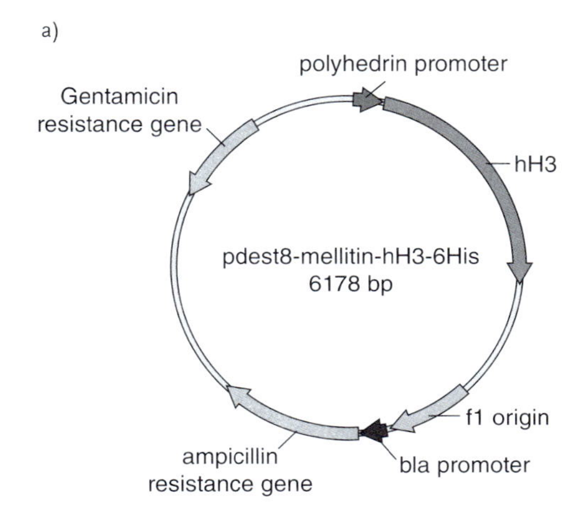

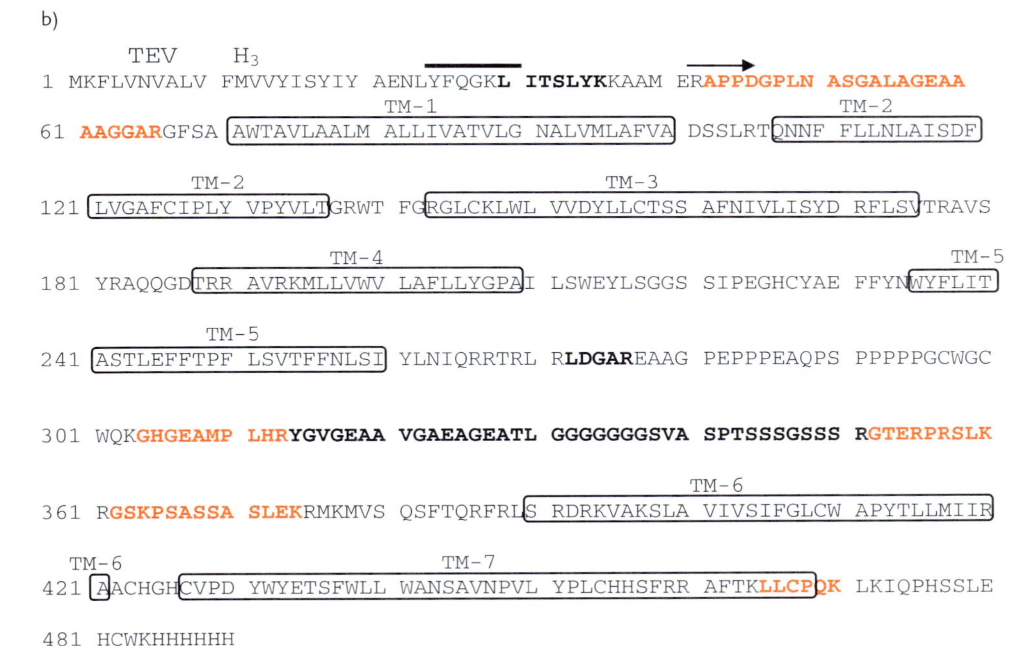

Figure 4.2 (a) Graphical representation of the pDEST™8 destination vector used for the baculovirus-infected insect cell expression of the histamine H_3 receptor. The construct contained the full-length histamine H_3 receptor gene (hH_3 [147]) with a N-terminal melittin signal sequence followed by a TEV protease site. The C-terminal portion of the expressed gene includes a His_6 tag. (b) Single amino acid protein sequence of the translated gene. Color coded in red and black are the peptides identified by liquid chromatography/mass spectroscopy analysis (Table 4.1) derived from the expressed gene product. The predicted transmembrane helices are boxed and labeled, and an arrow and a line indicate the start of the H_3 receptor sequence and the TEV protease site, respectively.

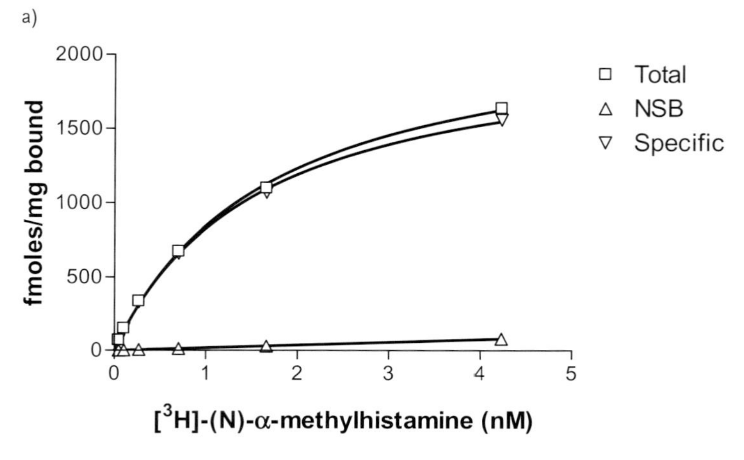

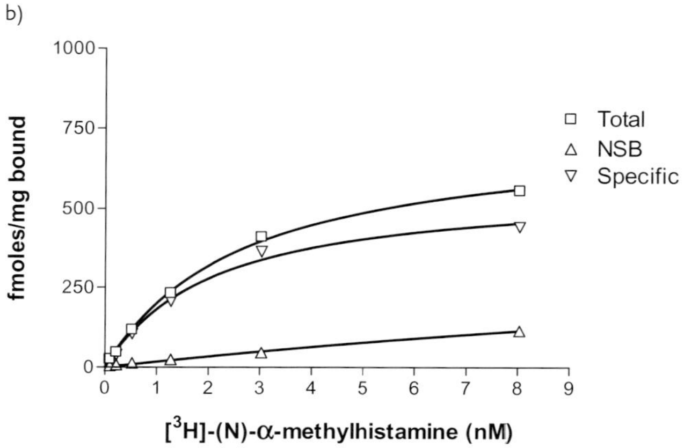

Figure 4.3 (a) [^{3}H]-(*N*)-α-Methylhistamine bound to untreated High Five insect membranes containing the histamine H_3 receptor. (b) [^{3}H]-(*N*)-α-Methylhistamine binding to H_3 receptor solubilized from High Five membranes using 0.1% (w/v) DDM. A representative example of three individual experiments is shown.

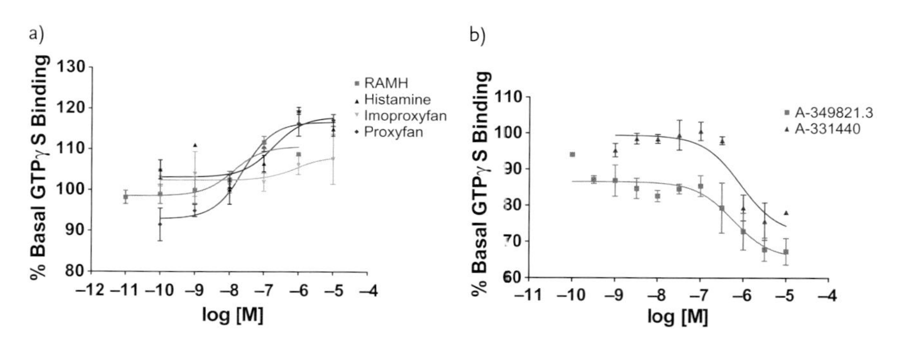

Figure 4.4 (a) Modulation of GTPγS binding by H_3 agonists utilizing High Five membranes expressing the human H_3 receptor. (b) Modulation of GTPγS binding by H_3 inverse agonists utilizing High Five membranes expressing the human H_3 receptor.

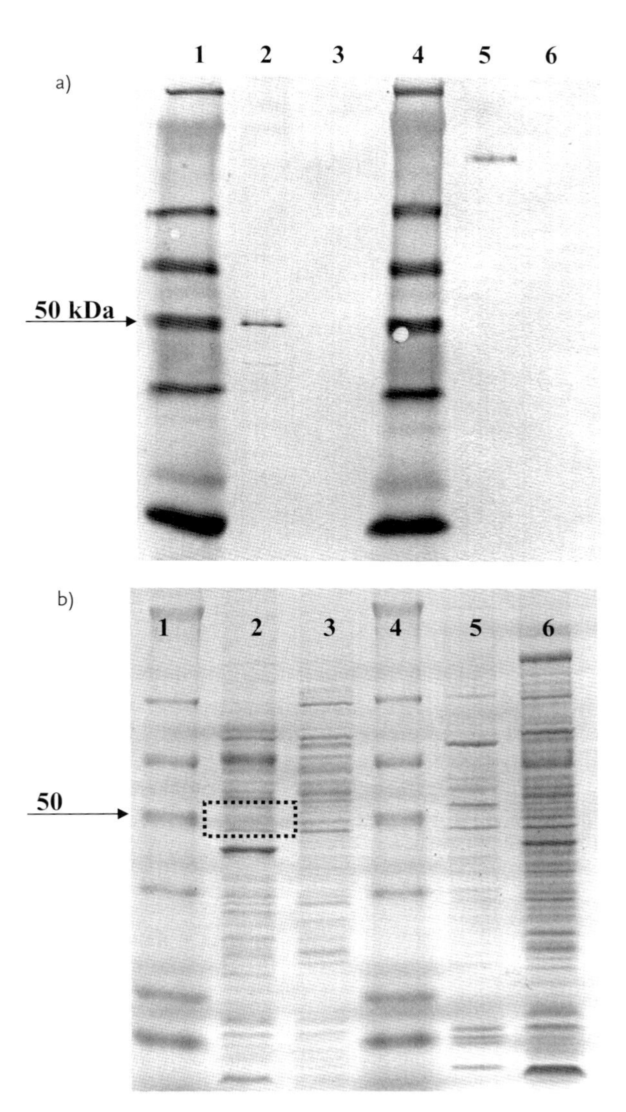

Figure 4.5 (a) Western blot and (b) SDS gel stained with Coomassie blue of the membrane (lane 2) and soluble (lane 3) fractions of High Five cell lysates infected with baculovirus containing the histamine H_3 receptor gene. Lane 5 contains an insect cell lysate known to have expressed a membrane protein with a C-terminal His_6 tag. Lane 6 is a total cell lysate of High Five cells known to not have expressed any histidine-tagged proteins. Identical samples shown in the SDS gel in (b) were used to generate the Western blot (a). The highlighted region on the SDS gel in lane 1 indicates the area where protein bands were excised for in-gel trypsin digestion followed by liquid chromatography/ mass spectroscopy analysis.

Table 4.2 Tryptic digests of human histamine H_3 receptor isolated from High Five expression.

Peptide identified	Molecular weight (calculated) Da	Molecular weight (experimental) Da	Color (Figure 4.2b)	Location on gene (construct numbering)
GSKPSASSASLEK	1247.64	1247.54	red	third cytoplasmic loop (residues 362–374)
YGVGEAAVGAEAGEATLGGGGGG GSVASPTSSSGSSSR	3242.29	3242.92	black	third cytoplasmic loop (residues 314–351)
GHGEAM*PLHR	1120.24	1120.23	red	third cytoplasmic loop (residues 304–313)
LLC*PQK	757.42	757.36	black	variable splice region (residues 465–470)
APPDGPLNASGALAGEAAAAGGAR	2062.20	2062.39	red	N-terminus (residues 43–66)
LDGAR	530.28	530.22	black	third cytoplasmic loop (residues 272–276)
GTERPRSLK	1042.59	1042.48	red	third cytoplasmic loop (residues 352–360)
LITSLYK	836.5	836.38	black	C-terminal to TEV site (residues 30–36)

The amino acid sequences of peptides derived from liquid chromatography/mass spectroscopy analysis of an in-gel trypsin digest of protein bands removed from an SDS PAGE containing the membrane fraction from an insect cell lysate expressing the histamine H_3 receptor. The table lists both measured and theoretical molecular weights for each peptide. For clarity, the peptides are color-coded alternatively red and black in Figure 2(b). The gene location of each peptide refers to the expressed sequence shown in Figure 2(b). * indicates skip in sequence information.

protease site (residues 22–28) to the histamine H_3 receptor. Assuming the signal sequence and TEV protease site are cleaved during *in vivo* processing, the theoretical molecular weight of the remaining sequence would be approximately 51 kDa. The total coverage of the sequence for all the peptides is 24% (117/490 possible residues). The combination of these results indicates the unambiguous presence of functional histamine H_3 receptor in the High Five membranes.

4.2.3.1 Solubilization of the Histamine H_3 Receptor

Solubilization experiments of the histamine H_3 receptor were conducted in the following five detergents: *n*-dodecyl-β-D-maltopyranoside (DDM), sucrose monododecanoate (SMD), fos-choline 12 (F-12), *n*-octyl-β-D-glucopyranoside (NOG), and dodecyl (lauryl) dimethyl amine oxide (LDAO). For appropriate comparison of the various detergents, all detergents were evaluated at 10X their respective critical micelle concentration values in water. For each of the five experiments, the samples were analyzed for percent total protein yield (Figure 4.6a) and remaining [^{3}H]-(*N*)-α-methylhistamine-binding activity (percent total ligand binding, Figure 4.6b). The results indicate that the detergents solubilize between 25 and 45% of the total protein in the membrane samples (Figure 4.6a), with LDAO solubilizing the most protein. Although LDAO was the best detergent for protein solubilization, the resulting soluble fraction had no measurable activity. However, greater than 60% of the total binding was maintained in the soluble fraction when DDM was used for solubilization (Figure 4.6b). The soluble histamine H_3 receptor was further characterized by saturation binding studies using [^{3}H]-(*N*)-α-methylhistamine (Figure 4.3b) in a buffer containing 0.01% (w/v) DDM. An average B_{max} of 532 ± 0.46 fmol/mg and an average K_D of 2.40 ± 0.51 nM were observed utilizing data obtained from three individual experiments.

4.2.3.2 Assay Validation

Typical filter-binding assays used to measure the ligand-binding activity of receptors often fail in the presence of detergents [47], complicating the characterization of detergent-solubilized receptors. To overcome this obstacle, gel filtration methods are often used to determine ligand-binding properties of detergent-solubilized receptors [125]. However, gel filtration methods are laborious and time-consuming. To validate the activity assays described and the observed binding properties for soluble histamine H_3 receptor (using DDM or digitonin), the activity of detergent-soluble histamine H_3 receptor, as measured by net disintegrations per minute (DPM), determined using the modified filter-binding assay was compared to the more laborious but well-characterized gel filtration assay (Figure 4.7a) to separate bound from free ligand. A comparison of the resulting activities measured for samples solubilized by identical methods and incubated with identical substrates is shown in Figure 4.7b. The *y*-axis indicates the net DPM (obtained by subtracting nonspecific binding from total binding) observed by each technique. Figure 4.7b shows there are no significant differences between the two methods when measuring ligand-binding activity for the detergent-soluble histamine H_3 receptor, indicating that the modified filter-binding assay is capable of accurately detecting receptor-bound radioligand.

a)

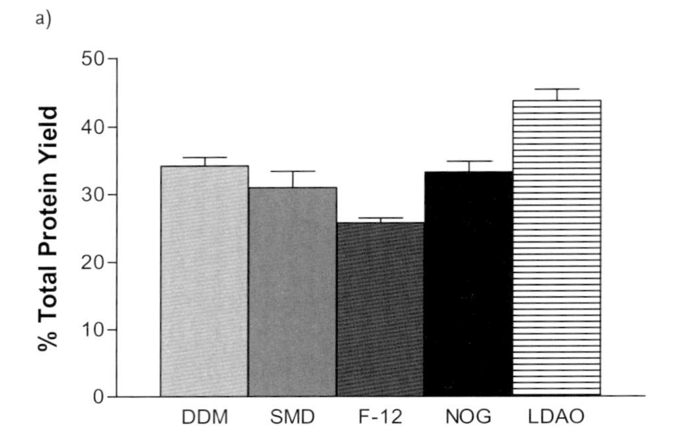

b)

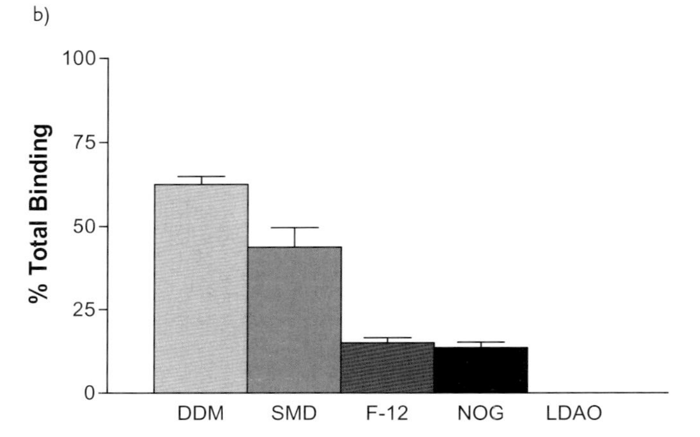

Figure 4.6 (a) Total percent protein yield and (b) total percent binding of High Five membranes containing the histamine H_3 receptor after solubilization with indicated detergents. Total and solubilized protein determinations were measured utilizing the BCA protein assay. Graphs show the average of four experiments, with a standard deviation for $n = 4$.

4.2.3.3 **Competition Analysis of Solubilized versus Membrane-Bound Receptor**

Competition binding studies using a panel of histaminergic ligands were conducted in order to evaluate their ability to displace (*N*)-α-methylhistamine from membrane-bound (Figure 4.8a) and detergent (DDM)-soluble histamine H_3 receptor (Figure 4.8b). For competition binding assays, [^{3}H]-(*N*)-α-methylhistamine (approximately 1.5 nM) was incubated at 25 °C for 30 min with the intact or solubilized membranes (0.5 ml reactions in 50 mM Tris, pH 7.4, 5 mM EDTA) either alone or in the presence of competing histaminergic ligands. Solubilized receptor reaction mixtures contained 0.01% (w/v) DDM. Nonspecific binding was defined with addition of thioperamide (10 μM). Radioligand binding was terminated and bound ligand was separated from free ligand utilizing vacuum filtration onto 0.3% (v/v) presoaked polyethylenimine GF/B filters, followed by thorough rinsing with

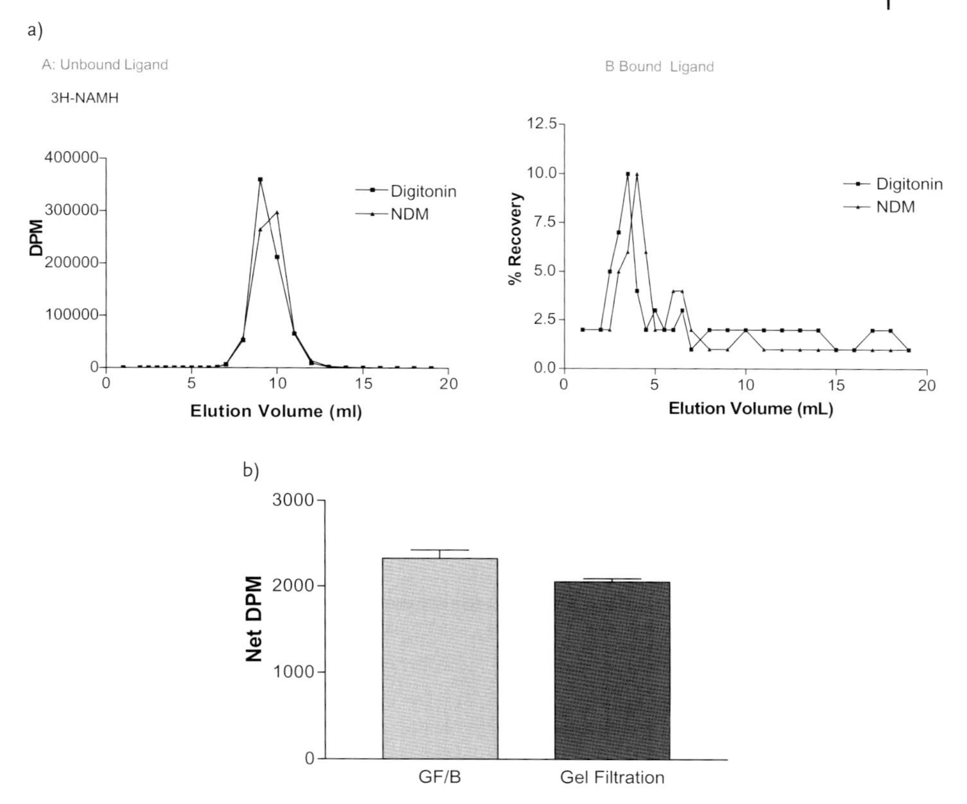

Figure 4.7 (a) Elution profiles of the free ligands and ligands bound to the detergent-solubilized protein. Detection of ^{3}H ligand was used. (b) Detergent-soluble (DDM) H_3 receptor was incubated with [^{3}H]-(N)-α-methylhistamine and binding was measured by either the filtration assay or by using Bio-Gel P-6DG polyacrylamide gel filtration as described in Section 4.2. Nonspecific binding was determined utilizing 10 μM thioperamide. Bar graphs represent specific binding for a single experiment performed in triplicate. The experiment was conducted at least three independent times with similar results.

50 mM Tris buffer (pH 7.4). GF/B filters used for intact membranes were soaked in 0.3% (v/v) polyethylenimine for 5 min; however, additional soaking time was required for their use with solubilized membranes (6 h soaking time). Bound radiolabel was determined by liquid scintillation counting utilizing the Packard Top Counter. Experiments were run in triplicate and data analysis performed [149] with pK_i values determined utilizing the Cheng–Prusoff equation [150]

The data show that the membrane-bound H_3 receptor expressed in insect cells and in C6 rat glioma cells show the same rank order of potencies for the compounds tested (Table 4.3), both for the agonists (imetit > (R)-α-methylhistamine > histamine) and for the antagonists (A-349821 > A-358239 > thioperamide). For the detergent-solubilized receptor, rank order of potencies is equivalent to that observed with the

a)

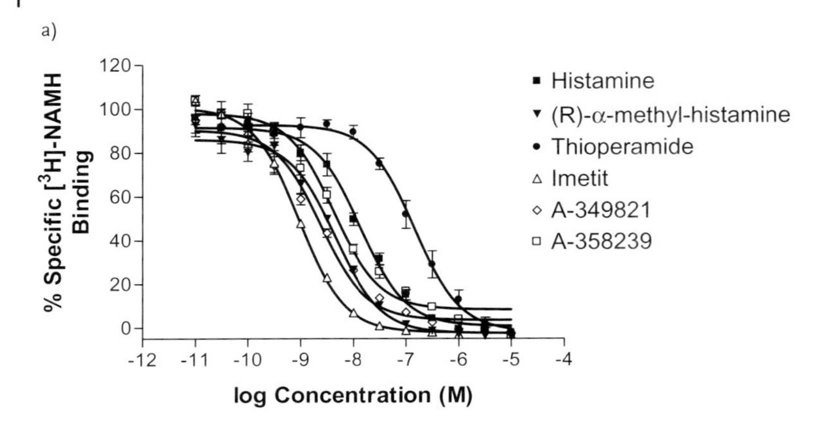

b)

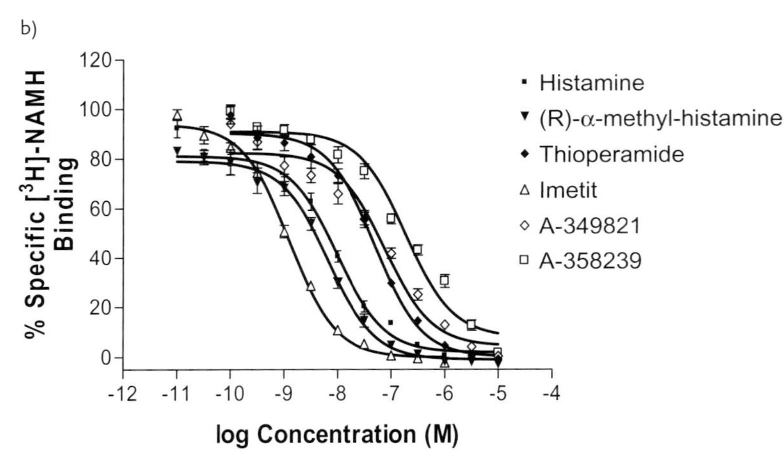

Figure 4.8 Competition of specific [^{3}H]-(*N*)-α-methylhistamine binding for membrane-bound (a, graph representative of $n = 4$) and 0.1% DDM-solubilized (b, graph representative of $n = 3$) H_3 receptor. Values for pK_i (nM) were determined from these graphs and are tabulated for comparison in Table 4.3.

membrane-bound receptors for the agonists. However, rank order of potencies for the antagonists differs from that observed for the membrane-bound receptors as thioperamide is essentially equivalent to A-349821 and A-358239 is the least potent.

4.3 Conclusions and Future Perspectives

The Bac to Bac method has been used to generate recombinant baculovirus containing the human histamine H_3 receptor. The heterologous expression via baculovirus infection of insect cells has generated histamine H_3 receptor that is functional

Table 4.3 Measured pK_i values of histaminergic ligands.

Ligand	Structure	Mean pK_i ± SEM (nM)		
		Insect cells	Detergent-solubilized	Mammalian cells
Histamine		8.34 ± 0.07	8.81 ± 0.03	8.56 ± 0.14 [145]
(*R*)-α-Methylhistamine		8.99 ± 0.07	9.03 ± 0.23	9.16 ± 0.08 [149]
Imetit		9.31 ± 0.07	9.26 ± 0.12	9.59 ± 0.09 [149]
Thioperamide		7.39 ± 0.19	7.88 ± 0.09	7.18 ± 0.06 [145]
A-349821		8.98 ± 0.12	7.78 ± 0.03	9.39 ± 0.82 [151]
A-358239		8.46 ± 0.01	6.99 ± 0.07	9.35 ± 0.04 [150]

The measured pK_i (nM) values for several histaminergic compounds determined for both detergent-solubilized and membrane-bound histamine H_3 receptor as described in Section 4.2. The values for different mammalian cell lines (C6 glioma cells; rat [145], human [149–151]) were previously published and are referenced accordingly. The experimental data used to generate these values are shown in Figure 4.8 (a and b).

in terms of ligand binding, rank-order competition with agonist and antagonist molecules, and G-protein coupling. Growth for 48 h was checked for expression via cell viability using the Trypan blue dye exclusion method and cell diameter size measurements. The expressed receptor was successfully solubilized under conditions that maintained ligand-binding activity, as shown by a validated filtration assay using [^{3}H]-(N)-α-methylhistamine – a known H_3 receptor radioligand. The current studies demonstrate that the solubilized histamine H_3 receptor maintains its structural integrity as evidenced by the results of a pharmacological comparison of histaminergic ligands against the soluble and membrane-bound receptors.

4.3.1
Executive Summary

- Baculovirus infection of insect cells can generate active recombinant GPCRs with post-translational modifications such as phosphorylation, glycosylation, and palmitoylation
- Recombinant baculovirus can be generated quickly, allowing a donor plasmid with the gene of interest that is cloned downstream of a baculovirus promoter to participate in site-specific transposition into a baculovirus DNA present in a plasmid or in the *E. coli* chromosome.
- Optimization of insect cell codon usage around the translation start site, and addition of signal sequences to the gene of interest can increase protein production levels. The addition of peptide tags (FLAG, His_6, HA, c-Myc) or protein domains such as GST and GFP can be useful tools to monitor protein expression and can be used as tools for purification.
- Monitoring cell viability and cell diameter size after baculovirus infection can serve as markers of the success of protein expression. Typically, most protein expression is best between 40 and 60 h postinfection.
- The addition of ligands, lipids, and protease inhibitors can increase protein stability and decrease protein degradation after baculovirus expression.
- Washing cell paste with protein-stability agents can protect protein viability during the freezing process.
- Assessment of ligand binding, rank-order competition using agonist and antagonist molecules, and G-protein activation are important tools to confirm the integrity of detergent-solubilized GPCRs.

4.3.2
Future Perspectives

Baculovirus-infected insect cells are useful tools for the production of GPCRs and other membrane proteins. Over the last few years, the addition of ligands and protease inhibitors during cell growth and purification has led to the emergence of new crystal structures. Nonetheless, there are still problems with saturation of the host cell protein-production machinery that can result in having proteins made with different levels of glycosylation, phosphorylation, palmitoylation, and proteolysis. This latter concern can result in clipped protein being produced not only in the plasma membrane, but also in the membranes of the Golgi and endoplasmic reticulum. Further advances with insect cell engineering to minimize proteolysis with controlled post-translational modification are being developed. In addition, the process to generate recombinant baculovirus still takes longer than that of bacterial and yeast cell production. Continued efforts to make the process faster are under way. If achievable, the resulting process to generate heterologous GPCRs and other membrane proteins for structural biology would be a powerful tool for the interpretation of pharmacological data *in vivo*.

Abbreviations

*Ac*MNPV	*Autographa californica* nuclear polyhedrosis virus
ATCC	American Type Culture Collection
BIIC	baculovirus-infected insect cell
DDM	*n*-dodecyl-β-D-maltopyranoside
DPM	disintegrations per minute
F-12	fos-choline 12
GFP	Green Fluorescent Protein
GPCR	G-protein-coupled receptor
GST	glutathione S-transferase
HEK	human embryonic kidney
LDAO	dodecyl (lauryl) dimethyl amine oxide
MOI	multiplicity of infection
NOG	*n*-octyl-β-D-glucopyranoside
PFU	plaque-forming unit
SDS	sodium dodecylsulfate
SMD	sucrose monododecanoate
TEV	tobacco etch virus
TIPS	titerless infected cell preservation and scale-up

References

1 Marinissen, M.J. and Gutkind, J.S. (2001) G protein-coupled receptors and signalling networks: emerging paradigms. *Trends Pharmacol. Sci.*, **22**, 368–376.

2 Cooper, M.A. (2004) Advances in membrane receptor screening and analysis. *J. Mol. Recognit.*, **17**, 286–315.

3 Lefkowitz, R.J. (2000) The superfamily of heptahelical receptors. *Nat. Cell Biol.*, **2**, E133–E136.

4 Sautel, M. and Milligan, G. (2000) Molecular manipulation of G protein coupled receptors: a new avenue into drug discovery. *Curr. Med. Chem.*, **7**, 889–896.

5 Chalmers, D.T. and Behan, D.P. (2002) The use of constitutively active GPCRs in drug discovery and functional genomics. *Nat. Rev. Drug Discov.*, **1**, 599–608.

6 Drews, J. (2000) Drug discovery. *Science*, **287**, 1960–1964.

7 Bywater, R.P. (2005) Location and nature of the residues important for ligand recognition in G-protein coupled receptors. *J. Mol. Recognit.*, **18**, 60–72.

8 Horn, F., *et al.* (2003) GPCRDB information system for G protein-coupled receptors. *Nucleic Acids Res.*, **31**, 294–297.

9 Civelli, O., Reinscheid, R.K., and Nothacker, H.P. (1999) Orphan receptors, novel neuropeptides and reverse pharmaceutical research. *Brain Res.*, **848**, 63–65.

10 Scussa, F. (2002) World's best selling drugs. *Med. Ad. News*, **21** (**5**), 1–46.

11 Behan, D.P. and Chalmers, D.T. (2001) The use of constitutively active receptors for drug discovery at the G protein-coupled receptor gene pool. *Curr. Opin. Drug Discov. Devel.*, **4**, 548–560.

12 Kenakin, T. (2005) New concepts in drug discovery: collateral efficacy and permissive antagonism. *Nat. Rev. Drug Discov.*, **4**, 919–927.

13 Ellis, C.R. (2004) State of GPCR research in 2004. *Nat. Rev. Drug Discov.*, **3**, 577–626.

14 Ghanouni, P., *et al.* (2001) Functionally different agonists induce distinct conformations in the G protein coupling domain of the beta 2 adrenergic receptor. *J. Biol. Chem.*, **276**, 24433–24436.

15 Vauquelin, G. and Van Liefde, I. (2005) G protein-coupled receptors: a count of 1001 conformations. *Fundam. Clin. Pharmacol.*, **19**, 45–56.

16 Ulloa-Aguirre, A., *et al.* (1999) Structure–activity relationships of G-protein coupled receptors. *Arch. Med. Res.*, **30**, 420–435.

17 Sarramegna, V., *et al.* (2006) Recombinant G protein-coupled receptors from expression to renaturation: a challenge towards structure. *Cell. Mol. Life Sci.*, **63**, 1149–1164.

18 Opekarova, M. and Tanner, W. (2003) Specific lipid requirements of membrane proteins – a putative bottleneck in heterologous expression. *Biochim. Biophys. Acta*, **1610**, 11–22.

19 Sarramegna, V., *et al.* (2003) Heterologous expression of G-protein-coupled receptors: comparison of expression systems from the standpoint of large-scale production and purification. *Cell. Mol. Life Sci.*, **60**, 1529–1546.

20 Chiu, M.L., *et al.* (2008) Over-expression, solubilization, and purification of G protein-coupled receptors for structural biology. *Comb. Chem. High Throughput Screen.*, **11**, 439–462.

21 Junge, F., *et al.* (2010) Modulation of G-protein coupled receptor sample quality by modified cell-free expression protocols: a case study of the human endothelin A receptor. *J. Struct. Biol.*, **172**, 94–106.11

22 Junge, F., *et al.* (2008) Large-scale production of functional membrane proteins. *Cell. Mol. Life Sci.*, **65**, 1729–1755.

23 Mouillac, B., *et al.* (1992) Agonist-modulated palmitoylation of beta 2-adrenergic receptor in Sf9 cells. *J. Biol. Chem.*, **267**, 21733–21737.

24 Bouvier, M., *et al.* (1995) Dynamic palmitoylation of G-protein-coupled receptors in eukaryotic cells. *Lipid Modifications Proteins*, **250**, 300–314.

25 King, L.A. and Possee, R.D. (1992) *The Baculovirus Expression System: A Laboratory Guide*, Chapman & Hall, London.

26 Griffiths, C.M. and Page, M.J. (1997) Production of heterologous proteins using the baculovirus/insect expression system. *Methods Mol. Biol.*, **75**, 427–440.

27 Maeda, S. (1989) Expression of foreign genes in insects using baculovirus vectors. *Annu. Rev. Entomol.*, **34**, 351–372.

28 Ruan, K.H., Cervantes, V., and Wu, J. (2008) A simple, quick, and high-yield preparation of the human thromboxane A_2 receptor in full size for structural studies. *Biochemistry*, **47**, 6819–6826.

29 Akermoun, M., *et al.* (2005) Characterization of 16 human G protein-coupled receptors expressed in

baculovirus-infected insect cells. *Protein Expr. Purif.*, **44**, 65–74.

30 Massotte, D. (2003) G protein-coupled receptor overexpression with the baculovirus-insect cell system: a tool for structural and functional studies. *Biochim. Biophys. Acta*, **1610**, 77–89.

31 Parker, E.M., *et al.* (1991) Reconstitutively active G protein-coupled receptors purified from baculovirus-infected insect cells. *J. Biol. Chem.*, **266**, 519–527.

32 Brillet, K., *et al.* (2008) Using EGFP fusions to monitor the functional expression of GPCRs in the *Drosophila* Schneider 2 cells. *Cytotechnology*, **57**, 101–109.

33 Perret, B.G., *et al.* (2003) Expression of EGFP-amino-tagged human mu opioid receptor in *Drosophila* Schneider 2 cells: a potential expression system for large-scale production of G-protein coupled receptors. *Protein Expr. Purif.*, **31**, 123–132.

34 Dong, G.Z., *et al.* (1995) Ligand binding properties of muscarinic acetylcholine receptor subtypes (m1–m5) expressed in baculovirus-infected insect cells. *J. Pharmacol. Exp. Ther.*, **274**, 378–384.

35 Hayashi, M.K. and Haga, T. (1996) Purification and functional reconstitution with GTP-binding regulatory proteins of hexahistidine-tagged muscarinic acetylcholine receptors (m2 subtype). *J. Biochem.*, **120**, 1232–1238.

36 Mazina, K.E., *et al.* (1996) Purification and reconstitution of a recombinant human neurokinin-1 receptor. *J. Recept. Signal Transduct. Res.*, **16**, 191–207.

37 Obermeier, H., Wehmeyer, A., and Schulz, R. (1996) Expression of mu-, delta- and kappa-opioid receptors in baculovirus-infected insect cells. *Eur. J. Pharmacol.*, **318**, 161–166.

38 Doi, T., *et al.* (1997) Characterization of human endothelin B receptor and mutant receptors expressed in insect cells. *Eur. J. Biochem.*, **248**, 139–148.

39 Massotte, D., *et al.* (1997) Characterization of delta, kappa, and mu human opioid receptors overexpressed in baculovirus infected insect cells. *J. Biol. Chem.*, **272**, 19987–19992.

40 Massotte, D., *et al.* (1999) Parameters influencing human mu opioid receptor over-expression in baculovirus-infected insect cells. *J. Biotechnol.*, **69**, 39–45.

41 Klaassen, C.H., *et al.* (1999) Large-scale production and purification of functional recombinant bovine rhodopsin with the use of the baculovirus expression system. *Biochem. J.*, **342**, 293–300.

42 Schiller, H., *et al.* (2001) Solubilization and purification of the human ETB endothelin receptor produced by high-level fermentation in *Pichia pastoris*. *Receptors Channels*, **7**, 453–469.

43 Furukawa, H., *et al.* (2002) Conformation of ligands bound to the muscarinic acetylcholine receptor. *Mol. Pharmacol.*, **62**, 778–787.

44 Bodor, E.T., *et al.* (2003) Purification and functional reconstitution of the human P2Y12 receptor. *Mol. Pharmacol.*, **64**, 1210–1216.

45 Gille, A. and Seifert, R. (2003) Co-expression of the β_2-adrenoceptor and dopamine D_1-receptor with $G_{s\alpha}$ proteins in Sf9 insect cells: limitations in comparison with fusion proteins. *Biochim. Biophys. Acta*, **1613**, 101–114.

46 Sen, S., *et al.* (2003) Functional expression and direct visualization of the human alpha 2B-adrenergic receptor and alpha 2B-AR–green fluorescent fusion protein in mammalian cell using Semliki Forest virus vectors. *Protein Expr. Purif.*, **32**, 265–275.

47 Sen, S., *et al.* (2005) Functional studies with membrane-bound and detergent-solubilized alpha2-adrenergic receptors expressed in Sf9 cells. *Biochim. Biophys. Acta*, **1712**, 62–70.

48 Stanasila, L., Pattus, F., and Massotte, D. (1998) Heterologous expression of G-protein coupled receptors. *Biochemie*, **80**, 563–571.

49 Ohtaki, T., *et al.* (1998) Expression, purification, and reconstitution of receptor for pituitary adenylate cyclase-activating polypeptide. Large-scale purification of a functionally active G protein-coupled receptor produced in Sf9 insect cells. *J. Biol. Chem.*, **273**, 15464–15473.

50 Asmann, Y.W., Dong, M., and Miller, L.J. (2004) Functional characterization and purification of the secretin receptor expressed in baculovirus-infected insect cells. *Regul. Pept.*, **123**, 217–223.

51 Panneels, V., *et al.* (2003) Pharmacological characterization and immunoaffinity purification of metabotropic glutamate receptor from *Drosophila* overexpressed in Sf9 cells. *Protein Expr. Purif.*, **30**, 275–282.

52 Aldecoa, A., *et al.* (2000) Mammalian calcitonin receptor-like receptor/receptor activity modifying protein complexes define calcitonin gene-related peptide and adrenomedullin receptors in *Drosophila* Schneider 2 cells. *FEBS Lett.*, **471**, 156–160.

53 Nekrasova, E., *et al.* (1996) Overexpression, solubilization and purification of rat and human olfactory receptors. *Eur. J. Biochem.*, **238**, 28–37.

54 Cherezov, V., *et al.* (2007) High-resolution crystal structure of an engineered human beta2-adrenergic G protein-coupled receptor. *Science*, **318**, 1258–1265.

55 Rosenbaum, D.M., *et al.* (2007) GPCR engineering yields high-resolution structural insights into beta2-adrenergic receptor function. *Science*, **318**, 1266–1273.

56 Rasmussen, S.G., *et al.* (2007) Crystal structure of the human beta2 adrenergic G-protein-coupled receptor. *Nature*, **450**, 383–387.

57 Warne, T., *et al.* (2008) Structure of a beta1-adrenergic G-protein-coupled receptor. *Nature*, **454**, 486–491.

58 Jaakola, V.P., *et al.* (2008) The 2.6 angstrom crystal structure of a human A2A adenosine receptor bound to an antagonist. *Science*, **322**, 1211–1217.

59 Sobolevsky, A.I., Rosconi, M.P., and Gouaux, E. (2009) X-ray structure, symmetry and mechanism of an AMPA-subtype glutamate receptor. *Nature*, **462**, 745–756.

60 Gonzales, E.B., Kawate, T., and Gouaux, E. (2009) Pore architecture and ion sites in acid-sensing ion channels and P2X receptors. *Nature*, **460**, 599–604.

61 Jasti, J., *et al.* (2007) Structure of acid-sensing ion channel 1 at 1.9 A resolution and low pH. *Nature*, **449**, 316–323.

62 Armstrong, N., *et al.* (2006) Measurement of conformational changes accompanying desensitization in an ionotropic glutamate receptor. *Cell*, **127**, 85–97.

63 Kawate, T., *et al.* (2009) Crystal structure of the ATP-gated $P2X_4$ ion channel in the closed state. *Nature*, **460**, 592–598.

64 Kurumbail, R.G., *et al.* (1996) Structural basis for selective inhibition of cyclooxygenase-2 by anti-inflammatory agents. *Nature*, **384**, 644–648.

65 Miller, L.K. (1997) *The Baculoviruses*, Plenum Press, New York.

66 Pennock, G.D., Shoemaker, C., and Miller, L.K. (1984) Strong and regulated expression of *Escherichia coli* beta-galactosidase in insect cells with a baculovirus vector. *Mol. Cell. Biol.*, **4**, 399–406.

67 Summers, M.D. (2006) Milestones leading to the genetic engineering of baculoviruses as expression vector systems and viral pesticides. *Adv. Virus Res.*, **68**, 3–73.

68 Jarvis, D.L. (2009) Baculovirus-insect cell expression systems. *Methods Enzymol.*, **463**, 191–222.

69 Invitrogen (2009) *Guide to Baculovirus Expression Vector Systems (BEVS) and Insect Cell Culture Techniques*, Invitrogen, Carlsbad, CA.

70 Jarvis, D.L., Weinkauf, C., and Guarino, L.A. (1996) Immediate-early baculovirus vectors for foreign gene expression in transformed or infected insect cells. *Protein Expr. Purif.*, **8**, 191–203.

71 Murhammer, D.W. (ed.) (2007) *Baculovirus and Insect Cell Expression Protocols*, 2nd edn, Methods in Molecular Biology, Humana Press, Totowa, NJ.

72 Smith, G.E., Summers, M.D., and Fraser, M.J. (1983) Production of human beta interferon in insect cells infected with a baculovirus expression vector. *Mol. Cell. Biol.*, **3**, 2156–2165.

73 Luckow, V.A. and Summers, M.D. (1989) High level expression of nonfused foreign genes with *Autographa californica* nuclear polyhedrosis virus expression vectors. *Virology*, **170**, 31–39.

74 Kitts, P.A. and Possee, R.D. (1993) A method for producing recombinant baculovirus expression vectors at high frequency. *Biotechniques*, **14**, 810–817.

75 Ciccarone, V.C., Polayes, D.A., and Luckow, V.A. (1998) Generation of recombinant baculovirus DNA in *E. coli* using a baculovirus shuttle vector. *Mol. Diagn. Infect. Dis.*, **13**, 213–235.

76 Ranjan, A. and Hasnain, S.E. (1995) Codon usage in the prototype baculovirus – *Autographa californica* nuclear polyhedrosis virus. *Indian J. Biochem. Biophys.*, **32**, 424–428.

77 Ranjan, A. and Hasnain, S.E. (1995) Influence of codon usage and translational initiation codon context in the *Ac*NPV-based expression system: computer analysis using homologous and heterologous genes. *Virus Genes*, **9**, 149–153.

78 Levin, D.B. and Whittome, B. (2000) Codon usage in nucleopolyhedroviruses. *J. Gen. Virol.*, **81**, 2313–2325.

79 Chang, M.J., Kuzio, J., and Blissard, G.W. (1999) Modulation of translational efficiency by contextual nucleotides flanking a baculovirus initiator AUG codon. *Virology*, **259**, 369–383.

80 Peakman, T.C., *et al.* (1992) Enhanced expression of recombinant proteins in insect cells using a baculovirus vector containing a bacterial leader sequence. *Nucleic Acids Res.*, **20**, 6111–6112.

81 Guan, X.M., Kobilka, T.S., and Kobilka, B.K. (1992) Enhancement of membrane insertion and function in a type IIIb membrane protein following introduction of a cleavable signal peptide. *J. Biol. Chem.*, **267**, 21995–21998.

82 Grunewald, S., *et al.* (1996) Glycosylation, palmitoylation, and localization of the human D_{2S} receptor in baculovirus-infected insect cells. *Biochemistry*, **35**, 15149–15161.

83 Korepanova, A., *et al.* (2009) Expression and purification of human TRPV1 in baculovirus-infected insect cells for structural studies. *Protein Expr. Purif.*, **65**, 38–50.

84 Suzuki, Y. and Nei, M. (2002) Origin and evolution of influenza virus hemagglutinin genes. *Mol. Biol. Evol.*, **19**, 501–509.

85 Tessier, D.C., *et al.* (1991) Enhanced secretion from insect cells of a foreign protein fused to the honeybee melittin signal peptide. *Gene*, **98**, 177–183.

86 Jarvis, D.L. and Garcia, A., Jr (1994) Biosynthesis and processing of the *Autographa californica* nuclear polyhedrosis virus gp64 protein. *Virology*, **205**, 300–313.

87 Murphy, C.I., *et al.* (1993) Enhanced expression, secretion, and large-scale purification of recombinant HIV-1 gp120 in insect cell using the baculovirus egt and p67 signal peptides. *Protein Expr. Purif.*, **4**, 349–357.

88 Loisel, T.P., *et al.* (1997) Recovery of homogeneous and functional beta 2-adrenergic receptors from extracellular baculovirus particles. *Nat. Biotechnol.*, **15**, 1300–1304.

89 Kawate, T. and Gouaux, E. (2006) Fluorescence-detection size-exclusion chromatography for precrystallization screening of integral membrane proteins. *Structure*, **14**, 673–681.

90 Peng, S., *et al.* (1993) One-step affinity isolation of recombinant protein using the baculovirus/insect cell expression system. *Protein Expr. Purif.*, **4**, 95–100.

91 Kwatra, M.M., *et al.* (1995) Immunoaffinity purification of epitope-tagged human beta 2-adrenergic receptor to homogeneity. *Protein Expr. Purif.*, **6**, 717–721.

92 Warne, T., Chirnside, J., and Schertler, G.F. (2003) Expression and purification of truncated, non-glycosylated turkey beta-adrenergic receptors for crystallization. *Biochim. Biophys. Acta*, **1610**, 133–140.

93 Kojima, K., *et al.* (2001) Tandem repetition of baculovirus ie1 promoter results in upregulation of transcription. *Arch. Virol.*, **146**, 1407–1414.

94 Douris, V., *et al.* (2006) Stably transformed insect cell lines: tools for expression of secreted and membrane-anchored proteins and high-throughput screening platforms for drug and insecticide discovery. *Adv. Virus Res.*, **68**, 113–156.

95 Cameron, I.R. and Possee, R.D. (1989) Conservation of polyhedrin gene

promoter function between *Autographa californica* and *Mamestra brassicae* nuclear polyhedrosis viruses. *Virus Res.*, **12**, 183–199.

96 Wasilko, D.J., *et al.* (2009) The titerless infected-cells preservation and scale-up (TIPS) method for large-scale production of NO-sensitive human soluble guanylate cyclase (sGC) from insect cells infected with recombinant baculovirus. *Protein Expr. Purif.*, **65**, 122–132.

97 Aloia, A.L., *et al.* (2009) GPCR expression using baculovirus-infected Sf9 cells. *Methods Mol. Biol.*, **552**, 115–129.

98 Luckow, V.A. and Summers, M.D. (1988) Signals important for high-level expression of foreign genes in *Autographa californica* nuclear polyhedrosis virus expression vectors. *Virology*, **167**, 56–71.

99 Lynn, D.E. (2007) Routine maintenance and storage of lepidopteran insect cell lines and baculoviruses. *Methods Mol. Biol.*, **388**, 187–208.

100 Lynn, D.E. (2007) Lepidopteran insect cell line isolation from insect tissue. *Methods Mol. Biol.*, **388**, 139–154.

101 Lynn, D.E. (2007) Available lepidopteran insect cell lines. *Methods Mol. Biol.*, **388**, 117–138.

102 Wickham, T.J., *et al.* (1995) Comparison of different cell lines for the production of recombinant baculovirus proteins. *Methods Mol. Biol.*, **39**, 385–395.

103 Davis, T.R., *et al.* (1993) Comparative recombinant protein production of eight insect cell lines. *In Vitro Cell. Dev. Biol. Anim.*, **29A**, 388–390.

104 Miyamoto, C., *et al.* (1985) Production of human c-Myc protein in insect cells infected with a baculovirus expression vector. *Mol. Cell. Biol.*, **5**, 2860–2865.

105 Smith, G.E., *et al.* (1985) Modification and secretion of human interleukin 2 produced in insect cells by a baculovirus expression vector. *Proc. Natl. Acad. Sci. USA*, **82**, 8404–8408.

106 Vaughn, J.L., *et al.* (1977) The establishment of two cell lines from the insect *Spodoptera frugiperda* (Lepidoptera; Noctuidae). *In Vitro*, **13**, 213–217.

107 Wickham, T.J., *et al.* (1992) Screening of insect cell lines for the production of recombinant proteins and infectious virus in the baculovirus expression system. *Biotechnol. Prog.*, **8**, 391–396.

108 Jardin, B.A., *et al.* (2007) High cell density fed batch and perfusion processes for stable non-viral expression of secreted alkaline phosphatase (SEAP) using insect cells: comparison to a batch Sf-9-BEV system. *Biotechnol. Bioeng.*, **97**, 332–345.

109 Kadwell, S.H. and Hardwicke, P.I. (2007) Production of baculovirus-expressed recombinant proteins in wave bioreactors. *Methods Mol. Biol.*, **388**, 247–266.

110 Murhammer, D.W. (2007) Useful tips, widely used techniques, and quantifying cell metabolic behavior. *Methods Mol. Biol.*, **388**, 3–22.

111 Palomares, L.A., Pedroza, J.C., and Ramirez, O.T. (2001) Cell size as a tool to predict the production of recombinant protein by the insect cell baculovirus expression system. *Biotechnol. Lett.*, **23**, 359–364.

112 Zeiser, A., *et al.* (2000) On-line monitoring of physiological parameters of insect cell cultures during the growth and infection process. *Biotechnol. Prog.*, **16**, 803–808.

113 Janakiraman, V., *et al.* (2006) A rapid method for estimation of baculovirus titer based on viable cell size. *J. Virol. Methods*, **132**, 48–58.

114 Janakiraman, V., Forrest, W.F., and Seshagiri, S. (2006) Estimation of baculovirus titer based on viable cell size. *Nat. Protoc.*, **1**, 2271–2276.

115 Sandhu, K.S., Naciri, M., and Al-Rubeai, M. (2007) Prediction of recombinant protein production in an insect cell-baculovirus system using a flow cytometric technique. *J. Immunol. Methods*, **325**, 104–113.

116 Horn, L.G. and Volkman, L.E. (1998) Preventing proteolytic artifacts in the baculovirus expression system. *Biotechniques*, **25**, 18–20.

117 Ikonomou, L., *et al.* (2001) Design of an efficient medium for insect cell growth and recombinant protein production. *In Vitro Cell. Dev. Biol. Anim.*, **37**, 549–559.

118 Ikonomou, L., Schneider, Y.J., and Agathos, S.N. (2003) Insect cell culture

for industrial production of recombinant proteins. *Appl. Microbiol. Biotechnol.*, **62**, 1–20.

119 Kobayashi, H., *et al.* (2009) Functional rescue of beta-adrenoceptor dimerization and trafficking by pharmacological chaperones. *Traffic*, **10**, 1019–1033.

120 Kobilka, B.K. (1995) Amino and carboxyl terminal modifications to facilitate the production and purification of a G protein-coupled receptor. *Anal. Biochem.*, **231**, 269–271.

121 Hampe, W., *et al.* (2000) Engineering of a proteolytically stable human beta 2-adrenergic receptor/maltose-binding protein fusion and production of the chimeric protein in *Escherichia coli* and baculovirus-infected insect cells. *J. Biotechnol.*, **77**, 219–234.

122 Reilander, H., *et al.* (1991) Purification and functional characterization of the human beta 2-adrenergic receptor produced in baculovirus-infected insect cells. *FEBS Lett.*, **282**, 441–444.

123 Bahia, D., *et al.* (2005) Optimisation of insect cell growth in deep-well blocks: development of a high-throughput insect cell expression screen. *Protein Expr. Purif.*, **39**, 61–70.

124 Withers, L.A. and King, P.J. (1979) Proline: a novel cryoprotectant for the freeze preservation of cultured cells of *Zea mays* L. *Plant Physiol.*, **64**, 675–678.

125 Weiss, H.M. and Grisshammer, R. (2002) Purification and characterization of the human adenosine A_{2a} receptor functionally expressed in *Escherichia coli*. *Eur. J. Biochem.*, **269**, 82–92.

126 White, J.F., *et al.* (2004) Automated large-scale purification of a G protein-coupled receptor for neurotensin. *FEBS Lett.*, **564**, 289–293.

127 Sarramegna, V., *et al.* (2005) Solubilization, purification, and mass spectrometry analysis of the human mu-opioid receptor expressed in *Pichia pastoris*. *Protein Expr. Purif.*, **43**, 85–93.

128 Ratnala, V.R., *et al.* (2004) Large-scale overproduction, functional purification and ligand affinities of the His-tagged human histamine H_1 receptor. *Eur. J. Biochem.*, **271**, 2636–2646.

129 Yeliseev, A.A., *et al.* (2005) Expression of human peripheral cannabinoid receptor for structural studies. *Protein Sci.*, **14**, 2638–2653.

130 Carpenter, J.W., *et al.* (2002) Configuring radioligand receptor binding assays for HTS using scintillation proximity assay technology. *Methods Mol. Biol.*, **190**, 31–49.

131 Wu, S. and Liu, B. (2005) Application of scintillation proximity assay in drug discovery. *BioDrugs*, **19**, 383–392.

132 Ott, D., *et al.* (2005) Engineering and functional immobilization of opioid receptors. *Protein Eng. Des. Sel.*, **18**, 153–160.

133 Navratilova, I., Sodroski, J., and Myszka, D.G. (2005) Solubilization, stabilization, and purification of chemokine receptors using biosensor technology. *Anal. Biochem.*, **339**, 271–281.

134 Harding, P.J., *et al.* (2006) Direct analysis of a GPCR-agonist interaction by surface plasmon resonance. *Eur. Biophys. J.*, **35**, 709–712.

135 Bruns, R.F., Lawson-Wendling, K., and Pugsley, T.A. (1983) A rapid filtration assay for soluble receptors using polyethylenimine-treated filters. *Anal. Biochem.*, **132**, 74–81.

136 Harms, A., *et al.* (2000) Development of a 5-hydroxytryptamine$_{2A}$ receptor binding assay for high throughput screening using 96-well microfilter plates. *J. Biomol. Screen.*, **5**, 269–278.

137 Filppula, S., *et al.* (2004) Purification and mass spectroscopic analysis of human CB2 cannabinoid receptor expressed in the baculovirus system. *J. Pept. Res.*, **64**, 225–236.

138 Takahashi, K., *et al.* (2002) Targeted disruption of H_3 receptors results in changes in brain histamine tone leading to an obese phenotype. *J. Clin. Invest.*, **110**, 1791–1799.

139 Hancock, A.A. (2003) H_3 receptor antagonists/inverse agonists as anti-obesity agents. *Curr. Opin. Investig. Drugs*, **4**, 1190–1197.

140 Esbenshade, T.A., Fox, G.B., and Cowart, M.D. (2006) Histamine H_3 receptor antagonists: preclinical promise for treating obesity and cognitive disorders. *Mol. Interv.*, **6**, 77–88.

141 Stark, H. (2003) Recent advances in histamine H_3/H_4 receptor ligands. *Expert Opin. Ther. Patents*, **13**, 851–865.

142 Leurs, R., Vollinga, R.C., and Timmerman, H. (1995) The medicinal chemistry and therapeutic potentials of ligands of the histamine H_3 receptor. *Prog. Drug Res.*, **45**, 107–165.

143 Hancock, A.A., *et al.* (2003) Genetic and pharmacological aspects of histamine H_3 receptor heterogeneity. *Life Sci.*, **73**, 3043–3072.

144 Ireland-Denny, L., *et al.* (2001) Species-related pharmacological heterogeneity of histamine H_3 receptors. *Eur. J. Pharmacol.*, **433**, 141–150.

145 Yao, B.B., *et al.* (2003) Cloning and pharmacological characterization of the monkey histamine H_3 receptor. *Eur. J. Pharmacol.*, **482**, 49–60.

146 Esbenshade, T.A. and Hancock, A.A. (2000) Characterization of histaminergic receptors. *Curr. Protoc. Pharmacol.*, **1**, 1.19.1–1.19.19.

147 Lovenberg, T.W., *et al.* (1999) Cloning and functional expression of the human histamine H_3 receptor. *Mol. Pharmacol.*, **55**, 1101–1107.

148 Cherifi, Y., *et al.* (1992) Purification of a histamine H_3 receptor negatively coupled to phosphoinositide turnover in the human gastric cell line HGT1. *J. Biol. Chem.*, **267**, 25315–25320.

149 Krueger, K.M., *et al.* (2005) G protein-dependent pharmacology of histamine H_3 receptor ligands: evidence for heterogeneous active state receptor conformations. *J. Pharmacol. Exp. Ther.*, **314**, 271–281.

150 Esbenshade, T.A., *et al.* (2005) Pharmacological properties of ABT-239 [4-(2-{2-[(2*R*)-2-methylpyrrolidinyl] ethyl}-benzofuran-5-yl)benzonitrile]: I. Potent and selective histamine H_3 receptor antagonist with drug-like properties. *J. Pharmacol. Exp. Ther.*, **313**, 165–175.

151 Esbenshade, T.A., *et al.* (2004) Pharmacological and behavioral properties of A-349821, a selective and potent human histamine H_3 receptor antagonist. *Biochem. Pharmacol.*, **68**, 933–945.

5
Membrane Protein Expression in Mammalian Cells

Deniz B. Hizal, Erika Ohsfeldt, Sunny Mai, and Michael J. Betenbaugh

5.1
Introduction

Mammalian systems are widely used in industry for the production of recombinant proteins because this expression system can provide a similar environment to the native one for higher-order eukaryotic membrane proteins [1]. Mammalian cells are often preferred in the production of biotherapeutics because of their ability to perform correct protein folding and post-translational modification [2–4]. Membrane proteins can be cell adhesion molecules, transduction receptors, and transport channels that require post-translational modifications such as glycosylation, phosphorylation, methylation, disulfide bond formation, proteolytic processing, and lipid addition for their functionality [5, 6]. Yeast cells have been used for expression of receptors; however, the enzymatic deficiency in their post-translational modification process can lead to poor glycosylation and misfolding of these proteins [3, 7]. A disadvantage of studying membrane proteins in bacterial systems is that the proteins can sometimes accumulate as insoluble aggregates, which limits the expression of the protein [3]. Drew *et al.* [8] monitored the aggregation by generating fusion constructs of Green Fluorescent Protein (GFP) and the protein of interest. The protein and GFP had coaggregated in inclusion bodies and no fluorescence was observed [8]. One of the other main differences between mammalian cells, insect cells, yeast, and bacteria is their lipid composition [3]. For example, overexpression of membrane proteins requires proper protein folding that necessitates the presence of cholesterol, which is abundant in mammalian expression systems. The absence of cholesterol can also affect the ability of ligands to bind to the receptors [9].

The first approval of a therapeutic protein from a mammalian cell line occurred in 1986 [1]. Since the mid-1980s protein expression in bioreactors has reached high levels, even gram per liter in some cases. There are numerous mammalian cell lines that have been used in protein expression including Chinese hamster ovary (CHO), human embryonic kidney (HEK), green monkey kidney (COS-1), and baby hamster kidney (BHK). CHO cells are the abundantly used cell lines in protein production because of their growth ability in suspension systems and bioreactors [10].

Production of Membrane Proteins: Strategies for Expression and Isolation, First Edition.
Edited by Anne Skaja Robinson.

Membrane proteins have essential roles in cell growth, differentiation, metabolism, flow of information, cell–cell communication, and migration [11]. Proteins that have a strong influence on the functionality of cellular processes can be good targets for pharmacological applications. Defects in some membrane proteins can lead to diseases, such as cancer [12]; therefore, it comes as no surprise that these proteins represent nearly 50% of the targets of therapeutic research [11, 13].

Membrane proteins have proven very difficult to study due to their natural low expression levels. It is desirable to study the structures of membrane proteins because this will help to better understand the function of these proteins. Since many membrane proteins have an effect on the progression of diseases this is of great importance. Many expression systems have arisen in recent years in an attempt to better study these delicate proteins. Studies that utilize a mammalian expression system may be highly useful for membrane protein production due to the superior nature of the folding and processing environment of these cells.

5.2 Mammalian Systems

5.2.1 Cell Culture Types and Media Optimization

Mammalian cells are relatively easy to grow and culture. Depending on the cell types, the cells can be grown in adherent or suspension culture using batch, fed-batch, or perfusion cultures.

5.2.1.1 Adherent Cell Culture

In order to grow and carry out metabolic functions, many mammalian cell types must adhere to the extracellular matrix (ECM) *in vivo* or on cell culture plates *in vitro* [14]. The cell membrane contains cell adhesion molecules such as integrins, cadherins, and selectins used to bind to surfaces. For this reason, manufacturers produce sterile pretreated single, multiwell plates or bottles to promote adhesion and growth of the cells. The type of media required depends on the cell type, but serum is often required for adherent cell cultures to facilitate binding to the surface. When the cells have grown to a confluent layer, the cells must be split and subcultured onto new plates. Trypsin is usually used to release cells for passing or cryopreservation since it can digest the membrane proteins that are adhered to the surface. The plates are stored in an incubator at a set temperature and a specific CO_2 level in order to maintain a constant pH. Cryopreservation of cell lines can be done to store the cells for future studies. The cells can be removed from the surface using trypsin; however, the enzyme must be removed subsequently by centrifugation of the cells. The cells can be resuspended in media that has been supplemented with a chemical, such as dimethyl sulfoxide, which will fix the cells in their current state. Freezing the cells slowly helps to keep the cells viable for when they are used later.

5.2.1.2 **Suspension Cell Culture**

Suspension cell culture is used for cells that do not require adhesion to a surface or to have cells with characteristics normally not expressed in an adherent culture. Since cell confluency depends on volume of media rather than the surface area of the container in suspension culture, more cells can be obtained in this type of culture. Furthermore, suspension culture is much more efficient for high-level protein production because of this; the cells can be adapted to suspension culture through the use of specially formulated media after transfection [2]. When the nutrient in the media is spent and when there is an accumulation of toxic substances, the suspended cells must be subcultured into fresh media.

5.2.1.3 **Batch and Fed-Batch Culture**

Batch culture involves growing the cells up in a single suspension or adherent culture process. The cells are typically collected after reaching a maximum cell density in which a nutrient such as glucose or an amino acid is exhausted. After the cells reach the maximum, there is often a decline in viability as the cells undergo apoptosis. In order to extend the culture periods, cells are often fed with additional nutrients at the end of the batch phase in a process called fed-batch or extended batch culture. The advantage of this process is that fresh medium can be added in batches or semicontinuously [2]. Fed-batch processing allows the expression scientists to reach higher final cell densities and delays the onset of the death phase. Fed-batch culturing is a process that is widely used for the manufacturing of protein therapeutics and potentially enables the production of target membrane proteins in much higher yields.

5.2.1.4 **Perfusion Process**

Perfusion culturing is a continuous process that requires the continual addition and removal of several reactor volumes of medium per day. In this case, the cells can be removed with the spent media or held in the bioreactor in order to reach highest possible cell densities. Perfusion culture allows the continual addition of nutrients, continual removal of toxins from the culture, and collection of cells or product prior to the completion of the run. One advantage of this technique is that harvesting can be done several times throughout the process in which the culture occurs, as seen in Figure 5.1. This is advantageous for proteins that are unstable and can be removed prior to degradation during extended culture times. Another advantage of this process is that the culturing can run for weeks and months at a time. The main disadvantage of perfusion culture is that it typically requires much larger medium volumes than fed-batch culture processing and medium costs can be problematic for mammalian cell culture processing.

5.2.1.5 **Media Optimization**

The optimization of cell culture medium plays an important role in increasing protein expression in mammalian cells. Different stages in cell growth and in the production process call for different media compositions. For instance, forming cell lines and stable pools in adherent cell cultures require media that promotes

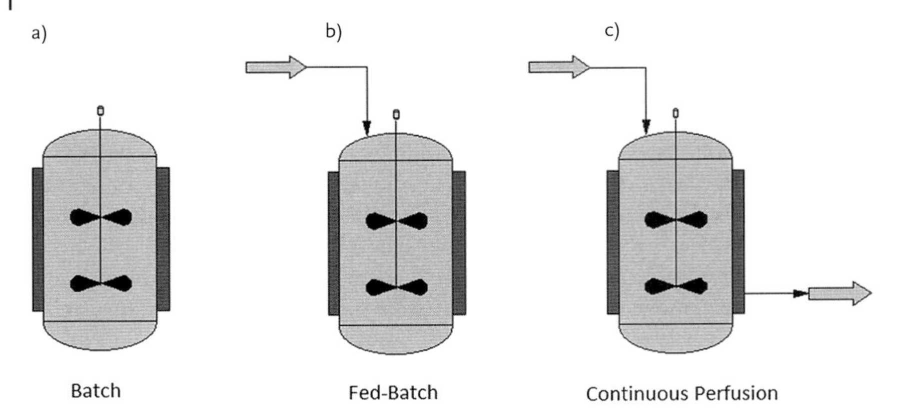

Figure 5.1 Schematic of cell culture process designs. (a) Batch process and (b) fed-batch process. The arrow entering the tank represents the influx of fresh medium. (c) Continuous perfusion process. New medium is added continuously while the product is removed.

adherence to culture plates. For this reason, media containing serum such as fetal bovine serum in Dulbecco's Modified Eagle's Medium is preferred during stable clone formation.

For suspension cells and bioreactors, suspension cell culture media must be used. During the final scale-up stage for producing large amounts of membrane and secreted proteins, serum-free media is used to yield products with a consistent composition and quality. Media containing serum is undesirable because serum is expensive and considered as chemically undefined because the composition of protein, carbohydrates, growth factors, lipids, and small molecules is unknown and always variable. This may lead to unwanted variability in the product from batch to batch as well as increasing production costs significantly. Chemically defined media with nonanimal proteins and growth factors are optimal for consistent product batches, but the formulations differ for various cells lines and must be determined for each specific cell line. These media include nutrients, pH buffers, and growth factors that can promote growth to high cell densities [15]. Many commercial vendors such as Invitrogen, Sigma, and others can provide chemically defined formulations for some widely used mammalian cell culture systems such as CHO or HEK-293. Unfortunately, the media often contain growth factors that can make the mammalian cell culture media expensive, at least in comparison to those used in bacterial or yeast cultures. As a result, one of the factors that have inhibited the greater use of mammalian cell culture for membrane protein production is the high costs of media for scaling-up production. Future efforts will require the development of compositions that provide for the large-scale production of membrane proteins using lower-cost additives and media.

5.2.2
Gene Delivery and Expression in Mammalian Systems

The successful delivery of the gene into the cells can be accomplished by transferring the gene to the surface of the cells, crossing the gene over the plasma membrane into the cytoplasm, and transporting it from cytoplasm to nucleus without any degradation. Viral vectors, and physical and chemical methods are used to overcome these hurdles [16]. The first nonviral DNA introduction to mammalian cells started with the study of Graham and van der Eb in 1973. They showed that transfer of DNA can be provided with calcium phosphate into mammalian cells [2]. Calcium phosphate transfection is now one of the most widely used techniques along with electroporation, lipofection, and cationic polymer transfection. These chemical methods are preferred in industrial applications since they are easy, reliable, and efficient for transient and stable transfection [2, 16].

In order to achieve the protein expression, DNA should be cloned into a suitable vector that has a promoter and polyadenylation signal. The common promoters in mammalian cell expression systems are mammalian elongation factor 1α (EF-1α), human cytomegalovirus (CMV), simian virus 40 (SV40), and Rous sarcoma virus (RSV) [17]. Transient transfection studies examine the protein expression level on the day following the transfection. This can provide evidence that the transfection procedure was successful in inserting the DNA into the cells. The mammalian expression vectors also include selectable markers such as neomycin, puromycin, and hygromycin, and these vectors are commonly available from commercial vendors. The transfected cells can be stably identified by selection of cells that are resistant to these antibiotics. In 2 weeks, the transfected cells can be isolated to form stable pools [18]. The stable pools can be also formed by pressure-induced gene amplification. Instead of using antibiotic resistances paired to the recombinant protein, a system using dihydrofolate reductase (DHFR) and glutamine synthetase (GS) genes is used. Media with methotrexate (MTX) and methionine sulfoximide (MSX), which are inhibitors of DHFR and GS, put selective pressure on cells. To survive, the cells must amplify the DHFR and GS, genes, which thus enhances the expression of the recombinant protein and increases productivity [2]. There are many other ways to improve gene amplification and expression [19]. These methods all provide the selection of the transfected cells to form a highly expressive stable pool. A number of different case studies are summarized in Table 5.1. The following is an example for high transfection efficiency with cationic liposome for membrane proteins.

5.2.2.1 High Transfection Efficiency in Adherent Cell Cultures with Cationic Liposome

High transfection efficiency is very important for the expression of membrane proteins since they are often difficult to express. One of the cationic liposomes that provide high transfection efficiency and high membrane expression is the Lipofectamine™ 2000 reagent from Invitrogen. Transfection can be done in

Table 5.1 Representative case studies of membrane proteins expressed in mammalian cells.

Membrane protein	Cell line	Vector construct	Selective marker	Promoter	Reference
ErbB2	CHO	pcDNA3.1/zeo	zeocin	P_{CMV}	[11]
92 GPCRs	BHK-21	pCR4Blunt-TOPO	kanamycin	P_{lac}	[20]
Fibroblast growth factor receptor 1 and 2 (FGFR1 and 2)	CHO	pCEP4	hygromycin	P_{CMV}	[21]
μ-opioid receptor	COS-7	pcDNA3	geneticin	P_{CMV}	[22]
Cellubrevin	BHK 21	pcDNA3	geneticin	P_{CMV}	[23]
Peroxisome proliferator-activated receptor-γ (PPAR-γ)	HEK-293T	$pLNCX_2$	geneticin	$P_{CMV\ IE}$	[24]
Caveolin-1 (Cav-1)	α1KO-Rec	pcDNA4.1	zeocin	P_{CMV}	[25]
ST8SiaII and ST8SiaIV	CHO-mutant 2A10 (PSA-negative cells)	pcDNA3	geneticin	P_{CMV}	[26]
$K_v1.1$	CHO/Lec (glycosylation-deficient cell line)	pZEM222	geneticin	MT	[27]
Shaker B	Pro5/Lec cells	pcDNA3.1/V5-His	geneticin	P_{CMV}	[28]

Synaptic adhesion-like molecule (SALM)	HEK-293T	pDisplay	kanamycin/geneticin	P_{CMV}	[29]
Gonadotropin releasing hormone type I receptor (GnRH-R)	HEK-293-HTR-8/SV neo	pcDNA3.1/hygro	hygromycin	P_{CMV}	[30]
Transient receptor potential melastatin type 6 (TRPM6)	HEK-293	pEBG	NA	EF-1α	[31]
Cytochrome P450 oxidoreductase (CPR)	Flp-In CHO cells	pcDNA5/FRT	hygromycin/zeocin	P_{CMV}	[32]
GluR2 ionotropic glutamate receptor	HEK-293T	pcDNA3.1	geneticin	P_{CMV}	[33]
Human calcium receptor (hCAR)	HEK-293	pCR3.1	geneticin	P_{CMV}	[34]
N-Methyl-D-aspartate (NMDA) receptors, NR2D, NR1	BHK-21	pTRE2/pCI-IRES-bla	geneticin/blasticidin	tetracycline-inducible promoter	[35]
Copper Transporter 1	COS-7	pAcGFP-N1	kanamycin/geneticin	P_{CMVIE}	[36]
Human olfactory receptor 17.4	HEK-293S	T-REx pcDNA 4/TO	blasticidin/zeocin	P_{CMV}	[37]
MUC13	SKOV-3	pcDNA3.1	geneticin	P_{CMV}	[38]

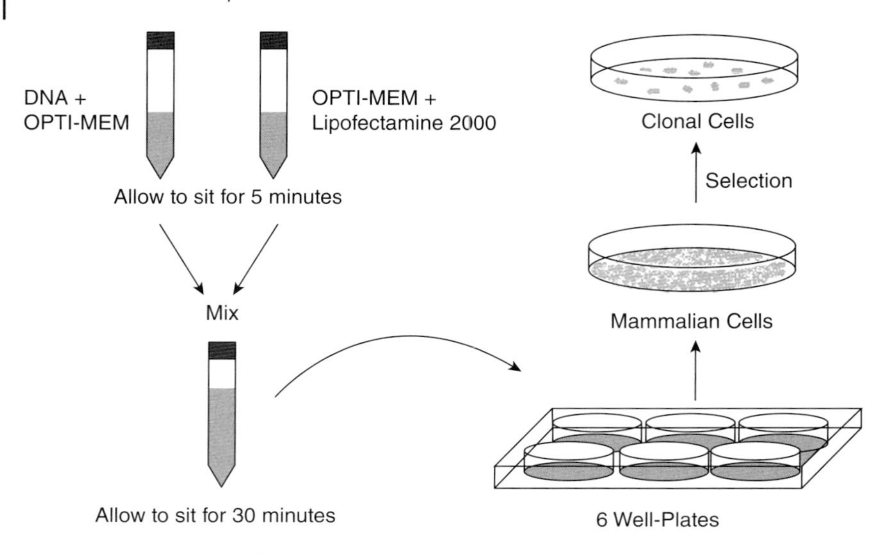

Figure 5.2 Transfection of gene of interest into mammalian cells and clonal selection.

six-well plates and does not require large amounts of this expensive reagent. The general process of transfection and selection is shown in Figure 5.2.

Day 1 The cells should be seeded with 2 ml growth media in each well 1 day before the transfection to provide the 80–85% confluency needed for the transfection.

Day 2 For each well, prepare 2 μg DNA in 250 μl of reduced serum medium OPTI-MEM® I from Invitrogen or in growth media without serum and prepare 10 μl Lipofectamine 2000 Reagent in 250 μl of OPTI-MEM I or in serum-free growth media. Let them sit for 5 min. Combine the tubes very gently after 5 min and let them sit for 30 min. Change the media of the six-well plates from the growth media to 2 ml of OPTI-MEM I media, and then add the 500 μl DNA and lipofectamine mixture onto the cells very carefully from the edge. After 5–7 h, remove the OPTI-MEM I media and put growth media into the wells. Keep the plates in an incubator at 37 °C containing 4–6% CO_2 for 24 h.

Day 3 The cells will be transiently transfected and they will be ready for stable transfection or to be assayed. For membrane proteins, special assays or Western blot analysis can be done for the first check of protein expression. The cells that should be stably transfected should be passed onto a 10-cm plate.

Day 4 Add the proper antibiotic for the selection of the transfected gene. The media should be changed every 3–4 days over the course of 15 days. After 15 days, the stable pools will be ready.

Stable pools consist of individual cell clones with different protein expression levels. Due to the nature of the random integration of the gene into one or more different sites on the chromosomes, thousands of different individual cell variants will be obtained, each with different expression levels and potentially other different traits such as growth rate or level of antibiotic resistance. The most tedious and time-consuming step in the transfection process is identifying the highest expressing cell line. The traditional method used to select the most promising expressing cell clone is to screen hundreds of cells. After estimating the necessary dilution, the cells can be divided into 96-well plates, each well having only one cell or the cells can be seeded very dilute in the 10-cm plates, and then the grown colonies can be picked and passed into one well of high-throughput plates. Then each well can be scaled-up for screening [39, 40]. In addition to the dilution method, sterile cloning rings can be used to isolate single cell line.

In the industrial setting, less labor-intensive techniques such as high-throughput automated selection systems and cell-sorting methods including flow cytometry approaches are often used. One of the most common ways to check the expression level of the cells is through the use of Western blot analysis, but flow cytometry is also possible. If a fluorescently conjugated antibody is bound to the membrane protein, then the cells can be scanned and sorted based on florescence intensity using a flow cytometer or immunofluorescent microscope. If the membrane protein has been fused with a fluorescent marker such as GFP, immunofluorescence can be used to examine both the location and expression of membrane proteins. Activity assays can also be used to evaluate the functionality of the receptor or transport channel. Due to the advances in techniques developed for secreted proteins such as monoclonal antibodies, mammalian systems are widely available and convenient for gene delivery and membrane protein expression. After checking the expression levels of the stable cell lines, the cells are also often screened for growth rate. The most robust cell lines in terms of high expression and efficient cell growth are then adapted to the suspension media for the scale-up.

5.2.3
Post-Translational Modifications in Mammalian Systems

One of the main advantages of mammalian cell culture for membrane protein production is its ability to perform numerous complex post-translational modifications. These post-translational modifications include glycosylation, lipid addition, phosphorylation, carboxylation, disulfide bond formation, amidation, and methylation that are sometimes either essential or helpful for membrane protein processing in the secretory compartment and functioning on the cell surface. The enzyme systems of the widely used cells and their modifications are shown in Table 5.2. It should be emphasized that even if a cell is capable of performing a post-translational modification, the efficiency or the process may be different in different organisms.

Glycosylation is a particularly complex modification that can often be important for folding and proper processing. Although all eukaryotes are able to perform

Table 5.2 Cell types and common post-translational modifications in membrane proteins.

	Glycosylation	Lipidation	Phosphorylation	Proteolytic processing	Disulfide bond formation
Mammalian cells	+	+	+	+	+
Insect cells	+	+	+	+	+
Yeast cells	+	+	+	+	+
Bacterial cells	−	−	+	−	+

glycosylation, the glycan moieties differ in each of them. Indeed, glycosylation processing in mammalian cells is very different compared to yeast and insect cells, and most bacteria are incapable of adding glycans to proteins [41–43]. The presence of *N*-acetylglucosamine (GlcNac) and mannose residues on *N*-glycan attachments is common in many cell types, but further modifications depend on the enzyme systems of the cells. For example, many yeast strains add numerous mannose residues to the sugar core for hypermannosylation. Alternatively, insect cells often generate paucimannose-type glycosylation with one to three mannose residues. In contrast, the presence other modifying enzymes such as sialyltransferase and galactosyltransferase in mammalian cells allows these membrane proteins to be further modified and function properly in cases where those sugars are required either for function or enhanced *in vivo* circulatory half-life [44].

5.2.3.1 Glycosylation

Mammalian cell proteins may be modified by glycosylation of an asparagine (*N*-linked glycosylation) and/or serine/threonine (*O*-linked glycosylation). The glycan group that is linked to asparagine or serine/threonine in mammalian cells typically includes a number of monosaccharide residues in human glycans including GlcNAc, mannose, glucose, fucose, galactose, *N*-acetylgalactosamine (GalNAc), *N*-acetylneuraminic acid (NANA or Neu5Ac), and *N*-glycolylneuraminic acid (NGNA). Each of these monosaccharides has different pharmacological roles on the glycan structures. Studies have shown that glycosylation can be controlled at some extent in mammalian cells by overexpression or deletion of certain genes. Limiting the level of glycosylation can sometimes be advantageous for structural analysis since consistent glycan patterns can sometimes be helpful in subsequent crystallization steps. Indeed, CHO *lec* mutants and an HEK cell line with deletions in genes that result in limited glycosylation are available [45, 46]. These cell lines can also be used for changing the glycosylation pattern on therapeutic proteins. For example, [47] engineered CHO cell lines and utilized Lec13, a fucosylation-deficient cell line, in order to produce nonfucosylated therapeutics. CHO, HEK, human retinal (PERC.6), mouse myeloma (NS0), and BHK cells are the cell lines often used for generating proteins in commercial applications with glycosylation similar to human glycoproteins [48, 49]. Figure 5.3 shows the typical glycan processing pattern in mammalian cell lines. However, this glycosylation pathway

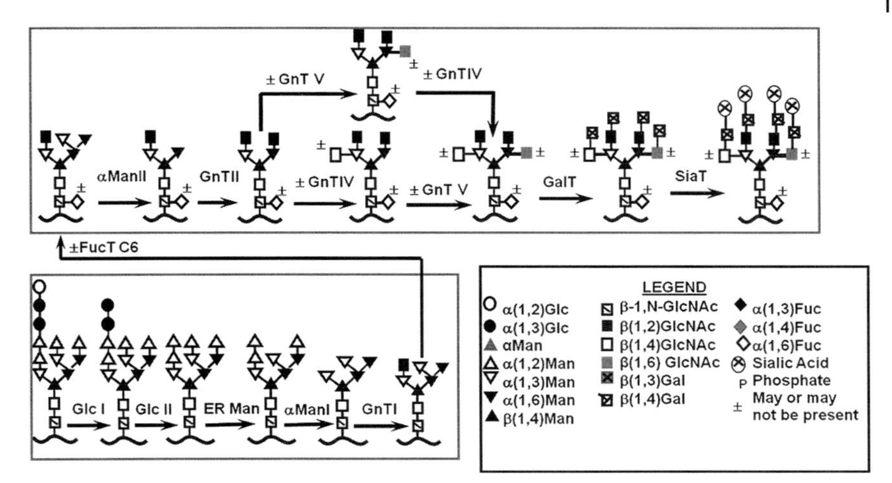

Figure 5.3 Glycan-processing in mammalian cells. Glc I, glucosidase I; Glc II, glucosidase II; Man I, α-mannosidase I, Man II, α-mannosidase II; GnT I, *N*-acetylglucosaminyltransferase I; GnT II, *N*-acetylglucosaminyltransferase II; GalT, galactosyltransferase; FucT C6, core α(1,6)-fucosyltransferase; SiaT, sialyltransferase [41].

is very different in the other expression systems because of the absence or presence of different glycosylating enzymes and substrates that will result in very different glycan patterns on membrane proteins.

5.2.3.2 **Protein Lipidation**

Lipid addition to proteins is an important post-translational modification that can be crucial for protein functioning and localization [50]. Improper lipidation can also cause physiological disorders including cancer or neurological diseases. There are many different ways in which the covalent attachment of lipids to proteins occurs [51].

N-terminal, C-terminal, and internal lipidation are the three major types found on membrane proteins. Modification of the N-terminal Gly residue with myristic acid is called *N*-myristoylation. The acylation of palmitic acid on cysteine residues of the N-terminus is *N*-palmitoylation. Internal lipidations occur on the Cys residues and Ser residues, and these are called *S*-palmitoylation and *O*-palmitoylation, respectively. Palmitoylation of proteins affect the localization and function of the proteins by enhancing the attraction between the protein and the membrane, whereas myristoylation has influence on the temporary interaction of the protein with the membrane. C-terminal lipidation can be found within three subgroups. Prenyltransferases, farnesyltransferases, and geranylgeranyltransferases are the three responsible enzymes that catalyze the lipid additions to the C-terminus. In this reaction, the isoprenoid is added to farnesyl or geranylgeranyl fatty acids and linked to the Cys residue of the C-terminus. Prenylation has the same role as the myristoylation; they both enhance the interaction with the membrane. The other type of lipidation that is seen in mammalian cells is the attachment of cholesterol to the C-terminus. Cholesterylation provides signaling properties to the protein and promotes the efficient trafficking of the protein. Addition of cholesterol can also affect the extracellular communication of the proteins. For example, the addition of cholesterol to Sonic Hedgehog provides its long-range signaling property and makes it one of the important proteins that govern tissue development. The other modification in the C-terminus is glycosylphosphatidylinositol (GPI) anchor addition. The GPI biosynthesis and the GPI addition to the protein occur in the endoplasmic reticulum (ER) by GPI transamidase. The GPI anchor has roles both in lateral movement and extracellular communication of the proteins [50–54].

Some proteins' activity and some protein–protein interactions are highly dependent on lipidation because the hydrophobic nature of these lipids can sometimes be important. While all of the lipid modifications can be observed in mammalian cells, some of them are lacking in other cell types. For instance, cholesterylation is specific to mammalian cells, and cholesterol is different from the other lipids in terms of size and hydrophobicity. Some proteins may need to undergo cholesterylation and thus their expression would not be appropriate in any system other than mammalian systems. Other cells often have different types of sterol; for example, yeasts have ergosterol and plants have sitosterol instead of cholesterol [53]. Furthermore, GPI anchors are composed of different oligosaccharides and different phospholipids in yeast, protozoa, and mammals, which means that the GPI-anchored proteins will vary widely from cell line to cell line. Indeed, a wide variety of oligosaccharides and phospholipids have been observed in GPI structures in mammalian cells [52].

Many other post-translational modifications such as phosphorylation, alkylation, acetylation, deamidation, ubiquitination, and formation of disulfide bridges, all observed in mammalian systems, may be what make mammalian systems preferable for membrane protein expression.

5.3 Case Studies

5.3.1 Increasing Membrane Protein Expression by Virus Vectors

Unfortunately, conventional transfection systems sometimes do not meet expectations in some cell lines in terms of expression levels for membrane proteins. Although transiently transfected cell lines sometimes show higher expression, the expression level can often decrease in stable pools and when they are scaled-up to suspension cultures. Alternatively, increased membrane protein expression is sometimes observed when viral vectors are used in mammalian systems. The two major concerns with viral gene transfer are the potential biosafety risks and the potential that overexpression can lead to improperly or incompletely processed target membrane proteins. Alphaviruses such as Venezuelan equine encephalitis virus (VEEV), Sindbus virus (SV), and Semliki Forest virus (SFV) are simple single-stranded RNA viruses used for gene delivery to mammalian systems [6]. The replication-deficient SFV gene delivery system was first developed in 1991 and increased expression of the transferrin receptor in BHK cells was the first proof that this system could be used to improve membrane protein production [3, 55].

Since 1991, numerous recombinant integral membrane proteins have been expressed in a variety of mammalian host cells through the SFV system without the generation of virus progeny [6]. Increased G-protein-coupled receptor (GPCR) and ion channel expression have been achieved using the SFV vector system. Eifler *et al.* [3] evaluated the expression levels of three types of membrane proteins – receptors, channels, and receptor channels – in three different expression systems – yeast, baculovirus/insect cells, and SFV/mammalian cells. Their study found that the SFV system achieved the best expression results for numerous membrane proteins in various mammalian cell lines [3]. However, this study also showed that some proteins can be severely degraded in either the SFV system or the baculovirus system. This example shows that although mammalian cells are similar to the native environment, the inclusion of a viral vector can sometimes lead to undesirable issues involving protein stability and processing.

GPCRs are important to the pharmaceutical industry, representing the target of numerous drug development projects; however, because of their transmembrane structure, it is hard to express and solubilize them [6, 13]. These membrane proteins have been studied in bacterial cells, like *Escherichia coli*, with poor results. The expression levels were low even when compensating for differences in codon bias between the bacterial and human genes [13]. The receptors that have been expressed in large quantities are usually aggregated into inclusion bodies so efforts were put forth to increase the proteins' stability by modifications [56]. However, these modifications made them inactive and could alter the structure seen in diffraction analysis. On the other hand, the SFV expression system has provided 10- to 100-fold higher expression of GPCRs in mammalian systems, and the assays proved that the expressed proteins were functional and stable [6, 57]. The SFV system has now been used on over 100 GPCRs in three different mammalian cell lines in a study by Hassaine *et al.* [20]. This expression system had yields near 1 mg/l, which is a level suitable for structural biology applications, for some of the GPCRs at the highest expression levels [20]. Interestingly, some of the GPCRs studied had no noticeable expression, others had expression but not at the levels of 1 mg/l, and some exhibited expression as high as 1 mg/l. The SFV expression system is very useful for structural analysis because functional proteins can be produced in high levels by this system.

Virus vectors for delivery have also been used on single transmembrane proteins [55], ion channels [58], potassium channels [59] and transporter proteins [60]. The functionality of the transfected ion channels can be evaluated by observing the movement of ions such as calcium, Ca^{2+}, after the mammalian cell has been stimulated with ATP [58]. The SFV expression system is also useful in neurobiological studies because of the ability of these vectors to infect the central nervous systems, and thus SFV is widely used in primary neuronal cultures for expression and localization studies. The infection level in primary neurons is generally as high as 75–95%, and the patch-clamp experiments for $K_v1.1$ and $K_v1.2$ potassium channels show high infection rates in mammalian systems [57].

Baculoviruses have been also used to transfect insect and mammalian cells with the genes of interest. For the mammalian systems, the insect cell-specific promoter is removed and a mammalian cell-active expression cassette replaced in these viruses [61]. Assays using the baculovirus vector have demonstrated that the baculovirus system can be more efficient in some cell lines than others [61].

Viral vectors can provide large quantities of protein within a relatively short period of time [62]. The SFV system provides almost 100% transfection efficiency of proteins into mammalian cells and uses very strong promoters for high expression levels [5, 6, 55], but there is a concern over the safety of the final product due to the virus component. To minimize the dangers, alphaviruses such as SFV are replication-deficient vectors, which improves the biosafety of the products. Alternatively, the baculovirus vector is not infectious to mammals and thus this vector transfects rather than infects mammalian cell lines. Finally, viral gene delivery can be applicable to the production of membrane and other proteins at large scales. Examples exist in which hundreds of milligrams of membrane protein production can be achieved in bioreactors by SFV system [63].

5.3.2
Anti-apoptosis Engineering for Increasing Membrane Protein Expression

One of the biggest problems in the production of proteins in mammalian cells is the stress placed on the cell by the use of bioreactors or the expression of a complex protein. Nutrient depletion, byproduct accumulation, and hydrodynamic stresses [64] can trigger the apoptosis pathway. Cells will tend to die in larger quantities near the end of the production process when the depletion of nutrients is having a greater impact on cell survival. Up to 80% of cell death in a serum-free bioreactor can be linked to apoptosis [64, 65]. Cells in the bioreactor that enter apoptosis will reduce the efficiency of the production process and the amount of proteins produced will be lowered. Apoptosis releases proteases into the media from lysed cells [66], which can reduce the quality of the protein product being produced.

Apoptosis, distinct from necrosis, is a programmed cell death cascade that is triggered by environmental stresses and other insults. Necrosis is a process in which chemical or physical injury to cells causes the cells to die through swelling and bursting. Apoptosis, in contrast, is a regulated process involving a series of cellular events that result in cell suicide. Cells can enter this process either when they are no longer needed or when environmental stimuli or other stresses such as glucose exhaustion or the overexpression of a complete protein force the cell to die. This process can be triggered from signals either inside or outside the cell that eventually causes the activation of a family of cysteine/aspartic proteases called caspases to be activated. Caspases are proteases that cleave their targeted protein at specific aspartic acids and contain a cysteine at their active site [67]. One of the pathways to apoptosis is through the mitochondria, which releases cytochrome *c* responsible for initiating the apoptosis cascade and the activation of proteases. Another cascade involves the binding of ligands to a death receptor on the surface of the cells, leading to apoptosis. Another potential cell death pathway may begin in the ER, but this pathway is not as well characterized. The caspase cascade is irreversible after a certain critical point and self-amplifying. There are eight members of the caspase family that are involved in apoptotic cell death: caspase-2, -3, -6, -7, -8, -9, -10, and -12 [1]. Caspase-8 and -9 are upstream initiator caspases, while caspase-3, -6, and -7 are downstream initiator caspases [66]. Caspase-3 plays a key role in the cell's commitment [1] to the irreversible apoptosis path way.

The permeability transition pore (PTP) of the mitochondria is regulated by the apoptotic signaling proteins. The formation of this pore allows cytochrome *c* to exit the mitochondria into the cytosol and start the caspase cascade. Anti-apoptosis proteins prevent the apoptosis pathway by inhibiting the activation of pro-apoptosis proteins and hinder the movement of cytochrome *c* to the cytosol from the mitochondria (see Figure 5.4) [64]. Pro-apoptosis proteins activate the release of cytochrome *c* by activating the formation of the PTP. Apoptosis can also be caused by stress to the ER from sources such as the unfolded protein response or improper protein folding and processing. The ER can cause changes in the cytosolic calcium levels that will signal the need for the cell to enter apoptosis. Another pathway to apoptosis is receptor-mediated, but this is unlikely to cause much apoptosis in cell cultures [64]. This pathway involves a death ligand, such as Fas or tumor necrosis factor, interacting with a death receptor. This will trigger internal signaling that will start the caspase cascade.

Anti-apoptosis engineering is a method for altering the apoptosis pathway either by increasing the expression of anti-apoptosis genes or proteins or by inhibiting expression or activity of a pro- apoptotic gene or protein [64]. By using anti-apoptosis engineering in the production process, the amount of recombinant protein produced can be increased and the cells will have increased cell viability. A number of different methods have been used for anti-apoptosis engineering by preventing the signaling pathway from being triggered through chemicals or other alterations to the extracellular environment and by using cell engineering to target components of the signaling pathway to alter their function. Extracellular alterations can include supplementation of growth factors, hydrolysates, and nutrients [64, 68, 69] to the media. These additions can reduce apoptosis in bioreactors by reducing the stress caused by nutrient depletion or byproduct accumulation. The inhibition of apoptosis proteins, including caspases, has been targeted using chemical reagents as a way to prevent the apoptosis signaling pathway from being initiated. Alternatively, expression of genes can be used to inhibit the apoptosis pathway and increase protein production [11, 70–72]. The Bcl-2 family of proteins contains both anti- (Bcl-x_L and Bcl-2) and pro-apoptosis (Bax, Bad, Bak, and Bim) proteins [64]. Bad functions by inhibiting the activity of Bcl-x_L and Bcl-2. The other proapoptotic members help the mitochondria to release cytochrome *c*. In one case, Bcl-x_L has

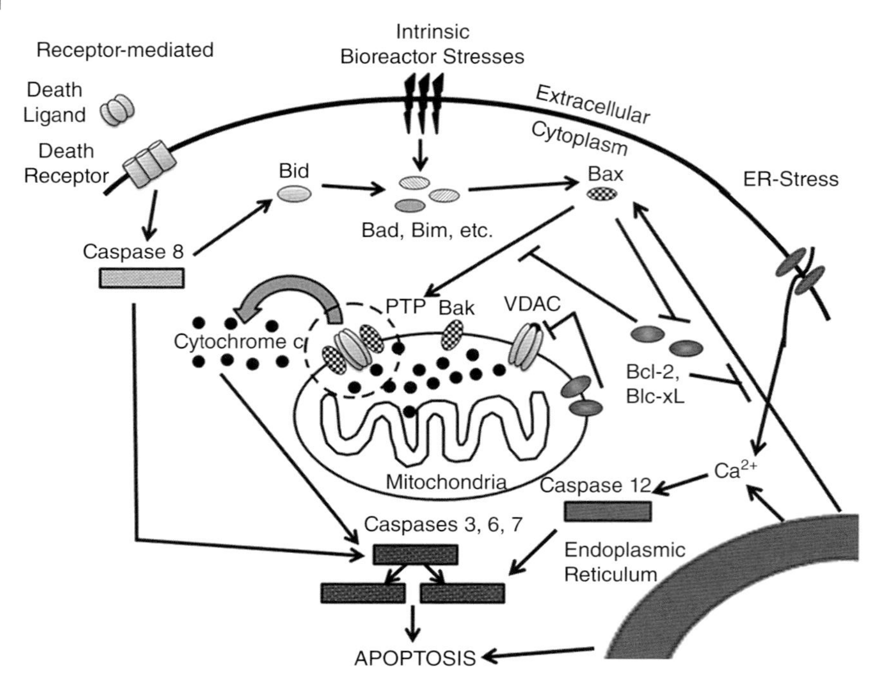

Figure 5.4 Pathways to apoptosis. The receptor-mediated pathway is triggered when a death ligand binds to a death receptor. The receptor sends an internal cellular signal to caspase-8 to trigger the caspase cascade. The intrinsic pathway can be triggered by the stresses felt by the cell during bioreactor growth. These stresses can activate the pro-apoptotic proteins such as Bad and Bim. These proteins will cause the PTP to form allowing the release of cytochrome *c* from the mitochondria. The cytochrome *c* activates the caspase cascade committing the cell to apoptosis. The anti-apoptosis proteins can prevent the release of cytochrome *c* by inhibiting the function of the pro-apoptosis proteins.

shown a better capability to protect cells from death compared to Bcl-2 in batch bioreactors under similar conditions [10]. Alternatively, viral inhibitors of apoptosis such as E1B-19K can be used instead of Bcl-2 inhibitors [72–74].

Anti-apoptosis engineering can attempt to reduce apoptosis in cell cultures by either increasing the expression of the anti-apoptosis proteins or reducing the expression of the pro-apoptosis proteins. Mammalian cell lines that have increased expression of either Bcl-2 or Bcl-x_L or viral anti-apoptosis genes have shown increases in production of recombinant proteins. The increase in protein production from Bcl-2 is due to the increase in cell viability, not an increase in specific productivity, at least in one case [10]. In this study by Meents *et al.*, Bcl-x_L showed a 2-fold higher increase in protein production than cell lines with Bcl-2. The use of non-Bcl-2 anti-apoptosis proteins such as Aven and E1B-19K has been used to increase the survival of cells in suspension cultures. The reduction in apoptosis can be seen by a decrease in caspase-3 activation [72]. In addition, these genes have been used to enhance production of monoclonal antibodies from CHO cells expressing both the target proteins and the anti-apoptosis genes [73, 74].

These anti-apoptosis genes also have the potential to improve production of membrane proteins. Indeed, O'Connor *et al.* [11] overexpressed the anti-apoptosis protein Bcl-x_L in concert with a receptor tyrosine kinase (RTK) membrane protein in CHO cells. RTKs are single-pass transmembrane proteins with cytoplasmic tyrosine kinase domains and extracellular ligand-binding domains [75]. CHO cells are widely used for protein production because they give high yields of secreted proteins, grow rapidly, can be easily adapted to the suspension conditions in a bioreactor, have defined media, and are safe for use in humans [1]. However, in some cases, expression of membrane proteins with CHO cells can be lower than from HEK cells.

In the O'Connor *et al.* [11] study, the specific RTK protein, ErbB2, was transfected into CHO cells both with and without the anti-apoptotic Bcl-x_L gene. Transient transfection Western blot analysis showed higher expression of ErbB2 in the anti-apoptosis CHO cell line compared to the CHO cell line without the gene. Expression levels in stable pools were detected by comparing the two cell lines in Western blots and flow cytometry experiments. The cells without the Bcl-x_L gene had decreased expression after several passages, but the cells containing the antiapoptosis gene had sustained expression over 33 days [11]. This method of increasing expression did not cause a change in the localization of the RTK protein that could alter its functionality. Finally, this effect of improved membrane protein production with overexpressed Bcl-x_L was seen in both pools of CHO cells and in CHO cell clonal isolates. Indeed, the researchers were unable to obtain any clones with significant detectable expression of the RTK in the CHO wild-type cells, but were able to obtain a number of clones with detectable RTK expression in the CHO Bcl-x_L cell lines. This anti-apoptotic engineering method may be extremely useful to improve the expression of other membrane proteins in CHO cell lines. The method by which an anti-apoptosis protein improves membrane production is not clear, however. Anti-apoptosis engineering may help protect the cells from apoptosis caused from the unfolded protein response [42, 76]. The unfolded protein response can occur when the overexpression of membrane proteins is sufficient to cause accumulation in the ER [77].

Another approach to anti-apoptosis engineering is to limit caspase activity, thereby preventing the caspase cascade from being activated. The advantage of engineering the caspase pathway is that this approach can protect the cells from both the intrinsic mitochondrial pathway to apoptosis and from the receptor-mediated pathway. The disadvantage of using this approach is that caspases are activated later in the apoptosis pathway and it will be more difficult to slow the cascade at this point. There are chemical inhibitors targeted to specific caspases, but adding these to bioreactors used for protein production can add to the already high cost associated with mammalian cell culture. Alternatively, genes that inhibit caspase activity can also be expressed in mammalian cell lines to limit apoptosis. Candidates include the inhibitor of apoptosis protein (IAP) family and its variants, p35 from *Autographa californica* nuclear polyhedrosis virus, and cytokine response modifier A (CrmA) [66, 78, 79]. Each of the inhibitors affects the apoptosis cascade in a different way because of the different caspases targeted. Therefore, the apoptosis insult that the cell will be protected from depends on the caspase inhibitor used.

Anti-apoptosis engineering and viral vector methods may potentially be combined to increase the viability of mammalian cells. One of the limitations of the viral vector method described above is that the infection by the virus can lead to cell death and in some cases the death may be potentially apoptotic. A way to limit the cell's death is to overexpress some of the anti-apoptosis genes. Indeed, in one study, Mastrangelo *et al.* [62] overexpressed Bcl-2 and Bcl-x_L in order to enhance cell viability and production of recombinant proteins for cells infected with the SFV viral vectors [62].This can be advantageous to the production process since inhibiting apoptosis can be used to extend the production stage following viral infection and perhaps improve yields of membrane protein targets. One limitation with anti-apoptosis engineering is that the increased production effects may vary in different cell lines or different conditions. Therefore, researchers must check initially to see if this method is helpful in the particular cell line considered.

5.3.3 Increasing Membrane Protein Expression by Chaperones

Aggregation, misfolding, and poor trafficking are some of the general problems encountered with the recombinant expression of membrane proteins, especially when they are overexpressed. The potential exists to limit this problem by increasing the folding and processing capacity of the host cells. Chaperones have the ability to increase the solubility of the proteins and to provide the folding or refolding of the proteins. For this reason, different chaperones have been coexpressed with various numbers of membrane proteins to promote the overexpression in different host cells [80, 81].

Ion channels are a very important group of proteins since they are widely produced in the brain, and they have significant roles in central and peripheral nervous system diseases such as Alzheimer's and Parkinson's diseases. Although mammalian cells are a potentially useful host for these proteins, their overexpression has always been problematic. Chaperones are known as the regulators of channel proteins and they have been used to increase the expression of them in mammalian cells. For instance, expression of α_7 nicotinic acetylcholine receptor (nAChR) protein has been attempted in different mammalian cell lines, but unfortunately functional expression has not been achieved. When this protein was coexpressed with the resistant to inhibitors of cholinesterase (Ric)-3 transmembrane chaperone, high protein expression of α_7 and α_8 nAChR was detected by Western blot [82–87]. Although the exact mechanism of interaction between Ric-3 and nAChRs is not known, it can be predicted from the experiments that Ric-3 has a specific chaperone activity towards nAChRs promoting the proper folding and assembly of nAChRs [82, 84, 87].

Furthermore, incubation of the mammalian cells with chemical chaperones may increase the proteins' expression. $Na_v1.8$ is a voltage-gated channel expressed from the dorsal root ganglion neurons and is important for the sensory nerve fibers. TsA201 (HEK-293 variant) cells were transfected with different sodium channels, but unfortunately the detected $Na_v1.8$ expression was very low. When the cells transfected with $Na_v1.8$ channel were incubated with lidocaine, the surface expression of the channel was dramatically increased. Lidocaine, a local anesthetic drug that provides loss of nociception, appears to be a chemical chaperone of $Na_v1.8$ expression. Lidocaine prevented the trapping of this membrane protein in the ER and provided high expression by demonstrating its chaperone activity [88].

5.3.4
Membrane Protein Expression in Cancer Cell Lines

Membrane proteins, such as those in the ErbB (epidermal growth factor receptors) family, have been targets of anticancer therapy studies. Cancer cells can overexpress their ErbB proteins that lead to the uncontrolled tumor growth [89]. The four receptors that comprise the ErbB family have varying specificity for ligand binding [90]. The ErbB2 receptor tyrosine kinase is overexpressed in 25% of breast cancer cells [89]. The mechanisms that cause the ErbB receptors to influence the malignant cellular processes of cancer have only recently become understood [12]. Cancer therapy has been developed against the overexpression of these receptors. The anti-cancer therapy can consist of either antibodies against the ligand-binding domain of the receptor or small molecules that limit kinase activity [90]. However, there has been a cardiac side-effect in some of these treatments that has limited its use due to the nonspecific cytotoxicity. It is unclear how the signaling pathway can cause cardiotoxic effects [89]. The issues with unintended side-effects illustrate the importance of continued research on membrane protein expression. Other RTKs can also have an impact on cancer progression. These RTK effects can be dependent on cell adhesion and activation of the epidermal growth factor receptors [91]. The phosphorylation of epidermal growth factor receptors is increased in attached cells versus cells grown in suspension.

Membrane proteins have been candidates for biomarker studies as well as targets for biotherapeutics due to the increased expression of these proteins in cancer cells. Cancer cell lines have been useful tools for studying these membrane protein targets. They are easy to culture since they can be grown in a similar manner to other mammalian cell lines. Different cancer cell lines will have different molecular properties such as the proteins expressed, the level of expression, and degrees of resistance to anticancer agents [92]. Probes, such as aptamers or antibodies, can be used to distinguish between these differences and assist in the purification of the desired membrane protein from the cancer cells [93]). The first step in developing a probe is to determine which membrane protein can be used as a potential biomarker. This requires comparing the cancer cell line of interest to noncancerous cells and other cancer cell lines, and evaluating the differences in membrane protein expression levels. Once the potential target has been identified, the probe has to be developed so that it will have specific binding to the protein of interest. Testing for specificity of the probe can be done by using immunofluorescence or flow cytometry and comparing the cancer cell line and a control cell line [93]. Once developed, the probes can be used to isolate either the protein from the cells or the cells with the protein from a mixture of cells.

Cancer cells can be transfected with membrane proteins to test for the effect of expression on the cells. For example, [94] examined the effect of a bound insulin-like growth factor-binding protein (IGFBP)-3, on the associated receptors' function in Ishikawa endometrial cancer cells. These cells naturally have between 15- and 20-fold more binding protein than the receptor [94]. Increasing the number of receptors on the cells through transfection allows functional assays to be completed more easily. Functional experiments can help illuminate the membrane proteins that may be responsible for some of the characteristics of cancer cells. The overexpression of some membrane proteins can lead to drug resistance of specific cancer cell lines [92]. Therefore, biotherapies that decrease the expression of these proteins may assist in increasing the efficiency of drug treatments. Alternatively, cancer cells in some cases may represent alternative systems to be considered for the overexpression of select membrane proteins that they produce as a result of their natural evolution or that can be transfected into the cells for high expression as a recombinant target.

5.3.5 Membrane Proteins as Biotherapeutics

Membrane proteins help regulate cellular processes necessary for cell survival. Abnormalities in membrane protein expression and functionality can affect the ability to control these processes and lead to the progression of diseases, including neurological disorders and cancer. For this reason, membrane proteins are studied as potential drug targets. For instance, blood–brain barrier membrane proteins play an important role in the transport and signaling characteristics of the blood–brain barrier [95]. Abnormalities in the functioning of the membrane proteins at this location can have an impact on the progression of neurological disorders. Understanding how membrane proteins involved in disease progression function in healthy individuals can help determine the appropriate targets for drug treatments for those afflicted with the disease. For the therapeutic use of membrane proteins, it is necessary to manufacture them in a manner that will create products safe for the consumer. In order to increase the biosafety of the production of recombinant proteins, it is important to remove the products from animal origin such as serum. The undefined products from animals found in serum and even serum-free media do not meet the stringent regulations. The contaminates that may be harmful to the consumers of the product should be reduced in the production process [96]. Therefore, there is a desire to create animal-free media by substituting the bovine albumin and other animal byproducts with recombinant proteins or chemically defined synthetic compounds. Some of these substitutions did not give the same level of cell viability due to deficiencies, such as lower zinc levels, in the supplements [96]. These deficits require additional supplements to supply the missing nutrients. On the other hand, the use of plant-derived hydrolyses has been found to allow as much cell growth as media containing serum [68]. The protein hydrolysate can be a cost-effective addition to bioreactor media since it has low cost and supplies amino acids and increases cell growth in serum-free media [97]. The hydrolysate addition can serve to increase cell growth and viability.

Furthermore, mammalian expression systems can be used for tissue engineering purposes. Immobilization of growth factors may provide a low-cost means to stimulate cell proliferation [98]. The immobilization involves increased stability of the growth factors' ability to bind to the extracellular portion of the growth factor receptors. The differences in degree of immobilization are due to the different strengths of the ionic and hydrophobic interactions during their formation [98]. Growth factors that have been immobilized in an oriented manner will lead to a stronger and more sustained growth factor receptor autophosyphorylation when compared to random immobilization [99]. The developments in the understanding of affinity of the growth factor family of membrane proteins to their receptors can be applied to wound-grafting treatments. Activation of the growth factor receptor will signal for the cell to grow and divide, allowing the wound to heal.

Membrane proteins have an important role in many cellular processes and an impact on diseases when these processes are interrupted. By studying the structure of membrane proteins it may be possible to determine how they function and gain an understanding of the molecular details of the protein's signal transduction pathway [100]. Determining these details of the membrane proteins will help to illustrate how they can be used as drug targets.

5.4 Conclusions

Mammalian expression systems represent an important host for membrane protein production because of established wide applicability to the therapeutic protein production processes and because their native intracellular environment can be helpful to mammalian protein expression. Furthermore, the capacity of mammalian cells to perform complex post-translational modifications such as glycosylation and lipid addition required for stability and function of some membrane proteins can make them the vehicle of choice for expression of some particularly complex mammalian membrane proteins. Indeed, many of the most relevant post-translational modifications critical to protein processing and function can be performed in mammalian systems, whereas some enzymes for these modifications are lacking in other organisms such as bacteria or yeast cells.

One of the major pitfalls to using mammalian cells is the high cost associated with the production of membrane proteins. Membrane proteins are expressed in low amounts within their native environments. The low expression and difficulties in isolating the proteins have made both biophysical and structural studies very difficult. Therefore, it is desirable to optimize the production process by increasing cell viability and protein production. Much future work should be dedicated to both lowering the costs of production of mammalian membrane proteins and increasing the yields of membrane proteins in mammalian cells.

There have been numerous attempts to create systems that will be helpful to increase the expression of membrane proteins in mammalian cells. Several of the approaches described in this chapter include using viral vectors, transient transfection, anti-apoptosis genes, chaperones, and using specialized cell lines. Although these techniques have been found to increase expression of some membrane proteins, the expression is still much lower than that obtained for some other secreted proteins. For example, secreted antibodies can reach yields of 10 g/l, while the maximum level of the most efficiently expressed membrane protein remains at about 1 mg/l. This illustrates the difficulties that still exist in expressing and studying membrane proteins. Membrane proteins only make up less than 1% of the proteins in the Protein Data Bank, illustrating the need to define ways to make studying these proteins easier [101]. In order to study the structures of membrane proteins it is essential to design methods that can improve the expression and purification of the proteins from the candidate mammalian hosts of choice [5, 6]. There is much work to be done in improving cell lines, expression vectors, and culture conditions, but many opportunities exist to make mammalian expression more effective and useful for the generation of the numerous membrane targets available.

Abbreviations

BHK	baby hamster kidney
CHO	Chinese hamster ovary
CMV	cytomegalovirus
CrmA	cytokine response modifier A
DHFR	dihydrofolate reductase
ECM	extracellular matrix
EF-1α	elongation factor 1α
ER	endoplasmic reticulum
GalNAc	*N*-acetylgalactosamine
GFP	Green Fluorescent Protein
GlcNac	*N*-acetylglucosamine
GPCR	G-protein-coupled receptor
GPI	glycosylphosphatidylinositol
GS	glutamine synthetase
HEK	human embryonic kidney

IAP	inhibitor of apoptosis protein
IGFBP	insulin-like growth factor-binding protein
MSX	methionine sulfoximide
MTX	methotrexate
nAChR	nicotinic acetylcholine receptor
NANA or Neu5Ac	*N*-acetylneuraminic acid
NGNA	*N*-glycolylneuraminic acid
PTP	permeability transition pore
Ric	resistant to inhibitors of cholinesterase
RSV	Rous sarcoma virus
RTK	receptor tyrosine kinase
SFV	Semliki Forest virus
SV	Sindbus virus
SV40	simian virus 40
VEEV	Venezuelan equine encephalitis virus

References

1 Mohan, C., Kim, Y.G., *et al.* (2008) Assessment of cell engineering strategies for improved therapeutic protein production in CHO cells. *Biotechnol. J.*, **3**, 624–630.

2 Wurm, F.M. (2004) Production of recombinant protein therapeutics in cultivated mammalian cells. *Nat. Biotechnol.*, **22**, 1393–1398.

3 Eifler, N., Duckely, M., *et al.* (2007) Functional expression of mammalian receptors and membrane channels in different cells. *J. Struct. Biol.*, **159**, 179–193.

4 Omasa, T., Onitsuka, M., *et al.* (2010) Cell engineering and cultivation of Chinese hamster ovary (CHO) cells. *Curr. Pharm. Biotechnol.*, **11**, 233–240.

5 Lundstrom, K. (2003) Semliki Forest virus vectors for large-scale production of recombinant proteins. *Methods Mol. Med.*, **76**, 525–543.

6 Lundstrom, K. (2003) Semliki Forest virus vectors for rapid and high-level expression of integral membrane proteins. *Biochim. Biophys. Acta*, **1610**, 90–96.

7 Lu, M., Echeverri, F., *et al.* (2003) Endoplasmic reticulum retention, degradation, and aggregation of olfactory G-protein coupled receptors. *Traffic*, **4**, 416–433.

8 Drew, D., Slotboom, D.J., *et al.* (2005) A scalable, GFP-based pipeline for membrane protein overexpression screening and purification. *Protein Sci.*, **14**, 2011–2017.

9 Tate, C.G. and Grisshammer, R. (1996) Heterologous expression of G-protein-coupled receptors. *Trends Biotechnol.*, **14**, 426–430.

10 Meents, H., Enenkel, B., *et al.* (2002) Impact of coexpression and coamplification of sICAM and antiapoptosis determinants bcl-2/bcl-x_L on productivity, cell survival, and mitochondria number in CHO-DG44 grown in suspension and serum-free media. *Biotechnol. Bioeng.*, **80**, 706–716.

11 O'Connor, S., Li, E., *et al.* (2009) Increased expression of the integral membrane protein ErbB2 in Chinese hamster ovary cells expressing the anti-apoptotic gene Bcl-x_L. *Protein Expr. Purif.*, **67**, 41–47.

12 Holbro, T., Civenni, G., *et al.* (2003) The ErbB receptors and their role in cancer progression. *Exp. Cell Res.*, **284**, 99–110.

13 Bane, S.E., Velasquez, J.E., *et al.* (2007) Expression and purification of milligram levels of inactive G-protein coupled receptors in *E. coli*. *Protein Expr. Purif.*, **52**, 348–355.

14 Mrksich, M. (2000) A surface chemistry approach to studying cell adhesion. *Chem. Soc. Rev.*, **29**, 267–273.
15 van der Valk, J., Brunner, D., *et al.* (2010) Optimization of chemically defined cell culture media – replacing fetal bovine serum in mammalian *in vitro* methods. *Toxicol. In Vitro*, 3 **24**, 1053–1063.
16 Heiser, W.C. (ed.) (2004) *Gene Delivery to Mammalian Cells*, Humana, Totowa, NJ.
17 Zheng, C. and Baum, B.J. (2008) Evaluation of promoters for use in tissue-specific gene delivery. *Methods Mol. Biol.*, **434**, 205–219.
18 Padmanabhan, R. and Thorgeirsson, S.S. (1997) Selection of transfected cells. Magnetic affinity cell sorting. *Methods Mol. Biol.*, **62**, 343–358.
19 Karlsson, G.B. and Liljestrom, P. (2004) Delivery and expression of heterologous genes in mammalian cells using self-replicating alphavirus vectors. *Methods Mol. Biol.*, **246**, 543–557.
20 Hassaine, G., Wagner, R., *et al.* (2006) Semliki Forest virus vectors for overexpression of 101 G protein-coupled receptors in mammalian host cells. *Protein Expr. Purif.*, **45**, 343–351.
21 Li, Y., Basilico, C., *et al.* (1994) Cell transformation by fibroblast growth factors can be suppressed by truncated fibroblast growth factor receptors. *Mol. Cell Biol.*, **14**, 7660–7669.
22 Rostami, A., Rabbani, M., *et al.*(2010) The role of N53Q mutation on the rat mu-opioid receptor function. *J. Biomol. Tech.*, **21**, 92–96.
23 Annaert, W.G., Becker, B., *et al.*(1997) Export of cellubrevin from the endoplasmic reticulum is controlled by BAP31. *J. Cell Biol.*, **139**, 1397–1410.
24 Wick, M., Hurteau, G., *et al.* (2002) Peroxisome proliferator-activated receptor-gamma is a target of nonsteroidal anti-inflammatory drugs mediating cyclooxygenase-independent inhibition of lung cancer cell growth. *Mol. Pharmacol.*, **62**, 1207–1214.
25 Chen, X., Whiting, C., *et al.* (2010) Integrin alpha1beta1 regulates epidermal growth factor receptor activation by controlling peroxisome proliferator-activated receptor gamma-dependent caveolin-1 expression. *Mol. Cell Biol.*, **30**, 3048–3058.
26 Bork, K., Reutter, W., *et al.*(2005) The intracellular concentration of sialic acid regulates the polysialylation of the neural cell adhesion molecule. *FEBS Lett.*, **579**, 5079–5083.
27 Thornhill, W.B., Wu, M.B., *et al.* (1996) Expression of Kv1.1 delayed rectifier potassium channels in Lec mutant Chinese hamster ovary cell lines reveals a role for sialidation in channel function. *J. Biol. Chem.*, **271**, 19093–19098.
28 Johnson, D. and Bennett, E.S. (2008) Gating of the shaker potassium channel is modulated differentially by N-glycosylation and sialic acids. *Pflugers Arch.*, **456**, 393–405.
29 Mah, W., Ko, J., *et al.* (2010) Selected SALM (synaptic adhesion-like molecule) family proteins regulate synapse formation. *J. Neurosci.*, **30**, 5559–5568.
30 Re, M., Pampillo, M., *et al.* (2010) The human gonadotropin releasing hormone type I receptor is a functional intracellular GPCR expressed on the nuclear membrane. *PLoS One.*, **5**, e11489.
31 Cao, G., Lee, K.P., *et al.* (2010) Methionine sulfoxide reductase B1 (MsrB1) recovers TRPM6 channel activity during oxidative stress. *J. Biol. Chem.*, **285**, 26081–26087.
32 Han, J.F., Wang, S.L., *et al.* (2006) Effect of genetic variation on human cytochrome p450 reductase-mediated paraquat cytotoxicity. *Toxicol. Sci.*, **91**, 42–48.
33 Huang, Z., Li, G., *et al.* (2005) Enhancing protein expression in single HEK 293 cells. *J. Neurosci. Methods*, **142**, 159–166.
34 Ray, K., Ghosh, S.P., *et al.*(2004) The role of cysteines and charged amino acids in extracellular loops of the human Ca(2+) receptor in cell surface expression and receptor activation processes. *Endocrinology*, **145**, 3892–3903.
35 Hansen, K.B., Mullasseril, P., *et al.* (2010) Implementation of a fluorescence-based screening assay

identifies histamine H3 receptor antagonists clobenpropit and iodophenpropit as subunit-selective N-methyl-D-aspartate receptor antagonists. *J. Pharmacol. Exp. Ther.*, **333**, 650–662.

36 Bertinato, J., Cheung, L., *et al.* (2010) Ctr1 transports silver into mammalian cells. *J. Trace. Elem. Med. Biol.*, **24**, 178–184.

37 Cook, B.L., Ernberg, K.E., *et al.* (2008) Study of a synthetic human olfactory receptor 17-4: expression and purification from an inducible mammalian cell line. *PLoS One*, **3**, e2920.

38 Chauhan, S.C., Vannatta, K., *et al.* (2009) Expression and functions of transmembrane mucin MUC13 in ovarian cancer. *Cancer Res.*, **69**, 765–774.

39 Dalby, B., Cates, S., *et al.* (2004) Advanced transfection with Lipofectamine 2000 reagent: primary neurons, siRNA, and high-throughput applications. *Methods*, **33**, 95–103.

40 Halterman, M.W., Giuliano, R., *et al.* (2009) In-tube transfection improves the efficiency of gene transfer in primary neuronal cultures. *J. Neurosci. Methods*, **177**, 348–354.

41 Betenbaugh, M.J., Tomiya, N., *et al.* (2004) Biosynthesis of human-type *N*-glycans in heterologous systems. *Curr. Opin. Struct. Biol.*, **14**, 601–606.

42 Jones, J., Krag, S.S., *et al.* (2005) Controlling *N*-linked glycan site occupancy. *Biochim. Biophys. Acta*, **1726**, 121–137.

43 Tomiya, N., Narang, S., *et al.* (2004) Comparing *N*-glycan processing in mammalian cell lines to native and engineered lepidopteran insect cell lines. *Glycoconj. J.*, **21**, 343–360.

44 Junge, F., Schneider, B., *et al.* (2008) Large-scale production of functional membrane proteins. *Cell. Mol. Life Sci.*, **65**, 1729–1755.

45 Campbell, C. and Stanley, P. (1984) A dominant mutation to ricin resistance in Chinese hamster ovary cells induces UDP-GlcNAc:glycopeptide beta-4-*N*-acetylglucosaminyltransferase III activity. *J. Biol. Chem.*, **259**, 13370–13378.

46 Reeves, P.J., Callewaert, N., *et al.* (2002) Structure and function in rhodopsin: high-level expression of rhodopsin with restricted and homogeneous *N*-glycosylation by a tetracycline-inducible *N*-acetylglucosaminyltransferase I-negative HEK293S stable mammalian cell line. *Proc. Natl. Acad. Sci. USA*, **99**, 13419–13424.

47 Shields, R.L., Lai, J., *et al.* (2002) Lack of fucose on human IgG1 N-linked oligosaccharide improves binding to human Fcgamma RIII and antibody-dependent cellular toxicity. *J. Biol. Chem.*, **277**, 26733–26740.

48 Durocher, Y. and Butler, M. (2009) Expression systems for therapeutic glycoprotein production. *Curr. Opin. Biotechnol.*, **20**, 700–707.

49 Hossler, P., Khattak, S.F., *et al.* (2009) Optimal and consistent protein glycosylation in mammalian cell culture. *Glycobiology*, **19**, 936–949.

50 Nadolski, M.J. and Linder, M.E. (2007) Protein lipidation. *FEBS J.*, **274**, 5202–5210.

51 Charron, G., Wilson, J., *et al.* (2009) Chemical tools for understanding protein lipidation in eukaryotes. *Curr. Opin. Chem. Biol.*, **13**, 382–391.

52 Ikezawa, H. (2002) Glycosylphosphatidylinositol (GPI)-anchored proteins. *Biol. Pharm. Bull.*, **25**, 409–417.

53 Epand, R.M. (2008) Proteins and cholesterol-rich domains. *Biochim. Biophys. Acta*, **1778**, 1576–1582.

54 Meinnel, T. and Giglione, C. (2008) Protein lipidation meets proteomics. *Front. Biosci.*, **13**, 6326–6340.

55 Liljestrom, P. and Garoff, H. (1991) A new generation of animal cell expression vectors based on the Semliki Forest virus replicon. *Biotechnology*, **9**, 1356–1361.

56 Martin, A., Damian, M., *et al.* (2009) Engineering a G protein-coupled receptor for structural studies: stabilization of the BLT1 receptor ground state. *Protein Sci.*, **18**, 727–734.

57 Lundstrom, K., Schweitzer, C., *et al.* (2001) Semliki Forest virus vectors: efficient vehicles for *in vitro* and *in vivo* gene delivery. *FEBS Lett.*, **504**, 99–103.

58 Lundstrom, K., Michel, A., *et al.* (1997) Expression of ligand-gated ion channels with the Semliki Forest virus expression

system. *J. Recept. Signal Transduct. Res.*, **17**, 115–126.

59 Shamotienko, O., Akhtar, S., *et al.* (1999) Recreation of neuronal Kv1 channel oligomers by expression in mammalian cells using Semliki Forest virus. *Biochemistry*, **38**, 16766–16776.

60 Lenhard, T., Marheineke, K., *et al.* (1998) Characterization of the human dopamine transporter heterologously expressed in BHK-21 cells. *Cell. Mol. Neurobiol.*, **18**, 347–360.

61 Ames, R., Nuthulaganti, P., *et al.* (2004) Heterologous expression of G protein-coupled receptors in U-2 OS osteosarcoma cells. *Receptors Channels*, **10**, 117–124.

62 Mastrangelo, A.J., Hardwick, J.M., *et al.* (2000) Part I. Bcl-2 and Bcl-x_L limit apoptosis upon infection with alphavirus vectors. *Biotechnol. Bioeng.*, **67**, 544–554.

63 Lundstrom, K. (2010) Expression of mammalian membrane proteins in mammalian cells using Semliki Forest virus vectors. *Methods Mol. Biol.*, **601**, 149–163.

64 Majors, B.S., Betenbaugh, M.J., *et al.* (2007) Links between metabolism and apoptosis in mammalian cells: applications for anti-apoptosis engineering. *Metab. Eng.*, **9**, 317–326.

65 Goswami, J., Sinskey, A.J., *et al.* (1999) Apoptosis in batch cultures of Chinese hamster ovary cells. *Biotechnol. Bioeng.*, **62**, 632–640.

66 Sauerwald, T.M., Oyler, G.A., *et al.* (2003) Study of caspase inhibitors for limiting death in mammalian cell culture. *Biotechnol. Bioeng.*, **81**, 329–340.

67 Alberts, B. (2002) *Molecular Biology of the Cell*. Garland Science, New York.

68 Burteau, C.C., Verhoeye, F.R., *et al.* (2003) Fortification of a protein-free cell culture medium with plant peptones improves cultivation and productivity of an interferon-gamma-producing CHO cell line. *In Vitro Cell. Dev. Biol. Anim.*, **39**, 291–296.

69 Rausch, J.M., Marks, J.R., *et al.* (2007) Beta-sheet pore-forming peptides selected from a rational combinatorial library: mechanism of pore formation in lipid vesicles and activity in biological membranes. *Biochemistry*, **46**, 12124–12139.

70 Arden, N. and Betenbaugh, M.J. (2006) Regulating apoptosis in mammalian cell cultures. *Cytotechnology*, **50**, 77–92.

71 Majors, B.S., Betenbaugh, M.J., *et al.* (2009) Mcl-1 overexpression leads to higher viabilities and increased production of humanized monoclonal antibody in Chinese hamster ovary cells. *Biotechnol. Prog.*, **25**, 1161–1168.

72 Nivitchanyong, T., Martinez, A., *et al.* (2007) Anti-apoptotic genes Aven and E1B-19K enhance performance of BHK cells engineered to express recombinant factor VIII in batch and low perfusion cell culture. *Biotechnol. Bioeng.*, **98**, 825–841.

73 Dorai, H., Kyung, Y.S., *et al.* (2009) Expression of anti-apoptosis genes alters lactate metabolism of Chinese Hamster Ovary cells in culture. *Biotechnol. Bioeng.*, **103**, 592–608.

74 Figueroa, B., Jr, Ailor, E., *et al.* (2007) Enhanced cell culture performance using inducible anti-apoptotic genes E1B-19K and Aven in the production of a monoclonal antibody with Chinese hamster ovary cells. *Biotechnol. Bioeng.*, **97**, 877–892.

75 He, L. and Hristova, K. (2008) Pathogenic activation of receptor tyrosine kinases in mammalian membranes. *J. Mol. Biol.*, **384**, 1130–1142.

76 Jones, J., Nivitchanyong, T., *et al.* (2005) Optimization of tetracycline-responsive recombinant protein production and effect on cell growth and ER stress in mammalian cells. *Biotechnol. Bioeng.*, **91**, 722–732.

77 Cudna, R.E. and Dickson, A.J. (2003) Endoplasmic reticulum signaling as a determinant of recombinant protein expression. *Biotechnol. Bioeng.*, **81**, 56–65.

78 Sauerwald, T.M., Betenbaugh, M.J., *et al.* (2002) Inhibiting apoptosis in mammalian cell culture using the caspase inhibitor XIAP and deletion mutants. *Biotechnol. Bioeng.*, **77**, 704–716.

79 Sauerwald, T.M., Figueroa, B., Jr, *et al.* (2006) Combining caspase and mitochondrial dysfunction inhibitors of apoptosis to limit cell death in mammalian cell cultures. *Biotechnol. Bioeng.*, **94**, 362–372.

80 Deuerling, E. and Bukau, B. (2004) Chaperone-assisted folding of newly synthesized proteins in the cytosol. *Crit. Rev. Biochem. Mol. Biol.*, **39**, 261–277.

81 de Marco, A., Deuerling, E., *et al.* (2007) Chaperone-based procedure to increase yields of soluble recombinant proteins produced in *E. coli*. *BMC Biotechnol.*, **17**, 32.

82 Millar, N.S. (2008) RIC-3: a nicotinic acetylcholine receptor chaperone. *Br. J. Pharmacol.*, **153** (Suppl. 1), S177–S183.

83 Banerjee, C., Nyengaard, J.R., *et al.* (2000) Cellular expression of alpha7 nicotinic acetylcholine receptor protein in the temporal cortex in Alzheimer's and Parkinson's disease – a stereological approach. *Neurobiol. Dis.*, **7B**, 666–672.

84 Lansdell, S.J., Gee, V.J., *et al.* (2005) RIC-3 enhances functional expression of multiple nicotinic acetylcholine receptor subtypes in mammalian cells. *Mol. Pharmacol.*, **68**, 1431–1438.

85 Roncarati, R., Seredenina, T., *et al.* (2008) Functional properties of alpha7 nicotinic acetylcholine receptors co-expressed with RIC-3 in a stable recombinant CHO-K1 cell line. *Assay Drug Dev. Technol.*, **6**, 181–193.

86 Valles, A.S., Roccamo, A.M., *et al.* (2009) Ric-3 chaperone-mediated stable cell-surface expression of the neuronal alpha7 nicotinic acetylcholine receptor in mammalian cells. *Acta Pharmacol. Sin.*, **30**, 818–827.

87 Williams, M.E., Burton, B., *et al.* (2005) Ric-3 promotes functional expression of the nicotinic acetylcholine receptor alpha7 subunit in mammalian cells. *J. Biol. Chem.*, **280**, 1257–1263.

88 Zhao, J., Ziane, R., *et al.* (2007) Lidocaine promotes the trafficking and functional expression of $Na_v1.8$ sodium channels in mammalian cells. *J. Neurophysiol.*, **98**, 467–477.

89 De Keulenaer, G.W., Doggen, K., *et al.* (2010) The vulnerability of the heart as a pluricellular paracrine organ: lessons from unexpected triggers of heart failure in targeted ErbB2 anticancer therapy. *Circ. Res.*, **106**, 35–46.

90 Sebastian, S., Settleman, J., *et al.* (2006) The complexity of targeting EGFR signalling in cancer: from expression to turnover. *Biochim. Biophys. Acta*, **1766**, 120–139.

91 Larsen, A.B., Stockhausen, M.T., *et al.* (2010) Cell adhesion and EGFR activation regulate EphA2 expression in cancer. *Cell Signal*, **22**, 636–644.

92 Yang, X. and Page, M. (1995) An M(r) 7-kDa membrane protein overexpressed in human multidrug-resistant ovarian cancer cells. *Cancer Lett.*, **88**, 171–178.

93 Shangguan, D., Cao, Z., *et al.* (2008) Cell-specific aptamer probes for membrane protein elucidation in cancer cells. *J. Proteome. Res.*, **7**, 2133–2139.

94 Karas, M., Danilenko, M., *et al.* (1997) Membrane-associated insulin-like growth factor-binding protein-3 inhibits insulin-like growth factor-I-induced insulin-like growth factor-I receptor signaling in ishikawa endometrial cancer cells. *J. Biol. Chem.* **272**, 16514–16520.

95 Agarwal, N. and Shusta, E.V. (2009) Multiplex expression cloning of blood–brain barrier membrane proteins. *Proteomics*, **9**, 1099–1108.

96 Zhang, J. and Robinson, D. (2005) Development of animal-free, protein-free and chemically-defined media for NS0 cell culture. *Cytotechnology*, **48**, 59–74.

97 Schlaeger, E.J. (1996) The protein hydrolysate, Primatone RL, is a cost-effective multiple growth promoter of mammalian cell culture in serum-containing and serum-free media and displays anti-apoptosis properties. *J. Immunol. Methods*, **194**, 191–199.

98 Boucher, C., St-Laurent, G., *et al.* (2008) The bioactivity and receptor affinity of recombinant tagged EGF designed for tissue engineering applications is defined by the nature and position of the tags. *Tissue Eng. A*, **14**, 2069–2077.

99 Boucher, C., Liberelle, B., *et al.* (2009) Epidermal growth factor tethered through coiled-coil interactions induces cell surface receptor phosphorylation. *Bioconjug. Chem.*, **20**, 1569–1577.

100 Marino, S.F. (2009) High-level production and characterization of a G-protein coupled receptor signaling complex. *FEBS J.*, **276**, 4515–4528.

101 Seddon, A.M., Curnow, P., *et al.* (2004) Membrane proteins, lipids and detergents: not just a soap opera. *Biochim. Biophys. Acta*, **1666**, 105–117.

6 Membrane Protein Production Using Photosynthetic Bacteria: A Practical Guide

Philip D. Laible, Donna L. Mielke, and Deborah K. Hanson

6.1 Introduction

6.1.1 The Membrane Protein Problem

The cell membrane serves as the interface between an organism and its environment, and internal membranes in eukaryotes separate functional compartments within cells. Proteins inserted in these membranes carry out many essential biological processes, including uptake of nutrients, excretion of wastes, signal transduction, and response to external stimuli. In addition, membrane proteins are used in elaborate bioenergetic schemes to fuel all normal cellular activities in healthy organisms. In this postgenomic era, about 35% of the genes in *any* genome encode membrane proteins. The fraction of proteins associated with the membrane in eukaryotes may be even higher (up to 40%). Notably, membrane proteins constitute the majority of drug targets; thus, knowledge of the structures of these proteins would contribute greatly to our understanding of biological processes. Unfortunately, structural information for membrane proteins is exceedingly scarce because it is notoriously difficult to purify quantities of native material that are sufficient for crystallization attempts.

Almost any viable strategy to obtain structural information for membrane proteins *must* rely on heterologous expression to produce enough of the target membrane protein before undertaking purification and crystallization attempts. Milligram quantities of pure membrane proteins are needed because the parameter space that must be searched in attempts of both purification and crystallization is quite large, much larger than the equivalent for a soluble protein. No precedent exists for a single "magic bullet" set of generic conditions (detergents, temperatures, incubation time, protein/surfactant ratios, crystallization-inducing agents, crystallization methods, etc.) under which membrane proteins can be purified in high yield in functional form and crystallized for structure determination.

Production of Membrane Proteins: Strategies for Expression and Isolation, First Edition.
Edited by Anne Skaja Robinson.

Heterologous expression of a candidate protein in *Escherichia coli* offers many advantages such as simplicity, low cost, and rapid growth. Eukaryotic protein expression systems offer the capability for post-translational modification of expressed proteins. Many have been employed successfully, but they are often cumbersome and extremely expensive for the preparation of the quantities of membrane proteins that are necessary for structure determination experiments. Other limitations of both prokaryotic and eukaryotic host organisms include inadequate membrane volume for accommodation of heterologously expressed membrane proteins [1, 2] and saturation of the secretory machinery for integration of the heterologous protein into the membrane [3]. Thus, overexpression strategies often fail, resulting in cell death or precipitation of aggregates of the heterologously expressed membrane protein as inclusion bodies.

The failure of many such expression systems with membrane proteins occurs because the space in the cell's membranes is already occupied by its own membrane proteins. Alternatively, several other expression systems that are tailored to the expression of membrane proteins are in use or in development [1, 2, 4–13]. A theme common to all of them is a proliferation of membranes that can address the problem of compartment space for the incorporation of heterologously expressed membrane proteins.

6.1.2 Exploiting the Physiology of Photosynthetic Bacteria

Photosynthetic bacteria are particularly capable of addressing the membrane protein expression problem since they feature an inducible intracytoplasmic membrane (ICM). Under conditions of light and/or lowered oxygen tension, the membrane surface in the native organism increases many-fold when the ICM is elaborated as invaginations of the cytoplasmic membrane (Figure 6.1; reviewed in [14]). Concomitantly, the same environmental cues induce synthesis of the photosynthetic apparatus. The new ICM in the native organism functions to sequester these complexes that are composed of transmembrane polypeptides and their

Figure 6.1 Under certain growth conditions, many new ICM vesicles (a; arrows in electron micrograph) are induced in *Rhodobacter*. This newly elaborated membrane houses the transmembrane protein–cofactor complexes of the photosynthetic unit (shown schematically in (b), excerpted from its specialized membrane environment in (c)). Aerobic conditions repress both the induction of this new membrane and synthesis of the photosynthetic unit (d). Since pigment biosysnthesis is regulated similarly, the color of the culture indicates that favorable conditions have been achieved for either induction (e and f) or repression (g) of membrane, protein, and pigment biosysnthesis. Maximal induction of these components occurs under anaerobic conditions where the organism is growing photosynthetically (h). The *Rhodobacter* membrane protein expression system takes advantages of key features of this physiology. The utility afforded by our engineered expression strains is that they can provide coordinated synthesis of foreign membrane proteins with synthesis of new membrane into which the nascent polypeptides can be incorporated.

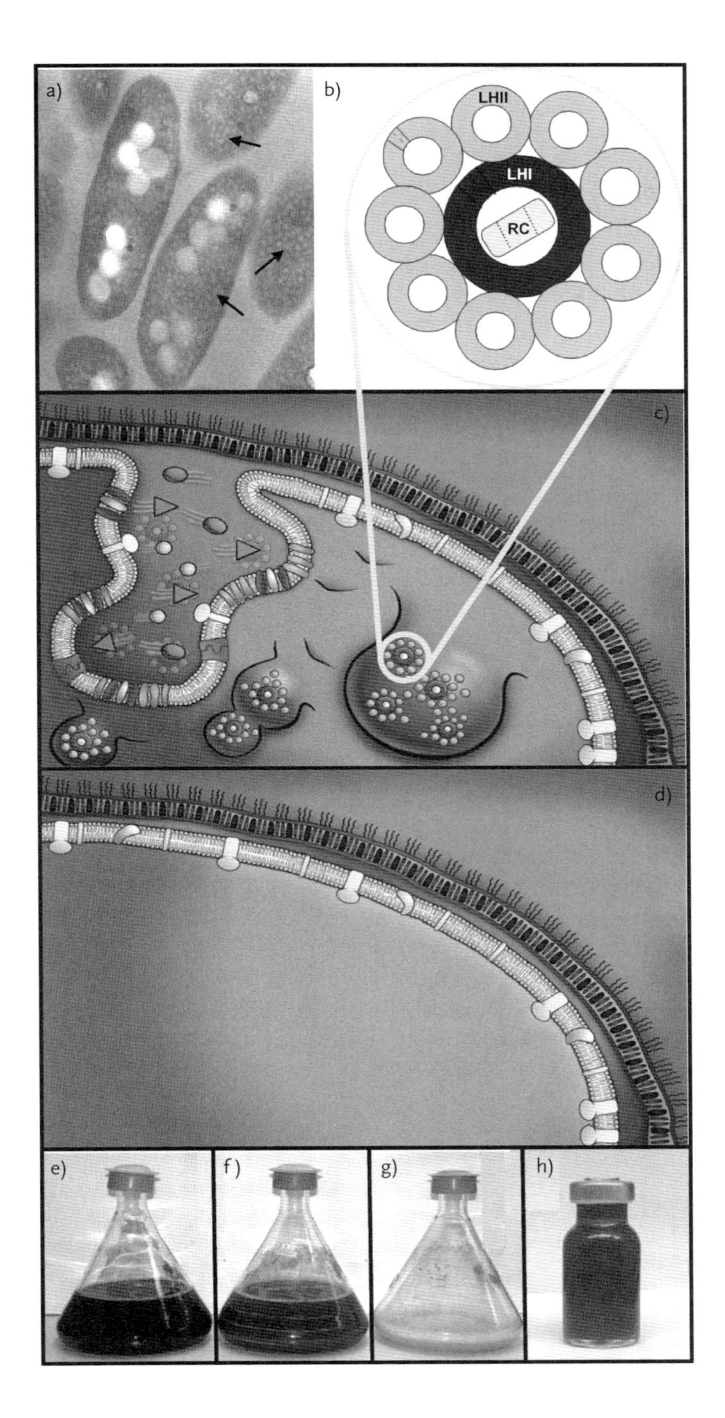
a)
b)
LHII
LHI
RC
c)
d)
e)
f)
g)
h)

associated hydrophobic redox and energy transfer cofactors (Figure 6.1b and c; reviewed in [15]). Upon cell disruption, the ICM invaginations break apart from the host's cytoplasmic membrane becoming sealed "inside-out" particles that, by virtue of their size, are easily isolated with differential centrifugation. Synthesis of the ICM and expression of its associated proteins, while coordinated, are not obligatorily linked. Mutant strains lacking some or all of the photosynthetic proteins still elaborate the ICM when induced appropriately.

We have exploited this inducible ICM of *Rhodobacter sphaeroides* to develop both *in vivo* and cell-free systems for the expression of foreign membrane proteins. We have employed the cell-based system for the heterologous expression of membrane proteins from a variety of prokaryotes, as well as fungi, plants, and mammals. The essential components of the *in vivo* system are summarized below and experimental protocols that enable adaptation of the *Rhodobacter* membrane protein expression system to the production of any target protein in any laboratory setting are described herein.

6.1.3 **Expression Strategies**

As in any *in vivo* expression strategy, the expression system consists of a host strain and an expression plasmid. Several hosts are available that carry deletions of genes encoding proteins of the photosynthetic apparatus. Likewise, a suite of plasmid vectors featuring cloning sites for insertion of foreign genes has been constructed. Heterologous expression is driven by a promoter that normally drives expression of genes encoding photosynthetic proteins. With any host/vector combination, genes encoding membrane proteins of interest can be induced by the environmental cues that also induce the ICM, and the overexpressed protein products are then incorporated into and purified from the ICM.

6.1.3.1 **Design of the Expression Plasmids**

Platform vectors (Figure 6.2) are based on the *puf* and *puc* operons that encode transmembrane components of the native photosynthetic apparatus. They utilize

Figure 6.2 The majority of the structural genes for the photosynthetic unit in *Rhodobacter* are encoded by the *puf* (a) and *puc* (b) operons. These operons have been utilized in the construction of platform vectors for the *Rhodobacter* expression system, based on the broad-host-range vector pRK404. Here, native genes are removed such that foreign genes can be inserted and placed under control of the oxygen- and/or light-regulated *puf* (P*puf*) or *puc* (P*puc*) promoters. In vector (c) and (d), foreign genes replace genes encoding the core light-harvesting antennae (LHI) or reaction center L and M genes in the *puf* operon. In vector (e), foreign genes replace genes encoding the peripheral light-harvesting antennae (LHII). While vectors (c) (pRKPLHT1), (d) (pRKPLHT4), and (e) (pRKPLHT7) use ligation-dependent cloning, vector (f) (pRKLICHT1) serves as an analog of vector (c) to facilitate ligation-independent cloning (LIC). Vectors encode a C-terminal heptahistidine tag (HT) that is fused in frame to two stop codons (*). Restriction sites in bold type are unique. A region of stable RNA secondary structure (hairpin) dictates the stability of the upstream transcript. Foreign genes can be inserted via the *SpeI*, *NdeI* and *BglII* sites in the multiple cloning site (MCS) of vectors (c, d, and e). Vector (f) carries a *PmlI* site that enables LIC.

either ligation-dependent (Figure 6.2c–e) or ligation-independent (Figure 6.2f and Figure 6.3a and b) cloning strategies and are designed to fuse an affinity tag to the target gene such that the foreign gene products can be purified efficiently by affinity chromatography. The platform vectors are based upon broad-host-range vector, pRK404 [16], which is maintained stably *in trans* in *R. sphaeroides* by selection for tetracycline resistance.

The plasmids shown in Figure 6.2 encode C-terminal heptahistidine tags. The inventory of expression vectors also includes plasmids that feature both cleavable and uncleavable N-terminal polyhistidine tags, as well as extended N- and C-terminal polyhistidine tags (10 × His, 13 × His) that enable increased availability and/or tighter binding of the tag to chromatography resin during purification

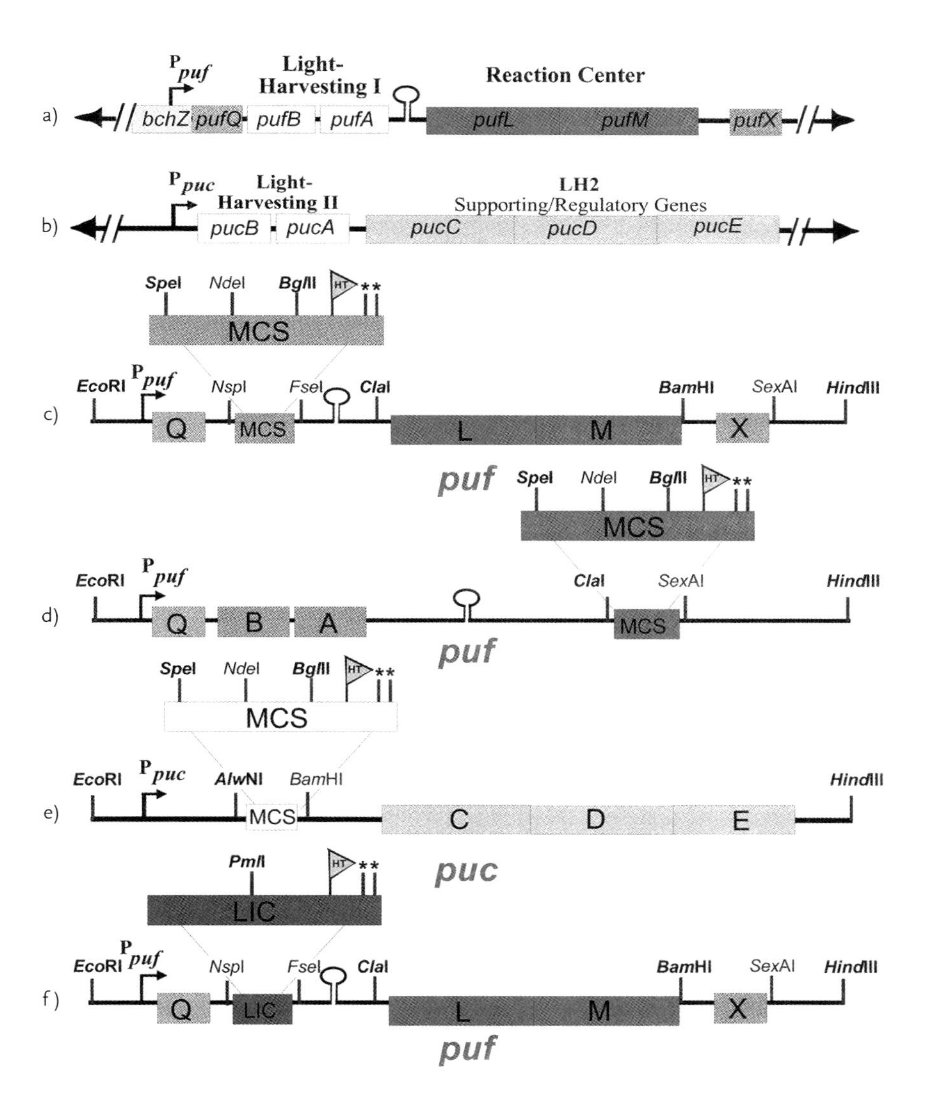

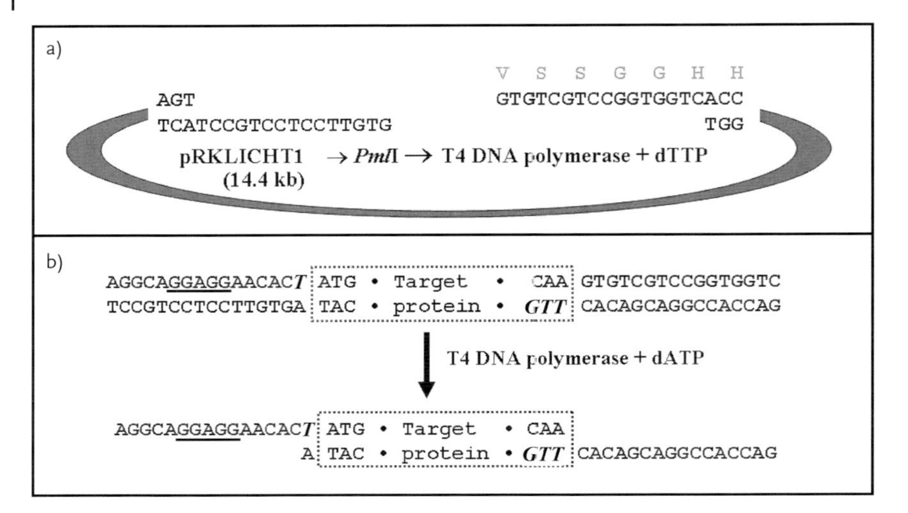

Figure 6.3 (a) LIC strategy employed to clone target membrane protein genes into pRKLICHT1. This vector is designed to fuse a C-terminal, seven-membered histidine tag (partially shown). The single-stranded overhangs are generated by *PmlI* digestion followed by 3′ → 5′ exonuclease activity of T4 DNA polymerase in the presence of dTTP. The "linker" residues between target gene and tag are also shown. (b) LIC overhangs for amplified target genes compatible for insertion into pRKLICHT1 are generated by T4 DNA polymerase in the presence of excess dATP. The resulting T_ms of the overhangs are sufficient to allow the transformation of competent *E. coli* to tetracycline resistance after a brief annealing process at room temperature. The single-stranded regions depict the LIC overhangs that are generated by T4 DNA polymerase digestion; underlined bases denote the RBS; and in bold-italic are noncomplementary, obligate bases that are necessary for generation of the LIC overhang.

by immobilized metal-affinity chromatography. Other platform vectors were developed that utilize the 1D4 epitope from bovine rhodopsin as a C-terminal tag or incorporate sequences that append an N-terminal signal peptide or membrane anchor domain. Expression plasmids are transferred to *R. sphaeroides* via conjugation from *E. coli* donor S17-1 [17]; its copy number in *Rhodobacter* strains is 4–6 per cell [18].

6.1.3.2 Design of Expression Hosts

Various strains of *R. sphaeroides* differing in the complement of native proteins present in their ICM have been evaluated as hosts for heterologous expression of membrane proteins. On the assumption that deletion of native ICM proteins could yield a strain that could accommodate more heterologously expressed membrane protein, engineered strains lack one or more of the three pigment–protein complexes of the photosynthetic apparatus (Figure 6.4). Some of these deletions render the organism incapable of photosynthetic growth, but *Rhodobacter* is a versatile organism that thrives under various growth regimes.

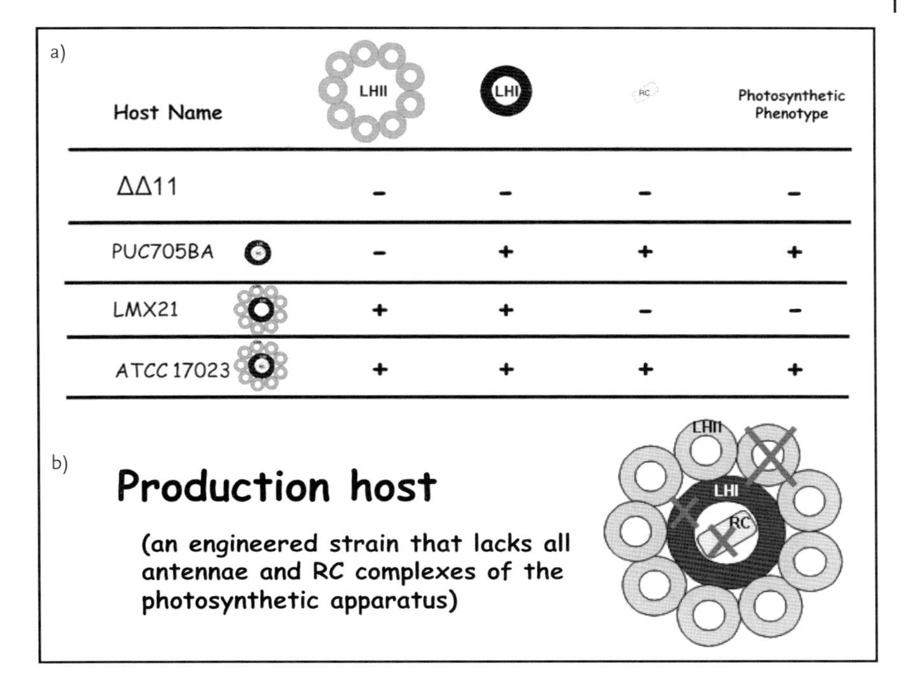

Host Name	LHII	LHI	RC	Photosynthetic Phenotype
ΔΔ11	-	-	-	-
PUC705BA	-	+	+	+
LMX21	+	+	-	-
ATCC 17023	+	+	+	+

Figure 6.4 (a) Several strains of *R. sphaeroides* that differ in the complement of native proteins present in the ICM have been evaluated. These strains were either acquired or constructed to test whether more foreign membrane protein might be incorporated into a partially depleted ICM. They range from a true wild-type (ATCC 17023) to an engineered strain deleted for the three pigment-protein complexes of the bacterial photosynthetic unit (ΔΔ11). The latter host strain is used most extensively in the production version of the *Rhodobacter* membrane protein expression system (b).

6.1.3.3 Autoinduction Conditions

Protein expression is controlled easily. By manipulating the culture conditions of the organism with regard to the induction cues (i.e., oxygen and/or light), expression of the desired protein is coordinated with the synthesis of the ICM. Depending upon the host strain employed, cultures can be grown either photosynthetically or chemoheterotrophically in the absence of light. In photosynthetic cultures (Figure 6.1h), anaerobiosis and/or light autoinduce the coupled synthesis. Repression of the *puf* and *puc* promoters can be achieved under aerobic conditions (e.g., with rapid shaking of flasks carrying a relatively small amount of medium; Figure 6.1g) when expression needs to be controlled tightly, as in the case of expression of a toxic protein or incorporation of selenomethionine (SeMet) or other labels into induced proteins [19].

In chemoheterotrophic cultures (dark, 34 °C, semi-aerobic; Figure 6.1e and f), concomitant synthesis of ICM and heterologous protein is autoinduced when the oxygen tension lowers as the cell density increases. No adjustment of culture conditions by the user is required. Protein expression is observed in early log-phase cultures and – for the majority of target proteins – reaches a maximum in late log-phase cultures when oxygen becomes limiting. At this point, cultures will have become highly pigmented as pigment biosynthesis is induced by the same cues (Figure 6.1e and f) .

6.1.4 Summary of Success Stories

Through extensive testing with foreign genes, the best yields of heterologous expression were obtained with plasmid pRKPLHT1 (Figure 6.2c) where foreign genes were inserted in the *puf* operon in place of the structural genes for LHI. Synthesis of the foreign protein is directed by the oxygen/light-sensitive *puf* promoter. The analysis has shown clearly that the highest heterologous expression was observed when a host strain with a modified ICM (Figure 6.4b) was employed. This host is devoid of the large number of native complexes of the photosynthetic apparatus that normally filled this membrane space.

In an initial case study, genes representing the entire *E. coli* membrane proteome (1030 members, 444 unique representatives chosen) were selected for cloning and expression studies. The majority of these targets were unannotated or "hypothetical" membrane proteins. Analysis of more than 400 of the expression strains has shown that approximately 60% of the *E. coli* membrane proteins were expressed in *R. sphaeroides* at levels that exceed 1 mg/l. Proteins that were expressed at or above this threshold level encompassed a wide range of size (14–86 kDa), isoelectric points (5.9–11.9), and transmembrane spans (1–14). Subcellular fractionation revealed that these expressed *E. coli* membrane proteins were localized within the *Rhodobacter* ICM, and many could be purified to yields of more than 10 mg/l of culture, rivaling the expression levels of native ICM proteins.

Subsequently, targets of increasing complexity have been introduced into the *Rhodobacter* expression system. This expanded target set included membrane proteins from a variety of functional classes and organisms – including eukaryotes. Among these targets are G-protein-coupled receptors (GPCRs) and ion channels that are of enormous interest in the development of pharmaceuticals and biomimetic devices. Success has been achieved in expressing GPCRs in *Rhodobacter*, including human HIV coreceptors CCR5 and CXCR4. *Rhodobacter* also excels in expression of multisubunit complexes, including those requiring protein maturation and attachment of complex cofactors [5, 6, 10, 20, 21].

Outlined below are the standardized protocols that have been established for cloning, expression, and purification of membrane proteins using *R. sphaeroides*. Details are provided for methods that are specific to this prokaryotic host. All media and buffer formulations are cataloged in the Appendix.

6.2
Preparation of Expression Constructs

While we realize that many different strategies can be used to perform the DNA manipulations that are necessary for the construction of expression plasmids, we present here an outline of the protocols that have been successful – and, therefore, have become standardized – in this laboratory. When pertinent, considerations for larger-scale experiments or *Rhodobacter*-specific considerations are brought to attention.

6.2.1
Platform Vector Preparation

In order to prepare the platform vectors for ligation-dependent (pRKPLHT1; Figure 6.2c) or ligation-independent (pRKLICHT1; Figure 6.2f) cloning, the following section outlines steps by which the relatively large vectors are linearized and compatible, cohesive ends are generated. The protocols for ligation-dependent cloning that use restriction enzymes are outlined separately from the protocols for LIC that employ the proofreading exonuclease activity of T4 DNA polymerase.

6.2.1.1 Large-Scale Vector Preparation Protocol for Ligation-Dependent Cloning

The platform vector, pRKPLHT1, used for ligation-dependent cloning has a simple multiple cloning site region with three unique restriction sites (*Spe*I, *Nde*I, and *Bgl*II) for target gene insertion (Figure 6.2c). Routine cloning employs *Spe*I and *Bgl*II, and protocols below are designed and written based on the assumption that these restriction endonucleases will be utilized.

Set up the following preparative-scale restriction enzyme digestion and allow it to incubate at 37 °C for at least 2 h to ensure complete digestion.

25 μl pRKPLHT1 (7.5–30 μg)
2.5 μl *Spe*I
2.5 μl *Bgl*II
8 μl 10 × Promega buffer B
42 μl sterile ddH_2O
For a total volume of 80 μl

Most commercial miniprep kits employ RNase during cell lysis, but RNase should be included in this reaction if plasmid DNA is prepared via simple alkaline lysis miniprep protocols.

When preparing to gel purify the DNA fragments, pour an 0.8% agarose gel with wide, preparative-scale lanes. Run the gel for at least 1.5 h at 60 V to help determine whether the digestion was complete and to be able to separate linear from circular uncut plasmid DNA. Excise the linear fragment. Extract the DNA from the excised agarose using a commercially available gel extraction kit. Before using the digested vector in an experimental ligation reaction, perform a control ligation/transformation experiment to determine the background of colonies in any experimental reaction that result from contamination of the digested vector with uncut or singly digested vector. Store the digested vector at 4 °C (or at −80 °C for long-term storage).

In the event that large amounts of digested platform vector are desired for the cloning of multiple target genes, set up multiple tubes using this protocol to generate a large supply of digested platform vector. We have found that the conditions that work well for small reactions do not scale well to large volumes, thus small reactions are preferred to one larger reaction to keep background levels of uncut plasmid low.

6.2.1.2 Large-Scale Vector Preparation Protocol for LIC

To generate the LIC overhangs, platform vector pRKLICHT1 (Figure 6.2f) is first linearized by digestion with *Pml*I and then treated with T4 DNA polymerase in the presence of dTTP. The exonuclease activity of the polymerase yields the single-stranded overhangs that are shown in red in Figure 6.3(a).

Vector Linearization with *Pml*I Digest 10 μl (around 1–2 μg) pRKLICHT1 (1/5 of the yield of plasmid DNA from a standard miniprep protocol) with *Pml*I in a 70-μl reaction volume for 1 h at 37 °C. *Pml*I is an unstable enzyme and retains maximal activity when stored at −80 °C in small aliquots. Best results are achieved by adding a second aliquot of enzyme half-way through the incubation. Following digestion, clean up the reaction with any standard purification kit or method that is suitable for plasmids larger than 10 kb.

Generation of LIC Overhangs One half of the *Pml*I-digested DNA should be used in generating the single-stranded overhangs with LIC-qualified T4 DNA polymerase.

pRKLICHT1/*Pml*I
1 μl 100 mM dTTP
2 μl 100 mM DTT
4 μl 10 × T4 polymerase reaction buffer
1 unit T4 DNA polymerase
Total volume of 40 μl

Incubate at room temperature for 30 min, then inactivate the polymerase at 75 °C for 20 min. This inactivated mixture can be used directly in annealing reactions or it can be cleaned up using a standard purification kit or method that is suitable for plasmids larger than 10 kb.

Before using the digested vector in an experimental reaction, perform a control ligation/transformation experiment to determine the background of colonies resulting from contamination of it by undigested vector. Again, since conditions that work well for small reactions often do not scale well to large volumes, the best results are achieved when multiple small reactions are performed and then combined following enzymatic digestion. Store the digested vector at 4 °C (long-term storage at −80 °C).

6.2.2 Design of Oligonucleotide Primers for Gene Amplification and Cloning

Determine whether the amplicon will be inserted by ligation-dependent cloning or LIC methodology. The genomic DNA of *Rhodobacter* is characterized by a high GC content (68%), therefore several codons are underutilized [22]. Therefore, another initial consideration is to determine whether there are any codons in the first 50 that are extremely rare in *Rhodobacter* (e.g., TTA). If so, one may want to consider cloning a paralog or homolog that lacks rare codons. For a higher-throughput approach to designing sets of oligonucleotides for the cloning of multiple target genes, primer generator tools [23] can be used *in lieu* of manual design.

6.2.2.1 Ligation-Dependent Cloning

For ligation-dependent cloning, decide which enzymes will be used to insert the gene into the expression vector, choosing those enzymes for which there are no sites in the target gene. The *Spe*I site in pRKPLHT1 is also compatible with *Xba*I and *Avr*II overhangs. The *Bgl*II site in pRKPLHT1 is compatible with *Bam*HI and *Bcl*I. Note that combining the *Bcl*I overhang with that of *Bgl*II produces an in-frame TGA stop codon; this may be desirable if it is preferable to express the target protein without the C-terminal polyhistidine tag.

Typical 5′- and 3′-oligonucleotides begin with four to six "dummy" bases at the 5′-end of each oligonucleotide to enable efficient digestion of the amplicon by the restriction enzyme. This sequence is followed in the "top" primer by the restriction site sequence and a ribosome-binding site (RBS, for *Rhodobacter* = GGAGG) placed 4–12 bases before the start codon; typically, the RBS is placed six bases before the start codon. The "bottom" primer incorporates the sequence for the second restriction enzyme site followed by the gene sequence on the complementary strand. Note that a polyhistidine tag and stop codons are encoded by the platform vectors (Figure 6.2), thus the native stop codon of the target gene should not be included in the amplicon. Oligonucleotides should be designed such that they have good GC content at the 3′-ends; at least three contiguous Gs or Cs are recommended.

Using any standard software, examine the oligonucleotide sequence to determine the melting temperature of the *complementary* region for use in determining annealing temperature for polymerase chain reaction (PCR) reactions. T_ms of the complementary regions of the primer sets should match within 5 °C.

6.2.2.2 LIC

Typical 5′- and 3′-oligonucleotides for use in LIC of a target gene are shown in Figure 6.3(b). The 5′-end of the top primer begins with the sequence that provides a LIC overhang which is complementary to that of the platform vector (Figure 6.3a), followed by the RBS placed 4–12 bases before the start codon. The 5′-end of the bottom primer begins with the other complementary LIC overhang, followed by the gene sequence. Both primers incorporate special noncomplementary bases that are required for proper T4 polymerase processing. Note that a polyhistidine tag and stop codons are encoded by the platform vectors, thus the native stop codon of the target gene should not be included in the amplicon. Oligonucleotides should be designed such that they have good GC content at the 3′-ends; at least three contiguous Gs or Cs are recommended.

Using any standard software, examine the oligonucleotide sequence to determine the melting temperature of the *complementary* region and to check for undesirable regions of stable secondary structure.

6.2.3 Target Gene Preparation

6.2.3.1 PCR Amplification of Target Gene

Perform a PCR reaction with the designed oligonucleotides to generate an amplicon carrying the gene encoding the target protein. Determine the yield of amplicon by running 10% of the PCR reaction mixture on an agarose gel. Remove salts/enzyme from the remainder of the PCR reaction by using a method or commercial kit of choice. Generate clonable fragments by either of the following methods:

6.2.3.2 Restriction Enzyme Digestion of PCR Amplicon

Perform a reaction to digest the amplicon with the enzymes whose restriction sites were included in the N-terminal and C-terminal oligonucleotide sequences (e.g., *Spe*I, *Xba*I, or *Avr*II for the 5′-end of the amplicon and *Bgl*II, *Bam*HI, or *Bcl*I for the 3′-end of the amplicon). Following the digestion reaction, remove enzymes and buffer by using a method or commercial reaction cleanup kit of choice.

6.2.3.3 Digestion of PCR Amplicon to Generate LIC Overhangs

Set up and perform a reaction to produce single-stranded overhangs (Figure 6.3b) that are complementary to the ones present in the platform vector (Section 6.2.1.2.). Use digestion conditions similar to those used to prepare the LIC platform vector (Section 6.2.1.2), in this case using dATP with LIC-qualified T4 DNA polymerase in the reaction. Remove enzymes and buffer following the digestion reaction by using a method or commercial reaction cleanup kit of choice.

6.2.4 Cloning of Digested Amplicons

6.2.4.1 Generation of Recombinant Plasmids

Recombinant molecules will be constructed by either ligation or annealing of LIC overhangs.

Ligation of Restriction Enzyme-Digested Amplicon and Platform Vector Set up a reaction to ligate the digested amplicon with the vector prepared according to the protocol in Section 6.2.1.1. A typical reaction utilizes 1–2 μl (around 50 ng) of the prepared vector stock and variable amounts of amplicon, depending on its concentration; the typical volume of the ligation reaction is 20 μl. Ensure that the background of colonies obtained following transformation with a control ligation of vector with no insert is minimal before using the vector stock for cloning of amplicons.

Annealing of Complementary LIC Platform Vector and Amplicon Mix the platform vector and amplicon to anneal complementary LIC overhangs produced by the exonuclease activity of T4 DNA polymerase (Sections 6.2.1.2 and 6.2.3.3). Incubate the reaction for 5 min at room temperature. A typical reaction utilizes 1–2 μl (around 50 ng) of the LIC platform vector and variable amounts of digested amplicon, depending on its concentration; the volume of the reaction is typically 20 μl. Again, ensure that the background of colonies obtained following transformation with a control reaction using vector with no insert is minimal.

6.2.4.2 Transformation of *E. coli* with Ligation or LIC Reactions

Use 5 μl of the reaction to transform an *E. coli* strain that is suitable for cloning (e.g., DH5α, XL1-Blue, etc.) according to any standard protocol, selecting transformants on LB agar containing 15 μg/ml tetracycline (LB/tet_{15}). Inoculate several colonies for overnight growth in 2xTY/tet_{15} medium for minipreps to screen for successful cloning of the desired insert. The number of colonies to screen is determined by knowledge of the background number of colonies obtained in transformations of control reactions utilizing vector with no insert.

6.2.5 Screening for Successful Insertion of Target Gene into Platform Vector

Enzymes that are conveniently used to screen for cloning of a PCR product into either platform vector are *Eco*RI and *Cla*I. These are unique within both platform vectors and flank the *Spe*I/*Bgl*II or LIC sites. In the absence of an insert, digestion with these enzymes will produce fragments of 13 116 and 1250 bp. Successful cloning of the PCR product will increase the size of the 1250-bp fragment according to the size of the PCR product (in the absence of *Eco*RI or *Cla*I sites within the PCR fragment). The relative positions of these sites are diagramed in Figure 6.2(c) for platform vector pRKPLHT1. It is recommended that the sequence of the expression construct be verified by using primers that flank the N-terminal elements and C-terminus of the inserted gene. Oligonucleotides that have yielded good-quality sequence data are: N-terminal primer 5′- GCGGCGGATTAATCG GG-3′, whose 3′-end is located 45 bp upstream (5′) of the *Spe*I site in the MCS of platform vector pRKPLHT1; and C-terminal primer 5′-TCCCTCTCCGCTTGCA GT-3′, whose 3′-end is located 97 bp from the second stop codon following the His_7 tag of either platform vector.

6.3
Transfer of Plasmid DNA to *Rhodobacter* via Conjugal Mating

Rhodobacter is not capable of being transformed with pure, double-stranded DNA containing sites for endogenous restriction enzymes. Therefore, the most efficient means for transfer of plasmid DNA to expression hosts is conjugation, whereby single-stranded DNA is transferred, thus bypassing the host's restriction system. Common *E. coli* strains used as hosts for cloning do not encode components necessary for mobilization and transfer of plasmids via conjugal mating. In order to enable transfer of the expression vector to *Rhodobacter* hosts, an *E. coli* host capable of conjugal mating (e.g., S17-1 [Tp^r Sm^r *recA*, *thi*, *pro*, *hsd*R^-M^+RP4: 2-Tc:Mu: Km Tn7]; ATCC 47055) must be utilized.

6.3.1
Transformation of *E. coli* S17-1

Prepare competent S17-1 cells according to any standard protocol. Transform 100 μl competent S17-1 with less than 100 ng of expression plasmid DNA according to a standard protocol, selecting transformants on LB/tet_{15} agar.

6.3.2
Conjugation of *E. coli* with *R. sphaeroides*

Two days prior to conjugation, inoculate 25 ml of MR26 medium (+ 1 × vitamins; 50-ml screw-cap Erlenmeyer flask) with the recipient strain ΔΔ11 and grow it, with shaking at 125 rpm, at 32 °C. The day before the conjugation, inoculate 3 ml of 2xTY/tet_{15} with the S17-1[pXYZ] donor strain; grow overnight at 37 °C.

The morning of the conjugation, dilute the stationary S17-1[pXYZ] overnight culture 1 : 50 into fresh 2xTY/tet_{15} and grow at 37 °C to early log phase. Prewarm a 2xTY plate at 37 °C; use one plate per conjugation. (*Note*: mini Petri dishes work well here.) In a sterile microfuge tube, mix 200–300 μl of the ΔΔ11 culture with 20 μl of the log-phase S17-1[pXYZ] culture and pellet cells in a microfuge. Decant the supernatant. Wash the cells two times with 1 ml of 2xTY. To prevent damage to pili, mix the cells *gently* by pipetting up and down; do not vortex. After the final wash, leave about 50 μl of supernatant above the pellet.

In a sterile hood, place a sterile nitrocellulose filter (13-mm diameter) on the surface of the prewarmed 2xTY plate using sterile forceps. Gently resuspend the cells in the 50 μl of remaining medium; filtered pipette tips are recommended to prevent contamination of one conjugation by another via aerosols. Place the cell suspension in one large drop on top of the filter and incubate at 37 °C for around 2 h. During the incubation period, the liquid will pass through the filter and be absorbed by the agar, leaving a paste of cells atop the filter.

At the end of the incubation period, return to the sterile hood and remove the filter from the plate using sterile forceps. Place it in a sterile snap-cap culture tube (e.g., Falcon 2059) with 1 ml of MR26 medium. Vortex to rinse the cell paste from the filter. Transfer the suspension to a microfuge tube, pellet the cells, and wash twice with 1 ml MR26 medium. Resuspend the final pellet in 1 ml MR26 medium and spread onto MR26/vit/$tet_{0.7}$ agar (platings of 100- and 900-µl aliquots are common). Incubate the plates at 32 °C.

After around 3–4 days, pigmented colonies (pink/brown/orange) will be visible. The S17-1 donor strain should not grow on minimal MR26 medium because it requires proline. However, this marker reverts with a measurable frequency, so a few tan colonies of the revertant S17-1 strain will appear. In addition, a virtual "lawn" of the original S17-1 strain lies dormant on the MR26 plate and will grow easily when transferred to rich medium. To purify the *R. sphaeroides* transconjugants from the *E. coli* donor (white/tan or dormant), pick pigmented colonies and streak them for single colonies onto YCC/tet_1 agar. Incubate the plates at 32 °C. *R. sphaeroides* transconjugants and contaminating *E. coli* colonies should be visible in 1–2 days. Repeat as necessary to ensure purity of *R. sphaeroides* transconjugants.

To initiate liquid culture, use a single colony of a pure *R. sphaeroides* transconjugant to inoculate a sterile, plastic, snap-cap culture tube (e.g., Falcon 2058) containing 1 ml YCC/tet_1. Shake at 32 °C for around 3 days until ready for subculturing and preparation of archival freezer stocks.

6.4
Small-Scale Screening for Expression and Localization of Target Protein in *Rhodobacter*

The following protocols (summarized schematically in Figure 6.5) describe initial experiments to screen for protein expression in small chemoheterotrophic cultures. In these cultures, coordinated synthesis of nascent membrane and target membrane protein is autoinduced by decreasing oxygen tension as the cell density increases during semi-aerobic culture. The level of oxygenation is determined by flask size, media volume, shaking speed, and, ultimately, the number of oxygen-consuming cells that are present in the culture. While the expression of the majority of target proteins is not highly sensitive to growth phase of the culture, the expression levels of some target proteins have been observed to exhibit definite maxima/minima throughout the culture period. Thus, it is recommended that the expression level of the target protein be determined at several different points during the maturation of the culture. An example is shown in Figure 6.6. At the extreme, if expression of a particular target proves to peak too early when cell density is inferior, alternate induction strategies could be employed. For example, culture turbidities could be allowed to increase under conditions that repress expression of the target gene (e.g., high oxygen tension), followed by strong induction once late log phase of the culture has been reached.

In initial screening, expression levels and cellular localization of the target protein are determined routinely by sodium dodecyl sulfate–polyacrylamide gel electrophoresis (SDS–PAGE) and immunoblotting as neither native

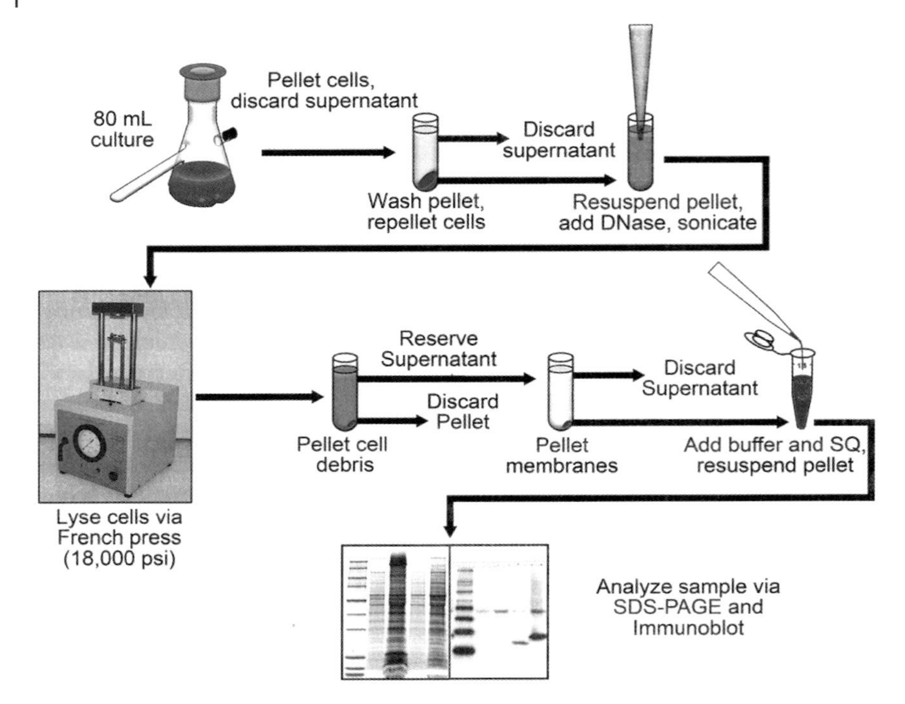

Figure 6.5 Screening for successful *Rhodobacter* expression and ICM insertion using 80-ml cell cultures grown semi-aerobically. Whole-cell lysates and membrane fractions are extracted and analyzed using SDS–PAGE and immunoblotting techniques. Overexpressed bands are not always clearly visible in Coomassie-stained gels. Therefore, immunoblots are used to visualize heterologously expressed membrane proteins, especially those expressed at low levels and those with anomalous mobility on gels.

photosynthetic proteins nor heterologously expressed membrane proteins are always visible in stained gels. This strategy also helps to identify membrane proteins that are characterized by anomalous mobility on SDS gels. In each experiment, signals obtained from cell fractions of expression strains are compared to those obtained from a negative control (Figure 6.7; expression host carrying an "empty" platform vector) and a positive control (host strain carrying a platform vector that expresses a protein at 1 mg/l of culture).

6.4.1 Small-Scale Growth and Preparation of Samples for SDS–PAGE

6.4.1.1 Growth and Harvest of Expression Strains

Steps in this protocol are diagramed in the flow chart shown in Figure 6.5. Inoculate 80 ml of YCC/tet$_1$ medium in a 125-ml baffled side-arm flask (see Figure 6.6). Incubate at 32–34 °C, shaking at 125 rpm, for 72–96 h.

Monitor the growth of the culture with regular measurements of turbidity. Side-arm flasks are especially convenient for determinations of culture turbidity with a

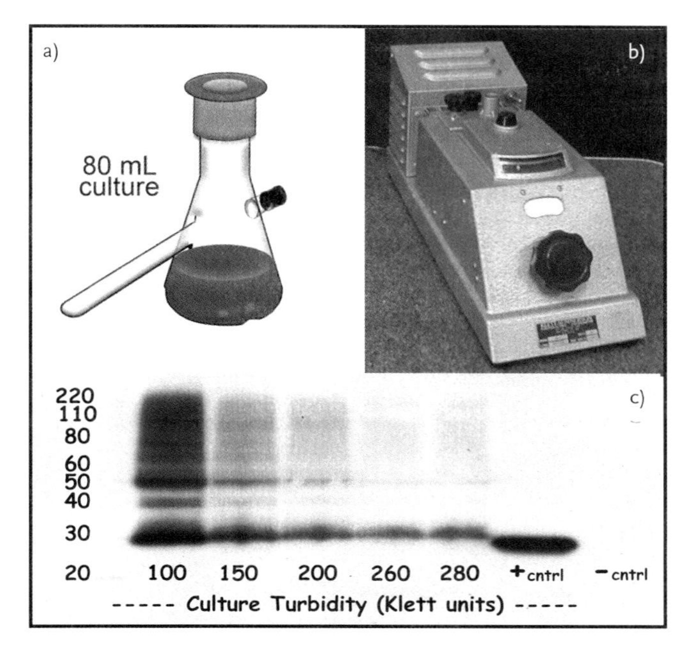

Figure 6.6 (a) Small-volume (80-ml) cultures of expression strains of *Rhodobacter* grown semi-aerobically are useful for small-scale screening to determine expression levels of target protein. Side-arm flasks facilitate noninvasive measurement of culture turbidity with a Klett–Summerson colorimeter (b). (c) Careful examination of the levels of expression of some target proteins throughout various stages of semi-aerobic culture suggests that expression levels can peak at different stages (GPCR EDG1 shown in this immunoblot example). These data, then, dictate that harvesting of such autoinduced cultures should occur earlier than late log phase.

Klett–Summerson colorimeter, eliminating the need to remove samples of the culture for repeated measurements. The equivalent OD_{600} may also be used. If side-arm flasks are not used, it is important to note that removal of successive aliquots of the culture for turbidity measurements will reduce the 80-ml culture volume significantly. Since media volume is a very important factor in determining the oxygen tension in shake flasks, the consequences of repeated sampling must be taken into consideration. When the culture reaches the desired turbidity, pellet the cells in aliquots of 5 and 75 ml.

Discard the supernatant, resuspend and wash the cell pellets with Buffer 1, and repellet cells. Following the protocols below for sample preparation for SDS–PAGE, the signal obtained for the whole-cell sample (derived from the 5-ml aliquot) will equal the combined signals obtained for the membrane and soluble fractions (derived from the 75-ml aliquot). Process them separately, as follows.

6.4.1.2 **Preparing Whole-Cell Samples for SDS–PAGE**

Treat the cell pellet from 5 ml of culture with Sample Quench/Tris-EDTA (SQ/TE) [24], as follows.

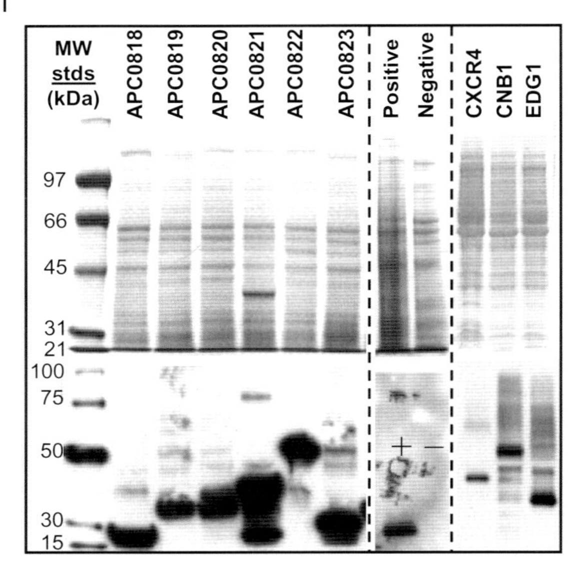

Figure 6.7 Quantification of heterologous expression of membrane in *Rhodobacter*. With the exception, here, of target APC0821, overexpressed proteins are not always clearly visible in Coomassie-stained gels (top). Immunoblots (bottom) with well-characterized controls are employed to probe the level of expression of membrane proteins from *E. coli* ("APC" numbers) in *Rhodobacter* membranes. Experimental membranes from expression strains are compared with membranes carrying a tagged protein (expressed at 1 mg/l culture; "+" control) and with membranes from a recombinant strain lacking a cloned gene ("–" control). Any target protein expressed at or above the "+" control level is considered a target worthy of purification attempts. The expression levels of some targets rival those of native ICM proteins that can be purified to yields of more than 10 mg/l culture. In these examples, the target proteins in the left panel and the controls were detected with an anti-polyhistidine antibody, while the GPCRs (CXCR4, CNB1, and EDG1) in the right panel were detected with an antibody to an epitope tag (1D4 of rhodopsin).

Resuspend the cell pellet to homogeneity in 167 μl 0.1 M Tris, pH 8.5, then add 167 μl SQ solution. Vortex for 2 min. Heat at 90 °C for 10 min. (*Note*: heating often causes aggregation of membrane proteins, so this step may be omitted in favor of incubation at room temperature.) Add 666 μl TE and vortex 30 s. Heat again at 90 °C for 5 min. Further incubation at room temperature is an option if heating is not desirable. If necessary to shear remaining DNA, sonicate the solution with a microtip. Total sample volume now equals 1 ml. Load 24 μl per gel lane. If not running the gel immediately, store the sample at 4 °C (–80 °C for long-term storage).

6.4.1.3 **Preparing Membranes and the Soluble Fraction for SDS–PAGE**

Resuspend the washed cell pellet from 75 ml of culture in 15 ml Buffer 1. Add 300 units of DNase (Sigma D-5025; stored as 20-μl aliquots in Buffer 1). Sonicate cells alternating 5 s on/ 5 s off, for a total of 30 s of sonication time. Lyse cells by passage through French press at 18 000 psi. Repeat, if necessary, to increase efficiency of breakage. Pellet cell debris at 16 500 g for 10 min. Discard the pellet and reserve the supernatant.

Transfer 1 ml of the supernatant to an ultracentrifuge tube and pellet membranes at 245 000 g for 90 min. Retain both the pellet and the supernatant. The supernatant represents the "soluble fraction" (total volume = 1 ml). Treat 24 μl with 6 μl 5 × SDS–PAGE sample buffer (5 × SB) and heat for 15 min at 90 °C. Load 30 μl per gel lane.

Treat the membrane pellet as follows. Resuspend the membrane pellet in 167 μl 0.1 M Tris, pH 8.5, and transfer the suspension to a 1.5-ml microfuge tube. Add 167 μl SQ solution and vortex for 2 min. Heat for 10 min at 90 °C (as mentioned above, heating is discretionary). Add 666 μl TE and vortex for 30 s. Heat again for 5 min at 90 °C or incubate at room temperature if heating is not desired. Total volume = 1 ml. Load 24 μl per gel lane. If not running gel immediately, store sample in the short term at 4 °C. Longer term storage requires freezing at −80 °C.

6.4.2
SDS–PAGE Followed by Electroblotting of Proteins to PVDF Membrane

For expression screening, replica gels – one for staining with Coomassie Brilliant Blue and one for immunoblotting – should be run in parallel. If heterologously expressed target proteins are not well-visualized on the stained gel, then target proteins can be detected on an immunoblot following electroblotting of the replica gel to poly(vinylidene fluoride) (PVDF) membrane.

Electroblotting of proteins to PVDF or nitrocellulose membranes can be performed with various apparati according to any one of numerous protocols that are generally available. The following methods are used routinely in this laboratory and have undergone some optimization for electroblotting of membrane proteins. In our experience, the hydrophobicity of PVDF (as compared to nitrocellulose) provides superior results, according to methods described below.

For gels that are to be electroblotted to PVDF membranes for immunoblots, prepare adequate quantities of the TGMS blotting buffer. This buffer contains 10% methanol to deter precipitation of hydrophobic membrane proteins in the gel before they are transferred onto the PVDF membrane. While the gel is running, prepare the PVDF membrane for transfer by first wetting it in a minimal amount of methanol, then placing it in 50 ml TGMS for further wetting (with rocking). When SDS–PAGE is complete, remove the gel and soak it in 50 ml TGMS for *only* 5 min, with rocking. This short soak ensures that some SDS remains to prevent membrane proteins from precipitating in the gel.

Assemble the components of the blotting cassette according to directions provided by the manufacturer, avoiding air bubbles between the gel and the PVDF membrane. All components of the cassette should be wetted thoroughly in TGMS at the time of assembly. Transfer the blotting cassette assembly to the tank filled with TGMS. Place a small stir bar in the bottom of the tank and use an ice reservoir to keep the initial transfer cold. Transfer at 300 mA for 1.5 h, initially, then overnight at 50 mA with slow stirring.

Separate the blotting cassette assembly layer by layer, taking care to note the orientation of the PVDF membrane, and place it in a container with *protein side up*. Either proceed immediately to development of the immunoblot or allow the PVDF membrane to air dry for later processing. If the PVDF is allowed to dry, it must be wetted again in methanol prior to transfer to any aqueous solution for further processing. Following transfer, stain the electroblotted gel to determine transfer efficiency.

6.4.3
Immunoblot Development

The development of immunoblots has become a standard technique, thus there are many choices available for protocols, detection schemes, and reagents. The following section outlines methods for chemiluminescent detection of tagged proteins that work well and have become routine in this laboratory. The example below will focus on detection of his-tagged membrane proteins with a primary anti-polyhistidine antibody that is conjugated to horseradish peroxidase (Penta-His HRP (horseradish peroxidase) Conjugate Kit; Qiagen 34460). Buffer formulations are outlined in the Appendix.

The following steps should be performed at room temperature, with gentle agitation or rocking during incubations. For the standard 5.5 cm × 8.5 cm pieces of PVDF that fit purchased mini-gels, flat plastic 6.5 cm × 9 cm trays are used for all incubations. Place the membrane in the tray with the protein side up, as determined by marking it or using colored molecular weight standards. Larger or smaller membranes (or containers) will require adjustment of the volumes.

If starting with a dried PVDF membrane, first re-wet it by soaking in methanol. Transfer the membrane to 15 ml 1 × TBS and perform two washes, each of 10 min. Discard the washes and incubate the membrane in 15 ml blocking solution (0.5% alkali-soluble casein (Pierce 70955) in 1 × TBST) for at least 1 h.

Remove 7.5 ml of blocking solution from the tray and save it for later. To the remaining 7.5 ml of blocking solution in the tray, add 0.5 μl of the Penta-His HRP conjugate antibody (thus diluted 1 : 15 000) and incubate for 1 h with rocking.

Wash 6 times, for 10 min each, with 20 ml 1 × TBSTT to remove unbound antibody. It is important to wash the membrane thoroughly at this point to achieve maximum signal-to-noise ratios. After the final washing step is complete, drain as much TBSTT from the membrane as possible.

Addition of the substrate: for a typical 5.5 cm × 8.5 cm membrane, use 3 ml each of Pierce Pico peroxide solution (Pierce 1856135) and Pierce Pico enhancer (1856136), and 0.25 ml each of Pierce Dura peroxide solution (1856158) and Pierce Dura enhancer (1856157) for a total volume of 6.5 ml. Incubate the membrane in the substrate at room temperature for 5 min with rocking.

Remove the membrane from the substrate. Drain any excess substrate from the membrane by touching the edge to a paper towel. Place the membrane in a clear plastic development folder and fold the plastic over the membrane. Remove any bubbles between the plastic and the membrane. Remove any liquid from the exterior of the plastic.

Use the membrane to expose X-ray film or a digital imaging system for identification of expressed target proteins. Be sure to orient the membrane such that the protein side faces the film or detector.

Typical results from a screening experiment of this type for *Rhodobacter* membranes derived from expression strains are shown in Figure 6.7. To determine the cellular localization of the expressed target membrane protein in *Rhodobacter* cells, expression in whole cells is compared (on an equal volume basis using immunoblot analysis) with the supernatant (soluble fraction) and pellet (membrane fraction) obtained from ultracentrifuge separation (245 000 g) of lysates that are devoid of cellular debris. Most of the target membrane proteins that have been studied are expressed predominantly in the *Rhodobacter* ICM. Very few target membrane proteins show any significant presence in the soluble fraction. An important note is that the sum of the signals from the soluble and membrane fractions should equal the total expression level observed in the cells. If this is not the case, one should investigate the debris pellet obtained from the lysate to test for the presence of target protein that may have aggregated as inclusion bodies – a phenomenon not yet observed with the expression of membrane proteins in *Rhodobacter*.

6.5 Large-Scale Culture

If small-scale screening has determined that the target protein is expressed at a level greater than 0.5 mg protein/l cell culture, then scale-up of target protein production is achieved efficiently by following the protocol below. Again, coordinated synthesis of nascent membrane and target membrane protein is autoinduced by decreasing oxygen tension as the cell density increases during semi-aerobic culture.

6.5.1 Growth and Harvest of Expression Culture

Inoculate 25 ml of YCC/tet_l medium in a 50-ml screw-cap Erlenmeyer flask with the desired heterologous expression strain of *R. sphaeroides*. Allow this culture to grow overnight in the dark at 32–34 °C while shaking at 125 rpm.

Subculture by transferring 0.5–25 ml of the 25-ml culture into 2 l of YCC medium containing tet_l. The volume of inoculum that is used for this subculture should be determined at the discretion of the experimenter based on turbidity of the 25-ml culture, the number of 2-l flasks that are to be inoculated, and time constraints for growth and harvesting. Incubate culture in the dark at 32–34 °C, shaking at

125 rpm, until the cell density reaches the desired value (see, e.g., considerations discussed in Section 6.4, Figure 6.6).

Harvest cells and wash the cell pellet in 500 ml of Buffer 1. Pellet the washed cells and resuspend pellet in 35 ml of Buffer 1, yielding a final volume of approximately 50 ml. Immediately add one crushed tablet of Protease Inhibitor Cocktail (e.g., cOmplete™, EDTA-free; Roche 1873580) to the resuspended cells and mix thoroughly. Cells may be frozen at this time. Add 300 units of DNase, then disrupt aggregates by sonicating the cells on ice for 1 min, alternating 5 s on and 5 s off. Repeat this 1-min cycle of sonication after 10 min on ice.

6.5.2 Cell Lysis

Lyse the cells mechanically with either a French press or a microfluidizer at 18 000 psi. Multiple passes (2–3) are recommended. The sample cell should be cold at the outset and the sample should be kept on ice throughout the procedure. Mechanical lysis (instead of lysozyme/osmotic shock) is necessary for efficient preparation of the membrane fraction.

Pellet the cell debris and unbroken cells for 10 min at 12 500 g. (If the volume of the pellet of unbroken cells is greater than 10% of the original pellet volume, then resuspend in Buffer 1 and repeat cell lysis following sonication to disrupt cell aggregates.) The supernatant (or "pressate") may be frozen at this step.

6.5.3 Membrane Isolation

Transfer the supernatant ("pressate") to ultracentrifuge tubes to pellet membranes at 240 000 g, 4 °C, for 1.5 h. The supernatant, which contains soluble proteins, is discarded. Weigh one pellet (as dry as possible) from each sample. Considering the mass of all pellets from a given sample to be approximately equal, calculate the total pellet weight for each sample. Use 12.5 ml of Buffer 1 to resuspend every 1 g of pellet. (*Note*: more concentrated membrane suspensions (up to 10 times) can be prepared by resuspension in a correspondingly smaller volume of Buffer 1.) Use a small-volume homogenizer, which may be connected to an electric tissue grinder to speed the process, to resuspend these membrane pellets in the buffer that has been added. Downstream processing can benefit from multiple rounds of washing with Buffer 1 and repelleting of the membranes. After final homogenization of the washed pellets, the membrane suspension can be stored indefinitely by freezing at −20 °C.

Membrane suspensions prepared in this way are comprised of inside-out vesicles (i.e., the cytoplasmic face of the expressed integral membrane protein is presented to the external, aqueous milieu.) These vesicles can be used as is in several types of biochemical assays to determine functionality and, with modification, can be used in others (Section 6.8). Proof of activity in membranes provides a benchmark to which the activity of detergent-purified forms of the same protein can be compared.

6.6 Detergent Solubilization and Chromatographic Purification of Expressed Membrane Proteins

For many applications, including crystallization, it is critical to remove *Rhodobacter*-expressed, target membrane proteins from the ICM lipid bilayer in functional form and isolate them subsequently by chromatographic separation to appropriate purity levels. Discovery of conditions that optimize solubilization of membrane proteins and stabilization of targets in detergent micelles prior to purification is a time-consuming, labor-intensive, and expensive process. Essential for success is the use of detergents that not only are gentle to the target membrane protein(s), but also are compatible with downstream chromatographic strategies. Rapid, efficient methods are critical for the production of samples that are stable, and retain the greatest level of native structural and functional integrity. Often, the solubilization step is a stage at which success or failure of a purification experiment is decided. At all stages of the purification process, one must balance the desire to increase purity with the time required to achieve such, often at the expense of activity for proteins that have very short half-lives when removed from the lipid bilayer.

Detergent-solubilized, tagged membrane proteins are notorious for exhibiting poor binding to affinity resins, especially metal-based resins. This can be made worse by the choice of detergent. In addition, detergent solubilization can be far from quantitative. From a survey of a variety of detergents used with a variety of different proteins, a set of general strategies has emerged that can be used for initial attempts at solubilization and purification of membrane proteins expressed heterologously in the *Rhodobacter* ICM. Our ability to generalize is because solubilization proceeds from a common lipid bilayer of defined chemical composition [25]. This is a significant advantage of expressing membrane proteins in the *Rhodobacter* ICM. Since the solubilization process involves penetration and disassembly of a lipid bilayer, understanding of the requirements for that process will then be applicable to the solubilization of most any protein that is contained within that bilayer. Subsequent testing for functionality will determine the effectiveness of this detergent set in stabilizing membrane proteins. Since the *Rhodobacter* ICM can be broken down with detergents that are considered to be fairly "gentle" (e.g., Deriphat 160 [26] and new tripod amphiphiles [27, 28]), these molecules are also likely to be effective in maintaining structural and functional integrity of the extracted membrane protein.

Herein, protocols are outlined specifically for purifications of target membrane proteins in native, functional form. Deriphat 160 has proven to be a workhorse in this laboratory as it is very effective, yet very gentle, in solubilizing membrane proteins from the *Rhodobacter* ICM.

Lauryldimethylamine-*N*-oxide (LDAO) has also been employed extensively but tends to be harsher for many proteins. Dodecyl maltoside and lauryl sarcosine have been employed successfully for work with expressed eukaryotic target membrane proteins. In the end, however, any detergent that is compatible with affinity chromatography may be substituted as long as the target membrane protein survives solubilization conditions. Detergent concentrations between 10 and 20 × critical micelle concentration (CMC) (100 × for dodecyl maltoside) are recommended for solubilization, followed by concentrations between 1 and 3 × CMC during all subsequent chromatographic steps.

The following illustration of protocols that work well in this laboratory utilizes Deriphat 160.

6.6.1 Solubilization of Membrane Proteins

Proceed with solubilization *only* if purification will follow immediately thereafter. Prewarm the membrane suspension (Section 6.5.3) to 23 °C in a water bath. Begin the solubilization of membrane proteins by stirring the mixture while slowly adding Deriphat 160 (dodecyliminodipropionate; Anatrace D345) to a final concentration of 1% (1 : 10 dilution from 10% stock, pH 7.8). Allow the mixture to continue stirring slowly (avoid foam) for 5–10 min to ensure homogeneity of the mixture. Return the solution to the 23 °C bath for 1 h with slow agitation. Again, avoid conditions that generate foam.

Remove membrane debris from the extract with ultracentrifugation at 240 000 g, 4 °C, for 2 h. This time the supernatant, which contains solubilized membrane proteins, will be used for subsequent chromatographic steps and the pellet will be discarded. To assess the efficiency of solubilization, aliquots of the supernatant and resuspended membrane debris pellet may be analyzed by SDS–PAGE. Ideally, one would like to see quantitative extraction of the target membrane protein. Since the presence of high detergent causes excessive streaking of sample lanes in SDS–PAGE, dialysis or treatment of the solution with Biobeads (BioRad 152-3920), is recommended to ensure optimal separation.

6.6.2 Chromatography

Affinity chromatography has become the standard, primary separation technique for purification of proteins. Thus, there are many choices available for protocols, chromatography media, and reagents. Choices depend on the scale of the operation, the availability of equipment, and the requirements for producing a pure, functional protein. Semiautomated systems have become popular and more readily available, but similar results can be obtained using good chromatographic technique at the bench. The following section outlines methods for purification of his-tagged proteins that work well and have become routine in this laboratory. Buffer formulations are outlined in the Appendix.

6.6.2.1 Bench-Top Affinity Chromatography

Whenever possible, keep all solutions on ice or perform some (or all) steps of chromatography in a refrigerated environment.

In a 50-ml Falcon tube, add Ni-NTA resin (e.g., Ni-NTA Superflow; Qiagen 30430) to the membrane extract supernatant from the ultracentrifuge spin following detergent extraction. Use 1 ml resin for membranes derived from 2 l of culture. Bind solubilized proteins to the resin at 4 °C, rotating, for at least 1 h and for up to 12 h. It is not uncommon to bind overnight.

Pack resin/protein suspension into column. Wash column with approximately 10 column volumes of Deriphat Buffer 2 until the flow-through is devoid of protein (e.g., $A_{280} \leq 0.05$). Elute the His-tagged protein using two column volumes each of Deriphat Buffers 3, 4, and 5. For each elution step (i.e., each different buffer), collect three fractions. Store fractions at –20 °C.

6.6.2.2 **Affinity Chromatography Using an ÄKTA-FPLC™**

To accommodate the specific requirements for running membrane protein samples through a semi-automated purification system designed for soluble proteins, the ÄKTA-FPLC in this laboratory has been highly modified [10]. Specific features have been incorporated (additional valves, a dedicated, low-head-volume sample pump, and a superloop) that accommodate the necessary large sample volumes (preserving key protein: detergent concentration ratios) and the slow binding characteristics that have been observed with many target membrane proteins.

To prepare the solubilized membrane protein slurry for the fast protein liquid chromatography (FPLC), decant the membrane extract (supernatant from the ultracentrifugation following solubilization; Section 6.1) into a precooled beaker. Discard the pellet. Filter the membrane extract into a second precooled beaker using 0.45-micron filters (Whatman PES filter 6904-2504; approximately one filter for each 50 ml of solution) and a 60-ml disposable syringe.

Pour the filtered mixture of membrane extract into an appropriate sample container. To limit sample aggregation during binding to the affinity resin, place the sample container in a beaker filled with wet ice. Binding of most heptahistidine-tagged target membrane proteins to Ni-charged columns is slow during the initial immobilized metal-affinity chromatography step, possibly due to restricted accessibility of the tag by the detergent micelle. Thus, the automated routines are set up to allow the protein to pass over the column at least twice to increase binding efficiency. Place sample inlet and outlet tubes in the appropriate sample containers to allow for one "recycle" run of the solution through the ÄKTA-FPLC. Low concentrations of imidazole (up to 10 mM) may be introduced during binding to limit the potential for copurification of impurities with weak affinity for the nickel resin.

Using standard software routines for affinity chromatography, modify run parameters (e.g., sample volume, recycling, etc.) where appropriate. Flow rates during the binding step should be low to facilitate maximum protein–resin interaction. Runs normally last approximately 2–3 h and are entirely dependent upon the initial sample volume. Detergents used in wash steps do not necessarily need to match those of the detergents used for solubilization, but most often they do. Detergent exchange during washing is straightforward and quantitative as long as wash volumes exceed 10 column volumes. Extra buffer (up to 20 column volumes) must be utilized when the CMC of the new detergent is greater than that of the previous detergent.

Collect fractions of the desired size. The ÄKTA-FPLC accommodates 96-well plates or racks for 15- or 50-ml Falcon tubes for larger fractions. Store elution fractions at 4 °C.

6.7
Protein Identification and Assessment of Purity

Concentrate elution fractions using centrifugal concentrators that are appropriate for the expected molecular weight of the target protein. Be careful to consider the consequences of the size of the complete protein-detergent complex (PDC) and the potential for concentrating empty detergent micelles at this stage. Determine the outcome of each purification experiment by analyzing elution fractions with SDS–PAGE and staining with Coomassie Brilliant Blue or silver. If the protein is not observed in stained gels of the elution fractions, proceed to protein detection via immunoblotting. If the protein purity is not sufficient for structural/functional studies, then additional chromatographic steps may be employed.

If the target protein contains a chromophore, UV/VIS spectroscopy can be used to compare the absorbance of bands unique to the target protein with that of the total protein absorbance at 280 nm. Circular dichroism (CD) spectroscopy can be used to monitor folding of detergent-purified integral membrane proteins [29]. The integrity of the purified protein can be assessed by observing changes in the CD spectra upon application of heat or chemical denaturing conditions.

The oligomeric state of the purified target protein can be assessed with size-exclusion chromatography, analytical ultracentrifugation, or PAGE using mildly denaturing conditions. The identity of the protein can be confirmed with N-terminal amino acid sequencing methods. If the protein is sufficiently pure, it can be sequenced as eluted from the column. Alternatively, the target protein can be separated by SDS–PAGE. The protein can then be excised from the gel and analyzed.

6.8
Preparations of Specialized *Rhodobacter* Membranes

Several types of membrane fractions can be produced from *Rhodobacter* (Figure 6.8) and they are ideal subjects for functional assays. The differences in membrane type depends not only on methods used to produce them, but in their source – several engineered strains are available that differ in the amounts of native proteins that are produced, which affects membrane morphology (Figure 6.4). A notable difference between these subpopulations is their "sidedness" and this feature can be exploited in the design of functional assays. The three major classes are inside-out vesicles, outside-out vesicles, and planar preparations.

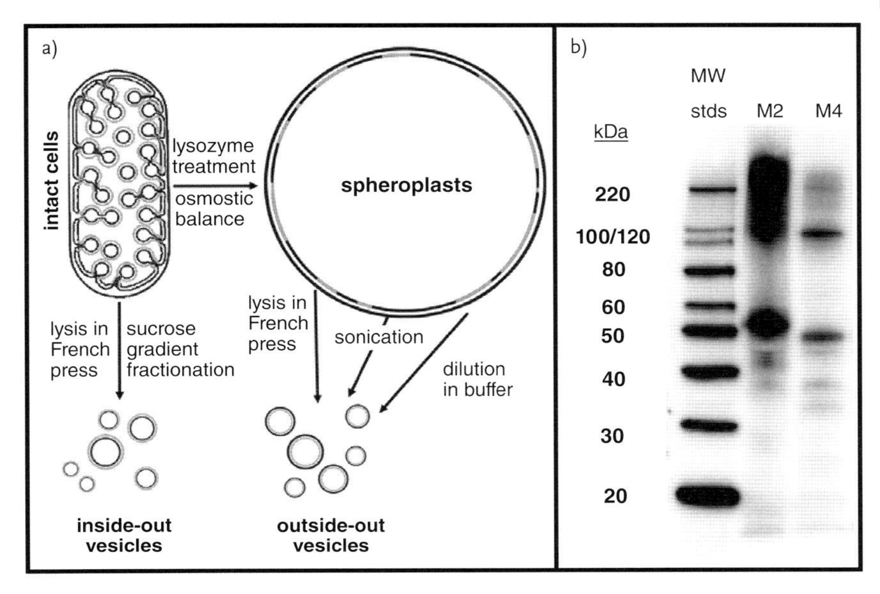

Figure 6.8 (a) Protocols exist for the preparation of different kinds of vesicles and fragments from ICMs in *Rhodobacter* cells. In each of these protocols, fractions specific to these specialized membranes can be isolated easily from other cellular components via differential centrifugation. As an alternative to detergent-purified proteins, these approaches provide oriented samples of membrane-bound target proteins that have proven utility in biochemical assays of function. (b) As an example, the right panel shows an immunoblot of membrane samples containing two *Rhodobacter*-expressed muscarinic receptors (M2 and M4; monomeric species around 48–52 kDa) that are oriented with the extracellular ligand-binding domains of the receptors displayed on the outside surface of the vesicle.

Separation of membrane subpopulations via sucrose density gradient centrifugation is a technique that has been used extensively in the preparation of membrane fragments that are enriched in native or foreign membrane proteins or membrane protein complexes expressed in *Rhodobacter*. Likewise, treatment of either inside-out or outside-out vesicles with sub-CMC concentrations of detergent (e.g., 0.03% LDAO) leaves the vesicles intact, but permeabilizes them such that trapped soluble proteins can diffuse out. These wash steps are a means of conditioning membranes for solubilization and making chromatographic steps "cleaner" by removing copurifying soluble components.

Mechanical cell lysis produces a membrane fraction that is composed of inside-out vesicles. In these membranes, the cytoplasmic domain of the expressed integral membrane protein is exposed to the external aqueous solution. Membrane vesicles oriented with the cytoplasmic face out (outside-out) can be prepared by a procedure that generates spheroplasts as a first step. Detergent (e.g., Brij 58) can be used to lyse spheroplasts leading to vesicle formation. Alternatively, outside-out vesicles can be prepared without detergent by lysis of spheroplasts with sonication, passage through a French press cell at 18 000 psi or 2-fold dilution of sucrose-stabilized spheroplasts with Buffer 1 (Figure 6.8). Each type of membrane preparation can then be pelleted by ultracentrifugation (245 000g, 1.5 h), washed, and resuspended by homogenization in Buffer 1 for biochemical characterization.

Appendix: Media and Buffer Formulations

Tetracycline

Prepare at 15 mg/ml in 70% EtOH and filter sterilize. Working concentrations are 15 μg/ml for *E. coli*, 1 μg/ml (or 0.7 μg/ml for agar) for *R. sphaeroides*.

YCC medium [30] (per liter)

5 g yeast extract
6 g casamino acids
5 ml Concentrated Base (see below)
adjust pH to 7.2 with NaOH

Concentrated Base for YCC (per liter)

$Na_2EDTA \cdot 2H_2O$	11.82 g
$CuSO_4 \cdot 5H_2O$	0.040 g
$CoCl_2 \cdot 6H_2O$	0.030 g
$(NH_4)_6Mo_7O_{24} \cdot 4H_2O$	0.020 g
$MgSO_4$	19.53 g
$FeSO_4 \cdot 7H_2O$	0.75 g
H_3BO_3	0.0125 g
$Ca(NO_3)_2 \cdot 4H_2O$	6.9 g

add water to around 950 ml; adjust pH to 7 using 5 N NaOH

MR26 medium

Modified Hutner's medium [31]. Use 20 ml of A, B, and C per liter of MR26 medium; add 1 ml D per liter *after* autoclaving.

A. *Potassium phosphate buffer*
   1 M, pH 6.8, adjusted with KOH or H_3PO_4
   K_2HPO_4 115 g/l
   KH_2PO_4 44.9 g/l

B. *Ammonium succinate*
   1 M, pH 6.8
   dissolve 118 g succinic acid in 500 ml H_2O
   adjust pH to 6.8 with ammonium hydroxide
   add H_2O to 1 l

C. *Concentrated base (per liter)*
add the following in order:

$Na_2EDTA \cdot 2H_2O$	11.16 g
$(NH_4)_6Mo_7O_{24} \cdot 4H_2O$	0.0093 g
$FeSO_4 \cdot 7H_2O$	0.099 g
"Metals 44"	50 ml
$MgSO_4$	14.5 g
$CaCl_2$	2.5 g

Metals 44 (per liter)

$FeSO_4 \cdot 7H_2O$	5.0 g
$Na_2EDTA \cdot 2H_2O$	6.5 g
$ZnSO_4 \cdot 7H_2O$	10.9 g
$MnCl_2 \cdot 4H_2O$	1.3 g
$CuSO_4 \cdot 5H_2O$	0.392 g
$CoCl_2 \cdot 6H_2O$	0.200 g
H_3BO_3	0.114 g

D. *Vitamins (per liter to prepare 1000 × stock solution)*

nicotinic acid	3.0 g
nicotinamide	3.0 g
thiamine-HCl	6.0 g
biotin	0.12 g

filter sterilize and store at 4 °C

Buffer 1
10 mM Tris, pH 7.8
100 mM NaCl

Deriphat Buffer 2
10 mM Tris, pH 7.8
0.1% Deriphat 160

Deriphat Buffer 3
40 mM imidazole
10 mM Tris, pH 7.8
0.1% Deriphat 160

Deriphat Buffer 4
100 mM imidazole
10 mM Tris, pH 7.8
0.1% Deriphat 160

Deriphat Buffer 5
1 M imidazole
10 mM Tris, pH 7.8
0.1% Deriphat 160

SQ [24]
6% SDS
0.24 M dithiothreitol
0.06% Bromphenol Blue
60% glycerol
20 mM Tris, pH 6.8
H_2O to 20 ml

TE
10 mM Tris, pH 7.4
1 mM EDTA, pH 8.0

10 × blotting buffer (per liter)
30.3 g Tris
144.1 g glycine

TGMS (per liter)
100 ml 10 × blotting buffer
200 ml methanol
1 ml 20% SDS

10 × TBS
200 mM Tris-Cl, pH 7.5
1.5 M NaCl
filter sterilize

10 × TBST
200 mM Tris-Cl, pH 7.5
1.5 M NaCl
0.5% Tween 20
filter sterilize

10 × TBSTT
200 mM Tris-Cl, pH 7.5
1.5 M NaCl
0.5% Tween 20
2% Triton X-100
filter sterilize

Abbreviations

CD	Circular dichroism
CMC	critical micelle concentration
FPLC	fast protein liquid chromatography
GPCR	G-protein-coupled receptor
HRP	horseradish peroxidase
ICM	intracytoplasmic membrane

LDAO Lauryldimethylamine-*N*-oxide
LIC ligation-independent cloning
MCS multiple cloning site
PAGE polyacrylamide gel electrophoresis
PCR polymerase chain reaction
PDC protein-detergent complex
PVDF poly(vinylidene fluoride)
RBS ribosome-binding site
SB SDS–PAGE sample buffer
SDS sodium dodecyl sulfate
SeMet selenomethionine
SQ Sample Quench
TE Tris-EDTA

References

1 Arechaga, I., *et al.* (2000) Characterisation of new intracellular membranes in *Escherichia coli* accompanying large scale over-production of the b subunit of F_1F_o ATP synthase. *FEBS Lett.*, **482**, 215–219.
2 Miroux, B. and Walker, J.E. (1996) Over-production of proteins in *Escherichia coli*: mutant hosts that allow synthesis of some membrane proteins and globular proteins at high levels. *J. Mol. Biol.*, **260**, 289–298.
3 Essen, L.-O. (2002) Structural genomics of "non-standard" proteins: a chance for membrane proteins? *Gene Funct. Dis.*, **3**, 39–48.
4 Collins, M.L.P. and Cheng, Y. (2004) Host/vector system for expression of membrane proteins, US Patent 20040086969.
5 De Smet, L., *et al.* (2001) A novel system for heterologous expression of flavocytochrome *c* in phototrophic bacteria using the *Allochromatium vinosum* rbcA promoter. *Arch. Microbiol.*, **176**, 19–28.
6 Graichen, M.E., *et al.* (1999) Heterologous expression of correctly assembled methylamine dehydrogenase in *Rhodobacter sphaeroides*. *J. Bacteriol.*, **181**, 4216–4222.
7 Gumpert, J. and Hoischen, C. (1998) Use of cell wall-less bacteria (L-forms) for efficient expression and secretion of heterologous gene products. *Curr. Opin. Biotechnol.*, **9**, 506–509.
8 Hoischen, C., *et al.* (2002) Novel bacterial membrane surface display system using cell wall-less L-forms of *Proteus mirabilis* and *Escherichia coli*. *Appl. Environ. Microbiol.*, **68**, 525–531.
9 Laible, P.D. and Hanson, D.K. (2002) Methods and constructs for expression of foreign proteins in photosynthetic organisms, US Patent 20020102655.
10 Laible, P.D., *et al.* (2004) Towards higher-throughput membrane protein production for structure genomics initiatives. *J. Struct. Funct. Genomics*, **5**, 167–172.
11 Nguyen, H.H., *et al.* (1998) The particulate methane monooxygenase from *Methylococcus capsulatus* (Bath) is a novel copper-containing three-subunit enzyme. Isolation and characterization. *J. Biol. Chem.*, **273**, 7957–7966.
12 Turner, G.J., *et al.* (1999) Heterologous gene expression in a membrane-protein-specific system. *Protein Expr. Purif.*, **17**, 312–323.
13 van Dijk, R., *et al.* (2000) The methylotrophic yeast *Hansenula polymorpha*: a versatile cell factory. *Enzyme Microb. Technol.*, **26**, 793–800.
14 Drews, G. and Golecki, J.R. (1995) Structure, molecular organization, and biosynthesis of membranes of purple

bacteria, in *Anoxygenic Photosynthetic Bacteria* (eds R.E. Blankenship, M.T. Madigan, and C.E. Bauer), Kluwer, Dordrecht, pp. 231–257.

15 Kiley, P.J. and Kaplan, S. (1988) Molecular genetics of photosynthetic membrane biosynthesis in *Rhodobacter sphaeroides*. *Microbiol. Rev.*, **52**, 50–69.

16 Scott, H.N., Laible, P.D., and Hanson, D.K. (2003) Sequences of versatile broad-host-range vectors of the RK2 family. *Plasmid*, **50**, 74–79.

17 Simon, R., Priefer, U., and Puhler, A. (1983) A broad host range mobilization system for *in vivo* genetic engineering: transposon mutagenesis in gram negative bacteria. *Biotechnology*, **1**, 37–45.

18 Donohue, T.J. and Kaplan, S. (1991) Genetic techniques in Rhodospirillaceae. *Methods Enzymol.*, **204**,. 459–485.

19 Laible, P.D., *et al.* (2005) Incorporation of selenomethionine into induced intracytoplasmic membrane proteins of *Rhodobacter* species. *J. Struct. Funct. Genomics*, **6**, 95–102.

20 Kappler, U. and McEwan, A.G. (2002) A system for the heterologous expression of complex redox proteins in *Rhodobacter capsulatus*: characterisation of recombinant sulphite:cytochrome *c* oxidoreductase from *Starkeya novella*. *FEBS Lett.*, **529**, 208–214.

21 Kirmaier, C., *et al.* (2002) Comparison of M-side electron transfer in *R. sphaeroides* and *R. capsulatus* reaction centers. *J. Phys. Chem. B*, **106**, 1799–1808.

22 Naylor, G., *et al.* (1999) The photosynthesis gene cluster of *Rhodobacter sphaeroides*. *Photosynth. Res.*, **62**, 121–139.

23 Yoon, J.R., *et al.* (2002) Express primer tool for high-throughput gene cloning and expression. *Biotechniques*, **33**, 1328–1333.

24 Blackshear, P.J. (1984) Systems for polyacrylamide gel electrophoresis. *Methods Enzymol.*, **104**, 237–255.

25 Benning, C. (1998) Membrane lipids in anoxygenic photosynthetic bacteria, in *Lipids in Photosynthesis: Structure, Function, and Genetics* (eds P.-A. Siegenthaler and N. Murata), Kluwer, Dordrecht, pp. 83–101.

26 Kirmaier, C., *et al.* (2003) Detergent effects on primary charge separation in wild-type and mutant *Rhodobacter capsulatus* reaction centers. *Chem. Phys.*, **294**, 305–318.

27 Chae, P.S., Laible, P.D., and Gellman, S.H. (2010) Tripod amphiphiles for membrane protein manipulation. *Mol. Biosyst.*, **6**, 89–94.

28 Chae, P.S., *et al.* (2008) Glycotripod amphiphiles for solubilization and stabilization of a membrane-protein superassembly: importance of branching in the hydrophilic portion. *ChemBioChem*, **9**, 1706–1709.

29 Lobley, A., Whitmore, L., and Wallace, B.A. (2002) DICHROWEB: an interactive website for the analysis of protein secondary structure from circular dichroism spectra. *Bioinformatics*, **18**, 211–212.

30 Taguchi, A.K.W., *et al.* (1992) Biochemical characterization and electron-transfer reactions of *sym1*, a *Rhodobacter capsulatus* symmetry mutant which affects the initial electron donor. *Biochemistry*, **31**, 10345–10355.

31 Gerhardt, P., *et al.* (ed.) (1994) *Methods for General and Molecular Bacteriology*, American Society for Microbiology, Washington, DC.

Part Two
Protein-Specific Considerations

7
Peripheral Membrane Protein Production for Structural and Functional Studies

Brian J. Bahnson

7.1
Introduction

The role that proteins play within and on the surface of membranes has been realized ever since the fluid mosaic model of membranes was presented [1]. Membrane proteins make up about a third of many genomes [2] and have functions that range from signal transduction, transport, structural, phospholipid remodeling, oxidative stress response, basic metabolic processes, to many yet uncovered functions. Some estimates suggest that greater than half of all proteins interact with membranes at some point during their functional lifetimes. Transmembrane proteins are typically the main focus when considering the challenges of membrane proteins for protein expression, purification, and characterization [3–5]. However, many challenges exist for proteins that interact with a membrane surface or are just partly imbedded within the membrane, and some of these challenges have been discussed in previous reviews [6, 7]. Three main groups of membrane proteins can be defined: peripheral, monotopic-integral, and transmembrane-integral. Peripheral membrane proteins are defined to include proteins that can exist in either an aqueous environment or associated to a membrane. Montopic-integral proteins face one side of a lipid bilayer and never leave the membrane. Transmembrane-integral proteins have hydrophilic domains on each side of a lipid bilayer. This chapter will cover the peripheral membrane class of proteins that can exist away from the membrane and then direct themselves to the membrane via a range of mechanisms as depicted in Figure 7.1.

Peripheral membrane proteins cover a range of protein folds, functions, and mediate their association to a membrane surface via a range of interactions. At least five separate classes can be defined, including: (i) nonspecific electrostatic, (ii) hydrophobic patch, (iii) covalent lipid anchor, (iv) specific lipid-binding domain, and (v) protein–protein interactions. Some proteins fall into more than one class combining two or more types of interactions in the association with a membrane or membrane-like surface. Despite the significance of proteins that interact with membranes via protein–protein interactions (class v), this chapter will not include a description of these interactions in order to focus on protein interactions with the

Production of Membrane Proteins: Strategies for Expression and Isolation, First Edition.
Edited by Anne Skaja Robinson.

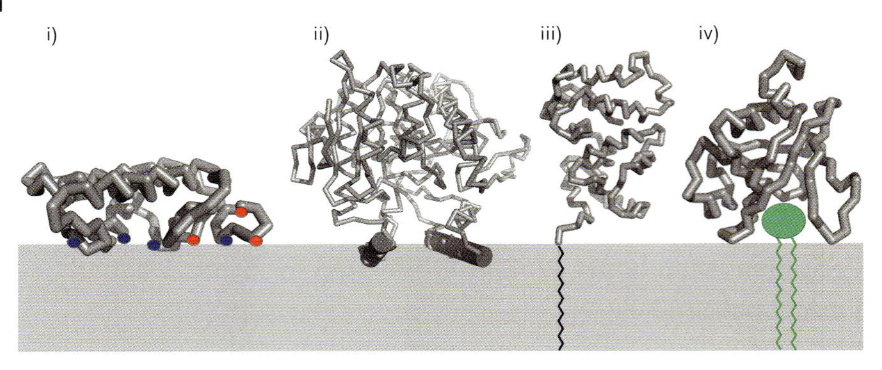

Figure 7.1 Peripheral membrane proteins are divided into four classes: (i) nonspecific electrostatic, (ii) hydrophobic patch, (iii) covalent lipid anchor, and (iv) specific lipid binding.

lipid components of membranes and membrane like particles.

Just as with another subset of proteins, peripheral membrane proteins have challenges in expression and study that range from disulfide bonds, glycosylation, lipid anchors, protein folding issues, aggregation, and so on. Previous high-throughput structural genomics projects like the National Institutes of Health-funded Protein Structure Initiative [8] mainly focused on aqueous-soluble proteins, and unfortunately often failed to troubleshoot challenges or relate the structure to the protein's function [9, 10]. As was true for those initiatives, choices now need to be made about what the challenges faced by membrane protein are [11], and what expression and purification approaches are best suited to give rise to structurally and functionally relevant forms of the protein. Furthermore, the yield and quality of the protein samples must satisfy the requirements for the experimental approaches that are to be performed. When targeting a larger number of membrane proteins, some will pose little difficulty, whereas others will require more creative approaches to express and study their structure and function. Toward that end several new tools exist to go after the "high-hanging fruit" subset of membrane proteins [3–5, 11–13].

Escherichia coli expression is often the system of choice if the protein does not require any significant post-translational modifications. Some common post-translational modifications that can be dealt with from *E. coli* expression systems include disulfide bond formation, phosphorylation of specific residues, and covalent lipid anchors. Disulfide bonds can sometimes be successfully made in thioredoxin reductase mutant strains of *E. coli* that have an altered redox poise [14] or periplasmic expression [15]. Another common approach is to express a disulfide-rich protein in inclusion bodies and refold in a redox buffer that will facilitate disulfide bond formation and proper disulfide shuffling. Later in this chapter (Section 7.2.3.1) an approach that was used to myristoylate proteins using coexpressed yeast myristoyltransferase enzyme will be described. The proper phosphorylation of eukaryotic proteins likewise requires the use of the appropriate eukaryotic kinases. Occasionally, post-translational modifications pose too great a difficulty and may require moving into eukaryotic cell expression (yeast, Sf9 insect cells, mammalian). The choice of expression system comes down to the particular challenges of the protein, cost, yield, purity, need for isotopic labeling, and ultimately what experiment the protein is being produced for.

The ability to fold a targeted eukaryotic protein and its solubility are critical issues to address for structural and functional studies. There are general considerations that apply to each of the classes of peripheral membrane proteins; however, the hydrophobic patch proteins, as the name would suggest, pose the most significant and frequent difficulties for protein expression, purification, and characterization. The common methods used to reduce folding and solubility difficulties include the use of fusion proteins with affinity tags, the coexpression of chaperones (heat-shock, protein disulfide isomerase, etc.), the expression of the protein as an inclusion body coupled with refolding trials, the use of detergents during all stages of purifications for solubilization from the membrane fraction, binding to vesicles or micelles, *in vitro* expression systems [12] that allow detergent, vesicle, or micelle use during protein translation, and a variety of other methods. Another approach that is gaining more popularity for the design and optimization of protein expression constructs is the expression of Green Fluorescent Protein (GFP) fusions for high-throughput expression and monodispersity analysis, where multiple constructs are evaluated in parallel and expression levels are qualitatively assessed by fluorescence comparison of the *E. coli* cell [16]. The modification of protein constructs can be optimized by techniques that rapidly define the limits of a protein domain such as protein proteolysis or H/D amide exchange coupled with mass spectrometry [17]. Alternatively, if structures of homologous proteins exist in the Protein Data Bank (PDB), then homology modeling becomes a powerful way to redesign a protein construct to optimize for expression and homogeneity qualities [18]. This comparative modeling approach will be described for a couple of systems that have a hydrophobic patch.

When a suspected peripheral membrane protein target has homology to a protein with known structure, then it is feasible to use these comparative models to predict the extent of interactions with membranes, as well as their orientation within a membrane. Several approaches have been developed [19–21] that give predictions of whether a protein interacts with a membrane and, if so, the orientation that it adopts. The method by Lomize *et al.* was applied to the peripheral membrane class of protein where they targeted 53 proteins from the PDB that are known to interact as peripheral membrane proteins [22]. The predicted interactions of these peripheral membrane proteins, together with integral membrane proteins are available in the Orientations of Proteins in Membranes (OPM) database [20]. The methods used in the OPM database allow the determination of the orientation of proteins in membranes by minimizing their transfer energy from water to a lipid bilayer [19]. The method can distinguish between membrane proteins and soluble proteins based on their transfer energy and membrane penetration depths. The OPM database method will be used as a visual tool in figures presented, and Table 7.1 summarizes the examples chosen to provide a description

Table 7.1 Peripheral membrane protein expression strategies.

Examples	Expression approach	PDB code
Nonspecific electrostatic		
Cytochrome c_2	expression from *R. sphaeroides* into the periplasmic space [23]	1cxa
Group IB $sPLA_2$	expression as inclusion bodies in *E. coli* and refolded in the presence of additives [24]	5p2p
Hydrophobic patch		
pPAFAH	*E. coli* expression, detergent solubilization of N- and C-terminally truncated enzyme [25, 26]	3d59
Human serum PON1	directed evolution to obtain soluble protein from *E. coli* expression [27, 28]	1v04
Cytochrome P450	*E. coli* expressed and detergent extracted from the membrane fraction using Cymal-5 [29]	2bdm
Group V and X $sPLA_2$	*E. coli* expression into inclusion bodies, refolding, purification by size exclusion [30]	2ghn, 1le6
Covalent lipid anchor		
Recoverin	*E. coli* expression with yeast myristoyl transfer enzymes coexpressed [31]	1iku, 1jsa
PAFAH-II	*E. coli* expression with myristoyl group added using intein chemistry (unpublished)	NA
Lipid-binding domain		
Phospholipase C PH domain δ1	*E. coli* expression with PH domain in soluble fraction [32]	1mai
Protein kinase C C2 α-domain	*E. coli* expression with Ni-affinity tag, binding to a water-soluble analog of phosphatidylserine [33]	1dsy

of strategies used in the expression, purification, and characterization of each of the classes of peripheral membrane proteins.

7.2 Case Studies of Peripheral Membrane Proteins

7.2.1 Electrostatic Interactions

Peripheral membrane proteins that interact with the interface through a combination of electrostatic and polar interactions are typically not plagued with the same difficulties of protein expression and purification that less-soluble membrane proteins are faced with. However, this does not mean that many of these systems do not have complex issues to resolve. There have been several recent studies geared at predicting the significance and magnitude of electrostatic interactions of peripheral membrane proteins [34–36]. Frequently, proteins that associate through electrostatics require the flexibility to come and go from a membrane when the need arises. Systems that are multiprotein complexes or even multidomain proteins are beyond the scope of this description. Here, we will focus on a few examples of relatively small proteins or protein domains that have been successfully expressed and characterized to demonstrate a range of considerations and challenges posed by the nonspecific electrostatic class of peripheral membrane proteins.

7.2.1.1 Case 1: Cytochrome c_2

Within the OPM database there are very few peripheral membrane proteins that have been identified to interact with the membrane solely by electrostatic interactions on the basis of structure alone [22]. Often electrostatic interactions with the polar head-groups of a lipid bilayer are an important initial step for membrane localization. However, the hydrophobic effect becomes the determining factor for each of the other classes discussed below, with the effect mediated by the hydrophobic patch provided by a portion of the protein, a covalently attached lipid or a specifically bound lipid. A notable exception to this comes from the protein cytochrome c_2. Due to its functional role of transferring electrons from one complex to another during the electron transport chain, it is expected that cytochrome c_2 would associate with a membrane surface. Figure 7.2 shows an electrostatic potential surface view and a cartoon view of the protein relative to the plane of the OPM database generated interface of a lipid bilayer. The surface predicted to contact the membrane is highly positively charged, whereas the opposite side is highly negatively charged. From an inspection of the structure, together with knowledge of the function, the prediction of how this protein contacts a membrane is straight forward. The cytochrome c_2 protein from *Rhodobacter sphaeroides* was obtained from an overproducing strain cycA1, harboring the plasmid pC2P404 and purified from periplasmic extracts containing cytochrome c_2 [23]. Peripheral membrane proteins of the electrostatic class are typically the easiest to express and characterize, as they come closest in solubility behavior to proteins that are cytosolic or non-membrane associated.

7.2.1.2 Case 2: Group IB Secreted Phospholipase A_2

The initial work with the small secreted class of phospholipase A_2 enzyme (sPLA_2) was done with cobra venom sPLA_2. The paradigm for study of the interfacial nature of a peripheral membrane enzyme is the group IB sPLA_2 enzyme. This mammalian enzyme is found in the small intestine and initial work used enzyme purified from porcine or bovine sources. The enzyme has also been shown to optimally function on anionic interface and is believed to be allosterically activated through a combination of charge interactions, together with desolvation of the interface

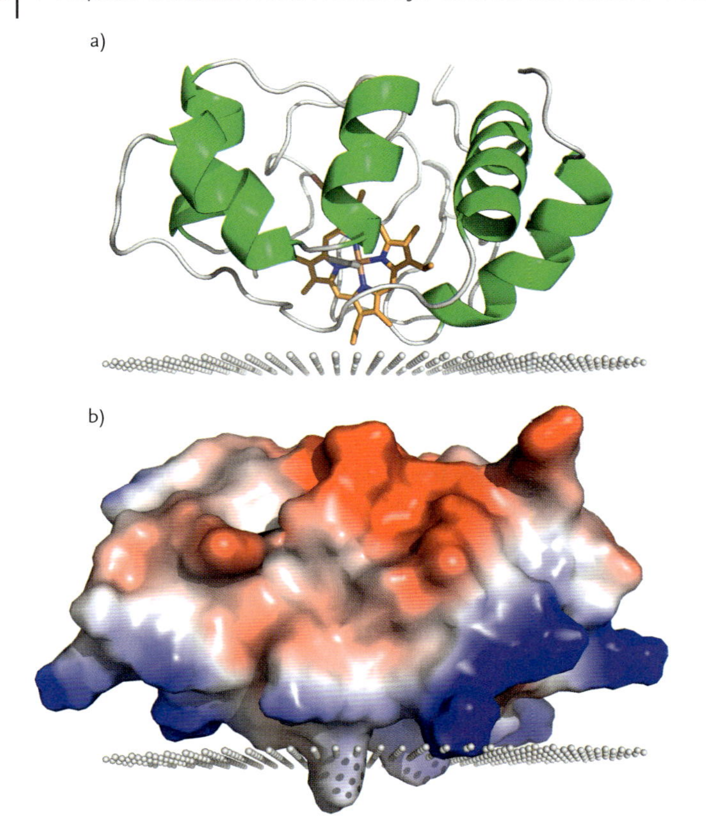

Figure 7.2 The protein cytochrome c_2 (PDB code 1cxa) is shown relative to its predicted membrane-binding orientation [22]. In each view, the plane of gray spheres represents the location of the ester carbonyls of lipids at the interface between nonpolar alkyl chain and polar head groups. (a) The protein is shown as a cartoon showing the proximity of the heme group relative to the predicted bound membrane. (b) Cytochrome c_2 is shown in the same orientation as (a) as an electrostatic potential surface. The positively charged region is facing the interface. Figures 7.2–7.6 were made using the program PyMol (www.pymol.org).

binding region [37]. The group IB enzyme was one of the first $sPLA_2$ enzymes to be cloned and over-expressed in an *E. coli* system [24]. The most challenging aspect to express homogeneous $sPLA_2$ enzyme is due to the large number of disulfide bonds for this secreted enzyme. The porcine group IB $sPLA_2$ enzyme has seven disulfide bonds. The enzyme has been studied extensively in the wild-type form, as well as numerous site-directed mutants. The enzyme has been studied on interfaces for function with the focus being kinetics and fluorescence spectroscopy. Additionally, many of the constructs used for functional studies have been used for nuclear magnetic resonance (NMR) or crystallographic studies. The original over-expression approaches have involved *E. coli* expression to inclusion bodies and refolding with an optimized balance of reduced to oxidized glutathione or an

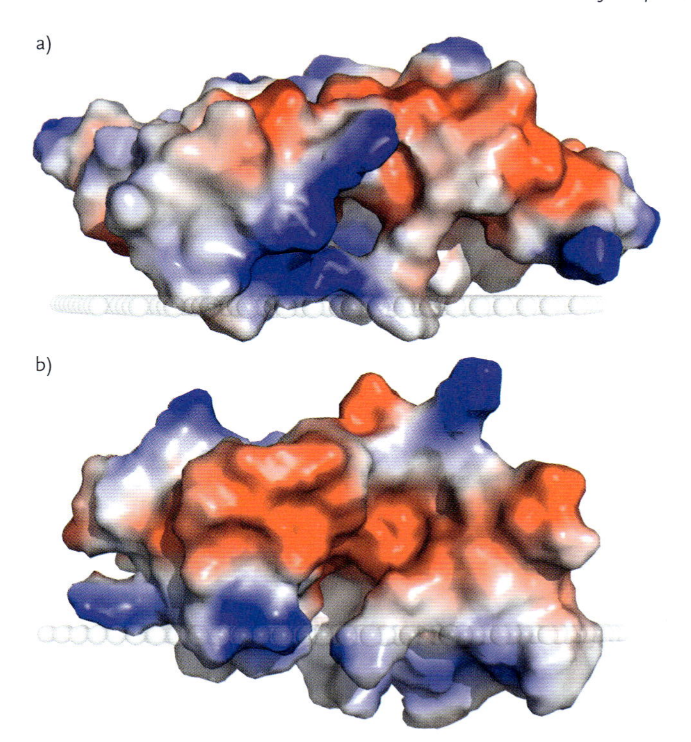

Figure 7.3 The enzyme $sPLA_2$ is shown as an electrostatic surface potential relative to its predicted membrane-binding orientation [22] (a) The membrane binding of group IB $sPLA_2$ enzyme (PDB code 5p2p) is dominated by electrostatic interactions made up of mainly Lys and Arg residues thought to interact with phospho head-groups of an anionic lipid interface. (b) The group-X $sPLA_2$ (PDB code 1le6) is known to interact with zwitterionic interfaces [40]. The increased penetration that is predicted for the hydrophobic resides of the group X versus the group IB enzyme is pronounced.

alternate reduced/oxidized thiol pair. This is still the method of choice for many $sPLA_2$ enzymes expressed and purified [38].

Other members of the $sPLA_2$ family behave differently on a membrane interface [39]. Figure 7.3 demonstrate the variation between group X and group IB $sPLA_2$ enzymes. The structure of the group X enzyme reveals that its mode of interface binding has shifted to be dominated by hydrophobic insertion [30]. This structural difference is correlated to the expected functional differences, along with the altered approach required for expression, folding, purification, and characterization [40].

7.2.2 Hydrophobic Patch

Proteins with a pronounced hydrophobic patch pose the greatest difficulty of the peripheral membrane proteins, yet they are often the easiest to pick out as membrane proteins using methods such as those used for the OPM database [20]. Even the mere extent of this patch can often muddle the choice of whether a protein should be defined as a monotopic-integral protein or a hydrophobic patch peripheral membrane protein. The examples described below will give context to some of the most popular methods used to reduce folding and solubility difficulties. The characterization of monodispersity, ability to pursue structural studies, and a demonstration of physiological relevant functions are each critical criteria to strive for.

7.2.2.1 **Case 1: Plasma Platelet-Activating Factor Acetylhydrolase**

The enzyme plasma platelet-activating factor acetylhydrolase (pPAFAH) hydrolyzes the *sn-2* ester of oxidatively fragmented phospholipids or the structurally similar GPCR ligand platelet-activating factor [41]. Physiologically, pPAFAH is associated to both low- and high-density lipoprotein particles, and is also referred to as lipoprotein-associated PLA_2 or group VIIA PLA_2 [42]. Early work with the enzyme [43] led to an *E. coli* expression system, and a detailed analysis of the limits of N- and C-terminal truncations. Also, the company ICOS Inc. had optimized the production [44] in a high-scale batch system. For routine lab analysis the protein is a challenging system, with *E. coli* expressions of the 42–441 construct on the order of 0.5–1 mg/l of *E. coli* culture [25].

Details of the expression and purification of pPAFAH that led to crystallographic work on both the ICOS-produced protein and protein produced in our lab have been reported previously [25, 44, 45]. Briefly summarized, the full-length clone of human pPAFAH was purchased from Invitrogen, the cDNA was amplified by polymerase chain reaction (PCR), and subcloned into the expression plasmid pGEX-4T3. The pPAFAH construct included residues 42–441 with an N-terminal extension of seven additional residues that remained after thrombin cleavage. The N-terminus start of pPAFAH found in human blood is heterogeneous, with a mixture of N-termini at positions S35, I42, or K55 [43]. The protein was expressed as a fusion of glutathione S-transferase (GST)–pPAFAH and was over-expressed in *E. coli* strain BL21. Cell lysis was followed by detergent extraction with 0.2% (w/v) Triton DF-16. The fusion protein was purified by affinity chromatography using glutathione Sepharose, and the GST tag was cleaved from the fusion protein via on-column cleavage using thrombin and purified protein was concentrated to a final concentrations of 1–2 mg/ml, always in the presence of a detergent.

Following the crystal structure of pPAFAH [26], a pronounced hydrophobic patch was fully elucidated first by simple inspection of the final structure, together with the OPM methods of Lomize *et al.* [19, 20]. The OPM methods were used to predict its association with a membrane surface as shown in Figure 7.4. The plane represents the interface between polar and nonpolar components. The polar headgroups of the phospholipids extend approximately 10 Å above the plane shown in Figure 7.4 [22]. Following this observation we compared the predicted membrane-binding orientation of pPAFAH by using the MAPAS method [21]. In contrast to the OPM method, MAPAS proposed several potential orientations, with one in agreement with the OPM method. In light of the functional data used to validate

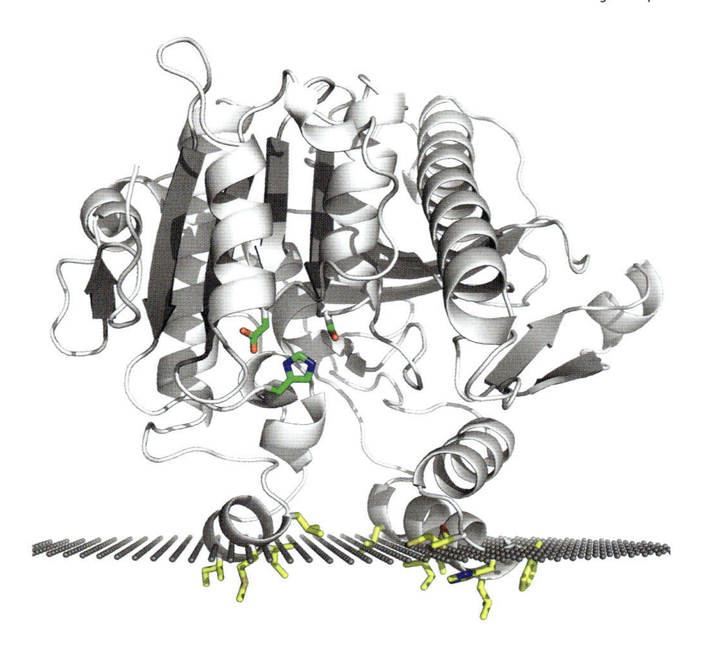

Figure 7.4 The enzyme plasma PAFAH (PDB code 3d59) is shown relative to its predicted membrane-binding orientation [22]. Residues penetrating into the hydrophobic region (yellow) are mainly nonpolar. The active-site catalytic triad (green) is poised above this plane, thereby providing an explanation of the enzymatic selectivity for oxidized and fragmented phospholipids [26].

the model of interface binding, it was concluded that the OPM prediction shown in Figure 7.4 serves as an initial model to understand and rationalize previously published data, as well as to direct future studies. This prediction was used as a basis to develop the model of pPAFAH associating with the phospholipid monolayer of low- and high-density lipoproteins [41]. The hydrophobic residues on helices 114–125 and 363–369 have been mutated to polar residues, and the lipoprotein binding properties are reduced. These mutations do not lead to a difference in catalytic activity judged from enzymatic assay using water-soluble substrates. A natural approach to improve expression yields and possibly open up certain avenues of investigation is to use a hydrophobic to polar site-directed mutation or even a combination of several mutants to convert pPAFAH into a soluble protein that expresses at higher yields. This solubilized protein would be expected to allow it to be concentrated to higher levels, to have a decreased need for detergents, and ultimately be a more homogeneous and monodisperse protein form. This approach has worked with pPAFAH, where a triple mutant of the hydrophobic patch has given a form that expressed with a higher yield and can be concentrated to a greater concentration in a monodisperse state (unpublished). Naturally, this approach is not recommended if the protein being produced is for functional studies related to its membrane bound function.

The work with pPAFAH has been adapted to a homologous protein, the intracellular human PAFAH type II, which shares 42% identity with the plasma enzyme. Using the program MODELER, a comparative model of PAFAH-II has allowed us to unequivocally identify the hydrophobic patch, which is likely to be at least partly responsible for membrane association [30]. The solubilization of PAFAH-II is being pursued as an avenue toward monodisperse protein for crystallographic structure analysis.

7.2.2.2 Case 2: Human Serum Paraoxonase 1

The human serum enzyme paraoxonase 1 (PON1) is expressed in human liver cells, secreted into the blood, and retains its N-terminal signal peptide of 22 amino acids. This N-terminal α-helix together with other hydrophobic helices direct the enzyme to bind to high-density lipoprotein particles. Early work with this enzyme has focused on samples obtained from purifications from human or rabbit blood [46]. Mammalian cell expression next allowed mutagenesis and other approaches to be pursued [47, 48]. Despite some success in Sf9 insect cell production [49], structural work required the improvement of the expression of this system. *E. coli* overexpression using fusions with maltose-binding protein, thioredoxin and GST had limited success. Once fusion partners were cleaved, the native PON1 enzyme did not stay monodisperse, even in a high concentration of detergents such as Triton X-100. Tawfik *et al.* at the Weizmann Institute used gene shuffling of PON1 sequences from several mammalian sources [27]. The gene-shuffled PON1 constructs were selected initially for increased solubility in *E. coli* expressions. Variants were also selected with increased activity and specificity of organophosphorus compound hydrolysis. The directed evolution approach soon led to the successful crystal structure determination by Harel *et al.* [28] of PON1, which had been the target of many structural biology labs.

7.2.3
Covalent Lipid Anchor

The attachment of a covalent lipid anchor is a common approach to direct particular proteins to a membrane surface. Covalent modifications vary from the myristoylation of the N-terminus, palmitoylation at cysteine residues, farnesylation of cysteine at the C-terminus of a protein and glycosyl phosphatidylinositol (GPI) anchors at the C-terminus as well. In each case, these particular post- or cotranslational modifications are directed to a particular site by a consensus sequence of the protein chain, which directs the auxiliary enzymes to attach these lipids. These modifications are specific to eukaryotic organisms, thereby requiring specialized approaches to express peripheral membrane proteins of this class. In the majority of cases, this limits the expression of these proteins to eukaryotic expression systems. However, in the first example, the protein recoverin was overexpressed in *E. coli* with the assistance of a coexpressed myristoyltransferase enzyme from yeast.

7.2.3.1 Case 1: Recoverin

The process of N-terminal myristoylation is a cotranslational process that targets the motif MGXXXS at the beginning of a protein chain, cleaves the first methionine, and conjugates a 14-carbon chain myristoyl group onto the glycine residue during protein translation [50]. Ray *et al.* [31] were successful at the *E. coli* expression of a myristoylated form of the bovine protein recoverin. Together with an expression vector containing bovine recoverin they coexpressed the *S. cerevisiae N*-myristoyltransferase enzyme. This enzyme transfers the C_{14} myristoyl group from myristoyl-CoA to the N-terminus of a targeted myristoyl site. This approach allowed the NMR structure determinations of a calcium bound and calcium free form of recoverin [51]. The protein recoverin is a 23-kDa protein that regulates the activity of guanylate cyclase by sensing calcium levels. In the calcium-free form the myristoyl tail is tucked inside a hydrophobic pocket. The NMR structure of Ames *et al.* provided a explanation of how calcium binding of recoverin causes the ejection of the myristoyl tail and thereby directs the protein to a cell membrane [51]. Figure 7.5 shows a snapshot from the calcium-free and calcium-bound forms of their respective NMR structures that demonstrate the basis for the calcium-myristoyl switch.

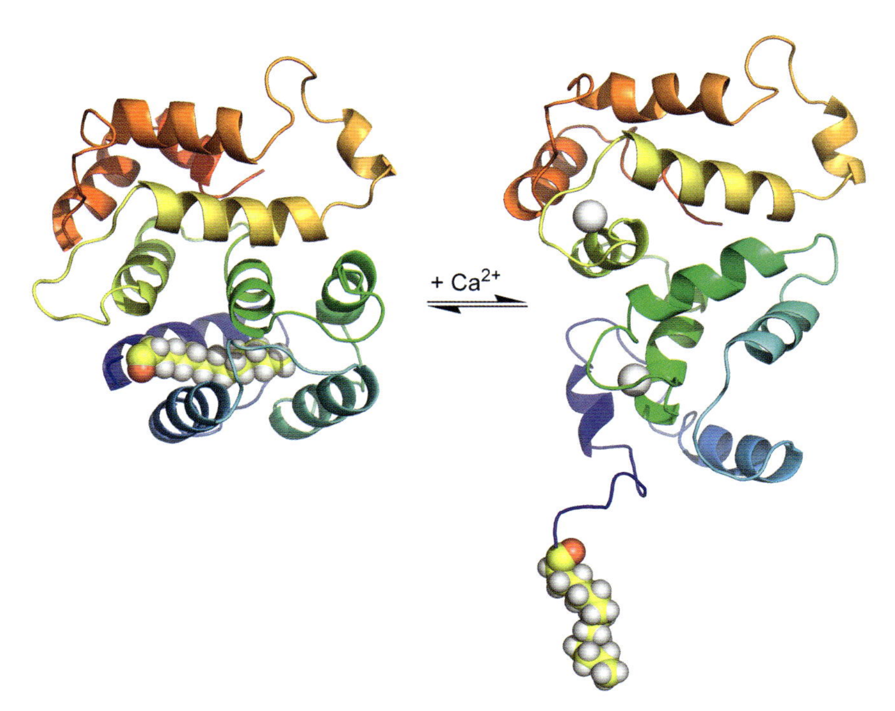

Figure 7.5 NMR structures of the enzyme bovine recoverin, which is N-terminally myristoylated, are shown in a calcium-free (left, PDB code 1iku, model 7) and calcium-bound form (right, PDB code 1jsa, model 9). The protein acts as a calcium-myristoyl switch, directing the calcium-bound form of the protein to a membrane surface.

7.2.3.2 Case 2: Intracellular Platelet-Activating Factor Acetylhydrolase Type II

The plasma PAFAH enzyme was described above as a protein with a hydrophobic patch made up of predominantly two α-helices. We have used the pPAFAH structure to build a comparative model of the homologous (42% identity) intracellular enzyme PAFAH-II. From this PAFAH-II model, we have identified a hydrophobic patch, which is likely to be at least partly responsible for membrane association [30]. However, unlike pPAFAH, the hydrophobic patch of PAFAH-II is not sufficient to direct the protein to the cell membrane. As described above for recoverin, PAFAH-II is a protein that is N-terminally myristoylated. It is hypothesized that the myristoyl group of PAFAH-II is critical for the localization of the enzymes to the inner leaflet membranes of human liver and kidney cells. Preliminary work on PAFAH-II has indicated that it exists in a myristoylated form in Madin-Darby bovine kidney cells [52], yet the exact role of the myristoyl group has not yet been elucidated. We are presently investigating the basis for this myristoyl-switch mechanism, which uses both the myristoyl group and a hydrophobic patch to direct the enzyme to the membrane surface following oxidative stress (unpublished).

7.2.4 Case 3: Palmitoylation of Human Proteins in Cell Culture

The examples of proteins that have been overexpressed for structural characterization is expectedly more extensive for proteins with a myristoyl group as discussed above. These are simpler systems due to a simple and predictable site of myristoylation that is limited to the N-terminus of proteins with a GXXXS consensus sequence. Other lipid modifications, such as palmitoylation, are harder to predict since cysteine residues are targeted. Also, whether a cysteine is palmitoylated is partly dependent on the structural context. Recently, a rapid screen of palmitoylation sites in a human cell line was performed using "click" chemistry analogs to ultimately bind biotin to palmitoylation sites [53]. The approach introduced into cells the compound 17-octadecynoic acid, which is a commercially available analog of palmitic acid. After its incorporation at palmitoylation sites, this compound was then readily coupled using click chemistry to biotin, which allowed for the identification of 125 palmitoylated protein sites. As more proteins are identified to have covalent lipid anchors, an increased number of structural and functional studies describing the membrane associated function will be elucidated.

7.2.5 Lipid-Binding Domain

The final group of peripheral membrane-binding proteins discussed includes proteins that are directed to bind a membrane surface through specific interactions with a highly specific lipid component or components of the membrane. Just as with the covalent lipid anchor proteins discussed above, the absence of a specific lipid in the membrane typically prevents any tight binding interaction of these peripheral membrane proteins or protein domains to the membrane. A wide range of lipid-binding domains have been identified in protein systems including: the pleckstrin homology (PH) domain [54], C1–C4 domains of protein kinase C [55], FYVE domains [56], PX domains [57] and others. Generally, these proteins and domains of larger proteins are water-soluble and do not pose the same difficulties as the hydrophobic patch or lipid anchored class of peripheral membrane proteins discussed above. Here, we will describe two examples of the structural and functional work that has been performed with the PH and C2 domains bound to their respective specific lipid ligand. In each case, the binding interactions to the specific lipid are believed to give the majority of the thermodynamic driving force for membrane interaction.

7.2.5.1 Case 1: Pleckstrin Homology Domain

The PH domain is a common motif made of about 120 amino acids with a β-sheet closed off with a C-terminal α-helix. The PH domain has a variable ligand-binding pocket that is distinguished by positively charged residues that are largely responsible for the specificity to bind specific anionic charged lipids. Surrounding this pocket are other positively charged residues that are believed to bind nonspecifically to the anionic phospho groups of other lipid components. Sigler *et al.* [58] were successful in solving the crystal structure of the PH domain from rat phospholipase C-δ1 complexed with the lipid inositol-1,4,5-trisphosphate (IP_3), as shown in Figure 7.6. The PH domain was expressed in the soluble fraction using an *E. coli* overexpression construct [32]. In addition to the expression of the PH domain, Lemmon *et al.* [32] characterized the specificity and demonstrated the modularity of PH domains in general.

The PH domain from the general receptor for phosphoinositides (phosphorylated phosphatidylinositols (PIPs)), isoform 1, was shown [59] to specifically bind the lipid phosphoinositol-3,4,5-triphosphate, also known as PIP_3. This work, done in the Falke lab, measured the K_d values for a range of PIP lipids to characterize their membrane targeting function. The truncated PH domain was also probed by protein-to-membrane fluorescence resonance energy transfer experiment to determine the magnitude of nonspecific interactions with this domain. Interestingly, phosphatidylserine lipids appear to direct a nonspecific electrostatic interaction of the PH domain to the membrane, and this step facilitates the docking of the specific lipid PIP_3 [59]. This PH domain highlights a general feature that the lipid-binding domain proteins first interact with membranes by a combination of nonspecific polar/electrostatic interactions with lipid head-groups, followed by a tight binding interaction with their specific lipid ligand.

7.2.5.2 Case 2: C2 Domain

Just as with the previous example described, the expression and purification of the C2 domains of many significant proteins are often attainable with a simple *E. coli* expression and purification facilitated by use of an affinity tag. Here, the challenge is to couple this expressed domain with the relevant experiments that elucidate their structure and function. The C2 α-domain of mammalian protein kinase C was expressed with a His tag [33]. The C2 α-domain provides the specificity of these protein kinases to target membrane interactions through a cooperative Ca^{2+}

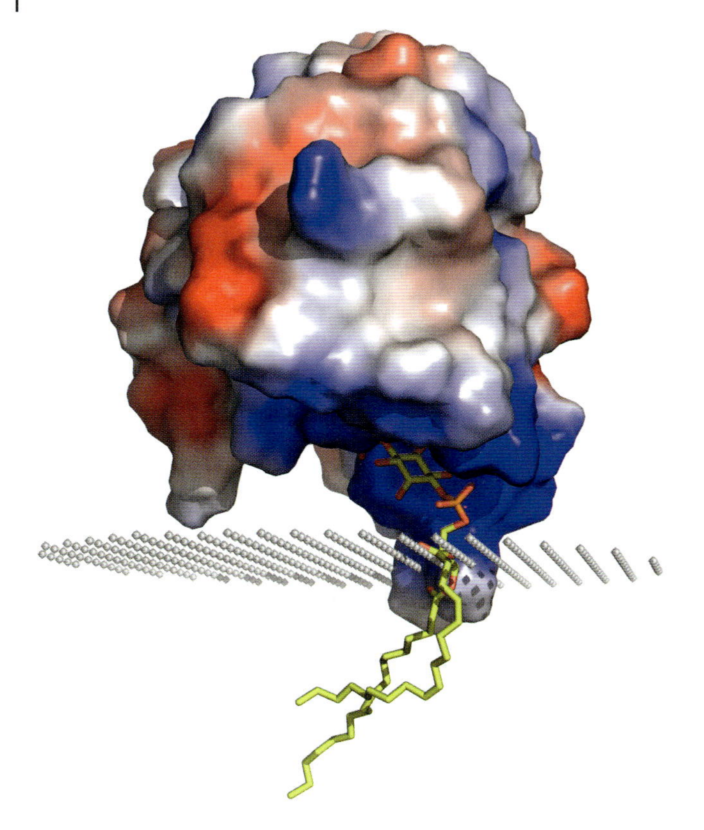

Figure 7.6 Phospholipase C-δ1 domain bound to the lipid second messenger IP_3 (PDB code 1mai) and shown relative to its predicted membrane-binding orientation [22].

and phosphatidylserine binding effect. The authors of this study were first able to demonstrate significant binding of a water-soluble version of the phosphatidylserine lipids. This 1,2-dicaproyl-*sn*-phosphatidyl-L-serine (DCPS) analog has two 6-carbon alkyl chains. The DCPS analog was then used to solve a crystal structure that provides a convincing model of how the C2 domain binds phosphoserine lipid and Ca^{2+}, as well as provides a starting point to understand membrane binding [33].

7.3
Conclusions

In studies that target peripheral membrane proteins, often the primary challenge is to identify the mode and significance of a membrane association in order to guide the design of the expression construct. Proteins of the nonspecific electrostatic and lipid-binding domain class, in general, pose the least technical challenges for protein expression and purification. In contrast, hydrophobic patch proteins pose the greatest challenges, and these systems have the most to gain from recent advances in technologies and strategies to express and characterize integral membrane proteins. Likewise, recent advances in the treatment of post-translational modifications, such as myristoylation and palmitoylation, will open up new opportunities to study this important class of lipid anchor proteins.

Acknowledgments

This work was supported in part by National Institute of Health grants 2P20 RR015588 from the National Center for Research Resources and 5R01 HL084366 from the National Heart, Lung, and Blood Institute.

Abbreviations

DCPS	1,2-dicaproyl-*sn*-phosphatidyl-L-serine
GFP	Green Fluorescent Protein
GPI	glycosyl phosphatidylinositol
GST	glutathione S-transferase
IP_3	inositol-1,4,5-trisphosphate
NMR	nuclear magnetic resonance
OPM	Orientations of Proteins in Membranes
PAFAH	platelet-activating factor acetylhydrolase
PCR	polymerase chain reaction
PDB	Protein Data Bank
PH	pleckstrin homology
PIP	phosphoinositide
PON1	paraoxonase 1
pPAFAH	plasma platelet-activating factor acetylhydrolase
$sPLA_2$	secreted phospholipase A_2

References

1 Singer, S.J. and Nicolson, G.L. (1972) The fluid mosaic model of the structure of cell membranes. *Science*, **175**, 720–731.

2 Wallin, E. and von Heijne, G. (1998) Genome-wide analysis of integral membrane proteins from eubacterial, archaean, and eukaryotic organisms. *Protein Sci.*, **7**, 1029–1038.

3 Edwards, A.M., Arrowsmith, C.H., Christendat, D., Dharamsi, A., Friesen, J.D., Greenblatt, J.F., and Vedadi, M. (2000) Protein production: feeding the crystallographers and NMR spectroscopists. *Nat. Struct. Biol.*, **7** (Suppl.), 970–972.

4 Midgett, C.R. and Madden, D.R. (2007) Breaking the bottleneck: eukaryotic membrane protein expression for

high-resolution structural studies. *J. Struct. Biol.*, **160**, 265–274.

5 Wagner, S., Bader, M.L., Drew, D., and de Gier, J.W. (2006) Rationalizing membrane protein overexpression. *Trends Biotechnol.*, **24**, 364–371.

6 Goni, F.M. (2002) Non-permanent proteins in membranes: when proteins come as visitors (Review). *Mol. Membr. Biol.*, **19**, 237–245.

7 Johnson, J.E. and Cornell, R.B. (1999) Amphitropic proteins: regulation by reversible membrane interactions (Review). *Mol. Membr. Biol.*, **16**, 217–235.

8 Dessailly, B.H., Nair, R., Jaroszewski, L., Fajardo, J.E., Kouranov, A., Lee, D., Fiser, A., Godzik, A., Rost, B., and Orengo, C. (2009) PSI-2: structural genomics to cover protein domain family space. *Structure*, **17**, 869–881.

9 Moore, P.B. (2007) Let's call the whole thing off: some thoughts on the protein structure initiative. *Structure*, **15**, 1350–1352.

10 Petsko, G.A. (2007) An idea whose time has gone. *Genome Biol.*, **8**, 107.

11 Mancia, F. and Love, J. (2010) High-throughput expression and purification of membrane proteins. *J. Struct. Biol.*, **172**, 85–93.

12 Katzen, F., Peterson, T.C., and Kudlicki, W. (2009) Membrane protein expression: no cells required. *Trends Biotechnol.*, **27**, 455–460.

13 Sonoda, Y., Cameron, A., Newstead, S., Omote, H., Moriyama, Y., Kasahara, M., Iwata, S., and Drew, D. (2010) Tricks of the trade used to accelerate high-resolution structure determination of membrane proteins. *FEBS Lett.*, **584**, 2539–2547.

14 Derman, A.I., Prinz, W.A., Belin, D., and Beckwith, J. (1993) Mutations that allow disulfide bond formation in the cytoplasm of *Escherichia coli*. *Science*, **262**, 1744–1747.

15 Luo, J., Choulet, J., and Samuelson, J.C. (2009) Rational design of a fusion partner for membrane protein expression in *E. coli*. *Protein Sci.*, **18**, 1735–1744.

16 Drew, D., Lerch, M., Kunji, E., Slotboom, D.J., and de Gier, J.W. (2006) Optimization of membrane protein overexpression and purification using GFP fusions. *Nat. Methods*, **3**, 303–313.

17 Pantazatos, D., Kim, J.S., Klock, H.E., Stevens, R.C., Wilson, I.A., Lesley, S.A., and Woods, V.L., Jr (2004) Rapid refinement of crystallographic protein construct definition employing enhanced hydrogen/deuterium exchange MS. *Proc. Natl. Acad. Sci. USA*, **101**, 751–756.

18 Goldschmidt, L., Cooper, D.R., Derewenda, Z.S., and Eisenberg, D. (2007) Toward rational protein crystallization: a web server for the design of crystallizable protein variants. *Protein Sci.*, **16**, 1569–1576.

19 Lomize, A.L., Pogozheva, I.D., Lomize, M.A., and Mosberg, H.I. (2006) Positioning of proteins in membranes: a computational approach. *Protein Sci.*, **15**, 1318–1333.

20 Lomize, M.A., Lomize, A.L., Pogozheva, I.D., and Mosberg, H.I. (2006) OPM: orientations of proteins in membranes database. *Bioinformatics*, **22**, 623–625.

21 Sharikov, Y., Walker, R.C., Greenberg, J., Kouznetsova, V., Nigam, S.K., Miller, M.A., Masliah, E., and Tsigelny, I.F. (2008) MAPAS: a tool for predicting membrane-contacting protein surfaces. *Nat. Methods*, **5**, 119–119.

22 Lomize, A.L., Pogozheva, I.D., Lomize, M.A., and Mosberg, H.I. (2007) The role of hydrophobic interactions in positioning of peripheral proteins in membranes. *BMC Struct. Biol.*, **7**, 44.

23 Axelrod, H.L., Feher, G., Allen, J.P., Chirino, A.J., Day, M.W., Hsu, B.T., and Rees, D.C. (1994) Crystallization and X-ray structure determination of cytochrome c_2 from *Rhodobacter sphaeroides* in three crystal forms. *Acta Crystallogr. D*, **50**, 596–602.

24 Thunnissen, M.M., Ab, E., Kalk, K.H., Drenth, J., Dijkstra, B.W., Kuipers, O.P., Dijkman, R., de Haas, G.H., and Verheij, H.M. (1990) X-ray structure of phospholipase A_2 complexed with a substrate-derived inhibitor. *Nature*, **347**, 689–691.

25 Samanta, U., Kirby, S.D., Srinivasan, P., Cerasoli, D.M., and Bahnson, B.J. (2009) Crystal structures of human group-VIIA phospholipase A_2 inhibited by organophosphorus nerve agents exhibit non-aged complexes. *Biochem. Pharmacol.*, **78**, 420–429.

26 Samanta, U. and Bahnson, B.J. (2008) Crystal structure of human plasma platelet-activating factor acetylhydrolase: structural implication to lipoprotein binding and catalysis. *J. Biol. Chem.*, **283**, 31617–31624.

27 Aharoni, A., Gaidukov, L., Yagur, S., Toker, L., Silman, I., and Tawfik, D.S. (2004) Directed evolution of mammalian paraoxonases PON1 and PON3 for bacterial expression and catalytic specialization. *Proc. Natl. Acad. Sci. USA*, **101**, 482–487.

28 Harel, M., Aharoni, A., Gaidukov, L., Brumshtein, B., Khersonsky, O., Meged, R., Dvir, H., Ravelli, R.B., McCarthy, A., Toker, L., Silman, I., Sussman, J.L., and Tawfik, D.S. (2004) Structure and evolution of the serum paraoxonase family of detoxifying and anti-atherosclerotic enzymes. *Nat. Struct. Mol. Biol.*, **11**, 412–419.

29 Zhao, Y., White, M.A., Muralidhara, B.K., Sun, L., Halpert, J.R., and Stout, C.D. (2006) Structure of microsomal cytochrome P450 2B4 complexed with the antifungal drug bifonazole: insight into P450 conformational plasticity and membrane interaction. *J. Biol. Chem.*, **281**, 5973–5981.

30 Winget, J.M. (2009) *In vitro* investigations into the structures of membrane-associated proteins augmented by the application of computational methods in chemistry and biochemistry. PhD Thesis. University of Delaware, Newark, NJ.

31 Ray, S., Zozulya, S., Niemi, G.A., Flaherty, K.M., Brolley, D., Dizhoor, A.M., McKay, D.B., Hurley, J., and Stryer, L. (1992) Cloning, expression, and crystallization of recoverin, a calcium sensor in vision. *Proc. Natl. Acad. Sci. USA*, **89**, 5705–5709.

32 Lemmon, M.A., Ferguson, K.M., O'Brien, R., Sigler, P.B., and Schlessinger, J. (1995) Specific and high-affinity binding of inositol phosphates to an isolated pleckstrin homology domain. *Proc. Natl. Acad. Sci. USA*, **92**, 10472–10476.

33 Verdaguer, N., Corbalan-Garcia, S., Ochoa, W.F., Fita, I., and Gomez-Fernandez, J.C. (1999) Ca^{2+} bridges the C2 membrane-binding domain of protein kinase Calpha directly to phosphatidylserine. *EMBO J.*, **18**, 6329–6338.

34 Murray, D. and Honig, B. (2002) Electrostatic control of the membrane targeting of C2 domains. *Mol. Cell.*, **9**, 145–154.

35 Lin, Y., Nielsen, R., Murray, D., Hubbell, W.L., Mailer, C., Robinson, B.H., and Gelb, M.H. (1998) Docking phospholipase A_2 on membranes using electrostatic potential-modulated spin relaxation magnetic resonance. *Science*, **279**, 1925–1929.

36 Murray, D., Ben-Tal, N., Honig, B., and McLaughlin, S. (1997) Electrostatic interaction of myristoylated proteins with membranes: simple physics, complicated biology. *Structure*, **5**, 985–989.

37 Berg, O.G., Gelb, M.H., Tsai, M.D., and Jain, M.K. (2001) Interfacial enzymology: the secreted phospholipase A_2 paradigm. *Chem. Rev.*, **101**, 2613–2654.

38 Rouault, M., Le Calvez, C., Boilard, E., Surrel, F., Singer, A., Ghomashchi, F., Bezzine, S., Scarzello, S., Bollinger, J., Gelb, M.H., and Lambeau, G. (2007) Recombinant production and properties of binding of the full set of mouse secreted phospholipases A_2 to the mouse M-type receptor. *Biochemistry*, **46**, 1647–1662.

39 Winget, J.M., Pan, Y.H., and Bahnson, B.J. (2006) The interfacial binding surface of phospholipase A_2s. *Biochim. Biophys. Acta*, **1761**, 1260–1269.

40 Pan, Y.H., Yu, B.Z., Singer, A.G., Ghomashchi, F., Lambeau, G., Gelb, M.H., Jain, M.K., and Bahnson, B.J. (2002) Crystal structure of human group X secreted phospholipase A_2. Electrostatically neutral interfacial surface targets zwitterionic membranes. *J. Biol. Chem.*, **277**, 29086–29093.

41 Srinivasan, P. and Bahnson, B.J. (2010) Molecular model of plasma PAF acetylhydrolase–lipoprotein association: insights from the structure. *Pharmaceuticals*, **3**, 541–557.

42 Schaloske, R.H. and Dennis, E.A. (2006) The phospholipase A_2 superfamily and its group numbering system. *Biochim. Biophys. Acta*, **1761**, 1246–1259.

43 Tjoelker, L.W., Eberhardt, C., Unger, J., Trong, H.L., Zimmerman, G.A.,

McIntyre, T.M., Stafforini, D.M., Prescott, S.M., and Gray, P.W. (1995) Plasma platelet-activating factor acetylhydrolase is a secreted phospholipase A_2 with a catalytic triad. *J. Biol. Chem.*, **270**, 25481–25487.

44 Cousens, L.S., Eberhardt, A., Gray, P., Tjoelker, L.W., Wilder, C.L., and Trong, H. (1996) Platelet-activating factor acetylhydrolase. US Patent 6045794.

45 Samanta, U., Wilder, C., and Bahnson, B.J. (2009) Crystallization and preliminary X-ray crystallographic analysis of human plasma platelet activating factor acetylhydrolase. *Protein Pept. Lett.*, **16**, 97–100.

46 Josse, D., Ebel, C., Stroebel, D., Fontaine, A., Borges, F., Echalier, A., Baud, D., Renault, F., Le Maire, M., Chabrieres, E., and Masson, P. (2002) Oligomeric states of the detergent-solubilized human serum paraoxonase (PON1). *J. Biol. Chem.*, **277**, 33386–33397.

47 Josse, D., Xie, W., Renault, F., Rochu, D., Schopfer, L.M., Masson, P., and Lockridge, O. (1999) Identification of residues essential for human paraoxonase (PON1) arylesterase/organophosphatase activities. *Biochemistry*, **38**, 2816–2825.

48 Yeung, D.T., Josse, D., Nicholson, J.D., Khanal, A., McAndrew, C.W., Bahnson, B.J., Lenz, D.E., and Cerasoli, D.M. (2004) Structure/function analyses of human serum paraoxonase (HuPON1) mutants designed from a DFPase-like homology model. *Biochim. Biophys. Acta*, **1702**, 67–77.

49 Draganov, D.I., Teiber, J.F., Speelman, A., Osawa, Y., Sunahara, R., and La Du, B.N. (2005) Human paraoxonases (PON1, PON2, and PON3) are lactonases with overlapping and distinct substrate specificities. *J. Lipid Res.*, **46**, 1239–1247.

50 Resh, M.D. (1999) Fatty acylation of proteins: new insights into membrane targeting of myristoylated and palmitoylated proteins. *Biochim. Biophys. Acta*, **1451**, 1–16.

51 Ames, J.B., Ishima, R., Tanaka, T., Gordon, J.I., Stryer, L., and Ikura, M. (1997) Molecular mechanics of calcium-myristoyl switches. *Nature*, **389**, 198–202.

52 Matsuzawa, A., Hattori, K., Aoki, J., Arai, H., and Inoue, K. (1997) Protection against oxidative stress-induced cell death by intracellular platelet-activating factor-acetylhydrolase II. *J. Biol. Chem.*, **272**, 32315–32320.

53 Martin, B.R. and Cravatt, B.F. (2009) Large-scale profiling of protein palmitoylation in mammalian cells. *Nat. Methods*, **6**, 135–138.

54 Lemmon, M.A. (2008) Membrane recognition by phospholipid-binding domains. *Nat. Rev. Mol. Cell. Biol.*, **9**, 99–111.

55 Cho, W. and Stahelin, R.V. (2005) Membrane–protein interactions in cell signaling and membrane trafficking. *Annu. Rev. Biophys. Biomol. Struct.*, **34**, 119–151.

56 Diraviyam, K., Stahelin, R.V., Cho, W., and Murray, D. (2003) Computer modeling of the membrane interaction of FYVE domains. *J. Mol. Biol.*, **328**, 721–736.

57 Sato, T.K., Overduin, M., and Emr, S.D. (2001) Location, location, location: membrane targeting directed by PX domains. *Science*, **294**, 1881–1885.

58 Ferguson, K.M., Lemmon, M.A., Schlessinger, J., and Sigler, P.B. (1995) Structure of the high affinity complex of inositol trisphosphate with a phospholipase C pleckstrin homology domain. *Cell*, **83**, 1037–1046.

59 Corbin, J.A., Dirkx, R.A., and Falke, J.J. (2004) GRP1 pleckstrin homology domain: activation parameters and novel search mechanism for rare target lipid. *Biochemistry*, **43**, 16161–16173.

8
Expression of G-Protein-Coupled Receptors

Alexei Yeliseev and Krishna Vukoti

8.1
Introduction

G-protein-coupled receptors (GPCRs) belong to a large class of integral membrane proteins found in eukaryotic organisms. GPCRs are localized in the cell membrane and transmit extracellular signals to the intracellular response elements. In mammalian cells, GPCRs are involved in a variety of signal transduction pathways. Therefore, it is not surprising that they are the targets of numerous pharmaceutical drugs – over 30% of drugs currently on the market, according to various estimates [1–5]. There are more than 900 different GPCRs encoded in mammalian genomes, yet structures of only a few of them, namely those of bovine and squid rhodopsin, β_1- and β_2-adrenergic receptors, and adenosine A_{2A} receptor, are currently available [6–15].

Much of the progress in the rational design of drugs targeting GPCRs will depend on the availability of structural information for various types of these receptors. Unlike rhodopsin, which is expressed at high density in the retina and can be obtained in significant (milligram) quantities in highly enriched form from natural sources, the vast majority of GPCRs are present in mammalian tissues at very low levels; therefore, isolation of GPCRs from these tissues in large quantities is not practical. A technically feasible alternative is the expression of these proteins either in a heterologous host or in a cell-free system. While the availability of homogenous preparations of GPCRs in milligram quantities is a prerequisite for initiating high-resolution structural studies by X-ray crystallography and nuclear magnetic resonance (NMR), valuable structural information can also be obtained by applying a wide range of biochemical and biophysical techniques including chemical cross-linking coupled with mass spectrometry [16], site-directed mutagenesis [17, 18], fluorescence spectroscopy [19–21], atomic force microscopy [22, 23], and circular dichroism spectroscopy [24], to name a few.

GPCRs perform a wide variety of biological functions and their sequence similarity is relatively low. Nevertheless, there are several important common structural features; namely, seven α-helical transmembrane domains, extracellular N-terminal domain, and intracellular C-terminus (Figure 8.1) [25]. The heptahelical transmembrane structure explains the high hydrophobicity of these proteins

Production of Membrane Proteins: Strategies for Expression and Isolation, First Edition.
Edited by Anne Skaja Robinson.

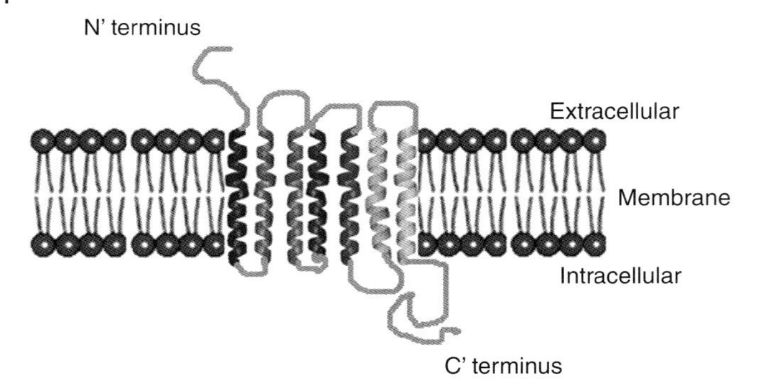

Figure 8.1 Schematic structure of a GPCR.

that can only be solubilized from membranes by the action of certain detergents or chaotropic agents.

The expression of GPCRs in a heterologous host is generally regarded as a complex task requiring careful consideration of several critical parameters. One of the most important is the selection of the host, which, in turn, may affect the cost of the production, yield, stability of the target protein, its functionality, and certain post-translational modifications. While many GPCR expression projects aim at obtaining large amounts of purified protein suitable for generating well-diffracting crystals, an ability to produce a stable isotope-labeled protein is an important requirement for studies by NMR and influences the choice of the expression strategy. Furthermore, the receptors have to be solubilized in detergent micelles, purified in order to obtain a homogenous preparation, and may need to be reconstituted into the lipid bilayer. Thus, successful GPCR sample preparation requires significant investments, labor-intensive technology development, and is commonly regarded as a high-risk project. Nevertheless, there are a number of recent publications that demonstrate the successful application of existing technologies and the development of new strategies for GPCR sample preparation, thus providing important contributions to this field [1, 6–9, 13, 26–32]. The objective of this chapter is to summarize these recent developments and provide the reader with the most up-to-date information available. A representative list of heterologously expressed GPCR including expression hosts and fusion tags used and expression levels achieved is provided in Table 8.1.

8.2
Bacterial Expression of GPCRs

The advantages of bacterial expression systems are numerous and have been discussed in several reviews [28, 31, 91–93]. They include relatively simple and inexpensive cultivation media, (generally) short time of fermentation, the availability of numerous (in the case of *Escherichia coli*) host strains and plasmids, and

Table 8.1 Heterologous expression of GPCRs.

Expression system	GPCR expressed	N-terminal tag	C-terminal tag	Expression level	Functional receptor	Purified	References
Bacteria							
E. coli	adenosine A_{2A}, human	MBP	TrxA-His_{10}	$1–2 \times 10^3$ receptors/cell	yes	yes	[33]
	adenosine A_{2A}, human		His_{10}-FLAG	2–3 mg/ml	yes	yes	[33]
	bradykinin B_2, human	FLAG	GFP-His_8	2 mg/l	yes	yes	[34]
	cannabinoid CB2, human	MBP	TrxA-His_{10}	1–2 mg/l	yes	yes	[35]
	cannabinoid CB2, human ([^{15}N]Trp)	MBP	TrxA-His_{10}	2 mg/l	yes	yes	[36]
	cannabinoid CB1, human	FLAG	GFP-His_8	0.5 mg/l	yes	yes	[34]
	leukotriene B_4, BLT1, human		His_6	IB	yes	yes	[37, 38]
	leukotriene B_4, BLT2, human	α_5 integrin	His_6	IB	yes	yes	[39]
	neurokinin NK_1, human		His_6	IB	no	yes	[40]
	serotonin 5-HT_4, mouse		ketosteroid isomerase, His_6	IB	yes	yes	[24]
	thyroid-stimulating hormone, human	His_6		2.5 mg/l		partly	[41]
	neurotensin 1, rat	MBP	TrxA-His_{10}	0.2–0.4 mg/l, 1.6 mg/200 l (PURE)	yes	yes	[42]
	neurotensin 1, rat	MBP	TrxA-His_{10}	0.8 mg/l	yes	yes	[43]
	neuropeptide Y2, human		His_8	20 mg/l	yes	yes	[44]
R. sphaeroides	angiotensin AT_{1A}, human		TEV-His_{10}	12 pmol/mg	yes	no	[45]
	adenosine A_{2A}, human		TEV-His_{10}	0.5 pmol/mg	yes	no	[45]
	bradykinin B2, human		TEV-His_{10}	0.7 pmol/mg	yes, low	no	[45]

(Continued)

Table 8.1 (Continued)

Expression system	GPCR expressed	N-terminal tag	C-terminal tag	Expression level	Functional receptor	Purified	References
Insect cells							
Sf9	adenosine A_1, human			46 pmol/mg	yes	no	[46]
	adenosine A_{2A}, human	His_6, FLAG		18.7 pmol/mg	yes	yes	[47]
	adrenergic α_2, human		GFP	11–64 pmol/mg	yes	no	[48]
	adrenergic α_2, human			31 pmol/mg	yes	no	[46]
	adrenergic β_2, turkey			1 pmol/mg			[49]
	adrenergic β_2, human			73 pmol/mg	yes	no	[46]
	adrenergic β_2, human			30 pmol/mg	yes	yes	[49]
	adrenergic β_2, human (mutant: N187E Δ(366–414))	HA, FLAG		30 pmol/mg	yes	yes	[8]
	adrenergic β_2, human (mutant: N187E Δ(366–414), intracytoplasmic loop 3 replaced with T4 lysozyme)	HA, FLAG		30 pmol/mg	yes	yes	[6, 7]
	endothelin ETB, human	His_6		19.6 pmol/mg	yes	yes	[50]
	histamine H_1, human		His_{10}	5–7 mg/l	yes	yes	[51]
	serotonin 5-HT_{1A}, human			63 pmol/mg	yes	no	[46]
	muscarinic M_1 acetylcholine, human			5 pmol/mg	yes	yes	[49]
	muscarinic M_2 acetylcholine, human			20–30 pmol/mg	yes	yes	[49, 52, 53]
	PACAP, human			82.6 pmol/mg	yes	yes	[54]
	rhodopsin, bovine		His_6	4 mg/l			[55]
	secretin, rat		HA	830 pmol/mg	yes	yes	[56]

Sf9 and *T. ni*	metabotropic glutamate, *D. melanogaster*			11–18 pmol/mg			[57]
	μ-opioid, human	HA-FLAG-His_{10}-YFP		12–40 pmol/mg	yes	yes	[58]
T. ni	adenosine A_2, human			270 pmol/mg	yes	no	[46]
	adrenergic β_1, human			79 pmol/mg	yes	no	[46]
	chemokine CXCR4, human			1 mg/8 l	yes	yes	[59]
Sf21	cannabinoid CB1, human			40 pmol/mg	yes	no	[46]
	cannabinoid CB1, human		His_6	5.4 pmol/mg	yes	yes	[60]
	cannabinoid CB2, human		His_6	9.3 pmol/mg	yes	yes	[61]
	adrenergic α_{2B}, human			95 pmol/mg	yes	no	[46]
	histamine H_1, human			85 pmol/mg	yes	no	[46]
Mammalian systems							
CHO	adenosine A_1, human	His_6, FLAG		30.7 pmol/mg	yes	yes	[47]
	muscarinic M_2 acetylcholine, porcine			1×10^6 receptors/cell	yes	yes	[62]
	PACAP, human			24 pmol/mg	yes	yes	[54]
	secretin, rat		HA	82 pmol/mg	yes	yes	[56]
	tachykinin NK_1, human	FLAG	His_6	10 pmol/mg	yes	yes	[63]

(Continued)

Table 8.1 (*Continued*)

Expression system	GPCR expressed	N-terminal tag	C-terminal tag	Expression level	Functional receptor	Purified	References
COS	adrenergic β_2, human		ID4	220 pmol/mg	yes	yes	[64]
	metabotropic glutamate R_6, human		ID4		yes	yes	[65]
	parathyroid hormone, human		ID4	0.25 nmol/mg	yes	yes	[66]
	rhodopsin, bovine (mutant)			5–10 μg/10^7 cells			[9]
	thromboxane A_2, human		His_6	24 pmol/mg	yes	yes	[67]
HEK-293	chemokine CCR1, human	StrepII/HA	His_8	0.1 mg/l	yes	yes	[68]
	melatonin MT_1, human		TAP	1 pmol/mg		yes	[69]
	melatonin MT_2, human		TAP	0.34 pmol/mg		yes	[69]
	μ-opioid, human	FLAG	V5, His_6	8.5 pmol/mg	yes	yes	[70]
	parathyroid hormone, human		ID4		yes	yes	[66]
	Ste2p, *S. cerevisiae*		GFP, His_8, StrepII	1–2 mg/l (pure)	yes	yes	[71]
HEK-293S	adrenergic β_2, human		ID4	220 pmol/mg	yes	yes	[64]
	bradykinin B_2, human	c-Myc/His_6	ID4	100 pmol/mg	no	yes	[72]
	opsin, bovine			9 mg/l		yes	[73]
	rhodopsin, human (^{13}C, ^{15}N)			2.5 mg/l			[74]
HEK-293T	chemokine CCR5, human		ID4	4×10^6 receptors/cell	yes	yes	[75]
BHK	κ-opioid, human		His_6-$G_{\alpha i1}$	0.4–0.5 mg/l	yes	yes	[76]

Yeast							
S. cerevisiae	adenosine A_{2A}, human		GFP	4 mg/l	yes	no	[77]
	adenosine A_{2A}, human		GFP	28 mg/l	yes	no	[78]
	adenosine A_{2A}, human		GFP-His_{10}	2 mg/l, ~1 mg/l purified	yes	yes	[79]
	adrenergic $\alpha 2C_2$, human			13 pmol/mg	yes	yes	[80]
	olfactory I7, human			327 pmol/mg	yes	no	[81]
P. pastoris	adenosine A_{2A}, human (V334 C-terminus truncated)			200 pmol/mg	yes	– –	[82]
	bradykinin B_2, human	FLAG-His_{10}	BRS	3.5 pmol/mg	yes	yes	[83]
	cannabinoid CB1, human	FLAG	c-Myc, His_6	3.6 pmol/mg	– –	yes	[84]
	cannabinoid CB2, human	FLAG	c-Myc, His_6	2.6 pmol/mg	– –	yes	[85]
	dG adenosine A_{2A}, human			8–12 pmol/mg (0.2 mg/l)		yes	[86]
	endothelin ETB, human		FLAG-His_{10}	45 pmol/mg	yes	yes	[87]
Other							
X. laevis	5-HT_{1A}	Rho_{15}	GFP	1–5 ng per tadpole	yes	yes	[88]
	EDG1	Rho_{15}	GFP		yes		[88]
Mouse	adenosine A_1	T7	Rho_{15}	2.1 pmol/mg	yes	yes	[89]
D. melanogaster	metabotropic glutamate, human			10.6 pmol/mg			[90]

BHK, baby hamster kidney; BRS, biotin recognition sequence; CHO, Chinese hamster ovary; COS, green monkey kidney; EDG, endothelial differentiation gene receptor; HA, hemagglutinin; ID4, C-terminal rhodopsin tag; PACAP, pituitary adenylate cyclase-activating polypeptide; TAP, tandem affinity purification; TEV, tobacco etch virus.

membrane-anchored AAA+ protease FtsH. Coexpression of this protease also enhanced bacterial growth. However, only a slight increase in the levels of active CB1 receptor was observed, suggesting that proper folding of the receptor was not achieved.

Among bacterial hosts besides *E. coli*, a photosynthetic bacterium *Rhodobacter sphaeroides* was used for production of GPCRs [45]. This organism can provide greater membrane surface compared to other hosts, due to its ability to develop, under oxygen-limiting conditions, an elaborate intracytoplasmic membrane system that houses components of photosynthetic apparatus. *R. sphaeroides* deletion strain DD13 lacking the LH1, LH2, and RC genes was used, and the expression of the recombinant proteins was conducted from a low-copy-number plasmid PRKEH10D under the control of the highly regulated photosynthetic promoter *pufQ*. This system was tested for the expression of several class A GPCRs, namely the human adenosine A_{2A} receptor ($A_{2A}R$), the human angiotensin AT_1 receptor ($AT_{1A}R$) and the human bradykinin B_2 receptor (B_2R). The best-expressed target, the $AT_{1A}R$, was produced at levels of 7–12 pmol/mg of functional protein. Several expression parameters still need to be optimized, the most critical being the development of a protease-deficient *R. sphaeroides* strains to increase yield of the full-length recombinant protein.

8.3
Expression of GPCRs in Inclusion Bodies, and Refolding

While heterologous expression is used for preparation of the correctly folded, functional GPCRs, an alternative strategy relies on production of target proteins accumulated in inclusion bodies. This approach circumvents the toxicity frequently caused by recombinant membrane proteins. Although there have been several examples of large-scale expression in *E. coli* of GPCRs as inclusion bodies, there is no single strategy that can be successfully applied to every representative of this protein family. In some instances, the use of the N-terminal T7 tag resulted in efficient expression [37, 111], while in other cases, the fusion of the GPCR to the protein partner was more beneficial, such as fusion to glutathione S-transferase [112] and ketosteroid isomerase (5-HT_{4A} receptor) [24]. At the same time, it was reported that use of ketosteroid isomerase fusion with other GPCRs did not result in high expression levels [24]. An efficient refolding of the purified receptors is one of the major problems in the production of high-quality homogenous proteins [24, 112–114]. In the majority of cases, refolding is performed using materials recovered from the inclusion bodies and since there is currently no detailed understanding of the various factors that affect this process, the commonly used experimental approach includes screening various refolding conditions [113]. The overall efficiency is mostly determined by the two competing processes – refolding and aggregation – and a very fine balance should be found in establishing conditions that favor refolding [113]. There have been several successes on this path including preparation of the leukotriene B_4 receptors (BLT1 and BLT2) and a serotonin

5-HT_{4A} receptor [24, 37, 44, 115, 116]. However, ligand affinity of the BLT1 receptor was somewhat lower than that of the native receptor expressed in mammalian tissues. High levels of expression were achieved upon codon optimization of the respective genes. Recent attempts to express, solubilize, and purify the receptor for neuropeptide Y2 have been reported, yielding large enough quantities of protein to initiate some biochemical and biophysical studies [44]. However, only a fraction of the refolded receptor was functional.

8.4 Expression of GPCRs in Yeast

Ease of genetic manipulation, ability to grow to high density in bioreactors, and the presence of eukaryotic secretion machinery make yeast an attractive host for expression of recombinant GPCRs. The majority of studies were performed using strains of *Saccharomyces cerevisiae* and *Pichia pastoris.* While many initial expression attempts focused on maximizing the yield of the target GPCRs through optimization of expression conditions, more recent studies have increasingly devoted significant attention to examining factors that affect correct folding and functionality of the recombinant GPCRs. We will consider several representative works dealing with these expression systems.

A series of successful studies on expression of human adenosine $A_{2A}R$ was reported [77–79, 117]. The $A_{2A}R$ was expressed with a C-terminal GFP fusion tag, and the quantification of the expression level was performed by fluorescence measurements, densitometry of the semiquantitative Western blots, and classical radioligand-binding assay. Using whole-cell GFP fluorescence, 340 000 A_{2A}-GFP molecules per cell were detected; approximately 70% of those were localized in plasma membranes. The yield of functional A_{2A}-GFP was as high as 4 mg/l of culture. The effects of cultivation conditions of *S. cerevisiae* on the total yield versus yield of the functional $A_{2A}R$ were studied [78]. A significant increase in the total levels of A_{2A} was achieved by performing induction at lower temperatures and at high cell densities. These optimized conditions resulted in the highest reported yield of 28 mg of A_{2A} per liter of culture. Content of the functional receptor generally increased with the decrease of the induction temperature. The authors suggest that the induction temperature affects the activity of the secretory pathway, causing incorrect folding of the target protein. The receptor was further purified in dodecyl maltoside detergent supplemented with CHAPS (3-(3-cholamidopropyl)-dimethylammoniopropane sulfonate)/CHS and was shown to retain its fold as measured by the content of α-helices [79].

A comparative study of the effects of cellular stress response on translocation, trafficking, expression levels, and functional activity of several GPCRs in *S. cerevisiae* was recently reported by the same group [117]. Twelve class A GPCRs have been expressed as fusions with C-terminal GFP followed by the histidine tag, at levels at least 1 mg/l or greater. Two of these receptors (i.e., A_{2A} and CB2) were expressed at close to 10 mg/l levels. High-level expression was reported for such receptors as hA_1, hA_{2B}, hA_3, hCCR5, hCXCR4, and hFSH receptor. However, only human adenosine $A_{2A}R$ located primarily at the plasma membrane was active for ligand binding. These findings suggest that membrane translocation is a critical limiting step in the production of active mammalian GPCRs in *S. cerevisiae.*

The C2 subtype of the human α_2-adrenergic receptor was successfully expressed in *S. cerevisiae* and purified by one-step affinity chromatography on a monoclonal antibody column [118]. The specific antagonist phentolamine stabilized the receptor during purification. However, only 10% of the purified receptor retained functionality.

Levels of total versus functional GPCRs expressed in *P. pastoris* vary significantly and seem to be protein target-dependent. For example, while the functional human μ-opioid receptor was expressed as a fusion with GFP at a level of 1 pmol/mg membrane protein, total levels of this protein determined by fluorescence were as high as 16 pmol/mg, suggesting incorrect folding of the bulk of the receptor [119, 120]. However, in another study, the recombinant human dopamine D_{2S} receptor expressed in *P. pastoris* at 3–13 pmol/mg of total protein was fully functional [121].

In another work, the activity of cannabinoid receptors CB1 and CB2 expressed in *P. pastoris* was significantly lower when compared to the native receptors from mammalian cells [84, 85]. The authors suggest that the compromise in the functional activity of these receptors might be due to the different lipid/sterol composition of mammalian and yeast cell membranes. Hence, it may be possible to increase the specific activity of the recombinant receptors by adding cholesterol/cholesterol derivatives during their solubilization and purification in detergent micelles [122].

In this respect, of interest is a thorough selection and optimization of the conditions for production of functional GPCRs in *P. pastoris* using a single expression screen [123]. An impact of several parameters affecting protein expression: temperature of expression, addition of dimethyl sulfoxide (DMSO), stabilizing ligands, and histidine, was examined. Among the factors beneficial for the expression of functional receptors were: (i) lowering of the temperature from 30 to 20 °C; (ii) supplementing the medium with a specific ligand at a concentration close to 100-fold K_d, (iii) addition of 2.5% DMSO, or (iv) 0.04 mg/ml histidine. For 10 GPCR targets, including human ADA1B, OPRK, and D2DR, up to a 6-fold increase in the expression of the functional receptor was observed at low induction temperature. Likewise, for most of the 20 receptors tested, the addition of the specific ligands to the medium resulted in increased production of functional receptors.

In another study from the same laboratory [124], the evaluation of expression levels of different GPCRs in *P. pastoris* was performed by quantitative dot-blot detection. Ten human and one mouse GPCR were expressed as fusions with the FLAG tag and a decahistidine tag (at the N-terminus), and biotin tag (at the C-terminus). The detection of as low as 100 pg of the recombinant GPCR was performed by the dot-blot using anti-FLAG M2 antibody. This semiquantitative protocol was also used to analyze the solubilization efficiency by using a panel of 16 different detergents.

Expression levels of the wild-type and truncated human adenosine $_{A2A}$ receptor in *P. pastoris* cultivated either in shake flasks or in a bioreactor were compared [82]. Cell density of OD_{600} around 75 was achieved in the bioreactor, 5-fold more than in the shake flasks, while the levels of the functional protein per mg of membrane protein were increased by 2-fold. Additionally, a stabilized and truncated version of the receptor V334 $A_{2A}R$ was expressed under similar conditions with yield of 200 pmol/mg of membrane protein.

Successful attempts to increase levels of functional human $A_{2A}R$ were made by coexpressing a transcriptional transactivator protein, addition of galactose, and introduction of a specific ligand into the yeast growth medium [125, 126]. However, when a similar approach was used to express the human D_{2S} dopamine receptor, no significant increase in yield was observed [127].

In some cases, the temperature shifts and pH variations have been used to increase the yield of the functional receptor. The increase in the pH values from 5.5 to 7.0 during the fermentation resulted in a 2- to 3-fold increase in expression levels of the human β_2- and α_2C-adrenergic receptors [128]. Likewise, the yield of the functional mouse 5-HT_{5A} serotonin receptor was doubled when the cells were subjected to a heat shock at 42 °C. This probably induced synthesis of the stress proteins acting as molecular chaperones [129]. On the other hand, lowering of the expression temperature may also assist in yielding high levels of the functional receptor, as was demonstrated by performing galactose induction of the VPAC1 and I7 olfactory receptors in *S. cerevisiae* at 15 °C [81, 130, 131]. In the latter case, fermentation was performed at 15 °C and protein levels were as high as 327 pmol/mg of membrane protein – one of the highest ever achieved for any GPCR expression in yeast.

Attempts have been made to solve the problem of the *in vivo* receptor degradation by expressing GPCRs in protease-deficient yeast strains [127, 132]. However, while there were seemingly positive effects such as decreased levels of degradation products and increased cellular levels of the full-length receptor, no significant increase in the amount of functional receptors in the plasma membranes was detected, suggesting that other factors involved in protein synthesis, folding, membrane targeting, and insertion play greater role in determining levels of functional receptor.

8.5 Expression of GPCRs in Insect Cells

Expression in insect cells has been a very popular approach for production of functional GPCRs [1, 46, 51, 133, 134]. It often yields high levels of receptors, although the proportion of active protein varies significantly depending on the individual GPCR. This expression system also allows, unlike a bacterial host, certain types of post-translational modifications of the recombinant proteins. However, the type and specificity of these post-translational modifications may not be exactly the same as in mammalian cells and this sometimes results in nonhomogenous receptor preparations. While there are a growing number of reports of the successful expression of homogenous and structurally intact receptors in insect cells, it appears that some caution has to be exercised when choosing this expression host. Depending on a goal of a particular project, different approaches have been tried with varied success: (i) expression in different cell lines and optimization of induction conditions; (ii) introduction of various N- and C-terminal tags; (iii) truncation of the target protein and introduction of stabilizing mutations; and (iv) coexpression with cognate G-proteins.

Coexpression with the cognate G-proteins was reported to aid production of some mammalian GPCRs. For example, the expression of two human olfactory receptors OR17-209 and OR17-210 in baculovirus Sf9 insect cells was performed by Matarazzo *et al* [133]. These proteins were coproduced with either $G_{\alpha off}$- or $G_{\alpha 16}$-proteins, taking advantage of the designed cassette baculovirus vectors containing FLAG/HA-tagged OR gene fused to an insect signal sequence. The leader sequence was used in order to target the recombinant protein to the secretory pathway. Each receptor was found to accumulate in a functional form at the cell surface and colocalized with both G_{α}-proteins. However, in the case of mouse δ-opioid receptor expressed in lepidopteran cell line Bm5, coexpression of exogenous $G_{\alpha 16}$ did not improve levels of functional receptor and reached only 1.2 pmol/mg membrane protein [134].

In several studies, the N-terminal tag was shown to significantly affect both the expression levels and functionality of the target receptor. The human peripheral type cannabinoid receptor CB2 was expressed in Sf21 cells with levels of functional protein 9.3 pmol/mg protein [61]. The addition of the N-terminal FLAG tag resulted in an approximately 30-fold decrease in the protein expression levels. In another publication from the same group an expression of the central-type cannabinoid receptor CB1 in Sf21 cells was achieved at levels as high as 24.5 pmol/mg protein in constructs carrying a N-terminal hexahistidine tag [60]. Its replacement with a FLAG tag resulted in a decrease of the B_{max} value by about 7-fold. Interestingly, the His-tagged protein was expressed at about the same level as the recombinant receptor without any tag.

The influence of the C-terminal tag on the expression level appears to be less crucial and a C-terminal polyhistidine tag has been frequently used for expression and purification of mammalian GPCRs. The large-scale production of the functional human histamine H_1 receptor was recently published by the W.J. DeGrip group [51]. The functional H_1-His-tagged receptor was accumulated at high levels (30–40 pmol/10^6 cells) in Sf9 cells. Scaling up to the 10-l bioreactor level resulted in 5–7 mg/l of functional receptor. Earlier, expression levels of up to 4 mg of the bovine rhodopsin per liter of the Sf9 cell culture were reported by the same group [55]. The C-terminal polyhistidine tag was used for the purification of the receptor and this rhodopsin preparation was successfully reconstituted into the lipid matrix.

Significant progress in structural studies of GPCRs was made in the past several years by several groups that succeeded in generating well-diffracting crystals of the recombinant proteins produced in insect cell systems [135–138]. The partial C-terminal deletion mutant of the β_2-adrenergic receptor, in fusion with HA and FLAG tags at the N-terminus, was efficiently expressed in Sf9 cells and purified by sequential antibody- and ligand-affinity chromatography [135]. In order to enhance conformational stability and increase the polar surface for the crystal contacts, crystallization was performed in the presence of monoclonal antibodies that bind to the intracytoplasmic loop 3 of the native β_2-adrenergic receptor [139].

In an effort to overcome the structural flexibility of the β_2-adrenergic receptor and to facilitate its crystallization, the same group has engineered the β_2-adrenergic receptor fusion protein in which T4 lysozyme replaces most of the third intracytoplasmic loop of the receptor. The β_2-adrenergic receptor–T4 lysozyme fusion protein was crystallized in the presence of the partial inverse agonist carazolol, in the cholesterol-doped lipidic cubic phase [6, 140, 141]. Both the T4 lysozyme fusion strategy and lipidic cubic phase crystallization were subsequently used to solve the structure of $A_{2A}R$ [13].

Introduction of stabilizing mutations greatly improved efficiency of crystallization of the turkey β_1-adrenergic receptor produced in the presence of the strong antagonist cyanopindolol [12, 142]. Likewise, thermostabilizing mutations were introduced into $A_{2A}R$ [143]. Thermostabilization of the neurotensin receptor NTS1 in detergent micelles in ligand-free as well as in neurotensin-bound form was achieved by combining four point mutations; only one of these was predicted to be localized in the transmembrane helix [144]. The mutated receptor bound neurotensin with an affinity similar to the wild-type receptor; however, agonist dissociation was slower and the receptor activated G-proteins poorly.

A recently published report described expression of three subtypes of human α_2-adrenergic receptor, including the truncated α_{2B}-adrenergic receptor carrying a deletion in the third intracellular loop, at levels between 11 and 64 pmol/mg of membrane protein [48]. The expressed receptor was localized in intracellular organelles, and the truncated form of α_{2B}-adrenergic receptor exhibited unaltered ligand affinities and enhanced stability.

Choice of a cell line is of particular importance in expressing GPCRs. In a study published by M. Akermoun *et al* [46], 16 human receptors, including A_{2A}, A_{1A}, α_{2A}-adrenergic, α_{2B}-adrenergic, α_{1B}-adrenergic, α_{1D}-adrenergic, β_1-adrenergic, β_2-adrenergic, 5-HT_{1A}, 5-HT_{2B}, BK_2, CB1, H_1, D_2, M_1, and μ-opioid, were expressed in three insect cell lines: Sf9 and Sf21 (derived from *Spodoptera frugiperda*) and High Five (derived from *Trichoplusia ni*) at levels ranging from 1 pmol/mg protein to 250 pmol/mg. No single set of expression conditions was found to be suitable for all GPCRs and significant differences were observed for individual GPCRs in different cell lines. Surprisingly, several closely related receptors did not share similar expression profiles. Remarkably, high levels (over 20 pmol/mg protein) were achieved for more than half of the GPCRs tested in the study. Generally, the best yield for all receptors was obtained at either 38 or 72 h postinfection.

8.6
Expression of GPCRs in Mammalian Cell Lines

While the expression in mammalian cell lines has been a method of choice for production of functional GPCRs, until recently it has not been widely used for high-level expression due to the high cost of the cultivation media and relatively low protein yields. Nevertheless, several studies have demonstrated the potential applicability of this host for generating large-scale protein samples suitable for structural investigations. High-level expression of the codon-optimized hamster β_2-adrenergic receptor gene in a functional form at levels 220 ± 40 pmol/mg in COS-1 cell line was reported [64]. Upon solubilization of β_2-adrenergic receptor in 1% (w/v) in decyl maltoside and purification on 1D4-Sepharose affinity column, about 70% of ligand-binding-competent receptor was obtained.

Successful GPCR production was also reported in the human embryonic kidney HEK-293 cell lines. A synthetic gene encoding human OR17-4 was expressed in adherent HEK-293 cells at levels of 30 mg per 150-mm tissue culture plate [145]. Scaling-up the suspension HEK-293S cells in a bioreactor yielded up to 3 mg/l of the receptor [146]. Functional activity of the recombinant receptor was confirmed by surface plasmon resonance experiments.

Opsin was expressed in a suspension culture of HEK-293S cells in a defined media containing [6-^{15}N]lysine and [2-^{13}C]glycine [147]. The authors reported a yield of 1.5–1.8 mg/l of culture. The functional rhodopsin was subsequently formed by binding to 11-*cis*-retinal and reconstituted into 1,2-dioleoyl-*sn*-glycero-3-phosphocholine proteoliposomes. It was estimated that expression in the HEK-293S cells resulted in as much as 50% of the amino acid residues in rhodopsin labeled with ^{15}N and ^{13}C, and the yield of the labeled receptor was as high as 1–2 mg of the purified protein per liter of culture [74].

Other examples of medium-to-high levels of production of GPCRs expressed in mammalian cell lines include functional yeast Ste2 [71, 148], a class B GPCR, human parathyroid hormone-1 receptor [149], human bradykinin B_2 [72], human adenosine A_1 and A_{2A} [47] (all in HEK-293), and thromboxane A_2 receptor (in COS-7 cells) [67].

8.7
Expression of GPCRs in Retina Rod Cells

Expression of functional GPCRs in retinal rod cells (rod outer segments) of transgenic animals has a potential of producing homogenous, highly concentrated preparations of these receptors. This approach was first applied to the production of metabotropic glutamate receptor (*Dm*GluRA) in photoreceptor cells of a transgenic fruit fly, *Drosophila melanogaster*, where its expression levels were at least 30-fold higher than in the conventional baculovirus expression system, primarily due to the higher membrane content of the photoreceptor cells [90].

A further development of this strategy was reported by K. Palczewski *et al.* [88] who used retina rod cells of transgenic *Xenopus laevis* for production of 20 different GPCRs. All recombinant proteins contained a C-terminal tag corresponding to 15 C-terminal amino acids of mouse rhodopsin that is required for vectorial transport of rhodopsin to rod outer segments. The receptors were accumulated in rod cells and homogeneously glycosylated. A C-terminal eGFP tag was used to demonstrate consistent expression of the target proteins across the entire retina. Fusion proteins were not detected in the endoplasmic reticulum or Golgi apparatus, suggesting that they were properly folded and transported to the rod cell disk membranes. The disadvantage of this expression system is that it is currently capable of producing only about 1–5 ng of fusion protein per transgenic tadpole and scaling-up is technically challenging.

8.8 Expression of GPCRs in a Cell-Free System

Application of the cell-free translational systems for the production of proteins for structural studies has become increasingly attractive, in particular to produce integral membrane proteins including GPCRs (see Table 8.2). The cell-free system allows production of stable isotope-labeled proteins [158–160] suitable for NMR structural studies and for protein engineering [161]. Detailed protocols for preparation of milligram quantities of membrane proteins of prokaryotic and eukaryotic origin using *E. coli* extracts in less than 24 h were described in [162]. Among the potential advantages of the cell-free expression are:

i) High specificity (the target protein is predominantly synthesized during the reaction).
ii) Short duration of the reaction, advantageous for preventing proteolytic degradation of the target protein.
iii) Small volume of the reaction and consequently low cost of isotope-labeled precursors.
iv) Possibility of using labeled precursors without inhibiting cell growth.

There are disadvantages of the cell-free system, such as relatively low expression levels and aberrant protein folding as compared to that of the native protein. Until recently, this has precluded its broader application for generating protein samples suitable for structural determination. The productivity of the cell-free translation system, in particular when applied to the synthesis of membrane proteins, was improved through optimization of the cell-free extracts and their supplementation with molecular chaperones that may be required to ensure correct folding of the nascent polypeptide chain [163, 164], addition of factors for membrane integration and translocation [165], as well as detergents and lipids that allow incorporation of newly synthesized polypeptides directly into preformed micelles [150, 151, 154, 166, 167].

Table 8.2 Expression of GPCRs and other integral membrane proteins in cell-free systems.

Protein expressed	Cell-free extract	Fusion/tag	Detergent/lipid (concentration)	Yield ("soluble" protein)	Functional activity detected	Reference
E. coli multidrug transporter EmrE	*E. coli* A19	T7-His_6	dilC_8PC (0.1%), decyl maltoside (0.2%), Brij98 (0.2%)	up to 3 g/ml in 20 h	yes	[150]
E. coli multidrug transporter EmrE	100 *E. coli* HY kit (Roche)	His	decyl maltoside (0.08%)	2–3 mg/ml	yes	[151]
E. coli nucleoside transporter Tsx	*E. coli* A19	T7-His_6	digitonin (0.4%), decyl maltoside (0.1%), Brij98 (0.2%), Triton X-100 (0.2%)	up to 3/ml to in 20 h	not reported	[150]
Odorant receptor OR5, rat	T7 TNT Quick *in vitro* expression (Promega)	VSV tag	PC/DMPE	not described	yes	[152]
Endothelin B, human	*E. coli* A19	T7-His_{10}	Brij58 (2%) or Brij78 (1%)	3 mg/ml	yes	[153]
Vasopressin type 2 (hV2R), human	*E. coli* A19	T7-His_6	Brij78 (1%)	3 mg/ml	not reported	[153]
Vasopressin type 2 (pV2R), porcine	*E. coli* A19	T7-His_6	Brij78 (1%) or Brij58 (1.5%)	up to 6 mg/ml in 20 h	yes	[150]
Corticotropin releasing factor, rat	*E. coli* A19	T7-His_6	Brij78 (1%)	3 mg/ml	not reported	[153]
Neuropeptide Y4 (NPY), human	*E. coli* A19	T7-His_{10}	Brij78 (1%)	0.8 mg/ml	not reported	[153]
Melatonin 1B (MTN), human	*E. coli* A19	T7-His_{10}	Brij78 (1%)	0.5 mg/ml	not reported	[153]
Muscarinic acetylcholine (M_2), human	S30 extract from BL21 Star (DE3)	TrxA-His_6	Brij35 (0.05–0.8 %)	0.15–0.2 mg/ml/h or 1 mg/ml in 6–8 h	not reported	[154]
β2-Adrenergic ($β_2$AR), human	S30 extract from BL21 Star (DE3)	TrxA-$G_{sα}$	Brij35 (≤1%) or digitonin (0.1–0.8%), PC/PS/ PE/1,2-dipalmitoyl-*sn*-glycero-3-phosphocholine	0.15 mg/ml/h or 1 mg/ml in 6–8 h	yes	[154]
β2-Adrenergic ($β_2$AR), human	S30 extract from BL21 Star (DE3)	TrxA-His_6	Brij35 (≤1%)	0.15 mg/ml/h or 1 mg/ml in 6–8 h	not reported	[154]
Neurotensin (NTS), rat	S30 extract from BL21 Star (DE)	TrxA-His_6	Brij35 (0.05–0.8%) or digitonin (0.1–0.8%)	0.1 mg/ml or 1 mg/ml in 6–8 h	not reported	[154]
Bacteriorhodopsin (bR), *Halobacterium salinarium*	Invitrogen's Expressway Maxi *E. coli* or Roche's RTS 500 ProteoMaster *E. coli*	C-terminal His_5	DMPC (2 mg/ml)	mg/ml, purified protein	yes	[155]
Bacteriorhodopsin (bR), *Halobacterium salinarium*	S30 extract from *E. coli*, continuous exchange dialysis	N-terminal His	6.7 mg/ml of phosphatidylcholine and 1% cholate or 1% CHAPS or 0.2% digitonin or 0.5% deoxycholate	0.3–0.7 mg/ml	yes	[156]
Olfactory receptor 17-4, human (hOR17-4)	wheat germ extract	N-terminus TrxA, C-terminus His_6	digitonin (0.2 % for expression) FC14 (3× critical micellar concentration for purification)	0.33 mg/ml	yes	[157]
Olfactory receptor OR23, mouse (mOR23)	wheat germ extract	N-terminus TrxA, C-terminus His_6	digitonin (0.6 %)	0.33 mg/ml	not determined	[157]
Olfactory receptor S51, mouse (mS51)	wheat germ extract	N-terminus TrxA, C-terminus His_6	Brij58 (0.1 %) or TD (1×)	0.33 mg/ml	not determined	[157]

Examples of production of integral membrane proteins using an *E. coli* cell-free coupled transcription–translation system include the fully functional multidrug transporters EmrE and SugE from *E. coli* as well as the larger transporter TehA and cysteine transporter YfiK [151]. Production in *E. coli* cell-free system of several GPCRs, including the β_2-adrenergic receptor, the muscarinic acetylcholine receptor M_2, and the neurotensin receptor fused to TrxA was reported (Table 8.2) [150, 151, 154, 168].

The reaction mixture often needs to be supplemented with either detergents or lipids to maintain solubility of the target membrane proteins. Of interest is the study in which suitability of several detergents supplemented in concentrations above their critical micellar concentrations for soluble expression of α-helical and β-barrel-type integral membrane proteins was evaluated [150]. The following parameters were assessed: (i) impact on productivity of the cell-free system, (ii) their efficiency in solubilizing newly synthesized membrane proteins, and (iii) their effect on the activity of the solubilized membrane proteins. It was found that both the type and concentration of a detergent have a profound impact on the yield, solubility, structural properties, and activity of the newly produced membrane proteins. Hence, the negative influence of detergents (particularly when used in high concentrations) on the reaction may limit their application. Of immediate practical interest, though, is that the long-chain polyoxyethylene derivatives Brij35, Brij58, Brij78, and Brij98 were found especially suitable for the high-level production of at least several different GPCRs [153].

While remarkable progress was achieved in high level (up to several milligrams per milliliter of reaction) cell-free production of several membrane receptors, including seven-transmembrane receptors, there are still significant difficulties in ensuring their functional activity. A few potentially useful approaches have been reported recently. One example is the production of the odorant receptor OR5 without the use of detergents [152]. The receptor was expressed as a fusion with a small vesicular stomatitis virus (VSV) affinity tag and incorporated into the tethered lipid matrix consisting of phosphocholine (PC) and 1,2-dimyristoyl-*sn*-glycero-3-phosphoethanolamine (DMPE). While the receptor appeared to be functionally active, its density in the lipid bilayer was rather small, varying from 20–100 molecules per μl of the reaction mixture.

A combination of the *E. coli*-based cell-free expression of bacteriorhodopsin and its rapid self-assembly into diskoidal nanolipoprotein particles was reported [155]. The coexpressed apolipoproteins sequestered lipid bilayer patches, solubilizing bacteriorhodopsin in the course of a single reaction. Addition of all-*trans*-retinal and 1,2-dimyristoyl-*sn*-glycero-3-phosphocholine (DMPC) to the reaction mixture resulted in a production of functional receptor.

Another group has synthesized bacteriorhodopsin integrated into liposomes using an *E. coli* cell-free system in a dialysis mode [156]. Supplementation of the reaction mixture with a combination of a detergent and phospholipid yielded up to 0.3–0.7 mg/ml of protein. The bacteriorhodopsin was shown to integrate into liposomes upon removal of detergents by dialysis.

A functional histamine H_1 receptor (hH_1R) was produced using *E. coli* cell-free expression system from Roche (Nutley, NJ), at levels of 3–5 mg/ml of reaction mixture [169]. Following immobilized-metal affinity chromatography purification, 0.2–0.7 mg of functional hH_1R was obtained and reconstituted into asolectin liposomes.

Use of a wheat germ extract for production of a functional human olfactory receptor 17-4 (hOR17-4) and two mouse olfactory receptors (mOR23 and MS51) was reported very recently [157]. Addition of digitonin in concentrations of 0.2% and 0.6% to the reaction mixture was optimal for production of hOR17-4 and mOR23, respectively, while 0.1% Brij58 was found better suited for production of the mS51 receptor.

8.9 Stabilization of GPCRs during Solubilization and Purification

It is crucial that the structure and activity of the recombinant receptors are preserved upon solubilization in the presence of detergents and during purification. An excellent review dealing with the latter issue was published not long ago [28]. It appears that the effects of a particular detergent or mixture of detergents on solubilization, purification and reconstitution of GPCR are highly receptor-specific. Therefore, solubilization conditions in most cases have to be determined empirically. Apart from comparing solubilization efficiency of various detergents, the effects of these detergents on the functionality of the target protein have to be taken into consideration [28, 29, 170].

For example, in the case of the μ-opioid receptor it was demonstrated that only a small number of detergents are suitable for both solubilization and preservation of its function [170]. A combination of 0.5% (w/v) CHAPS and 0.1% (w/v) CHS was found to be optimal for efficient extraction from cellular membranes and for preservation of activity in a micellar state. Likewise, CHS was found extremely beneficial for preservation of functional activity of other GPCRs [109, 122].

The use of specific ligands during receptor solubilization and purification has beneficial effects on increasing the yield of functionally active and structurally unperturbed protein as was reported for the dopamine D_4 receptor [171], vasopression V_2 receptor [172], rhodopsin [173], δ-opioid receptor [174], human chemokine receptor CXCR4 [59], and α_2-adrenergic receptor [118], to name just a few.

8.10 Conclusions

The development of a robust technology for the production of structurally unperturbed, functional GPCRs, in quantities sufficient for structural investigations, is prerequisite for understanding their structural features and mechanisms of action. Choice of the expression strategy is influenced by many factors such as yield of the target receptor, availability of post-translational modifications, and requirements of the particular biophysical technique utilized in subsequent structural characterization. One of the approaches focuses on achieving high yields rather than production of a functional and correctly folded receptor, and relies on expression of the recombinant protein as inclusion bodies in bacterial cells. Subsequent solubilization, refolding, and purification of the recombinant receptors are necessary to produce correctly folded homogenous samples suitable for structural investigations.

Expression of GPCRs in mammalian and insect cell lines or in rod outer segments of transgenic animals, on the other hand, takes advantage of adequate machinery for synthesis, membrane insertion, and post-translational modifications of the membrane receptors that is offered by these systems. Performance of these heterologous hosts is, however, receptor- and cell line-specific, and the protein yield is sometimes quite low. Yet another strategy takes advantage of the bacterial expression of GPCRs as fusions with MBPs that allows production of receptors in a functional form. Bacterial systems offer the potential for high expression levels of the recombinant GPCRs, and are well suited for production of stable isotope-labeled proteins, which is prerequisite for subsequent NMR studies. Several research groups have demonstrated its potential for generating protein samples suitable for the large-scale structural characterization. Likewise, several class A GPCRs have been produced in large quantities using yeast expression systems.

One of the promising novel approaches to production of GPCRs utilizes cell-free systems that offer such advantages as high specificity of protein synthesis and opportunity to use labeled precursors in small reaction volume. This technology can potentially be well adapted to the preparation of labeled proteins for high-resolution studies and currently experiences rapid growth.

Although it is evident that expression method is extremely important, the success of the particular strategy in application to a particular GPCR cannot be reliably forecast at present. The studies discussed here demonstrate rapid progress in the field of GPCR production. It is likely to result in the development of more reliable and efficient technologies, allowing for production of large quantities of receptors suitable for structural investigations that would, in turn, provide the basis for further improvement in rational drug design.

Acknowledgments

This work was supported by the Intramural Research Program of the National Institute on Alcohol Abuse and Alcoholism, National Institutes of Health.

Abbreviations

$A_{2A}R$	A_{2A} receptor
$AT_{1A}R$	AT_1 receptor
B_2R	B_2 receptor

CHAPS	3-(3-cholamidopropyl)-dimethylammoniopropane sulfonate
CHS	cholesteryl hemisuccinate
DMPE	1,2-dimyristoyl-*sn*-glycero-3-phosphoethanolamine
DMSO	dimethyl sulfoxide
eCFP	enhanced Cyan Fluorescent Protein
eGFP	enhanced Green Fluorescent Protein
eYFP	enhanced Yellow Fluorescent Protein
GFP	Green Fluorescent Protein
GPCR	G-protein-coupled receptor
HEK	human embryonic kidney
MBP	maltose-binding protein
NMR	nuclear magnetic resonance
PC	phosphocholine
Tig	peptydyl-prolyl isomerase trigger factor
TrxA	thioredoxin
VSV	vesicular stomatitis virus

References

1 Lundstrom, K. (2005) Structural biology of G protein-coupled receptors. *Bioorg. Med. Chem. Lett.*, **15**, 3654–3657.

2 Jacoby, E., *et al.* (2006) The 7 YM G-protein-coupled receptor target family. *Chem. Med. Chem.*, vol. **1**, pp. 760–782.

3 Palczewski, K., *et al.* (2000) Crystal structure of rhodopsin: a G protein-coupled receptor. *Science*, **289**, 739–745.

4 Li, J., *et al.* (2004) Structure of bovine rhodopsin in a trigonal crystal form. *J. Mol. Biol.*, **343**, 1409–1438.

5 Salom, D., *et al.* (2006) Crystal structure of a photoactivated deprotonated intermediate of rhodopsin. *Proc. Natl. Acad. Sci. USA*, **103**, 16123–16128.

6 Cherezov, V., *et al.* (2007) High-resolution crystal structure of an engineered human beta$_2$-adrenergic G protein-coupled receptor. *Science*, **318**, 1258–1265.

7 Rosenbaum, D.M., *et al.* (2007) GPCR engineering yields high-resolution structural insights into beta$_2$-adrenergic receptor function. *Science*, **318**, 1266–1273.

8 Rasmussen, S.G., *et al.* (2007) Crystal structure of the human beta2 adrenergic G-protein-coupled receptor. *Nature*, **450**, 383–387.

9 Standfuss, J., *et al.* (2007) Crystal structure of a thermally stable rhodopsin mutant. *J. Mol. Biol.*, **372**, 1179–1188.

10 Park, J.H., *et al.* (2008) Crystal structure of the ligand-free G-protein-coupled receptor opsin. *Nature*, **454**, 183–187.

11 Scheerer, P., *et al.* (2008) Crystal structure of opsin in its G-protein-interacting conformation. *Nature*, **455**, 497–502.

12 Warne, T., *et al.* (2008) Structure of a beta$_1$-adrenergic G-protein-coupled receptor. *Nature*, **454**, 486–491.

13 Jaakola, V.P., *et al.* (2008) The 2.6 angstrom crystal structure of a human A_{2A} adenosine receptor bound to an antagonist. *Science*, **322**, 1211–1217.

14 Gloriam, D.E., Fredriksson, R., and Schioth, H.B. (2007) The G protein-coupled receptor subset of the rat genome. *BMC Genomics*, **8**, 338.

15 Takeda, S., *et al.* (2002) Identification of G protein-coupled receptor genes from the human genome sequence. *FEBS Lett.*, **520**, 97–101.

16 Son, C.D., *et al.* (2005) Analysis of ligand–receptor cross-linked fragments by mass spectrometry. *J. Pept. Res.*, **65**, 418–426.

17 Raitio, K.H., *et al.* (2005) Targeting the cannabinoid CB2 receptor: mutations, modeling and development of CB2 selective ligands. *Curr. Med. Chem.*, **12**, 1217–1237.

18 Jacobsen, R.B., *et al.* (2006) Structure and dynamics of dark-state bovine rhodopsin revealed by chemical cross-linking and high-resolution mass spectrometry. *Protein Sci.*, **15**, 1303–1317.

19 Scherrer, G., *et al.* (2006) Knockin mice expressing fluorescent delta-opioid receptors uncover G protein-coupled receptor dynamics *in vivo*. *Proc. Natl. Acad. Sci. USA*, **103**, 9691–9696.

20 McVey, M., *et al.* (2001) Monitoring receptor oligomerization using time-resolved fluorescence resonance energy transfer and bioluminescence resonance energy transfer – the human delta-opioid receptor displays constitutive oligomerization at the cell surface, which is not regulated by receptor occupancy. *J. Biol. Chem.*, **276**, 14092–14099.

21 Terrillon, S. and Bouvier, M. (2004) Roles of G-protein-coupled receptor dimerization – from ontogeny to signalling regulation. *EMBO Rep.*, **5**, 30–34.

22 Liang, Y., *et al.* (2003) Organization of the G protein-coupled receptors rhodopsin and opsin in native membranes. *J. Biol. Chem.*, **278**, 21655–21662.

23 Fotiadis, D., *et al.* (2003) Atomic-force microscopy: rhodopsin dimers in native disc membranes. *Nature*, **421**, 127–128.

24 Baneres, J.L., *et al.* (2005) Molecular characterization of a purified 5-HT_4 receptor – a structural basis for drug efficacy. *J. Biol. Chem.*, **280**, 20253–20260.

25 Lundstrom, K. (2006) Latest development in drug discovery on G protein-coupled receptors. *Curr. Protein Pept. Sci.*, **7**, 465–470.

26 Horn, F., *et al.* (2003) GPCRDB information system for G protein-coupled receptors. *Nucleic Acids Res.*, **31**, 294–297.

27 Ratnala, V.R.P. (2006) New tools for G-protein coupled receptor (GPCR) drug discovery: combination of baculoviral expression system and solid state NMR. *Biotechnol. Lett.*, **28**, 767–778.

28 Sarramegna, V., *et al.* (2006) Recombinant G protein-coupled receptors from expression to renaturation: a challenge towards structure. *Cell. Mol. Life Sci.*, **63**, 1149–1164.

29 Grisshammer, R. (2006) Understanding recombinant expression of membrane proteins. *Curr. Opin. Biotechnol.*, **17**, 337–340.

30 Grisshammer, R., *et al.* (2005) Large-scale expression and purification of a G-protein-coupled receptor for structure determination – an overview. *J. Struct. Funct. Genomics*, **6**, 159–163.

31 Lundstrom, K. (2005) Structural genomics of GPCRs. *Trends Biotechnol.*, **23**, 103–108.

32 Sarramegna, V., *et al.* (2003) Heterologous expression of G-protein-coupled receptors: comparison of expression systems from the standpoint of large-scale production and purification. *Cell. Mol. Life Sci.*, **60**, 1529–1546.

33 Weiss, H.M. and Grisshammer, R. (2002) Purification and characterization of the human adenosine A_{2a} receptor functionally expressed in *Escherichia coli*. *Eur. J. Biochem.*, **269**, 82–92.

34 Link, A.J., *et al.* (2008) Efficient production of membrane-integrated and detergent-soluble G protein-coupled receptors in *Escherichia coli*. *Protein Sci.*, **17**, 1857–1863.

35 Yeliseev, A.A., *et al.* (2005) Expression of human peripheral cannabinoid receptor for structural studies. *Protein Sci.*, **14**, 2638–2653.

36 Berger, C., *et al.* (2010) Preparation of stable isotope-labeled peripheral cannabinoid receptor CB2 by bacterial fermentation. *Protein Expr. Purif.*, **70**, 236–247.

37 Baneres, J.L., *et al.* (2003) Structure-based analysis of GPCR function: conformational adaptation of both agonist and receptor upon leukotriene B-4 binding to recombinant BLT1. *J. Mol. Biol.*, **329**, 801–814.

38 Baneres, J.L. and Parello, J. (2003) Structure-based analysis of GPCR function: evidence for a novel pentameric assembly between the dimeric leukotriene B_4 receptor BLT1 and the G-protein. *J. Mol. Biol.*, **329**, 815–829.

39 Arcemisbehere, L., *et al.* (2010) Leukotriene BLT2 receptor monomers activate the G_{i2} GTP-binding protein more efficiently than dimers. *J. Biol. Chem.*, **285**, 6337–6347.

40 Bane, S.E., Velasquez, J.E., and Robinson, A.S. (2007) Expression and purification of milligram levels of inactive G-protein coupled receptors in *E. coli*. *Protein Expr. Purif.*, **52**, 348–355.

41 Busuttil, B.E., Turney, K.L., and Frauman, A.G. (2001) The expression of soluble, full-length, recombinant human TSH receptor in a prokaryotic system. *Protein Expr. Purif.*, **23**, 369–373.

42 White, J.F., *et al.* (2004) Automated large-scale purification of a G protein-coupled receptor for neurotensin. *FEBS Lett.*, **564**, 289–293.

43 Attrill, H., *et al.* (2009) Improved yield of a ligand-binding GPCR expressed in *E. coli* for structural studies. *Protein Expr. Purif.*, **64**, 32–38.

44 Schmidt, P., *et al.* (2009) Prokaryotic expression, *in vitro* folding, and molecular pharmacological characterization of the neuropeptide Y receptor type 2. *Biotechnol. Prog.*, **25**, 1732–1739.

45 Roy, A., *et al.* (2008) Employing *Rhodobacter sphaeroides* to functionally express and purify human G protein-coupled receptors. *Biol. Chem.*, **389**, 69–78.

46 Akermoun, M., *et al.* (2005) Characterization of 16 human G protein-coupled receptors expressed in baculovirus-infected insect cells. *Protein Expr. Purif.*, **44**, 65–74.

47 Robeva, A.S., *et al.* (1996) Double tagging recombinant A_1- and A_{2A}-adenosine receptors with hexahistidine and the FLAG epitope. Development of an efficient generic protein purification procedure. *Biochem. Pharmacol.*, **51**, 545–555.

48 Sen, S., *et al.* (2005) Functional studies with membrane-bound and detergent-solubilized alpha$_2$-adrenergic receptors expressed in Sf9 cells. *Biochim. Biophys. Acta Biomembr.*, **1712**, 62–70.

49 Parker, E.M., *et al.* (1991) Reconstitutively active G protein-coupled receptors purified from baculovirus-infected insect cells. *J. Biol. Chem.*, **266**, 519–527.

50 Doi, T., *et al.* (1997) Characterization of human endothelin B receptor and mutant receptors expressed in insect cells. *Eur. J. Biochem.*, **248**, 139–148.

51 Ratnala, V.R.P., *et al.* (2004) Large-scale overproduction, functional purification and ligand affinities of the His-tagged human histamine H_1 receptor. *Eur. J. Biochem.*, **271**, 2636–2646.

52 Park, P.S. and Wells, J.W. (2003) Monomers and oligomers of the M_2 muscarinic cholinergic receptor purified from Sf9 cells. *Biochemistry*, **42**, 12960–12971.

53 Park, P., *et al.* (2001) Nature of the oligomers formed by muscarinic M_2 acetylcholine receptors in Sf9 cells. *Eur. J. Pharmacol.*, **421**, 11–22.

54 Ohtaki, T., *et al.* (1998) Expression, purification, and reconstitution of receptor for pituitary adenylate cyclase-activating polypeptide. Large-scale purification of a functionally active G protein-coupled receptor produced in Sf9 insect cells. *J. Biol. Chem.*, **273**, 15464–15473.

55 Klaassen, C.H., *et al.* (1999) Large-scale production and purification of functional recombinant bovine rhodopsin with the use of the baculovirus expression system. *Biochem. J.*, **342**, 293–300.

56 Asmann, Y.W., Dong, M., and Miller, L.J. (2004) Functional characterization and purification of the secretin receptor expressed in baculovirus-infected insect cells. *Regul. Pept.*, **123**, 217–223.

57 Panneels, V., *et al.* (2003) Pharmacological characterization and immunoaffinity purification of metabotropic glutamate receptor from *Drosophila* overexpressed in Sf9 cells. *Protein Expr. Purif.*, **30**, 275–282.

58 Kuszak, A.J., *et al.* (2009) Purification and functional reconstitution of monomeric mu-opioid receptors: allosteric modulation of agonist binding by G_{i2}. *J. Biol. Chem.*, **284**, 26732–26741.

59 Dukkipati, A., *et al.* (2006) *In vitro* reconstitution and preparative purification of complexes between the chemokine receptor CXCR4 and its ligands SDF-1 alpha, gp120-CD4 and AMD3100. *Protein Expr. Purif.*, **50**, 203–214.

60 Xu, W., *et al.* (2005) Purification and mass spectroscopic analysis of human CB1 cannabinoid receptor functionally expressed using the baculovirus system. *J. Pept. Res.*, **66**, 138–150.

61 Filppula, S., *et al.* (2004) Purification and mass spectroscopic analysis of human CB2 cannabinoid receptor expressed in the baculovirus system. *J. Pept. Res.*, **64**, 225–236.

62 Peterson, G.L., *et al.* (1995) Purification of recombinant porcine m2 muscarinic acetylcholine receptor from Chinese hamster ovary cells. Circular dichroism spectra and ligand binding properties. *J. Biol. Chem.*, **270**, 17808–17814.

63 Alves, I.D., *et al.* (2007) Analysis of an intact G-protein coupled receptor by MALDI-TOF mass spectrometry: molecular heterogeneity of the tachykinin NK-1 receptor. *Anal. Chem.*, **79**, 2189–2198.

64 Chelikani, P., *et al.* (2006) The synthesis and high-level expression of a beta$_2$-adrenergic receptor gene in a tetracycline-inducible stable mammalian cell line. *Protein Sci.*, **15**, 1433–1440.

65 Weng, K., *et al.* (1997) Functional coupling of a human retinal metabotropic glutamate receptor (hmGluR6) to bovine rod transducin and rat Go in an *in vitro* reconstitution system. *J. Biol. Chem.*, **272**, 33100–33104.

66 Shimada, M., *et al.* (2002) Purification and characterization of a receptor for human parathyroid hormone and parathyroid hormone-related peptide. *J. Biol. Chem.*, **277**, 31774–31780.

67 Pawate, S., *et al.* (1998) Expression, characterization, and purification of C-terminally hexahistidine-tagged thromboxane A_2 receptors. *J. Biol. Chem.*, **273**, 22753–22760.

68 Allen, S.J., *et al.* (2009) Expression, purification and *in vitro* functional reconstitution of the chemokine receptor CCR1. *Protein Expr. Purif.*, **66**, 73–81.

69 Daulat, A.M., *et al.* (2007) Purification and identification of G protein-coupled receptor protein complexes under native conditions. *Mol. Cell. Proteomics*, **6**, 835–844.

70 Christoffers, K.H., *et al.* (2003) Purification and mass spectrometric analysis of the mu opioid receptor. *Brain Res. Mol. Brain Res.*, **118**, 119–131.

71 Shi, C.H., *et al.* (2005) Purification and characterization of a recombinant G-protein-coupled receptor, *Saccharomyces cerevisiae* Ste2p, transiently expressed in HEK293 EBNA1 cells. *Biochemistry*, **44**, 15705–15714.

72 Camponova, P., *et al.* (2007) High-level expression and purification of the human bradykinin B_2 receptor in a tetracycline-inducible stable HEK293S cell line. *Protein Expr. Purif.*, **55**, 300–311.

73 Reeves, P.J., Kim, J.M., and Khorana, H.G. (2002) Structure and function in rhodopsin: a tetracycline-inducible system in stable mammalian cell lines for high-level expression of opsin mutants. *Proc. Natl. Acad. Sci. USA*, **99**, 13413–13418.

74 Werner, K., *et al.* (2008) Isotope labeling of mammalian GPCRs in HEK293 cells and characterization of the C-terminus of bovine rhodopsin by high resolution liquid NMR spectroscopy. *J. Biomol. NMR*, **40**, 49–53.

75 Mirzabekov, T., *et al.* (1999) Enhanced expression, native purification, and characterization of CCR5, a principal HIV-1 coreceptor. *J. Biol. Chem.*, **274**, 28745–28750.

76 Marino, S.F. (2009) High-level production and characterization of a G-protein coupled receptor signaling complex. *FEBS J.*, **276**, 4515–4528.

77 Niebauer, R.T. and Robinson, A.S. (2006) Exceptional total and functional yields of the human adenosine (A2a) receptor expressed in the yeast *Saccharomyces cerevisiae. Protein Expr. Purif.*, **46**, 204–211.

78 Wedekind, A., *et al.* (2006) Optimization of the human adenosine A_2a receptor

yields in *Saccharomyces cerevisiae*. *Biotechnol. Prog.*, **22**, 1249–1255.

79 O'Malley, M.A., *et al.* (2007) High-level expression in *Saccharomyces cerevisiae* enables isolation and spectroscopic characterization of functional human adenosine A2a receptor. *J. Struct. Biol.*, **159**, 166–178.

80 Kapat, A., *et al.* (2000) Production and purification of recombinant human alpha 2C2 adrenergic receptor using *Saccharomyces cerevisiae*. *Bioseparation*, **9**, 167–172.

81 Minic, J., *et al.* (2005) Functional expression of olfactory receptors in yeast and development of a bioassay for odorant screening. *FEBS J.*, **272**, 524–537.

82 Singh, S., *et al.* (2008) Large-scale functional expression of WT and truncated human adenosine A2A receptor in *Pichia pastoris* bioreactor cultures. *Microb. Cell Fact.*, **7**, 28.

83 Shukla, A.K., *et al.* (2007) Heterologous expression and characterization of the recombinant bradykinin B_2 receptor using the methylotrophic yeast *Pichia pastoris*. *Protein Expr. Purif.*, **55**, 1–8.

84 Kim, T.K., *et al.* (2005) Expression and characterization of human CB1 cannabinoid receptor in methylotrophic yeast *Pichia pastoris*. *Protein Expr. Purif.*, **40**, 60–70.

85 Feng, W., *et al.* (2002) Expression of CB2 cannabinoid receptor in *Pichia pastoris*. *Protein Expr. Purif.*, **26**, 496–505.

86 Fraser, N.J. (2006) Expression and functional purification of a glycosylation deficient version of the human adenosine 2a receptor for structural studies. *Protein Expr. Purif.*, **49**, 129–137.

87 Schiller, H., *et al.* (2001) Solubilization and purification of the human ETB endothelin receptor produced by high-level fermentation in *Pichia pastoris*. *Receptors Channels*, **7**, 453–469.

88 Zhang, L., *et al.* (2005) Expression of functional G protein-coupled receptors in photoreceptors of transgenic *Xenopus laevis*. *Biochemistry*, **44**, 14509–14518.

89 Li, N., *et al.* (2007) Heterologous expression of the adenosine A1 receptor in transgenic mouse retina. *Biochemistry*, **46**, 8350–8359.

90 Eroglu, C., *et al.* (2002) Functional reconstitution of purified metabotropic glutamate receptor expressed in the fly eye. *EMBO Rep.*, **3**, 491–496.

91 Grisshammer, R. and Tate, C. (2003) Preface: overexpression of integral membrane proteins. *Biochim. Biophys. Acta Biomembr.*, **1610**, 1.

92 Grisshammer, R. and Tate, C.G. (1995) Overexpression of integral membrane proteins for structural studies. *Quart. Rev. Biophys.*, **28**, 315–422.

93 Tate, C.G. and Grisshammer, R. (1996) Heterologous expression of G-protein-coupled receptors. *Trends Biotechnol.*, **14**, 426–430.

94 Grisshammer, R., Averbeck, P., and Sohal, A.K. (1999) Improved purification of a rat neurotensin receptor expressed in *Escherichia coli*. *Biochem. Soc. Trans.*, **27**, 899–903.

95 Grisshammer, R. and Tucker, J. (1997) Quantitative evaluation of neurotensin receptor purification by immobilized metal affinity chromatography. *Protein Expr. Purif.*, **11**, 53–60.

96 Tucker, J. and Grisshammer, R. (1996) Purification of a rat neurotensin receptor expressed in *Escherichia coli*. *Biochem. J.*, **317**, 891–899.

97 Luca, S., *et al.* (2003) The conformation of neurotensin bound to its G protein-coupled receptor. *Proc. Natl. Acad. Sci. USA*, **100**, 10706–10711.

98 Harding, P.J., *et al.* (2009) Constitutive dimerization of the G-protein coupled receptor, neurotensin receptor 1, reconstituted into phospholipid bilayers. *Biophys. J.*, **96**, 964–973.

99 Niebauer, R.T., *et al.* (2006) Characterization of monoclonal antibodies directed against the rat neurotensin receptor NTS1. *J. Recept. Signal Transduct. Res.*, **26**, 395–415.

100 Mancia, F., *et al.* (2007) Production and characterization of monoclonal antibodies sensitive to conformation in the $5HT_{2c}$ serotonin receptor. *Proc. Natl. Acad. Sci. USA*, **104**, 4303–4308.

101 Bokoch, M.P., *et al.* (2010) Ligand-specific regulation of the extracellular surface of a G-protein-coupled receptor. *Nature*, **463**, 108–112.

102 Ahuja, S., *et al.* (2009) Location of the retinal chromophore in the activated state of rhodopsin. *J. Biol. Chem.*, **284**, 10190–10201.

103 Werner, K., *et al.* (2007) Combined solid state and solution NMR studies of alpha,epsilon-N-15 labeled bovine rhodopsin. *J. Biomol. NMR*, **37**, 303–312.

104 Crocker, E., *et al.* (2004) Dipolar assisted rotational resonance NMR of tryptophan and tyrosine in rhodopsin. *J. Biomol. NMR*, **29**, 11–20.

105 Klein-Seetharaman, J., *et al.* (2004) Differential dynamics in the G protein-coupled receptor rhodopsin revealed by solution NMR. *Proc. Natl. Acad. Sci. USA*, **101**, 3409–3413.

106 Etzkorn, M., *et al.* (2007) Secondary structure, dynamics, and topology of a seven-helix receptor in native membranes, studied by solid-state NMR spectroscopy. *Angew. Chem. Int. Ed.*, **46**, 459–462.

107 Krepkiy, D., *et al.* (2006) Bacterial expression of functional, biotinylated peripheral cannabinoid receptor CB2. *Protein Expr. Purif.*, **49**, 60–70.

108 Krepkiy, D., Yeliseev, A., and Gawrisch, K. (2006) Towards NMR structural studies on the peripheral cannabinoid receptor CB2 in micellar solution. *FASEB J.*, **20**, A920–A920.

109 Yeliseev, A.A., Zoubak, L., and Gawrisch, K. (2007) Use of dual affinity tags for expression and purification of functional peripheral cannabinoid receptor. *Protein Expr. Purif.*, **53**, 153–163.

110 Ren, H., *et al.* (2009) High-level production, solubilization and purification of synthetic human GPCR chemokine receptors CCR5, CCR3, CXCR4 and CX3CR1. *PLoS ONE*, **4**, e4509.

111 Tian, C.L., *et al.* (2005) Solution NMR spectroscopy of the human vasopressin V2 receptor, a G protein-coupled receptor. *J. Am. Chem. Soc.*, **127**, 8010–8011.

112 Kiefer, H., Maier, K., and Vogel, R. (1999) Refolding of G-protein-coupled receptors from inclusion bodies produced in *Escherichia coli*. *Biochem. Soc. Trans.*, **27**, 908–912.

113 Baneres, J.L. (2006) Refolding of G-Protein-Coupled Receptors, in *Structural Biology of Membrane Proteins* (eds R.G. Grisshammer and S.K. Buchanan), RSC, London, pp. 3–14.

114 Kiefer, H., Vogel, R., and Maier, K. (2000) Bacterial expression of G-protein-coupled receptors: prediction of expression levels from sequence. *Receptors Channels*, **7**, 109–119.

115 Michalke, K., *et al.* (2009) Mammalian G-protein-coupled receptor expression in *Escherichia coli*: I. High-throughput large-scale production as inclusion bodies. *Anal. Biochem.*, **386**, 147–155.

116 Martin, A., *et al.* (2009) Engineering a G protein-coupled receptor for structural studies: stabilization of the BLT1 receptor ground state. *Protein Sci.*, **18**, 727–734.

117 O'Malley, M.A., *et al.* (2009) Progress toward heterologous expression of active G-protein-coupled receptors in *Saccharomyces cerevisiae*: linking cellular stress response with translocation and trafficking. *Protein Sci.*, **18**, 2356–2370.

118 Liitti, S., *et al.* (2001) Immunoaffinity purification and reconstitution of human alpha$_2$-adrenergic receptor subtype C2 into phospholipid vesicles. *Protein Expr. Purif.*, **22**, 1–10.

119 Sarramegna, V., *et al.* (2002) Optimizing functional versus total expression of the human mu-opioid receptor in *Pichia pastoris*. *Protein Expr. Purif.*, **24**, 212–220.

120 Sarramegna, V., *et al.* (2002) Green fluorescent protein as a reporter of human mu-opioid receptor overexpression and localization in the methylotrophic yeast *Pichia pastoris*. *J. Biotechnol.*, **99**, 23–39.

121 de Jong, L.A.A., *et al.* (2004) Purification and characterization of the recombinant human dopamine D_{2S} receptor from *Pichia pastoris*. *Protein Expr. Purif.*, **33**, 176–184.

122 Grisshammer, R. (2009) Purification of recombinant G-protein-coupled receptors. *Methods Enzymol.*, **466**, 631–645.

123 Andre, N., *et al.* (2006) Enhancing functional production of G protein-coupled receptors in *Pichia pastoris* to levels required for structural studies via

a single expression screen. *Protein Sci.*, **15**, 1115–1126.

124 Zeder-Lutz, G., *et al.* (2006) Dot-blot immunodetection as a versatile and high-throughput assay to evaluate recombinant GPCRs produced in the yeast *Pichia pastoris*. *Protein Expr. Purif.*, **50**, 118–127.

125 King, K., *et al.* (1990) Control of yeast mating signal transduction by a mammalian beta-2-adrenergic receptor and Gs alpha-subunit. *Science*, **250**, 121–123.

126 Minic, J., *et al.* (2005) Yeast system as a screening tool for pharmacological assessment of G protein coupled receptors. *Curr. Med. Chem.*, **12**, 961–969.

127 Sander, P., *et al.* (1994) Heterologous expression of the human D-2s dopamine-receptor in protease-deficient *Saccharomyces cerevisiae* strains. *Eur. J. Biochem.*, **226**, 697–705.

128 Sizmann, D., *et al.* (1996) Production of adrenergic receptors in yeast. *Receptors Channels*, **4**, 197–203.

129 Bach, M., *et al.* (1996) Pharmacological and biochemical characterization of the mouse $5HT_{5A}$ serotonin receptor heterologously produced in the yeast *Saccharomyces cerevisiae*. *Receptors Channels*, **4**, 129–139.

130 Hansen, M.K., *et al.* (1999) Functional expression of rat $VPAC_1$ receptor in *Saccharomyces cerevisiae*. *Receptors Channels*, **6**, 271–281.

131 Pajot-Augy, E., *et al.* (2003) Engineered yeasts as reporter systems for odorant detection. *J. Recept. Signal Transduct. Res.*, **23**, 155–171.

132 Sander, P., *et al.* (1994) Constitutive expression of the human D-2s-dopamine receptor in the unicellular yeast *Saccharomyces cerevisiae*. *Biochim. Biophys. Acta Biomembr.*, **1193**, 255–262.

133 Matarazzo, V., *et al.* (2005) Functional characterization of two human olfactory receptors expressed in the baculovirus Sf9 insect cell system. *Chem. Senses*, **30**, 195–207.

134 Swevers, L., *et al.* (2005) Functional expression of mammalian opioid receptors in insect cells and high-throughput screening platforms for receptor ligand mimetics. *Cell. Mol. Life Sci.*, **62**, 919–930.

135 Kobilka, B.K. (1995) Amino and carboxyl-terminal modifications to facilitate the production and purification of a G-protein-coupled receptor. *Anal. Biochem.*, **231**, 269–271.

136 Kenakin, T. (2003) Ligand-selective receptor conformations revisited: the promise and the problem. *Trends Pharmacol. Sci.*, **24**, 346–354.

137 Kobilka, B.K. and Deupi, X. (2007) Conformational complexity of G-protein-coupled receptors. *Trends Pharmacol. Sci.*, **28**, 397–406.

138 Kobilka, B.K. (2007) G protein coupled receptor structure and activation. *Biochim. Biophys. Acta Biomembr.*, **1768**, 794–807.

139 Day, P.W., *et al.* (2007) A monoclonal antibody for G protein-coupled receptor crystallography. *Nat. Methods*, **4**, 927–929.

140 Hanson, M.A., *et al.* (2008) A specific cholesterol binding site is established by the 2.8 angstrom structure of the human $beta_2$-adrenergic receptor. *Structure*, **16**, 897–905.

141 Roth, C.B., Hanson, M.A., and Stevens, R.C. (2008) Stabilization of the human $beta_2$-adrenergic receptor TM4–TM3–TM5 helix interface by mutagenesis of Glu122(3.41), a critical residue in GPCR structure. *J. Mol. Biol.*, **376**, 1305–1319.

142 Serrano-Vega, M.J., *et al.* (2008) Conformational thermostabilization of the beta 1-adrenergic receptor in a detergent-resistant form. *Proc. Natl. Acad. Sci. USA*, **105**, 877–882.

143 Magnani, F., *et al.* (2008) Co-evolving stability and conformational homogeneity of the human adenosine A_{2a} receptor. *Proc. Natl. Acad. Sci. USA*, **105**, 10744–10749.

144 Shibata, Y., *et al.* (2009) Thermostabilization of the neurotensin receptor NTS1. *J. Mol. Biol.*, **390**, 262–277.

145 Cook, B.L., *et al.* (2008) Study of a synthetic human olfactory receptor 17-4: expression and purification from an inducible mammalian cell line. *PLoS ONE*, **3**, e2920.

146 Cook, B.L., *et al.* (2009) Large-scale production and study of a synthetic G protein-coupled receptor: human olfactory receptor 17-4. *Proc. Natl. Acad. Sci. USA*, **106**, 11925–11930.

147 Eilers, M., *et al.* (1999) Magic angle spinning NMR of the protonated retinylidene Schiff base nitrogen in rhodopsin: expression of N-15-lysine- and C-13-glycine-labeled opsin in a stable cell line. *Proc. Natl. Acad. Sci. USA*, **96**, 487–492.

148 Yin, D.Z., *et al.* (2005) Successful expression of a functional yeast G-protein-coupled receptor (Ste2) in mammalian cells. *Biochem. Biophys. Res. Commun.*, **329**, 281–287.

149 Gan, L., *et al.* (2006) Large-scale purification and characterization of human parathyroid hormone-1 receptor stably expressed in HEK293S $GnTI^-$ cells. *Protein Expr. Purif.*, **47**, 296–302.

150 Klammt, C., *et al.* (2005) Evaluation of detergents for the soluble expression of alpha-helical and beta-barrel-type integral membrane proteins by a preparative scale individual cell-free expression system. *FEBS J.*, **272**, 6024–6038.

151 Elbaz, Y., *et al.* (2004) *In vitro* synthesis of fully functional EmrE, a multidrug transporter, and study of its oligomeric state. *Proc. Natl. Acad. Sci. USA*, **101**, 1519–1524.

152 Robelek, R., *et al.* (2007) Incorporation of *in vitro* synthesized GPCR into a tethered artificial lipid membrane system. *Angew. Chem. Int. Ed.*, **46**, 605–608.

153 Klammt, C., *et al.* (2007) Cell-free production of G protein-coupled receptors for functional and structural studies. *J. Struct. Biol.*, **159**, 194–205.

154 Ishihara, G., *et al.* (2005) Expression of G protein coupled receptors in a cell-free translational system using detergents and thioredoxin-fusion vectors. *Protein Expr. Purif.*, **41**, 27–37.

155 Cappuccio, J.A., *et al.* (2008) Cell-free co-expression of functional membrane proteins and apolipoprotein, forming soluble nanolipoprotein particles. *Mol. Cell. Proteomics*, **7**, 2246–2253.

156 Shimono, K., *et al.* (2009) Production of functional bacteriorhodopsin by an *Escherichia coli* cell-free protein synthesis system supplemented with steroid detergent and lipid. *Protein Sci.*, **18**, 2160–2171.

157 Kaiser, L., *et al.* (2008) Efficient cell-free production of olfactory receptors: detergent optimization, structure, and ligand binding analyses. *Proc. Natl. Acad. Sci. USA*, **105**, 15726–15731.

158 Ozawa, K., *et al.* (2006) ^{15}N-labelled proteins by cell-free protein synthesis. Strategies for high-throughput NMR studies of proteins and protein–ligand complexes. *FEBS J.*, **273**, 4154–4159.

159 Sobhanifar, S., *et al.* (2010) Cell-free expression and stable isotope labelling strategies for membrane proteins. *J. Biomol. NMR*, **46**, 33–43.

160 Vinarov, D.A., Newman, C.L.L., and Markley, J.L. (2006) Wheat germ cell-free platform for eukaryotic protein production. *FEBS J.*, **273**, 4160–4169.

161 Shimuzu, Y., *et al.* (2006) Cell-free translation systems for protein engineering. *FEBS J.*, **273**, 4133–4140.

162 Schwarz, D., *et al.* (2007) Preparative scale expression of membrane proteins in *Escherichia coli*-based continuous exchange cell-free systems. *Nat. Protoc.*, **2**, 2945–2957.

163 Hartl, F.U. and Hayer-Hartl, M. (2002) Molecular chaperones in the cytosol: from nascent chain to folded protein. *Science*, **295**, 1852–1858.

164 Kigawa, T., *et al.* (1999) Cell-free production and stable-isotope labeling of milligram quantities of proteins. *FEBS Lett.*, **442**, 15–19.

165 Koch, H.G., Moser, M., and Muller, M. (2003) Signal recognition particle-dependent protein targeting, universal to all kingdoms of life. *Rev. Physiol. Biochem. Pharmacol.*, **146**, 55–94.

166 Berrier, C., *et al.* (2004) Cell-free synthesis of a functional ion channel in the absence of a membrane and in the presence of detergents. *Biochemistry*, **43**, 12585–12591.

167 Klammt, C., *et al.* (2006) Cell-free expression as an emerging technique for the large scale production of integral

membrane protein. *FEBS J.*, **273**, 4141–4153.

168 Klammt, C., *et al.* (2004) High level cell-free expression and specific labeling of integral membrane proteins. *Eur. J. Biochem.*, **271**, 568–580.

169 Sansuk, K., *et al.* (2008) GPCR proteomics: mass spectrometric and functional analysis of histamine H-1 receptor after baculovirus-driven and *in vitro* cell free expression. *J. Proteome Res.*, **7**, 621–629.

170 Ott, D., *et al.* (2005) Engineering and functional immobilization of opioid receptors. *Protein Eng. Des. Sel.*, **18**, 153–160.

171 Van Craenenbroeck, K., *et al.* (2005) Folding efficiency is rate-limiting in dopamine D_4 receptor biogenesis. *J. Biol. Chem.*, **280**, 19350–19357.

172 Morello, J.P., *et al.* (2000) Pharmacological chaperones functionally rescue misfolded mutant V2 vasopressin receptors that cause nephrogenic diabetes insipidus. *FASEB J.*, **14**, A1352–A1352.

173 Noorwez, S.M., *et al.* (2003) Pharmacological chaperone-mediated *in vivo* folding and stabilization of the P23H-opsin mutant associated with autosomal dominant retinitis pigmentosa. *J. Biol. Chem.*, **278**, 14442–14450.

174 Petaja-Repo, U.E., *et al.* (2002) Ligands act as pharmacological chaperones and increase the efficiency of delta opioid receptor maturation. *EMBO J.*, **21**, 1628–1637.

9
Structural Biology of Membrane Proteins

David Salom and Krzysztof Palczewski

9.1
Introduction

Membrane proteins represent about a third of the proteins in living organisms [1], but knowledge of their various functions is hampered by the scarcity of structural information. Owing to their central role in basically all physiological processes, membrane proteins constitute around 60% of approved drug targets [2] and, therefore, their experimentally determined three-dimensional structures are eagerly sought to assist in structure-based drug design. Fortunately, the number of high-resolution membrane protein structures has grown exponentially since the first membrane protein crystal structure was solved [3], although most membrane protein crystal structures solved to date are from bacteria. To explain this bias one might assume that eukaryotic membrane proteins are just more difficult to crystallize. However, the fact that half of all eukaryotic membrane proteins crystallized were purified from native sources speaks of difficulties encountered with their heterologous expression. Of those eukaryotic membrane proteins expressed heterologously, less than one-quarter were expressed in *Escherichia coli* [4]. By contrast, the vast majority of soluble proteins and prokaryotic membrane proteins crystallized thus far were produced as recombinant proteins in bacterial systems [4, 5].

9.2
Folding and Structural Analysis of Membrane Proteins

9.2.1
Folding

The hydrophobic effect is a dominant contributor to the structural stabilization of soluble proteins and extramembranous regions of membrane proteins, but water is essentially absent in the hydrocarbon core of lipid bilayers. Consequently, the relative importance of other forces, such as van der Waals packing and hydrogen bonding, increase in the apolar environment of the membrane core. There are

Production of Membrane Proteins: Strategies for Expression and Isolation, First Edition.
Edited by Anne Skaja Robinson.

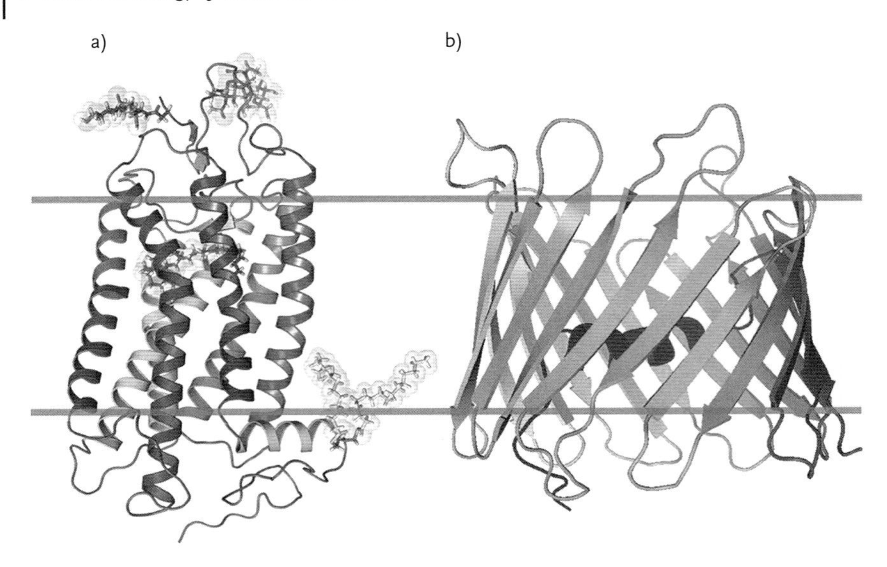

Figure 9.1 Examples of the two structural motives found in membrane proteins: the α-helical bundle (a, bovine rhodopsin) and a β-barrel protein (b, human mitochondrial voltage-dependent anion channel), in their approximate positions in the membrane bilayer. Retinal, carbohydrates, and palmitoyl chains are also shown in the rhodopsin illustration.

virtually no unfolded segments in transmembrane domains of membrane proteins because of the high thermodynamic cost of burying polar peptide bonds in a membrane, which also explains why the membrane interface is a potent catalyst of secondary structure formation. To date, two structural motifs – α-helices and β-barrels – have been found in membrane proteins (Figure 9.1). β-Barrels, made up of β-strands, are found in Gram-negative bacterial outer membranes as well as in mitochondrial and chloroplast membranes, and these structures function as channels or transporters for nutrients, proteins, hydrophobic toxic substances, and other molecules. α-Helical bundles are structurally and functionally more versatile, serving as receptors, channels, transporters, electron transporters, and redox facilitators.

Owing to their quasi-two-dimensional environment, the conformational space of membrane proteins is severely restricted. Over the past few years, the steady increase of published membrane protein structures has turned up numerous examples of structural similarities among apparently unrelated protein families. For example, structural similarity was found between the short acid transporter FocA and aquaporins [6], and between the transmembrane domain of the ionotropic glutamate receptor (iGluR) and bacterial K^+ channels [7]. However, it is still unclear whether these structural similarities reflect a divergence from common ancestral proteins so ancient that sequence similarities have been erased by genetic drift or whether they just represent a manifestation of the restricted folding space available to membrane proteins [8].

9.2.2
Prediction Methods

Early methods for predicting transmembrane helices were based on the premise that a protein segment would partition into a membrane if it was sufficiently hydrophobic and long enough to cross it. Starting with the method of Kyte and Doolittle [9], various algorithms for detecting transmembrane segments were proposed from experimental and computational data. These relied on an amino acid hydrophobicity scale based on the free energy needed to transfer them from hydrophilic to hydrophobic media. Typically, the algorithms searched for stretches of around 20 amino acid residues that could cross the roughly 30-Å hydrophobic core of a membrane. Sequence stretches scoring above a certain propensity threshold would be predicted to constitute transmembrane helices. However, these methods and others developed in the 1990s that relied on statistical inference provide only moderate success rates, and are unable to identify "irregular" structures, such as half helices and re-entrant loops found in the K^+ channel and aquaporins [10]. More sequences and structures then provided new insights, such as: aromatic Trp and Tyr residues tend to cluster near the ends of transmembrane segments; loops connecting transmembrane helices have different amino acid compositions depending on whether they face the inside or outside of the cell; and certain sequence motifs in transmembrane segments exhibit a higher-than-random occurrence [11]. Recently, von Heijne *et al.* proposed two methods based on first principles that employ an experimental scale of position-specific amino acid contributions to the free energy of membrane insertion; when coupled with the "positive-inside" rule [12], this combination predicts the topology of α-helical membrane proteins with performance levels rivaling the best machine learning methods [13]. Finally, a computational approach for optimizing the spatial arrangement of protein structures in lipid bilayers has been developed by minimizing their transfer energies from water to the lipid bilayer. The last method has been applied to all unique transmembrane domains and hundreds of peripheral proteins in the Protein Data Bank (PDB), and the results are available online [14].

Prediction of β-barrel motifs from primary sequences is still difficult because of the presence of short transmembrane stretches. Furthermore, the interior of the barrel is not always hydrophilic. Therefore, searching for alternating polar and nonpolar residues is not a promising approach. The C- and N-terminal β-barrel signature sequences are only moderately successful in identifying proteins with β-barrel motifs [15].

9.2.3
Membrane Insertion

With few exceptions, membrane protein synthesis on ribosomes and their membrane insertion, led by a signal sequence, are concerted steps. In many cases the signal sequence of eukaryotic membrane proteins is removed in the endoplasmic reticulum where their maturation is completed by different post-translational modifications before export to their final cellular destinations. By contrast, the fold of soluble proteins is normally defined entirely within the sequence itself. It is now clear that the translocon complex itself plays an important role in determining the final membrane protein topology. von Heijne's group developed an experimental method for quantifying sequence-dependent translocon selection of transmembrane helices, which allows us to determine the membrane insertion efficiency of suspected transmembrane sequences [16]. The fact that important charged residues are not conserved among translocons of different species could be one of the reasons why it is so difficult to express eukaryotic membrane proteins in heterologous systems [17].

9.2.4 Estimating the Molecular Weight of Membrane Proteins

Most methods for protein structural characterization were first developed for soluble proteins and then adapted to membrane proteins. Thus, in many cases, cautious interpretation of experimental results is advisable when a membrane protein is investigated. For example, one of the most used biophysical methods for protein isolation is sodium dodecyl sulfate–polyacrylamide gel electrophoresis (SDS–PAGE), which generally provides useful information about protein purity and molecular weight. However, this method often provides misleading information when used to analyze membrane proteins. In contrast to the relatively unstructured SDS-induced unfolded states of most water-soluble proteins, unfolded states of membrane proteins contain a significant percentage of secondary structure [18]. In addition, native (e.g., glycophorin A) or irreversible (e.g., G-protein-coupled receptors (GPCRs)) oligomerization may occur in SDS micelles. Also, if not boiled prior to electrophoresis, OmpA migrates to different positions on SDS–PAGE depending on the compactness of its structure. Native OmpA migrates with an apparent molecular weight of around 30 kDa, whereas completely unfolded OmpA migrates as an around 35-kDa protein [19]. In some cases, differential binding of SDS to membrane proteins during electrophoresis has been suggested to cause deviations up to ±50% in their apparent molecular weights [20]. Similar artifacts can arise when size-exclusion chromatography (SEC) is used to separate membrane proteins.

9.2.5 Amino Acid Composition

From the first crystal structure of a membrane protein [21] it was clear that the distribution of protein residues along the lipid bilayer depended on their depth in the membrane. Leu, Ile, and Phe residues were the most abundant in the acyl chain areas of membrane lipids, Trp and Tyr residues in the interface, and polar residues in the aqueous zone. A decade later, analysis of available membrane protein crystal structures, as well as work with model peptides, confirmed these results [22, 23]. This observed hydrophobic distribution applies only to those residues facing the lipid acyl chains, because the interiors of α-helix membrane proteins are not significantly more hydrophobic than those of soluble proteins, although transmembrane domains bury smaller residues on average [24].

Trp and Tyr residues have such a marked tendency to locate in the interfacial area that most membrane proteins have what is known as an "aromatic belt." Whenever a hydrophobic mismatch occurs, the protein and/or the membrane can change their structures in order to match the length of their respective hydrophobic areas [25–28]. Lys and Arg residues, with their long and flexible side-chains, can "snorkel" towards the lipid/water interface region, where their positive charge can interact with the negatively charged phosphate groups of phospholipid [23]. Similarly, lipid molecules located close to a transmembrane helix can adapt to the presence of polar residues and water molecules can help solvate polar groups located well within the bilayer plane [29–32].

9.2.6
Transmembrane Helix Association Motifs and Membrane Protein Oligomerization

Consideration of individual transmembrane α-helices as independent folding units has facilitated a "divide-and-conquer" approach to the study of membrane protein assembly, where interactions between peptides recapitulate membrane-embedded portions of helical membrane proteins [33]. The utility of peptides as tools for investigating membrane protein folding originates from a series of early experiments that revealed that the native protein fold of bacteriorhodopsin could be regenerated from various fragments – an indication that transmembrane helix–helix interactions alone have sufficient specificity to generate tertiary and quaternary structures [34, 35]. The "divide-and-conquer" strategy has been successfully applied to the structural analysis of transmembrane segments of a number of membrane proteins [36, 37], such as the structural determination of the M2 proton channel from influenza A virus [38]. In addition, valuable structural information can be obtained in some instances from small transmembrane peptides or membrane protein fragments in organic solvents.

Work with model peptides has allowed investigating the thermodynamics and sequence dependence of transmembrane helix–helix association, for example, showing that polar residues can drive oligomerization of transmembrane helices [39–41]. Classic work by Engelman's group with glycophorin A showed that the GX_3G sequence also drives the association of transmembrane helices [42]; subsequently, this motif was found in both water-soluble [43] and membrane proteins [44]. In glycophorin the helices cross with a right-handed crossing angle of around 40°. Since then, a number of transmembrane helix self-association motifs have been determined experimentally [36] and an exhaustive analysis of helix-packing motifs in membrane proteins was published in 2006 [45]. The later study revealed that the most frequent association motifs showed a tendency to segregate small residues to the helix–helix interface every seven or four residues in the sequence, driving tight helix–helix associations. Thus, the universe of common transmembrane helix-pairing motifs is relatively simple [46].

Crystallographic and functional studies have demonstrated that a large number of membrane proteins can form homo- and hetero-oligomeric structures or even may require oligomerization for function. Some cases are obvious, such as tetrameric or pentameric ion channels where the pore runs through the symmetry axis of the helix bundle. In extreme cases, such as the photosystem II/light harvesting complex II in grana membranes, bacteriorhodopsin in purple membranes, and rhodopsin in rod outer segments (ROS), membrane proteins can be packed at high density in para-crystalline arrays, in order to increase the surface area capable of harvesting photons. There is mounting evidence for many GPCRs that homo- and heterodimers (and even larger oligomers) play an important role in the GPCR activity cycle. In some cases, GPCR hetero-oligomers give rise to pharmacological properties that differ from those of their individual GPCR components.

In the case of rhodopsin, the size of the minimum *in vivo* functional unit is still a matter of debate. Our laboratory, using transmission electron microscopy to study the effect of *n*-alkyl-β-D-maltoside (C_nM) chain length on the oligomerization state of rhodopsin, found that in micelles containing $C_{12}M$, rhodopsin exists as a mixture of monomers and dimers in equilibrium depending on the protein/detergent ratio. In $C_{14}M$ and $C_{16}M$ rhodopsin forms higher-ordered structures and, especially in $C_{16}M$, most of the particles are present in tightly packed rows of dimers similar to those seen in ROS membranes. The fact that the activity of rhodopsin increased with its oligomerization state in detergents suggests that oligomerization of GPCRs may be crucial for signal transduction [47]. However, in subsequent experiments we reconstituted monomeric bovine rhodopsin into an apolipoprotein A-I phospholipid particle derived from high-density lipoprotein. Photoactivation of rhodopsin in these particles resulted in rapid activation of transducin at a rate comparable with that found in native ROS and 20-fold faster than rhodopsin in detergent micelles. These data suggest that monomeric rhodopsin is the minimal functional unit for G-protein activation and that oligomerization is not absolutely required for this process [48]. Two points about these studies with nanoparticles, however: (i) they do not answer the real question (i.e., how does rhodopsin activate transducin in its native para-crystalline arrangement) and (ii) the structures of these nanoparticles are much more complex than initially thought [49].

9.2.7 Post-Translational Modifications

There are more than 150 known post-translational modifications in proteins [50]. Reversible and regulated post-translational modifications include phosphorylation, acetylation, *S*-palmitoylation, and ubiquitination. Prokaryotic systems lack the cellular machinery needed for post-translational modifications. Heterogeneous modification is a potential problem because the protein may run as several bands or a smear during electrophoresis, making it difficult to assess its identity, molecular weight and purity. In addition, heterogeneous modifications may inhibit the growth of protein crystals. For example, heterogeneous glycosylation would prevent growth of crystal lattices involving specific carbohydrate–carbohydrate contacts, as was the case for two rhodopsin crystal forms grown in our laboratory (Figure 3 in [51]).

9.2.7.1 Glycosylation

The most prevalent post-translational modification of plasma membrane and secretory proteins is *N*-linked glycosylation [52]. This linkage participates in many important biological processes such as protein folding, intracellular targeting, immune responses, cell adhesion, and protease resistance. However, such glycosylation is also the most frequent source of heterogeneity in membrane proteins heterologously expressed in eukaryotic systems. In our laboratory, many GPCRs have been stably expressed in HEK-293 cells, but all, with the exception of the CB2 cannabinoid receptor, exhibited markedly heterogeneous (hyper)glycosylation (unpublished results).

Several strategies have been used to minimize or avoid heterogeneous glycosylation, the most common being enzymatic or mutational removal of the carbohydrates, or prior addition of the glycosylation inhibitor, tunicamycin. However, although many successful experiments have been performed with deglycosylated proteins, deglycosylation may negatively affect the expression, stability, folding, and/or and function of membrane proteins.

Several glycosylation-deficient expression systems also have been developed to address this problem [53]. Our laboratory developed a GPCR expression system that produces homogeneous glycosylated GPCRs by expressing them in mouse rod cells under the influence of the rhodopsin promoter [54, 55].

Rhodopsin's glycosylation pattern in rod cells is exquisitely homogeneous (Figure 9.2). However, when rhodopsin was heterologously expressed in liver [55], yeast [56], or mammalian cell cultures [57], heterogeneous glycosylation was observed. Rhodopsin in bovine retina is glycosylated at residues N2 and N15 [58]. Mutations of these residues or their surroundings (especially T4 and T17) can cause different degrees of retinal degeneration. Disruption of the normal N-terminal structure could result in loss of disk structure, potentially attributable to poor packing efficiency of rhodopsin within these membranes or decreased ability of rhodopsin to self-associate [52].

9.2.7.2 Palmitoylation

Palmitoylation consists of the reversible addition of a 16-carbon saturated fatty acyl chain, usually to cytoplasmic Cys residues via a thioester linkage (*S*-palmitoylation). Many other acyl lipids can be added to proteins as well [59].

Palmitoylation of soluble cytoplasmic proteins has been well documented to regulate their interaction with specific membranes or membrane domains. Less is known about the consequences of palmitoylation in transmembrane proteins, because of the difficulty of working with membrane proteins and the complexity of palmitoylation-induced behavior. Palmitoylation has been reported to be involved in regulation of membrane protein folding and targeting, trafficking, and protein–protein interactions [60]. It also has been suggested that palmitoylation

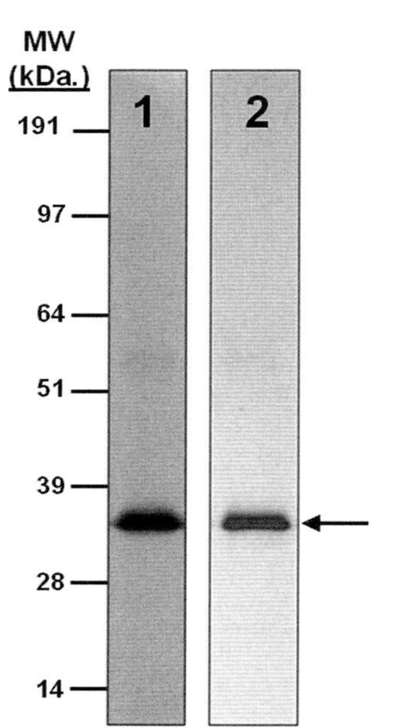

Figure 9.2 Bovine rhodopsin purified with immobilized 1D4 antibody. Lane 1: immunoblot with 1D4 used as the primary antibody followed by an alkaline phosphatase-conjugated secondary antibody. Signals were visualized by treatment with BCIP/NBT Color Development Substrate (Promega). Lane 2: silver-stained SDS–PAGE gel of purified rhodopsin. Arrow shows the position of rhodopsin. Mobility and molecular weight (MW) of markers (SeeBlue® Plus2; Invitrogen) are shown at the left.

determines the orientation of the transmembrane domain with respect to the plane of lipid bilayers [59].

Many GPCRs have been experimentally shown to be mono-, bi-, or tripalmitoylated at conserved C-terminal Cys residues [59]. This trend indicates that these residues and post-translational modification are important for GPCR function and/or trafficking, but there is no consistency among GPCRs about the role that palmitoylation plays in receptor structure and function. We analyzed engineered mice lacking both C-terminal palmitoylated Cys residues of rhodopsin (Cys322 and Cys323) and found that absence of palmitoylation generally only had minor effects [61]. Not surprisingly, the palmitoyl chains adopted different conformations in the seven crystal structures of bovine rhodopsin solved so far.

9.2.8 Sequence Modifications

Most eukaryotic membrane proteins crystallized so far were purified from native sources [4]. However, heterologous expression of membrane proteins is sometimes needed or preferable because of their low expression levels in native sources, instability, difficulties with purification, and so on. As membrane proteins are translocated into membranes starting with their N-terminus during translation, modifications and additions to the C-terminus are in general less dysfunctional than those at the N-terminus. For example, placing a polyhistidine tag at the N-terminus adds several positive charges that may adversely affect membrane insertion. This said, modifications of the N-terminus constitute a common strategy to improve expression levels. Also, C-terminal length variants such as splicing variants are common and these do not seem to affect function in many cases. This is why the C-terminus is generally preferred over the N-terminus to add tags and fusion proteins for different purposes (e.g., to facilitate purification, detection, crystallization, etc.).

We successfully expressed many different GPCRs in rod cells of *Xenopus* under the rhodopsin promoter with Green Fluorescent Protein (GFP) fused to their C-terminal for easy detection [62]. Recently we also expressed a 5-HT_{4R}–DsRed fusion protein in a *Sf*9/baculovirus system and found, despite a reasonable expression level, that most of the fusion protein was degraded by cleavage between its two protein components (unpublished results).

Fluorescence-activated cell sorting has been used to sort and enrich the best cell lines expressing membrane protein–GFP fusions [63]. The Gouaux group has used GFP fusions combined with fluorescence-detection SEC quite successfully to rapidly evaluate the expression level, stability, optimal detergent composition, and monodispersity of a number of membrane proteins expressed in HEK-293 cells [64]. Before crystallization trials, these constructs are cleaved with a specific protease and the GFP removed by SEC.

One strategy designed to enhance plasma membrane expression of GPCRs in heterologous cells consists of adding leader sequences to these receptors. The first example of this approach was the engineering of an artificial signal sequence onto the N-terminus of the β_2-adrenoceptor [65], that resulted in a several-fold increase in insertion of this receptor into the plasma membrane. Similarly, the addition of an N-terminal signal peptide or the partial truncation of the exceptionally long N-terminus of the cannabinoid CB1 receptor increased its expression level, with no effect on ligand binding. In the same line, truncation of 79 amino acids from the N-terminus of the α_{1D}-adrenergic receptor enhanced its expression. By contrast, grafting of the α_{1D}-adrenergic receptor N-terminus onto α_{1A}-adrenergic receptor or α_{1B}-adrenergic receptor reduced surface expression of these receptors in heterologous cells [66]. However, addition of a signal sequence provides no assurance that the protein will be expressed correctly and targeted to the plasma membrane [67]. In summary, N- and C-terminal modifications can be neutral in terms of membrane protein folding and function, but N-terminal modifications generally have the highest impact on membrane protein expression.

A more important and underappreciated problem occurs when researchers introduce sequence modifications in a membrane protein to stabilize it, reduce polydispersity, and eliminate heterogeneity as a strategy to increase the probability of obtaining crystals. Although important structural information can be gained from such engineered membrane proteins, caution should be exercised in extrapolating it to the wild-type protein. For example, a truncated, minimally functional acid-sensing ion channel ASIC1 was recently crystallized that provided more information about this channel's gating and ion selectivity properties than a previous structure of an inactive, more truncated version of ASIC1 [68]. In the case of GPCRs, thermostabilizing mutations, deletions, chemical modifications, deglycosylation, and/or substitution of a loop with a soluble protein have allowed the crystallization of β-adrenergic receptors and adenosine A_{2A} receptors at the cost of losing their native-like ligand-binding properties [69–71].

9.2.9
Lipids and Water

It is generally known that addition of lipids is sometimes necessary for proper functioning of detergent-solubilized membrane proteins. For example, we have found that phospholipids are needed for the proper formation, stability, and function of the photoactivated rhodopsin–transducin complex [72]. Nevertheless, membrane protein crystallographers had initially assumed that lipids were a detrimental contaminant that should be removed as completely as possible before crystallization trials. However, it is now becoming clear that some proteins crystallize more readily in the presence of a number of specifically bound lipids, whereas others, such as the porins, only crystallize after delipidation [73]. For example, well-diffracting crystals of cytochrome b_6f can be obtained only by adding phospholipids that stabilize the intact complex [74]. Successful crystallization of a heterologously expressed rabbit sarcoplasmic–endoplasmic reticulum Ca^{2+}-ATPase isoform also required addition of native phospholipid [75]. Moreover, we obtained the first high-resolution crystals of rhodopsin in mixed phospholipid/detergent micelles [76].

However, excess lipid may be detrimental for crystal growth. Therefore, purification procedures leading to crystallization need to include optimization of the detergent:lipid:protein ratio. Examples of lipids found in crystal structures include those in photosystem I and II, bacteriorhodopsin, archaerhodopsin-2, spinach major light harvesting complex, ATP-binding cassette transporter, cytochrome b_6f and bc_1 complexes ([77] and references therein) and rhodopsin [78]. In addition, two crystal structures of the β_2-adrenergic receptor bound to cholesterol along with a sequence analysis have led to the hypothesis that most family A GPCRs have a cholesterol-binding motif [69, 79].

This is why it is hardly surprising that addition of lipids during or after chromatographic purification of membrane proteins is sometimes necessary or helpful in order to maintain their function and/or grow crystals. In the case of glycerol-3-phosphate transporter (GlpT) from *E. coli*, the extent of delipidation, that increases with successive purification steps, is critical for its crystallization. Two chromatography steps were found to be optimal wherein GlpT was found to copurify with about equal amounts of phosphatidylethanolamine, phosphatidylglycerol, and cardiolipin, at a total phospholipid/protein molar ratio of around 23 [80].

A systematic study of the effect of phospholipids on crystallization of lactose permease (LacY) resulted in three different crystal forms that diffracted to increasingly better resolution in a manner that correlated with the concentration of copurified phospholipid. Consistently, progressive addition of *E. coli* phospholipids to delipidated LacY led to different crystal forms. Tetragonal crystals were obtained with improved diffraction quality for a stable mutant by carefully adjusting phospholipid content. Furthermore, crystals of good quality from wild-type LacY, a particularly difficult protein to crystallize, were also obtained by using the same approach [81, 82]. Thus, proper adjustment of the phospholipid environment is a good strategy for crystallizing membrane proteins.

Comparisons of high-resolution crystals of bacterial and mammalian cytochrome *c* oxidases have revealed that the positions occupied by native membrane lipids and detergent substitutes are highly conserved, along with amino acid residues in their vicinity. Well-defined detergent head-groups (maltose) were found associated with aromatic residues in a manner similar to phospholipid head groups, and this likely contributes to the success of alkyl-glycoside detergents in supporting membrane protein activity and crystallization [83].

Especially interesting are structures obtained by electron diffraction of two-dimensional crystals, which are grown in phospholipid planar bilayers faithfully mimicking cell membranes. Although technical difficulties usually prevent the high resolution expected for three-dimensional crystals, it has been possible to solve the structure of lens-specific aquaporin-0 (AQP0) at 1.9-Å resolution. This allowed atomic modeling of the lipid bilayer surrounding AQP0 tetramers and a description of the lipid–protein interactions. Indeed, close inspection of this model revealed that lipids bridge all the contacts between AQP0 tetramers within a layer and that these tetramers have virtually no direct lateral interaction [84].

Finally, an alternative to crystallization *in surfo* (in surfactant micelles) is to crystallize membrane proteins in bicelles (diskoidal phospholipid/detergent micelles) or to use *in cubo* methods wherein the protein is purified in detergent micelles and then embedded into a lipidic cubic phase [85]. However, the last method has the disadvantage of a high viscosity that makes crystal detection and collection challenging.

On the other side of the polarity spectrum, water is basically absent from the hydrophobic interior of phospholipid bilayers and also until recently was believed to be absent from the interior of membrane proteins, with the obvious exception of channel lumens. However, high-resolution crystal structures of a number of membrane proteins clearly show electron densities corresponding to water molecules that can act as cofactors or prosthetic groups.

High-resolution crystals of bacteriorhodopsin in different signaling states along with spectroscopic studies have provided information about the cofactor role of several water molecules in this light-activated proton pump [86]. We analyzed the distribution of water molecules in the transmembrane region of all available GPCR structures and found conserved contacts with microdomains shown to be involved in receptor activation [87]. Armed with this knowledge, we used radiolytic labeling to identify structural water molecules in different GPCRs. Radiolytic labeling (or footprinting) involves generating hydroxyl radicals with synchrotron X-ray white light and then using liquid chromatography coupled to mass spectrometry to identify which residues react with those radicals. This method also can provide valuable information about the role of structural water in different conformational states of a membrane protein, either in bilayers or membrane mimetic systems [88].

9.2.10 Purity and Contaminants

Common experience dictates that proteins need to be purified to homogeneity for their crystallization in the presence of maximally purified chemicals. However, several exceptions to this precept exist wherein impurities not only do not seem to affect crystal growth, but may even be essential for it.

For the crystallization of urea transporter at 3.8 Å, the last purification step consists of gel filtration on a Superdex 200 10/300 (GE Health Sciences) while changing the detergent to 40 mM low-purity C_8M (Sol-Grade from Anatrace). Curiously, crystals grown from protein purified in high-purity C_8M (AnaGrade) failed to diffract to better than 4.5–5 Å [89].

As noted below with retinal pigmented epithelium-specific protein 65 kDa (RPE65), high protein purity is not always essential for growth of high-resolution crystals. Obtaining high-purity preparations can be especially difficult when the protein is obtained from native sources and purified by nonaffinity chromatographic methods. In a systematic study it was found that the photosynthetic reaction center was able to form crystals at 75% purity by using the vapor diffusion method (and just 50% purity for lipid cubic phase crystallization) [90].

Finally, gramicidin (gA′ or gD) is commercially available as a mixture of gA (Trp in position 11) and a small amount of gB (Phe11) and gC (Tyr11). Analysis of a high-resolution crystal structure of gA suggested that a small percentage of gC present in the sample acted as a nucleation agent for gA crystallization [91].

9.2.11 Current Trends in the Crystallization of α-Helical Membrane Proteins

In 2008, Iwata's group published an exhaustive study on the detergents and conditions used for crystallization of α-helical [92] and β-barrel [93] membrane proteins. The most successful detergents for α-helical membrane protein crystallization were, in this order, *n*-alkyl-β-D-maltosides ($C_{12}M$ and $C_{10}M$), *n*-alkyl-β-D-glucosides (C_8G and C_9G), lauryldimethylamine-*N*-oxide (LDAO), and polyoxyethylene glycol detergents ($C_{12}E_8$ and $C_{12}E_9$). This statistical analysis was facilitated by the Membrane Protein Data Bank [94] and Stephen White's data base of membrane proteins of known three-dimensional structure [3], both available online. High-resolution structures of native human membrane proteins in particular are intensely pursued because of their potential to serve as templates for structure-based drug design. Unfortunately, vertebrate proteins in general seem especially difficult to overexpress and crystallize [95]. Here, we describe a minisurvey performed to discern the latest trends in crystallizing vertebrate α-helical membrane proteins, including only those *unique* structures published in 2008 and 2009. In this order, the most successful detergents for crystallization (usually exchanged during purification) were $C_{12}M$ (three structures), C_9G (two structures), plus C_8G, $C_{11}M$, C_{11}-thio-M, and $C_{12}E_8$ with one structure. Several of these crystals were grown with variable amounts of phospholipids; $C_{12}M$ includes one GPCR structure crystallized in lipidic cubic phase. The most used expression systems were *Sf*9 (four) cells, *Pichia pastoris* (three), plus *E. coli* and native tissue each with one structure. In conclusion, alkyl-glycosides are still the most successful detergents, and insect cells and yeast are being increasingly successful as expression system for the crystallization of vertebrate membrane proteins.

9.3 Test Cases

9.3.1 Rhodopsin

Protocol 1

Protein type	seven-transmembrane, family A GPCR
Crystal space group	tetragonal $P4_1$
Source	native (bovine retina)
Purification steps	organelle (ROS) fractionation, zinc acetate precipitation of apoprotein
Detergent	*n*-nonyl-β-D-glucoside (C_9G)

Rhodopsin is a photoreceptor membrane protein responsible for visual signal transduction [96]. As such, it is also the pharmacological target for drugs used to treat certain retinal diseases and age-related visual dysfunction. Notably, rhodopsin is a prototypical GPCR [97], by far the most important family of pharmacological receptors because about half of the therapeutic drugs in the market today target GPCRs [98].

Rhodopsin is glycosylated at Asn2 and Asn15 residues [99], palmitoylated at Cys322 and Cys323 residues [100], and acetylated at the N-terminal Met. It also contains eight Ser and Thr residues that are partially phosphorylated in a light-dependent manner [101]. These post-translational modifications are not completely homogeneous, so it was feared that they would negatively affect crystallization of this membrane protein. We addressed this challenge by purifying bovine rhodopsin from native sources without including a chromatographic step. The first and most important purification step was actually accomplished by the cow, because only properly folded rhodopsin is transported from the rod cell's endoplasmic reticulum to ROS where it forms para-crystalline arrays at millimolar concentrations [102]. Once in the laboratory, ROS were mechanically separated from bovine retinas in phosphate buffer (67 mM KP_i, pH 7.0, 1 mM $Mg(OAc)_2$, 0.1 mM EDTA, 1 mM dithiothreitol (DTT)). Then a stepwise sucrose gradient centrifugation allowed enrichment of the ROS at the interface between densities 1.11 and 1.13 g/ml (method adapted from [103]). The ROS suspension was washed several times with water to remove loosely attached proteins. Then, ROS were solubilized with C_9G to yield a final concentration of around 10 mg rhodopsin/ml, 50 mM 2-(*N*-morpholino)ethanesulfonic acid (MES), pH 6.3, and around 100 mM

$Zn(OAc)_2$. After 5–10 h incubation at room temperature, solubilized rhodopsin was separated from insoluble Zn^{2+}-opsin complexes by centrifugation. Solubilization of rhodopsin was optimal at a detergent/rhodopsin ratio of around 2.2 (w/w) and, after centrifugation, the sample of rhodopsin in its ground state was pure enough (around 98%) to grow diffracting crystals [104], which led to solving the three-dimensional structure of a GPCR for the first time [76]. All experimental procedures were done in the dark or under dim red light because these crystals were unstable under white light. Such instability most probably occurs because conformational changes upon photoactivation involve areas of rhodopsin engaged in protein–protein contacts that stabilize the crystalline array. Addition of heptanetriol, one of the most used additives for membrane protein crystallization, helped improve the resolution to 2.8 Å. Resolution was later enhanced to 2.2 Å by switching the detergent from C_9G to C_7-thio-G [105].

Protocol 2

Crystal space group	hexagonal $P6_4$
Source	native (bovine retina) and COS-1 cells
Purification steps	organelle (ROS) fractionation, and affinity (concanavalin A), gel-filtration, and anion-exchange chromatographies
Detergent	*n*-octyltetraoxyethylene (C_8E_4), LDAO

After our initial structure was published, the Schertler group independently solved the ground-state rhodopsin structure by using a different purification strategy and crystallization conditions [106, 107]. ROS membranes were solubilized in LDAO and the protein was purified by successive lectin-affinity, SEC and anion-exchange chromatographic steps, while the detergent was partially exchanged to C_8E_4. Furthermore, heterologous expression in COS-1 cells of a thermostable rhodopsin mutant and a similar purification scheme also yielded well-diffracting crystals [108]. In the last case, the cofactor, 11-*cis*-retinal, had to be added at the first step of purification. The space group was reinterpreted from trigonal $P3_1$ to hexagonal $P6_4$ [109].

Protocol 3

Crystal space group	trigonal $P3_112$ and rhombohedral R32
Source	native (bovine retina)
Purification steps	organelle (ROS) fractionation, zinc acetate precipitation of apoprotein, affinity (antibody) chromatography, NH_4SO_4-induced phase separation
Detergent	*n*-nonyl-β-D-glucoside (C_9G)

We designed a new purification scheme with the hope of solving the structure of photoactivated rhodopsin. $C_9G/Zn(OAc)_2$ extraction of rhodopsin from ROS membranes, as described above, was initially omitted, but was added later as an initial purification step in an attempt to improve crystal resolution.

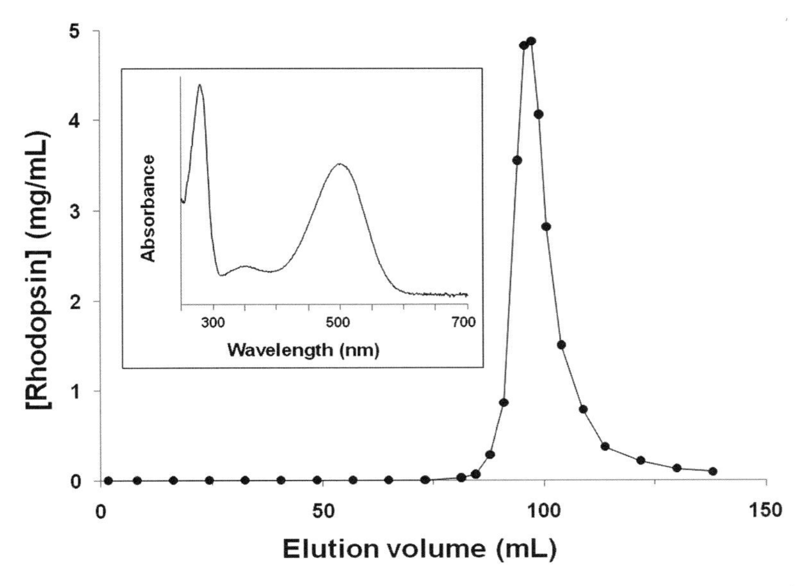

Figure 9.3 Chromatographic purification of bovine rhodopsin on a 2.5-cm × 20-cm column filled with 95 ml of Sepharose-immobilized 1D4 antibody. Inset: absorption spectrum of purified rhodopsin.

An affinity chromatographic support was prepared by immobilizing 1D4 antibody, which recognizes rhodopsin's C-terminus [110], in CNBr-activated Sepharose [111]. A detergent-solubilized ROS sample was loaded onto this affinity column at a ratio of about 0.6 mg rhodopsin/ml 1D4 gel. Then, the column was washed at a linear flow rate of around 18 cm/h with 10 column volumes of 50 mM Tris, pH 7.4, 280 mM NaCl, and 6 mM KCl containing 50 mM C_9G, and eluted with 0.75 mg/ml of a competing peptide (TETSQVAPA) at 7 cm/h (Figure 9.3). The rhodopsin concentration in each fraction was determined by diluting an aliquot into 10 mM Tris, pH 7.2, 1 mM $C_{12}M$, and 10 mM NH_4OH, and measuring the absorbance at 500 nm. Peak fractions reached around 5 mg/ml and the $A_{280\ nm}/A_{500\ nm}$ ratio of the fractions containing ground-state rhodopsin was 1.58 ± 0.2 (Figure 9.3, inset), indicating a purity of around 99% in agreement with electrophoresis results (Figure 9.2).

The most highly purified fractions were pooled together to achieve 1–2 mg/ml rhodopsin. Then, 0.25 volumes of 0.5 M MES, pH 6.3, were added and 0.69 g solid $(NH_4)_2SO_4$/ml of total solution was dissolved by stirring at room temperature. Addition of a high concentration of $(NH_4)_2SO_4$ induces phase separation in many detergent solutions, yielding a top phase rich in detergent where nearly all the rhodopsin partitioned. Treatment of purified rhodopsin with solid $(NH_4)_2SO_4$ resulted in a 12- to 15-fold increase in the concentration of this protein. A typical final yield was 0.2–0.3 mg of purified ground-state rhodopsin per bovine retina.

Crystals obtained from this purification scheme diffracted up to around 4 Å, allowing us to solve the structure of both the ground-state and photoactivated states of rhodopsin in a putatively physiological, parallel dimer orientation [51]. Crystals grown without the $(NH_4)_2SO_4$-induced phase separation step were larger, but diffracted poorly and lost integrity after photo-activation, much like our previous tetragonal crystals [104].

9.3.2 RPE65

Protein type	monotopic enzyme, seven-bladed β-propeller
Crystal space group	hexagonal $P6_5$
Source	native (bovine RPE); crystallization from heterologous expression systems failed
Purification steps	differential centrifugation and anion-exchange chromatography
Detergent	*n*-octyltetraoxyethylene (C_8E_4)

Vertebrate vision is maintained by a complex cyclic enzymatic pathway that continuously operates in the retina to regenerate the visual chromophore, 11-*cis*-retinal. A key enzyme in this pathway is the microsomal membrane protein RPE65, which catalyzes the conversion of all-*trans*-retinyl esters to 11-*cis*-retinol in the retinal pigmented epithelium (RPE). Mutations in RPE65 are known to be responsible of a subset of cases of the most common form of childhood blindness, Leber congenital amaurosis. Like other enzymes with hydrophobic substrates, RPE65 was suspected to be associated with the membrane, but the strength of this association was a matter of debate [112, 113].

9.3.2.1 Expression in *E. coli*

RPE65 was expressed as a fusion protein with maltose-binding protein and purified by metal-affinity chromatography. However, crystals obtained by this method corresponded to GroEL, which copurified with the RPE65 construct [114].

9.3.2.2 Expression in *Sf*9 Cells

RPE65 was purified by metal-affinity chromatography and, although the protein was more than 99% pure after gel-filtration chromatography, crystal trials failed to produce diffracting crystals.

9.3.2.3 Purification from Native Sources

Bovine RPE microsomes were solubilized for 1 h in 10 mM Tris acetate, pH 7.0, containing 1 mM DTT and 24 mM C_8E_4. The mixture was centrifuged at 100 000 g for 1 h to pellet insoluble material. RPE65 was purified from the supernatant by anion-exchange chromatography on a DEAE-Macroprep column (Bio-Rad) eluted with a 0–500 mM linear NaCl gradient. The pooled RPE65 sample was dialyzed against 10 mM Tris acetate, pH 7.0, containing 1 mM DTT and 19.2 mM C_8E_4. The

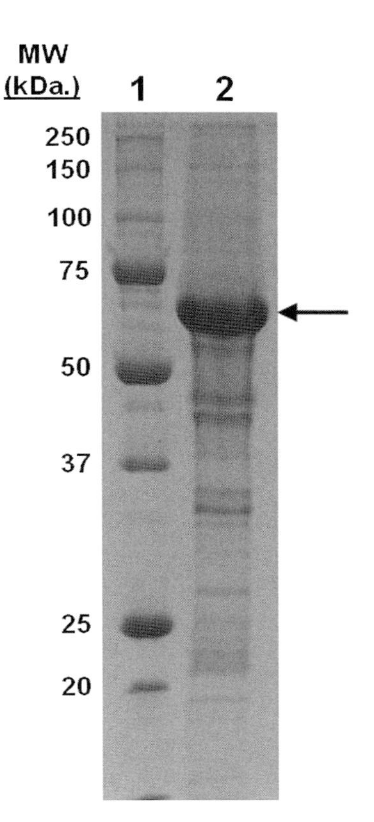

Figure 9.4 Coomassie-stained SDS–PAGE of purified RPE65 used for crystallization trials. Arrow shows the position of RPE65. Lane 1: molecular weight (MW) standards; lane 2: RPE65 sample.

resulting protein preparation was used directly for crystallization trials and biochemical experiments. On average, 150 µl of protein solution was obtained at a concentration of 10–15 mg/ml and a purity of around 90%, as judged by Coomassie-stained gels (Figure 9.4) from 300 bovine eyes. Crystals, obtained by a vapor diffusion method, diffracted to less than 2 Å. The structure of RPE65 revealed a β-propeller architecture, the membrane-binding domain [115], and bound product in the active site.

9.3.3 Transmembrane Domain of M2 Protein from Influenza A Virus

Protein type	proton channel, four-helix bundle
Crystal space group	primitive monoclinic and orthorhombic, $P2_1$ and $P2_12_12$
Source	chemical synthesis
Purification steps	C_4 reverse-phase chromatography
Detergent	*n*-octyl-β-D-glucoside (C_8G)

Flu virus kills hundreds of thousands of people in nonpandemic years worldwide and millions in pandemic years. M2, a proton channel essential for

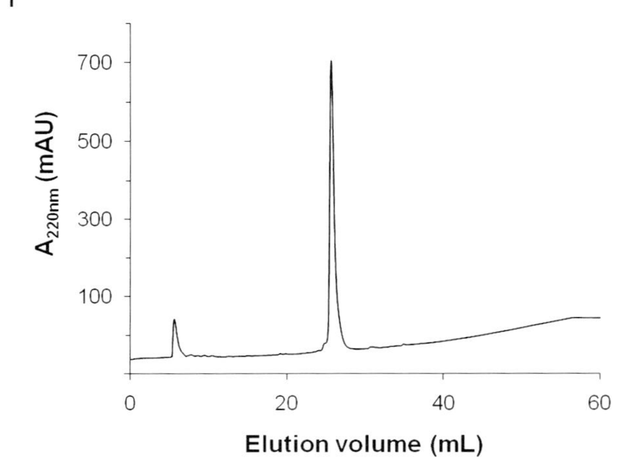

Figure 9.5 Analytical HPLC of purified M2TM on a reverse-phase C_4 column. Elution occurred in a nonlinear gradient of 60–100% buffer B, at 1 ml/min (see text).

viral replication, is the target for the related drugs, amantadine and rimantadine, although resistance to these drugs is now generalized. Both the 97-residue M2 protein (prepared by native chemical ligation of two synthetic peptide segments) and a synthetic 50-residue variant (representing a stable tryptic digest of M2) [116] failed to grow crystals.

Then, we attempted the crystallization of a 25-residue peptide containing the only transmembrane helix of M2 (M2TM) able to recapitulate the tetramerization, drug binding/resistance, low-pH activation, gating, and proton transport properties of the full-length protein [116–118]. Udorn strain wild-type plus multiple natural and artificial single-point mutants of M2TM were prepared by solid-phase peptide synthesis with standard Fmoc chemistry. Purification proceeded by reverse-phase high-performance liquid chromatography (HPLC) with a preparative C_4 column (Vydac) and gradients of buffer B (6 : 3 : 1 2-propanol/acetonitrile/H_2O containing 0.1% trifluoroacetic acid (TFA)) versus 0.1% aqueous TFA. Most peptide variants eluted at 80–84% buffer B. M2TM purity was around 99%, as assessed by analytical HPLC on a C_4 column (Figure 9.5) and its molecular mass was confirmed by matrix-assisted laser desorption/ionization-time of flight mass spectrometry. Purified peptide fractions were lyophilized and the peptide powder was stored at −20 °C or in organic solvent at −80 °C. To prepare samples for biophysical studies, an aliquot of M2TM stock in methanol, ethanol, or trifluoroethanol was transferred to a glass vial, and the organic solvent was evaporated under a N_2 stream and overnight under high vacuum. Finally, a detergent solution in the appropriate buffer (typically 50 mM C_8G in water for crystal trials, to achieve a final peptide concentration of around 1 mM) was added and the sample was vortexed vigorously [38, 116]. Crystals grew robustly from C_8G-solublized samples under different conditions. Crystals of M2TM, free and bound to amantadine, diffracted at 2.0 and 3.5 Å, respectively, revealing the 4-helix bundle structure of the proton channel and the amantadine-binding site [38].

Acknowledgments

This research was supported in part by grants EY008061, EY019478, GM079191 and P30 EY11373 from the National Institutes of Health. We want to thank to Drs. Leslie T. Webster, Jr, David T. Lodowski, and Ismael Mingarro for valuable comments, Philip Kiser for help in preparing Figure 9.4, and Dr. Amanda L. Stouffer for help in preparing Figure 9.5.

Abbreviations

DTT	dithiothreitol
GFP	Green Fluorescent Protein
GPCR	G-protein-coupled receptor
HPLC	high-performance liquid chromatography
LDAO	lauryldimethylamine-*N*-oxide
M2TM	transmembrane helix of M2
MES	2-(*N*-morpholino)ethanesulfonic acid
PAGE	polyacrylamide gel electrophoresis
PDB	Protein Data Bank
ROS	rod outer segments
RPE	retinal pigmented epithelium
SDS	sodium dodecyl sulfate
SEC	size-exclusion chromatography
TFA	trifluoroacetic acid

References

1 Krogh, A., Larsson, B., von Heijne, G., and Sonnhammer, E.L. (2001) Predicting transmembrane protein topology with a hidden Markov model: application to complete genomes. *J. Mol. Biol.*, **305**, 567–580.

2 Yildirim, M.A., Goh, K.I., Cusick, M.E., Barabasi, A.L., and Vidal, M. (2007) Drug-target network. *Nat. Biotechnol.*, **25**, 1119–1126.

3 White, S.H. (2009) Biophysical dissection of membrane proteins. *Nature*, **459**, 344–346.

4 Carpenter, E.P., Beis, K., Cameron, A.D., and Iwata, S. (2008) Overcoming the challenges of membrane protein crystallography. *Curr. Opin. Struct. Biol.*, **18**, 581–586.

5 Loll, P.J. (2003) Membrane protein structural biology: the high throughput challenge. *J. Struct. Biol.*, **142**, 144–153.

6 Wang, Y., Huang, Y., Wang, J., Cheng, C., Huang, W., Lu, P., Xu, Y.N., Wang, P., Yan, N., and Shi, Y. (2009) Structure of the formate transporter FocA reveals

a pentameric aquaporin-like channel. *Nature*, **462**, 467–472.

7 Sobolevsky, A.I., Rosconi, M.P., and Gouaux, E. (2009) X-ray structure, symmetry and mechanism of an AMPA-subtype glutamate receptor. *Nature*, **462**, 745–756.

8 Theobald, D.L. and Miller, C. (2010) Membrane transport proteins: surprises in structural sameness. *Nat. Struct. Mol. Biol.*, **17**, 2–3.

9 Kyte, J. and Doolittle, R.F. (1982) A simple method for displaying the hydropathic character of a protein. *J. Mol. Biol.*, **157**, 105–132.

10 Fleishman, S.J. and Ben-Tal, N. (2006) Progress in structure prediction of alpha-helical membrane proteins. *Curr. Opin. Struct. Biol.*, **16**, 496–504.

11 Elofsson, A. and von Heijne, G. (2007) Membrane protein structure: prediction versus reality. *Annu. Rev. Biochem.*, **76**, 125–140.

12 Nilsson, J., Persson, B., and von Heijne, G. (2005) Comparative analysis of amino acid distributions in integral membrane proteins from 107 genomes. *Proteins*, **60**, 606–616.

13 Bernsel, A., Viklund, H., Falk, J., Lindahl, E., von Heijne, G., and Elofsson, A. (2008) Prediction of membrane-protein topology from first principles. *Proc. Natl. Acad. Sci. USA*, **105**, 7177–7181.

14 Lomize, M.A., Lomize, A.L., Pogozheva, I.D., and Mosberg, H.I. (2006) OPM: Orientations of Proteins in Membranes database. *Bioinformatics*, **22**, 623–625.

15 Brosig, A., Nesper, J., Boos, W., Welte, W., and Diederichs, K. (2009) Crystal structure of a major outer membrane protein from *Thermus thermophilus* HB27. *J. Mol. Biol.*, **385**, 1445–1455.

16 Hessa, T., Kim, H., Bihlmaier, K., Lundin, C., Boekel, J., Andersson, H., Nilsson, I., White, S.H., and von Heijne, G. (2005) Recognition of transmembrane helices by the endoplasmic reticulum translocon. *Nature*, **433**, 377–381.

17 Bowie, J.U. (2005) Solving the membrane protein folding problem. *Nature*, **438**, 581–589.

18 Booth, P.J. and Curnow, P. (2009) Folding scene investigation: membrane proteins. *Curr. Opin. Struct. Biol.*, **19**, 8–13.

19 Kleinschmidt, J.H., Wiener, M.C., and Tamm, L.K. (1999) Outer membrane protein A of *E. coli* folds into detergent micelles, but not in the presence of monomeric detergent. *Protein Sci.*, **8**, 2065–2071.

20 Rath, A., Glibowicka, M., Nadeau, V.G., Chen, G., and Deber, C.M. (2009) Detergent binding explains anomalous SDS–PAGE migration of membrane proteins. *Proc. Natl. Acad. Sci. USA*, **106**, 1760–1765.

21 Deisenhofer, J., Epp, O., Miki, K., Huber, R., and Michel, H. (1985) Structure of the protein subunits in the photosynthetic reaction centre of *Rhodospeudomonas viridis* at 3 Å resolution. *Nature*, **318**, 618–624.

22 von Heijne, G. (1996) Principles of membrane protein assembly and structure. *Prog. Biophys. Mol. Biol.*, **66**, 113–139.

23 Killian, J.A. and von Heijne, G. (2000) How proteins adapt to a membrane–water interface. *Trends Biochem. Sci.*, **25**, 429–434.

24 Oberai, A., Joh, N.H., Pettit, F.K., and Bowie, J.U. (2009) Structural imperatives impose diverse evolutionary constraints on helical membrane proteins. *Proc. Natl. Acad. Sci. USA*, **106**, 17747–17750.

25 Killian, J.A. (1998) Hydrophobic mismatch between proteins and lipids in membranes. *Biochim. Biophys. Acta*, **1376**, 401–415.

26 Salom, D., Perez-Paya, E., Pascal, J., and Abad, C. (1998) Environment- and sequence-dependent modulation of the double-stranded to single-stranded conformational transition of gramicidin A in membranes. *Biochemistry*, **37**, 14279–14291.

27 Lee, A.G. (2003) Lipid–protein interactions in biological membranes: a structural perspective. *Biochim. Biophy. Acta*, **1612**, 1–40.

28 Orzaez, M., Lukovic, D., Abad, C., Perez-Paya, E., and Mingarro, I. (2005) Influence of hydrophobic matching on association of model transmembrane

fragments containing a minimised glycophorin A dimerisation motif. *FEBS Lett.*, **579**, 1633–1638.

29 Johansson, A.C. and Lindahl, E. (2006) Amino-acid solvation structure in transmembrane helices from molecular dynamics simulations. *Biophys. J.*, **91**, 4450–4463.

30 Dorairaj, S. and Allen, T.W. (2007) On the thermodynamic stability of a charged arginine side chain in a transmembrane helix. *Proc. Natl. Acad. Sci. USA*, **104**, 4943–4948.

31 Freites, J.A., Tobias, D.J., von Heijne, G., and White, S.H. (2005) Interface connections of a transmembrane voltage sensor. *Proc. Natl. Acad. Sci. USA*, **102**, 15059–15064.

32 Krepkiy, D., Mihailescu, M., Freites, J.A., Schow, E.V., Worcester, D.L., Gawrisch, K., Tobias, D.J., White, S.H., and Swartz, K.J. (2009) Structure and hydration of membranes embedded with voltage-sensing domains. *Nature*, **462**, 473–479.

33 White, S.H., Ladokhin, A.S., Jayasinghe, S., and Hristova, K. (2001) How membranes shape protein structure. *J. Biol. Chem.*, **276**, 32395–32398.

34 Liao, M.J., London, E., and Khorana, H.G. (1983) Regeneration of the native bacteriorhodopsin structure from two chymotryptic fragments. *J. Biol. Chem.*, **258**, 9949–9955.

35 Kahn, T.W., and Engelman, D.M. (1992) Bacteriorhodopsin can be refolded from two independently stable transmembrane helices and the complementary five-helix fragment. *Biochemistry*, **31**, 6144–6151.

36 Rath, A., Tulumello, D.V., and Deber, C.M. (2009) Peptide models of membrane protein folding. *Biochemistry*, **48**, 3036–3045.

37 Bordag, N. and Keller, S. (2010) Alpha-helical transmembrane peptides: a "divide and conquer" approach to membrane proteins. *Chem. Phys. Lipids*, **163**, 1–26.

38 Stouffer, A.L., Acharya, R., Salom, D., Levine, A.S., Di Costanzo, L., Soto, C.S., Tereshko, V., Nanda, V., Stayrook, S., and DeGrado, W.F. (2008) Structural basis for the function and inhibition of an influenza virus proton channel. *Nature*, **451**, 596–599.

39 Gratkowski, H., Lear, J.D., and DeGrado, W.F. (2001) Polar side chains drive the association of model transmembrane peptides. *Proc. Natl. Acad. Sci. USA*, **98**, 880–885.

40 Zhou, F.X., Merianos, H.J., Brunger, A.T., and Engelman, D.M. (2001) Polar residues drive association of polyleucine transmembrane helices. *Proc. Natl. Acad. Sci. USA*, **98**, 2250–2255.

41 Lear, J.D., Gratkowski, H., Adamian, L., Liang, J., and DeGrado, W.F. (2003) Position-dependence of stabilizing polar interactions of asparagine in transmembrane helical bundles. *Biochemistry*, **42**, 6400–6407.

42 Lemmon, M.A., Treutlein, H.R., Adams, P.D., Brunger, A.T., and Engelman, D.M. (1994) A dimerization motif for transmembrane alpha-helices. *Nat. Struct. Biol.*, **1**, 157–163.

43 Kleiger, G., Grothe, R., Mallick, P., and Eisenberg, D. (2002) GXXXG and AXXXA: common alpha-helical interaction motifs in proteins, particularly in extremophiles. *Biochemistry*, **41**, 5990–5997.

44 Senes, A., Gerstein, M., and Engelman, D.M. (2000) Statistical analysis of amino acid patterns in transmembrane helices: the GxxxG motif occurs frequently and in association with beta-branched residues at neighboring positions. *J. Mol. Biol.*, **296**, 921–936.

45 Walters, R.F. and DeGrado, W.F. (2006) Helix-packing motifs in membrane proteins. *Proc. Natl. Acad. Sci. USA*, **103**, 13658–13663.

46 Oberai, A., Ihm, Y., Kim, S., and Bowie, J.U. (2006) A limited universe of membrane protein families and folds. *Protein Sci.*, **15**, 1723–1734.

47 Jastrzebska, B., Maeda, T., Zhu, L., Fotiadis, D., Filipek, S., Engel, A., Stenkamp, R.E., and Palczewski, K. (2004) Functional characterization of rhodopsin monomers and dimers in detergents. *J. Biol. Chem.*, **279**, 54663–54675.

48 Whorton, M.R., Jastrzebska, B., Park, P.S., Fotiadis, D., Engel, A., Palczewski, K., and Sunahara, R.K. (2008) Efficient

coupling of transducin to monomeric rhodopsin in a phospholipid bilayer. *J. Biol. Chem.*, **283**, 4387–4394.

49 Wu, Z., Gogonea, V., Lee, X., Wagner, M.A., Li, X.M., Huang, Y., Undurti, A., May, R.P., Haertlein, M., Moulin, M., Gutsche, I., Zaccai, G., Didonato, J.A., and Hazen, S.L. (2009) Double superhelix model of high density lipoprotein. *J. Biol. Chem.*, **284**, 36605–36619.

50 Aitken, A. (1996) Protein chemistry methods, post-translational modification, consensus sequences, in *Proteins LabFax* (ed. N.C. Price), BIOS/Academic Press, Oxford, pp. 253–285.

51 Salom, D., Lodowski, D.T., Stenkamp, R.E., Le Trong, I., Golczak, M., Jastrzebska, B., Harris, T., Ballesteros, J.A., and Palczewski, K. (2006) Crystal structure of a photoactivated deprotonated intermediate of rhodopsin. *Proc. Natl. Acad. Sci. USA*, **103**, 16123–16128.

52 Tam, B.M. and Moritz, O.L. (2009) The role of rhodopsin glycosylation in protein folding, trafficking, and light-sensitive retinal degeneration. *J. Neurosci.*, **29**, 15145–15154.

53 Reeves, P.J., Callewaert, N., Contreras, R., and Khorana, H.G. (2002) Structure and function in rhodopsin: high-level expression of rhodopsin with restricted and homogeneous *N*-glycosylation by a tetracycline-inducible *N*-acetylglucosaminyltransferase I-negative HEK293S stable mammalian cell line. *Proc. Natl. Acad. Sci. USA*, **99**, 13419–13424.

54 Li, N., Salom, D., Zhang, L., Harris, T., Ballesteros, J.A., Golczak, M., Jastrzebska, B., Palczewski, K., Kurahara, C., Juan, T., Jordan, S., and Salon, J.A. (2007) Heterologous expression of the adenosine A1 receptor in transgenic mouse retina. *Biochemistry*, **46**, 8350–8359.

55 Salom, D., Wu, N., Sun, W., Dong, Z., Palczewski, K., Jordan, S., and Salon, J.A. (2008) Heterologous expression and purification of the serotonin type 4 receptor from transgenic mouse retina. *Biochemistry*, **47**, 13296–13307.

56 Mollaaghababa, R., Davidson, F.F., Kaiser, C., and Khorana, H.G. (1996) Structure and function in rhodopsin: expression of functional mammalian opsin in *Saccharomyces cerevisiae*. *Proc. Natl. Acad. Sci. USA*, **93**, 11482–11486.

57 Reeves, P.J., Thurmond, R.L., and Khorana, H.G. (1996) Structure and function in rhodopsin: high level expression of a synthetic bovine opsin gene and its mutants in stable mammalian cell lines. *Proc. Natl. Acad. Sci. USA*, **93**, 11487–11492.

58 Hargrave, P.A. (1977) The amino-terminal tryptic peptide of bovine rhodopsin. A glycopeptide containing two sites of oligosaccharide attachment. *Biochim. Biophys. Acta*, **492**, 83–94.

59 Chini, B. and Parenti, M. (2009) G-protein-coupled receptors, cholesterol and palmitoylation: facts about fats. *J. Mol. Endocrinol.*, **42**, 371–379.

60 Charollais, J. and Van Der Goot, F.G. (2009) Palmitoylation of membrane proteins (Review). *Mol. Membr. Biol.*, **26**, 55–66.

61 Park, P.S., Sapra, K.T., Jastrzebska, B., Maeda, T., Maeda, A., Pulawski, W., Kono, M., Lem, J., Crouch, R.K., Filipek, S., Muller, D.J., and Palczewski, K. (2009) Modulation of molecular interactions and function by rhodopsin palmitylation. *Biochemistry*, **48**, 4294–4304.

62 Zhang, L., Salom, D., He, J., Okun, A., Ballesteros, J., Palczewski, K., and Li, N. (2005) Expression of functional G protein-coupled receptors in photoreceptors of transgenic *Xenopus laevis*. *Biochemistry*, **44**, 14509–14518.

63 Mancia, F., Patel, S.D., Rajala, M.W., Scherer, P.E., Nemes, A., Schieren, I., Hendrickson, W.A., and Shapiro, L. (2004) Optimization of protein production in mammalian cells with a coexpressed fluorescent marker. *Structure*, **12**, 1355–1360.

64 Kawate, T. and Gouaux, E. (2006) Fluorescence-detection size-exclusion chromatography for precrystallization screening of integral membrane proteins. *Structure*, **14**, 673–681.

65 Guan, X.M., Kobilka, T.S., and Kobilka, B.K. (1992) Enhancement of membrane insertion and function in a type IIIb membrane protein following

introduction of a cleavable signal peptide. *J. Biol. Chem.*, **267**, 21995–21998.

66 Hague, C., Chen, Z., Pupo, A.S., Schulte, N.A., Toews, M.L., and Minneman, K.P. (2004) The N terminus of the human alpha1D-adrenergic receptor prevents cell surface expression. *J. Pharmacol. Exp. Ther.*, **309**, 388–397.

67 O'Malley, M.A., Mancini, J.D., Young, C.L., McCusker, E.C., Raden, D., and Robinson, A.S. (2009) Progress toward heterologous expression of active G-protein-coupled receptors in *Saccharomyces cerevisiae*: linking cellular stress response with translocation and trafficking. *Protein Sci.*, **18**, 2356–2370.

68 Gonzales, E.B., Kawate, T., and Gouaux, E. (2009) Pore architecture and ion sites in acid-sensing ion channels and P2X receptors. *Nature*, **460**, 599–604.

69 Cherezov, V., Rosenbaum, D.M., Hanson, M.A., Rasmussen, S.G., Thian, F.S., Kobilka, T.S., Choi, H.J., Kuhn, P., Weis, W.I., Kobilka, B.K., and Stevens, R.C. (2007) High-resolution crystal structure of an engineered human beta2-adrenergic G protein-coupled receptor. *Science*, **318**, 1258–1265.

70 Jaakola, V.P., Griffith, M.T., Hanson, M.A., Cherezov, V., Chien, E.Y., Lane, J.R., Ijzerman, A.P., and Stevens, R.C. (2008) The 2.6 angstrom crystal structure of a human A2A adenosine receptor bound to an antagonist. *Science*, **322**, 1211–1217.

71 Warne, T., Serrano-Vega, M.J., Baker, J.G., Moukhametzianov, R., Edwards, P.C., Henderson, R., Leslie, A.G., Tate, C.G., and Schertler, G.F. (2008) Structure of a beta1-adrenergic G-protein-coupled receptor. *Nature*, **454**, 486–491.

72 Jastrzebska, B., Goc, A., Golczak, M., and Palczewski, K. (2009) Phospholipids are needed for the proper formation, stability, and function of the photoactivated rhodopsin–transducin complex. *Biochemistry*, **48**, 5159–5170.

73 Pebay-Peyroula, E., Garavito, R.M., Rosenbusch, J.P., Zulauf, M., and Timmins, P.A. (1995) Detergent structure in tetragonal crystals of OmpF porin. *Structure*, **3**, 1051–1059.

74 Zhang, H., Kurisu, G., Smith, J.L., and Cramer, W.A. (2003) A defined protein–detergent–lipid complex for crystallization of integral membrane proteins: the cytochrome b_6f complex of oxygenic photosynthesis. *Proc. Natl. Acad. Sci. USA*, **100**, 5160–5163.

75 Jidenko, M., Nielsen, R.C., Sorensen, T.L., Moller, J.V., Maire, M., Nissen, P., and Jaxel, C. (2005) Crystallization of a mammalian membrane protein overexpressed in *Saccharomyces cerevisiae*. *Proc. Natl. Acad. Sci. USA*, **102**, 11687–11691.

76 Palczewski, K., Kumasaka, T., Hori, T., Behnke, C.A., Motoshima, H., Fox, B.A., Le Trong, I., Teller, D.C., Okada, T., Stenkamp, R.E., Yamamoto, M., and Miyano, M. (2000) Crystal structure of rhodopsin: a G protein-coupled receptor. *Science*, **289**, 739–745.

77 Yoshimura, K. and Kouyama, T. (2008) Structural role of bacterioruberin in the trimeric structure of archaerhodopsin-2. *J. Mol. Biol.*, **375**, 1267–1281.

78 Ruprecht, J.J., Mielke, T., Vogel, R., Villa, C., and Schertler, G.F. (2004) Electron crystallography reveals the structure of metarhodopsin I. *EMBO J.*, **23**, 3609–3620.

79 Hanson, M.A., Cherezov, V., Griffith, M.T., Roth, C.B., Jaakola, V.P., Chien, E.Y., Velasquez, J., Kuhn, P., and Stevens, R.C. (2008) A specific cholesterol binding site is established by the 2.8 A structure of the human beta2-adrenergic receptor. *Structure*, **16**, 897–905.

80 Lemieux, M.J., Song, J., Kim, M.J., Huang, Y., Villa, A., Auer, M., Li, X.D., and Wang, D.N. (2003) Three-dimensional crystallization of the Escherichia coli glycerol-3-phosphate transporter: a member of the major facilitator superfamily. *Protein Sci.*, **12**, 2748–2756.

81 Guan, L., Smirnova, I.N., Verner, G., Nagamori, S., and Kaback, H.R. (2006) Manipulating phospholipids for crystallization of a membrane transport protein. *Proc. Natl. Acad. Sci. USA*, **103**, 1723–1726.

82 Guan, L., Mirza, O., Verner, G., Iwata, S., and Kaback, H.R. (2007) Structural

determination of wild-type lactose permease. *Proc. Natl. Acad. Sci. USA*, **104**, 15294–15298.

83 Qin, L., Hiser, C., Mulichak, A., Garavito, R.M., and Ferguson-Miller, S. (2006) Identification of conserved lipid/detergent-binding sites in a high-resolution structure of the membrane protein cytochrome *c* oxidase. *Proc. Natl. Acad. Sci. USA*, **103**, 16117–16122.

84 Gonen, T., Cheng, Y., Sliz, P., Hiroaki, Y., Fujiyoshi, Y., Harrison, S.C., and Walz, T. (2005) Lipid–protein interactions in double-layered two-dimensional AQP0 crystals. *Nature*, **438**, 633–638.

85 Seddon, A.M., Curnow, P., and Booth, P.J. (2004) Membrane proteins, lipids and detergents: not just a soap opera. *Biochim. Biophys. Acta*, **1666**, 105–117.

86 Hirai, T., Subramaniam, S., and Lanyi, J.K. (2009) Structural snapshots of conformational changes in a seven-helix membrane protein: lessons from bacteriorhodopsin. *Curr. Opin. Struct. Biol.*, **19**, 433–439.

87 Angel, T.E., Chance, M.R., and Palczewski, K. (2009) Conserved waters mediate structural and functional activation of family A (rhodopsin-like) G protein-coupled receptors. *Proc. Natl. Acad. Sci. USA*, **106**, 8555–8560.

88 Angel, T.E., Gupta, S., Jastrzebska, B., Palczewski, K., and Chance, M.R. (2009) Structural waters define a functional channel mediating activation of the GPCR, rhodopsin. *Proc. Natl. Acad. Sci. USA*, **106**, 14367–14372.

89 Levin, E.J., Quick, M., and Zhou, M. (2009) Crystal structure of a bacterial homologue of the kidney urea transporter. *Nature*, **462**, 757–761.

90 Kors, C.A., Wallace, E., Davies, D.R., Li, L., Laible, P.D., and Nollert, P. (2009) Effects of impurities on membrane-protein crystallization in different systems. *Acta Crystallogr. D*, **65**, 1062–1073.

91 Burkhart, B.M., Gassman, R.M., Langs, D.A., Pangborn, W.A., Duax, W.L., and Pletnev, V. (1999) Gramicidin D conformation, dynamics and membrane ion transport. *Biopolymers*, **51**, 129–144.

92 Newstead, S., Ferrandon, S., and Iwata, S. (2008) Rationalizing alpha-helical membrane protein crystallization. *Protein Sci.*, **17**, 466–472.

93 Newstead, S., Hobbs, J., Jordan, D., Carpenter, E.P., and Iwata, S. (2008) Insights into outer membrane protein crystallization. *Mol. Membr. Biol.*, **25**, 631–638.

94 Raman, P., Cherezov, V., and Caffrey, M. (2006) The Membrane Protein Data Bank. *Cell. Mol. Life Sci.*, **63**, 36–51.

95 Muller, D.J., Wu, N., and Palczewski, K. (2008) Vertebrate membrane proteins: structure, function, and insights from biophysical approaches. *Pharmacol. Rev.*, **60**, 43–78.

96 Palczewski, K. (2006) G protein-coupled receptor rhodopsin. *Annu. Rev. Biochem.*, **75**, 743–767.

97 Mirzadegan, T., Benko, G., Filipek, S., and Palczewski, K. (2003) Sequence analyses of G-protein-coupled receptors: similarities to rhodopsin. *Biochemistry*, **42**, 2759–2767.

98 Lundstrom, K. (2006) Latest development in drug discovery on G protein-coupled receptors. *Curr. Protein Pept. Sci.*, **7**, 465–470.

99 Fukuda, M.N., Papermaster, D.S., and Hargrave, P.A. (1979) Rhodopsin carbohydrate. Structure of small oligosaccharides attached at two sites near the NH_2 terminus. *J. Biol. Chem.*, **254**, 8201–8207.

100 Ovchinnikov Yu, A., Abdulaev, N.G., and Bogachuk, A.S. (1988) Two adjacent cysteine residues in the C-terminal cytoplasmic fragment of bovine rhodopsin are palmitylated. *FEBS Lett.*, **230**, 1–5.

101 Ohguro, H., Palczewski, K., Ericsson, L.H., Walsh, K.A., and Johnson, R.S. (1993) Sequential phosphorylation of rhodopsin at multiple sites. *Biochemistry*, **32**, 5718–5724.

102 Fotiadis, D., Liang, Y., Filipek, S., Saperstein, D.A., Engel, A., and Palczewski, K. (2003) Atomic-force microscopy: rhodopsin dimers in native disc membranes. *Nature*, **421**, 127–128.

103 Papermaster, D.S. (1982) Preparation of retinal rod outer segments. *Methods Enzymol.*, **81**, 48–52.

104 Okada, T., Le Trong, I., Fox, B.A., Behnke, C.A., Stenkamp, R.E., and Palczewski, K. (2000) X-ray diffraction analysis of three-dimensional crystals of bovine rhodopsin obtained from mixed micelles. *J. Struct. Biol.*, **130**, 73–80.

105 Okada, T., Sugihara, M., Bondar, A.N., Elstner, M., Entel, P., and Buss, V. (2004) The retinal conformation and its environment in rhodopsin in light of a new 2.2 A crystal structure. *J. Mol. Biol.*, **342**, 571–583.

106 Li, J., Edwards, P.C., Burghammer, M., Villa, C., and Schertler, G.F. (2004) Structure of bovine rhodopsin in a trigonal crystal form. *J. Mol. Biol.*, **343**, 1409–1438.

107 Edwards, P.C., Li, J., Burghammer, M., McDowell, J.H., Villa, C., Hargrave, P.A., and Schertler, G.F. (2004) Crystals of native and modified bovine rhodopsins and their heavy atom derivatives. *J. Mol. Biol.*, **343**, 1439–1450.

108 Standfuss, J., Xie, G., Edwards, P.C., Burghammer, M., Oprian, D.D., and Schertler, G.F. (2007) Crystal structure of a thermally stable rhodopsin mutant. *J. Mol. Biol.*, **372**, 1179–1188.

109 Stenkamp, R.E. (2008) Alternative models for two crystal structures of bovine rhodopsin. *Acta Crystallogr. D*, **D64**, 902–904.

110 MacKenzie, D., Arendt, A., Hargrave, P., McDowell, J.H., and Molday, R.S. (1984) Localization of binding sites for carboxyl terminal specific anti-rhodopsin monoclonal antibodies using synthetic peptides. *Biochemistry*, **23**, 6544–6549.

111 Salom, D., Le Trong, I., Pohl, E., Ballesteros, J.A., Stenkamp, R.E., Palczewski, K., and Lodowski, D.T. (2006) Improvements in G protein-coupled receptor purification yield light stable rhodopsin crystals. *J. Struct. Biol.*, **156**, 497–504.

112 McBee, J.K., Kuksa, V., Alvarez, R., de Lera, A.R., Prezhdo, O., Haeseleer, F., Sokal, I., and Palczewski, K. (2000) Isomerization of all-*trans*-retinol to *cis*-retinols in bovine retinal pigment epithelial cells: dependence on the specificity of retinoid-binding proteins. *Biochemistry*, **39**, 11370–11380.

113 Travis, G.H., Golczak, M., Moise, A.R., and Palczewski, K. (2007) Diseases caused by defects in the visual cycle: retinoids as potential therapeutic agents. *Annu. Rev. Pharmacol. Toxicol.*, **47**, 469–512.

114 Kiser, P.D., Lodowski, D.T., and Palczewski, K. (2007) Purification, crystallization and structure determination of native GroEL from *Escherichia coli* lacking bound potassium ions. *Acta Crystallogr. F*, **63**, 457–461.

115 Kiser, P.D., Golczak, M., Lodowski, D.T., Chance, M.R., and Palczewski, K. (2009) Crystal structure of native RPE65, the retinoid isomerase of the visual cycle. *Proc. Natl. Acad. Sci. USA*, **106**, 17325–17330.

116 Kochendoerfer, G.G., Salom, D., Lear, J.D., Wilk-Orescan, R., Kent, S.B., and DeGrado, W.F. (1999) Total chemical synthesis of the integral membrane protein influenza A virus M2: role of its C-terminal domain in tetramer assembly. *Biochemistry*, **38**, 11905–11913.

117 Ma, C., Polishchuk, A.L., Ohigashi, Y., Stouffer, A.L., Schon, A., Magavern, E., Jing, X., Lear, J.D., Freire, E., Lamb, R.A., DeGrado, W.F., and Pinto, L.H. (2009) Identification of the functional core of the influenza A virus A/M2 proton-selective ion channel. *Proc. Natl. Acad. Sci. USA*, **106**, 12283–12288.

118 Salom, D., Hill, B.R., Lear, J.D., and DeGrado, W.F. (2000) pH-dependent tetramerization and amantadine binding of the transmembrane helix of M2 from the influenza A virus. *Biochemistry*, **39**, 14160–14170.

Part Three
Emerging Methods and Approaches

10
Engineering Integral Membrane Proteins for Expression and Stability

Igor Dodevski and Andreas Plückthun

10.1
Introduction

Natural evolution provides integral membrane proteins (IMPs) with the necessary structural and biophysical properties to fulfill their function in the lipid bilayer of the cell. However, having evolved under the selective pressures of the cell and organism, IMPs have not been particularly adapted for high-level overproduction in laboratory expression hosts, let alone for high stability in detergent-solubilized form. These properties, as important as they appear in the daily work of the biochemist, have no relevance in the natural context of cellular function. These properties may be irrelevant for IMPs and they may even be selected against – some IMPs might need to be degraded and thus a limited stability might even be a desired natural trait.

To improve the process of producing a membrane protein at sufficient amounts and quality for structural studies, an ever-growing number of techniques for screening, selecting, and generating variations and mutations are being developed and implemented. The high pace of methodological developments reflects the observation that most wild-type IMPs are difficult to study structurally and biophysically – sometimes it is virtually impossible.

Currently, there are two different experimental strategies that are routinely used for increasing the chances of producing a well-expressed, soluble, and active membrane protein. The first strategy, which is the most intuitive – and probably part of every structure determination effort – relies on the extensive screening of a large number of experimental conditions for expression and solubilization of a given construct. Typically, many experimental parameters such as plasmid design, expression host, expression media, and temperatures, detergent, and buffer for solubilization have to be empirically optimized for any given target protein. These "classical" alterations may also include changes of fused tags, domains, or truncations and the use of a host of natural ligands binding to the protein. Nonetheless, it is frequently found that *none* of these conditions achieves the desired goal, as the protein itself is the limiting factor. In other words, for some proteins simply no conditions might exist in which this particular protein with its given sequence

Production of Membrane Proteins: Strategies for Expression and Isolation, First Edition.
Edited by Anne Skaja Robinson.

would show the required characteristics that would allow it to be studied in a purified, detergent-solubilized state.

The second strategy thus relies on changing the protein under study. Most popular has been the screening of protein homologs of a given target protein in order to identify a homolog that shows favorable biophysical properties. The homologs are usually screened in the context of a few globally optimized conditions for expression and solubilization. The most promising candidates for this strategy are proteins of prokaryotic origin, because the source of homologous sequences is large and diverse. Interestingly, many homolog screens tend to converge to sequences that derive from a small number of extremophile organisms, such as *Ralstonia metallidurans*, *Pyrococcus horikoshii*, or *Thermus thermophilus*. Clearly, extremophile organisms have evolved their protein repertoire to sustain extreme conditions (such as high temperature in the case of the thermophiles). Therefore, these proteins naturally show higher stability. The crystal structures of IMPs that derive from extremophile origin are over-represented among the set of solved atomic-resolution structures (http://blanco.biomol.uci.edu/Membrane_Proteins_xtal.html). However, for many eukaryotic IMPs, including proteins of special medical relevance, such as the G-protein-coupled receptors (GPCRs), the strategy of homolog screening is not very useful, because there simply are no homologous sequences found in prokaryotic organisms.

Despite the strong research efforts put into these two strategies, they still often fail to establish appropriate conditions for expression and solubilization of a given IMP. Thus, the structural or biophysical characterization of many biologically relevant IMPs has remained a formidable challenge.

In such cases, the techniques of protein engineering can offer an alternative, third strategy. Protein engineering allows one to alter the biophysical properties of a given IMP itself, in order to increase its expression, solubility, stability, or crystal-forming propensity. It puts the focus on the protein sequence itself as an experimental parameter to be optimized. It is, in essence, a much "milder" approach than the search for homologs from thermophilic bacteria, but follows the same logic: protein engineering seeks to find a sequence *as close as possible* to the given starting sequence that bestows the desired properties on the IMP of interest.

By a process consisting of modifying the underlying amino acid sequence of a protein and screening or selecting by evolutionary strategies for improved protein versions, many roadblocks inherent to the production of IMPs can be removed. In this chapter we review the current experimental approaches for improving the biophysical properties of IMPs by protein engineering to produce better expressed and more stable IMPs.

10.2
Engineering Higher Expression

There are basically two ways of improving the biophysical properties of a given target protein by protein engineering. The first relies on modifying the amino acid sequence of a protein by rational design. This approach usually requires a detailed structural description as well as a mechanistic understanding of the protein to be engineered. For many if not most IMPs, however, this information is not available. Almost by definition, a protein that is to be engineered in order to make it stable enough to have its structure determined will not have such information available.

However, even if a good structural model and a general mechanistic understanding of the protein was available, the sequence modifications required to rationally design better biophysical properties and higher stability are too complex to be predicted. Especially in IMPs, the detailed understanding of mutations that influence biosynthesis, folding, and aggregation is almost completely lacking.

The second protein engineering approach relies on a combination of random mutagenesis and selection or screening (see below for a definition of both terms and their distinction). This at first seemingly "irrational" approach is very powerful, provided sampling is very wide, and selection or screening is efficient and accurate. In other words, mutations tested must really include those that make the decisive difference, and selection or screening must be powerful enough to discern small improvements and to handle the commensurate number of mutants to be tested.

This strategy exploits the evolutionary principle of nature (i.e., random mutagenesis and selection, ideally in an iterative fashion) to generate a desired molecular property. This concept is known as directed protein evolution. Its strength lies in the fact that a desired molecular property can be obtained based on very low input information, provided the generation of diversity and the selection of the desired protein mutants is efficient. Since the iteration between diversification and selection allows one to explore the evolutionary potential of a protein in a combinatorial fashion, it is possible to evolve rather complex biophysical properties, such as the ones defining protein biosynthesis, membrane insertion, and folding in the membrane, usually summarized under the heading "expression."

The process of evolving such properties of a given IMP thus involves two experimental steps. In the first step, a diverse set of protein variants (a library) is generated by introducing mutations in the gene coding for the membrane protein of interest. Mutations can in principle comprise point mutations, insertions, or deletions. There are many standard molecular biology techniques by which such mutations can be introduced [1, 2] and those will not be reviewed here. Genetic diversity can be either concentrated to particular regions of the protein, to particular amino acid types [3] or the gene can be mutagenized randomly at a predefined error rate. Clearly, a strategy different from a random mutagenesis of the whole gene will require that additional information is available that justifies focusing of the efforts to particular regions.

In the second step, the library of mutant protein variants is analyzed with a functional assay in order to identify candidates showing the desired molecular property (e.g., higher expression and/or stability), termed "selection" or "screening." A technical distinction can be made between these two terms. *Screening* defines the analysis of individual mutants, which are kept separately, and whose sequence identity is usually known before the experiment. As samples need to be handled in parallel, screening is restricted to cases where the number of mutant protein variants in the library is relatively small (in the range of hundreds). While several parameters can be determined in each sample, the challenges of sample handling puts a practical limit on sample numbers. Furthermore, individual samples of mutants will inevitably have some error in amounts, media, and concentrations, limiting the possibility of reliably identifying small differences.

If the sequence space that needs to be analyzed is much larger (e.g., in an attempt to evolve a particular protein), with no structural insight which regions to mutagenize preferentially, a screening of individually grown mutants is no more practical. With a library of millions of randomly mutagenized protein variants, it is simply not possible to screen each variant individually even with robotics equipment. In this case the methods of directed evolution, which are based on *selection* rather than screening, are more appropriate. The term "selection" defines methods in which all mutants are handled as a pool, in a single tube, and either a genetic selection (e.g., based on growth characteristics) or a physical selection (e.g., based on a direct measurement of receptor levels in an individual cell) is used to identify the clones with improved characteristics. In a *genetic* selection, the desired mutants have a growth advantage; in a *physical* selection, they are sorted or enriched by some physical separation.

Increasing the expression level of a target IMP represents a highly complex design task. We will not concentrate on promoter, transcription, or tags – even though they could in principle also be subjected to directed evolution – since they can usually be taken from other well-characterized working systems as a reasonable starting point. We will concentrate on the protein sequence itself. The protein sequence itself influences all the steps along the biosynthesis of an IMP: its translation rate (which might be influenced by the chosen codons), its incorporation into the membrane, its misfolding and aggregation in- and outside of the membrane during biosynthesis, as well as its susceptibility to aggregation or degradation after successful membrane insertion. All these steps influence the steady-state "expression level" of native protein. At the current state of knowledge, it is close to impossible to predict which mutations in a given amino acid sequence would influence any of these steps. We are therefore not able to engineer a molecular property called "functional expression level" by a rational, structure-based approach.

In the absence of such information it is necessary to include in the screening or selection as many mutants as possible to identify the ones that increase expression. In recent years two different approaches have been published that have proven to be very effective in evolving well-expressed IMPs starting from weakly expressed wild-type protein templates [4, 5].

10.2.1
Directed Evolution of a GPCR for Higher Expression

To directly address the importance of the protein sequence as an experimental parameter in IMP expression, we have developed a selection method in our laboratory based on the principles of directed evolution [5]. The method allows one to isolate well-expressed and functional GPCRs from libraries as big as 10^7–10^8 individual mutants, and it involves an iteration between diversification and selection. To demonstrate the power of the method we chose as the first example the neurotensin receptor NTR1, which expresses at about 800 functional receptors per *Escherichia coli* cell [6]. After several rounds of random mutagenesis and selection a receptor variant could be isolated that expresses about 10-fold more functional protein, and it also turned out to be more stable in detergents.

E. coli is the most convenient expression host for such experiments, since the transformation of large libraries is straightforward. As in the directed evolution approach, a diversified library will have to be brought into cells repeatedly; this ease of transformation is a very important consideration. Additionally, the handling of *E. coli* is very convenient on a large scale such that the final purification of the GPCR from *E. coli* is more easily implemented than with many other hosts.

The general method is schematically shown in Figure 10.1 for NTR1. (1) The wild-type NTR1 cDNA is randomly mutagenized by error-prone PCR (epPCR) to generate a library of NTR1 mutants. The mutant DNA is cloned into an expression vector, which previously had been optimized for the functional expression of NTR1 in *E. coli* [6]. The GPCR sequence is genetically fused to an N-terminal maltose-binding protein (MBP) and a C-terminal thioredoxin A (TrxA). The MBP, whose signal sequence will direct it to the Sec pathway, may enforce a periplasmic localization of the N-terminus of the GPCR, while the C-terminal TrxA may enforce a cytoplasmic localization of the C-terminus. (2) *E. coli* is transformed with the gene library and the proteins are expressed in the inner membrane of *E. coli* in liquid culture. Expression takes place at 20 °C for 20 h. (3) The cells are incubated with fluorescently labeled ligand BODIPY-neurotensin (BODIPY-NT), which shows high specificity and affinity for the neurotensin receptor. To allow binding of the fluorescent ligand to the receptors expressed in the inner membrane of *E. coli*, the outer membrane has to be partially permeabilized by an appropriate permeabilization buffer. (4) The cells expressing the highest number of functional receptors, which therefore exhibit the greatest fluorescence, are sorted by fluorescence activated cell sorting (FACS). The cells are sorted directly into growth medium and can directly be cultivated for a next round of sorting of the highest expressing mutants. By repeating the cycle of receptor expression and sorting, the highest expressing mutants can be strongly enriched from the large pool of initial mutants. Since all mutants are in the same test tube, any small variation in permeabilization efficiency of the buffer or other variations in concentrations and times will affect all mutants equally. Therefore, this approach can be thought of as a competitive experiment between different cells, amplifying small differences by serial repetition.

Whenever additional genetic diversity is desired after any FACS round, the plasmid DNA of the selected mutants is isolated and the GPCR sequence is further randomized by epPCR. The FACS selection is then repeated.

The procedure for evolving the expression level of NTR1 is outlined in Figure 10.2. The initial randomized NTR1 library was subjected to four rounds of FACS.

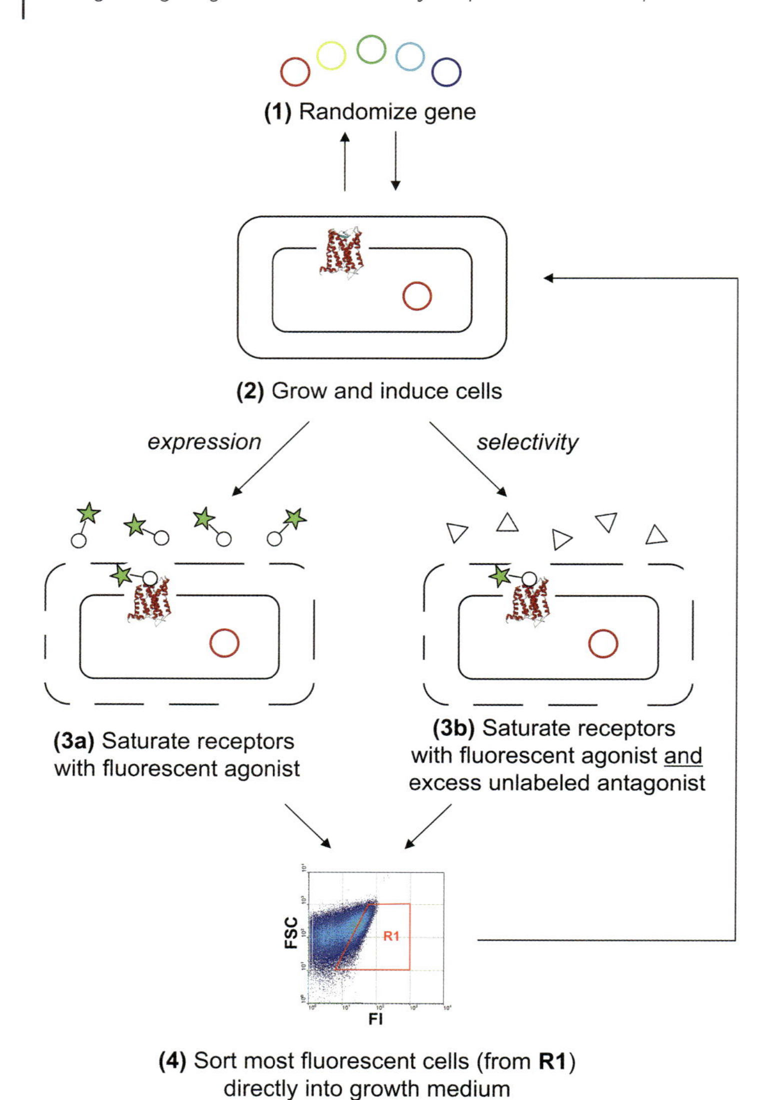

Figure 10.1 General selection scheme for increasing expression level (steps 1, 2, 3a, 4, back to 2) or altering ligand selectivity (steps 1, 2, 3b, 4, back to 2).

In each round, only the most fluorescent 0.1–1% of the cells were collected. Nonetheless, after these rounds, the evolved pool had a mean fluorescence intensity (MFI) no greater than that of the wild-type sequence. epPCR was used to overlay another set of random mutations on top of those that were enriched after the first four rounds of FACS and this rerandomized library was again subjected to four rounds of sorting. In this second set of sorts, the MFI of the pool overtook that of

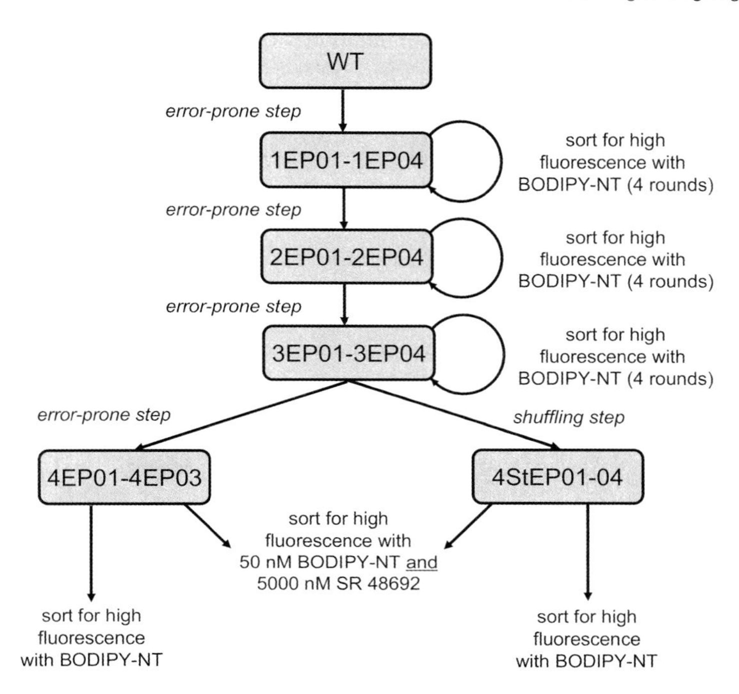

Figure 10.2 Flowchart for selections of NTR1 variants leading to increased expression level or altered ligand selectivity. WT = wild-type.

wild-type NTR1. After a third randomization step followed by four more rounds of FACS, the evolved pool was split into two. One half was randomized by epPCR a fourth time and the other half was subjected to DNA shuffling, using the staggered extension process [7]. After these selections, the MFI was approximately 5 times that of wild-type NTR1.

From the enriched "error-prone" pool (4EP03 pool), 48 single clones were analyzed for receptor expression level. Figure 10.3 shows that the clone with the best functional receptor expression level per cell, D03, exhibited approximately a 10-fold increase in specific signal, as assayed by [^{3}H]neurotensin binding. This shows that the receptor has not "adapted" to the fluorescent dye, as binding was assessed with unmodified ligand. The receptor shows nine amino acid mutations compared to wild-type NTR1. Analysis of the expression level of D03 in eukaryotic expression hosts shows that it also expressed about 12-fold better in *Pichia pastoris* and 3-fold better in HEK293T cells. These increased expression levels show that the receptor has also not "adapted" to a prokaryotic expression host. With respect to protein purification, D03 can functionally and quantitatively be solubilized in detergent micelles from the inner membrane of *E. coli* and yields about 0.5 mg of solubilized and functional GPCR per liter of shaking flask expression culture (around 4–5 g cells wet weight). Therefore, the increased functional level seen in whole cells is maintained after solubilization and purification.

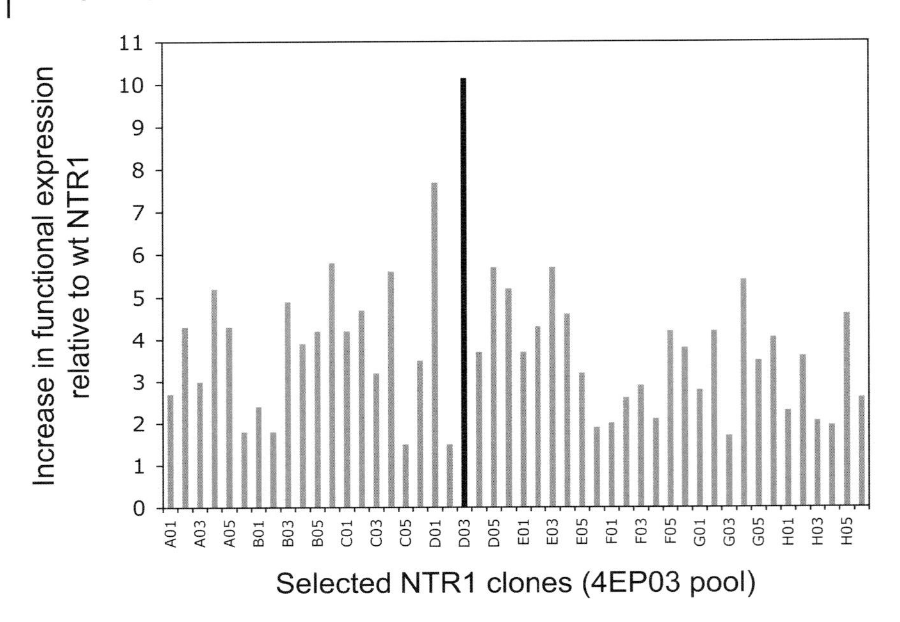

Figure 10.3 Radioligand binding analysis of 48 clones isolated from selections for maximum receptor expression level (4EP03 pool, four rounds of epPCR, each followed by three to four rounds of sorting by FACS). The bars represent receptor expression level per cell, relative to wild-type NTR1 expression level. D03 (black bar) is the clone showing the highest expression level (10-fold that of wild-type NTR1).

The evolved receptor D03 also retains the biochemical and pharmacological properties of wild-type NTR1. As D03 had been evolved under selective pressure to conserve a functional ligand-binding site, it retains the same affinity for the high-affinity agonist neurotensin as wild-type NTR1 ($K_d \sim 0.1$ nM in intact *E. coli* cells). Note that selection also has not improved the affinity. This can be explained: since the fluorescent dye was present at 50 nM during the selection, the receptor should be essentially saturated such that higher affinity mutants would not be rewarded. Neurotensin binding can also be competed by a molar excess of the antagonist SR48692, indicating that the ligand-binding site has indeed been conserved during the selections.

In addition to the selections performed for increasing the expression level, the selection method has also been applied to altering the binding selectivity of wild-type NTR1 (Figure 10.2). A motivation for such an experiment can be to favor one conformation of the GPCR over others. These selections were done by incubating a mutant library with a 100-fold excess of unlabeled antagonist SR48692, in addition to the labeled agonist BODIPY-NT, and selecting the cells showing the highest fluorescence. Those receptor mutants are thus expected to still bind the labeled agonist, but not the antagonist competitor. The selected mutants showed one strong consensus mutation in TM7 (F358S). The effects of mutation at Phe358 had been studied by others [8, 9] and reveal that a substitution to alanine at this position results not only in decreased antagonist affinity, but also in spontaneous basal inositol phosphate production in a receptor-dependent manner. Thus, more generally, the selection method by FACS has the ability to rapidly isolate mutations that may trap receptors in the active or inactive state. This approach is thus complementary to the engineering of GPCR stability, reviewed in Section 10.3.2.

There are several technical advantages inherent to this FACS-based selection method. Most importantly, for the first time IMPs can be evolved under laboratory conditions by harnessing the full power of the natural selection process. By physically linking the genotype to the phenotype – by containing the gene encoding the corresponding receptor in an *E. coli* cell – it is possible to move from a low-throughput *screening* set-up to a high-throughput *selection* setup. This allows one to sample a much larger sequence space in the range of 10^7–10^8 individual mutants per selection round as opposed to only thousands of mutants, when they have to be screened one-by-one. Being able to sample the largest sequence space possible strongly increases the probability of finding rare expression enhancing mutations. Moreover, since directed evolution is an iterative process, selection is clearly advantageous over screening. The selection from a large pool has to be done after each library diversification – a prohibitive amount of work when screening single mutants.

A second advantage, which makes the application of the selection method very appealing, is that there is a strong selection pressure for correctly folded, functional receptors inherent in the selection for the number of active receptors. Well-expressing receptor variants are selected only if the ligand-binding site remains conserved throughout the evolutionary process, because in order to be selected the receptor variants must bind the fluorescent ligand. Receptor variants adopting non-native receptor folds are selected against, even if they are well expressed. For GPCRs, ligand binding is a very strong indicator of a correctly folded protein. The ligand-binding sites of GPCRs recognizing small ligands are mostly contributed by several amino acids located on different transmembrane helices. These different helices must therefore be in a wild-type-like conformation to provide a functional high-affinity ligand-binding site.

The third advantage in implementing this method relies on the fact that the expression level of a given IMP can be improved without having to make assumptions about the mechanism of the expression process or about the amino acid substitutions that might influence it, or even the regions in the sequence that would affect it. The combinatorial approach inherent in the method allows evolving the desired property in a nearly assumption-free manner. This is a very important consideration, as many sequence correlations probably have yet to be discovered and others may be incomplete or even incorrect.

The method's potential to select for functional receptors by using fluorescent ligands is – at the moment – compromised by the fact that fluorescent ligands are not yet available for every GPCR. To date, the chemical synthesis of fluorescent ligands has been described in the literature for maybe 30 GPCRs or so [10, 11]. Fortunately, the ever-increasing research activity in the field of GPCRs and IMP biochemistry in general makes it very likely that new fluorescently labeled ligands will be developed as powerful tools in membrane protein research. Moreover, many GPCR ligands do not require complicated chemistry for their derivatization with fluorophores. Some can be synthesized in any laboratory that is equipped with a basic equipment for chemical synthesis; this is especially true for peptide ligands, which provide a variety of possibilities for chemical derivatization. The N-terminal amino groups, ε-amino groups of lysine residues, thiol groups of cysteine residues, or the C-terminal carboxyl groups can often be coupled to fluorophores without considerably compromising the binding affinity of the peptide ligand.

While the selection method by FACS is able to evolve well-expressed GPCRs that conserve their functional ligand-binding site (a very strong criterion for a correctly folded GPCR), other functional characteristics of these proteins are more difficult to enforce to be conserved during the directed evolution process. Most importantly, the ability to transmit binding signals (i.e., to couple to G-proteins) is very difficult to select as a property to be retained, as the selection process takes place in *E. coli*. Nevertheless, in the example of the NTR1 receptor studied, the ability to signal via G-protein *was* retained [5]. This was determined experimentally by recloning the selected mutants to mammalian cells and measure the ability to signal via G-proteins directly. Thus, it is important to characterize evolved mutants very carefully after the selection experiment, as is generally the case for every protein engineering method presented in this chapter.

To prove the usefulness of evolved mutants for structural studies their phenotypes should be carefully compared to their wild-type progenitor. When we performed this analysis for D03, we found that the signaling capability was retained, but somewhat more agonist was needed, compared to wild-type NTR1. This was found in mammalian cells by measuring Ca^{2+} mobilization via coupling to $G_{q/11}$. When one of the evolved mutations in D03 (R167L in the conserved (D/E)R(W/Y) motif in TM3) was reverted to wild-type Arg167, D03 would show a similar signaling behavior as wild-type NTR1 [5]. This analysis underlines the importance of performing a careful functional characterization of evolved mutants. Nonetheless, it also emphasizes that the basic functions, such as agonist binding, antagonist binding, and signaling via the G-proteins, can be maintained in these evolved GPCRs.

10.2.2
Increasing Expression by Random Mutagenesis and Dot-Blot Based Screening

A second screening method for increasing expression levels of IMPs has been developed by the Nordlund lab [4]. The colony filtration (CoFi) blot method is well suited to screen somewhat smaller mutant libraries for expression, since thousands of single mutants can be screened simultaneously. However, since single colonies must be discernable on plates, it is difficult to extend this to very large libraries. By applying the CoFi blot method to several membrane proteins their expression level could be increased. After one round of random mutagenesis and screening, mutants could be identified that express 1.5- to 40-fold better than the corresponding wild-type protein, depending on the protein under study and the initial expression level.

The CoFi blot has originally been developed for screening the expression level of soluble proteins in *E. coli* [12]. It has proven very efficient in identifying mutants of eukaryotic proteins that show higher soluble expression yields than the wild-type proteins. Recently, the method has been adapted to allow screening of mutant libraries of membrane proteins [4]. The strategy consists of four steps:

i) A library of mutants is generated by random mutagenesis of the wild-type open reading frame used as template DNA, cloning the library into an expression vector (containing an N-terminal FLAG tag and a C-terminal His_6 tag), and transforming *E. coli* cells. The cells are plated onto an LB-agar plate and grown until medium-sized colonies are visible.

ii) The colonies are picked up onto a Durapore filter membrane by overlaying the colonies with the membrane and peeling them off the agar, as they will stick to the membrane. Expression is then induced by placing the membrane onto an LB-agar plate containing isopropyl-β-D-thiogalactopyranoside.

iii) The membrane is placed on top of a nitrocellulose membrane and a Whatman 3MM paper soaked in lysis buffer containing detergents. Solubilized membrane proteins will diffuse through the filter membrane (Durapore) and are captured on the nitrocellulose membrane.

iv) The amount of expressed and solubilized membrane protein is quantified by probing the nitrocellulose membrane with a His-tag-specific reagent using standard equipment for Western blotting. The mutant colonies' expression level can easily be compared to a wild-type reference and the best-expressing mutants can be identified.

The method was benchmarked by subjecting nine membrane proteins to one round of random mutagenesis and screening the mutant libraries for expression. The set of target proteins consisted of eight prokaryotic proteins and one human protein, and they were from different functional classes. They were classified as showing either no, low-, or medium-level expression. For five of the nine target proteins (all classified as low or medium expressing proteins) the method was able to identify better expressing mutants. For an *E. coli* glycosyl transferase the expression yield was improved by an impressive 40-fold over the wild-type level (from about 25 μg/l to 1 mg/l of purified protein) as a result of three amino acid mutations. For the human microsomal glutathione S-transferase 2, several single amino acid mutations could be identified that increase the expression level 2-fold. These results are very encouraging because they are based on only one round of mutagenesis and screening.

The ability to apply the method to nine different proteins from different functional classes in a benchmark test exemplifies the biggest advantage of the CoFi blot method – its generality. In principle, it can be applied to any membrane protein with no prior structural or functional information about the target protein. Moreover, the method is easy to implement and the costs are low. However, the biggest strength of the method – its simplicity and generality – represents at the same time its greatest potential weakness. Random mutagenesis of a target gene is expected to generate mostly misfolded and nonfunctional protein mutants. As the C-terminal His_6 tag is used as the sole indicator of expression, the structural and functional state of the protein mutants is completely neglected in the screen. Reducing the screening criteria to His_6 tag detection entails a considerable risk of identifying well-expressed, but misfolded proteins. Since there is no functional screening criterion directly implemented into the method, it is crucial to carefully characterize the functional state of improved mutants after the screen. With regard to this risk, the developers of the CoFi blot analyzed the catalytic activity of selected mutants of one of their target proteins – the human microsomal glutathione S-transferase 2. The two analyzed mutants both showed only one amino acid mutation and their catalytic activities were comparable to the wild-type protein. Despite the encouraging results on these single-amino-acid mutants of one membrane protein, the risk of losing functionality, when proteins are evolved in the absence of selective pressure for functionality, remains considerably high. It is very likely that this risk increases when multiple mutations start to accumulate in experiments in which several rounds of mutagenesis and selection are performed.

10.3 Engineering Higher Stability

Establishing high expression levels of correctly folded IMPs is only the first critical step in the process of producing sufficient amounts of functional protein for biophysical and structural studies. Equally critical is the protein purification process as well as the stability of the purified protein in the solubilized state, for example, for structural studies, where the detergent-solubilized protein will be studied at high concentrations for extended times. Even though stability and functional expression yield show some correlation (Schlinkmann *et al.*, unpublished; Dodevski *et al.*, unpublished), the properties are not identical and, thus, stability must be tested separately.

Unlike soluble globular proteins, the purification process of IMPs and most methods of biophysical analysis absolutely require the help of detergent molecules to extract them from their natural physical environment – the phospholipid membrane – and to transfer them into an isotropic solution, established by detergent micelles. This process is problematic from a thermodynamic point of view because the physical environment provided by the detergent micelle is very different from the phospholipid membrane. Membrane protein solubilization by detergents therefore frequently leads to protein unfolding, aggregation, and loss of function.

For GPCRs, for example, the solubilization process turns out to be particularly destructive because of their marginal biophysical stability. GPCRs naturally exist as structurally flexible molecules, as conformational changes are required to exert their function in transducing extracellular signals across the phospholipid bilayer. The observation that GPCRs can activate G-proteins even in the absence of an activating agonist ligand – a phenomenon called basal activity – implies that the receptors can continuously adopt different receptor conformations that are of similar thermodynamic energy and separated by rather small energetic barriers. This structural and thermodynamic heterogeneity is most likely responsible for the rapid receptor unfolding observed upon receptor solubilization with detergents. The marginal stability of GPCRs in detergent micelles represents the rule for IMPs, rather than the exception. In fact, the activity of most IMPs is linked to conformational flexibility in certain regions of the proteins. Most IMPs tend to quickly lose their activity when solubilized in detergent micelles, because of limited biophysical stability.

The protein engineer's contribution to solve this problem is to identify stability-enhancing mutations in the amino acid sequence of a target protein. Here, we define "stability" as the molecular property of the protein to maintain a correctly folded and active conformation in detergent-solubilized form. We therefore set out to identify mutations that increase the half-life of a protein in detergent micelles. At the current state of knowledge and in the absence of high-resolution structural information for the great majority of IMPs, it is close to impossible to predict in advance what mutations will improve the stability of a given protein target. Stated more practically, the identification of stability-enhancing mutants relies on setting up appropriate screening experiments for rapidly and reliably analyzing collections of many mutants.

10.3.1 Stabilizing a Prokaryotic IMP by Cysteine-Scanning, Random Mutagenesis, and Screening in a 96-Well Assay Format

The approach of using protein engineering for stabilizing IMPs in detergent micelles gained momentum with the findings by the Bowie lab published in 1999, namely that stability-enhancing mutations were not rare at all in IMPs [13]. In the analysis of 20 single cysteine-substituted mutants of the *E. coli* diacylglycerol kinase (DGK), two mutants showed significantly higher resistance to thermal inactivation. Combining the two mutations in a single protein revealed a partly additive effect and the stability was further increased. Strikingly, while the half-life of the double mutant at 70 °C is 51 min, the half-life of wild-type DGK is less than 1 min. In a follow-up study on DGK, a collection of 1560 random mutants was screened for thermal stability [14]. Twelve different single mutants of DGK showed higher stability in detergent solution. The four most stabilizing mutations were combined to construct the quadruple mutant CLLD-DGK, which showed a half-life of 35 min at 80 °C. This is about 18 times that of the most stable single DGK mutant. Most importantly, the stabilized mutant showed similar catalytic activity as the wild-type.

How were the 1560 single clones screened to identify stability-enhancing mutations? The screening setup is relatively simple. The mutants were expressed in

96-deep-well plates and the protein was solubilized with detergents. In a 96-well format an aliquot was directly assayed for DGK enzymatic activity in a colorimetric assay. A second aliquot was assayed after it had been exposed to heat for a defined time period to inactivate the protein. Mutants showing higher activity than wild-type after the inactivation step were selected and further characterized. The finding that stability-enhancing mutations in IMPs may be identified relatively easily has inspired other research groups, including our own, to explore the potential of protein engineering for improving the biophysical properties of IMPs, and to devise additional methods to screen for stability for IMPs without enzymatic activity. In this respect, GPCRs have been of special interest.

10.3.2
Stabilizing GPCRs by Alanine-Scanning and Single-Clone Screening

The Tate lab has applied an alanine-scan to identify mutants of GPCRs showing increased thermal stability in detergent solution. This strategy has successfully been implemented for three different GPCRs [15–17] – a class of membrane proteins that is notoriously unstable in detergent solution. The most significant example of their work is the engineering of a β_1-adrenergic receptor mutant, which is highly stable in the detergent octylglucoside. The combination of alanine mutations rendered the receptor stable enough to be crystallized and to determine its atomic resolution structure [18].

Their strategy for identifying stability-enhancing mutations in a GPCR is based on a relatively simple methodology. A collection of mutants was prepared by mutating each position in a receptor to an alanine residue. If the wild-type residue was already alanine, leucine was introduced, as this is a helix-forming residue compatible with a membrane location. The single-point mutants were then analyzed for their potential to increase thermal stability of the receptor in detergent solution. The thermal stability of individual receptor mutants was determined on unpurified samples after detergent solubilization of whole cells. To measure the stability of each mutant, one receptor aliquot was heated for a fixed period of time, a second aliquot was kept on ice. Both of the samples were then assayed for their content of folded receptor by a radioligand binding assay (LBA). The best single mutants were then combined to test for additivity and to produce more stabilized receptor variants.

The alanine-screening methodology was successfully applied to three different GPCRs. In the case of the adenosine receptor A_{2A}, two different thermostabilized mutants were constructed [17]. For the mutant A2a-rant21, the melting temperature (T_m) was 17 °C higher than that of the wild-type, for the mutant A2a-rag23 the improvement was 9 °C. Each receptor shows four mutations. Interestingly, the two mutants seem to be stabilized in different conformations: A2a-rant21 preferentially adopts an antagonist-binding state, A2a-rag23 prefers an agonist-binding state. As a second receptor, NTR1 was stabilized as a result of four mutations [15]. The NTR1 mutant termed "NTS1-7m" was 17 °C more stable than wild-type NTR1. Lastly, the turkey β_1-adrenergic receptor was stabilized by combining six mutations

[16]. The mutant b_1AR-m23 was 23 °C more stable than the wild-type protein. This mutant was then crystallized in octylglucoside and its atomic resolution structure was determined in the presence of the antagonist cyanopindolol.

The high success rate of identifying stabilizing mutations in both scanning mutagenesis strategies, with alanine scanning or cysteine scanning, are encouraging in that stabilizing mutations appear to be very frequent. However, both strategies suffer form a basic limitation. Their application is limited to those membrane proteins that show a relatively high wild-type expression level. This excludes many membrane proteins that fail to be expressed above a certain threshold level. Furthermore, the linear scanning against one type of amino acid will only detect positions where an unfavorable amino acid needs to be removed. Since a full screen is not carried out against all substitutions, amino acids able to make a new interaction, or those filling out a cavity better, would not be discovered. Since these methods are screening methods where each mutant is expressed and solubilized separately, it would hardly be feasible to increase the throughput to the required scale.

10.3.3
Stabilizing GPCRs by Random Mutagenesis and Screening in a 96-Well Assay Format

In Section 10.2.1 we described a high-throughput selection method with the main goal of identifying *expression*-enhancing mutations in GPCRs. An advantage of this strategy is that it generates large sets of well-expressing and functional mutants. We wondered if there was a quick and reliable way of screening these collections of mutants for their potential of increasing thermal stability in addition to increasing expression. As we wanted to perform the stability screen on as many receptor mutants as possible, we had to revise the LBA, which is the rate-limiting step of the conventional stability screening method for GPCRs (which had conventionally been employed in the alanine-scanning method by Tate [15–17] and previously by us [5]). In the conventional LBA method, each sample has to be processed by a small size-exclusion column (typically a spin column) to separate bound from unbound ligand and assess the ligand-binding signal, since solubilized receptor cannot be quantitatively bound to filters. This spin-column step cannot be performed in a 96-well assay format and therefore strongly limits the assay throughput. The key feature of the newly developed method [19] presented below is the immobilization of biotinylated receptor on streptavidin-coated paramagnetic beads. By immobilizing the receptor, all essential experimental steps of the stability screen – purification, exposure to heat, and LBA – can be performed with small receptor amounts and in a highly parallelized 96-well format. Immobilized receptor can easily be separated from detergent-solubilized lysates of whole cells by magnetic force, which yields highly concentrated and purified receptor preparations. Most importantly, magnetic capturing also allows for a convenient separation of bound from unbound ligand in the LBA, which avoids the handling of size exclusion spin columns. All essential steps can therefore be performed in a 96-well assay format [19].

The method for screening thermal stability of GPCRs in a 96-well assay format [19] consists of four steps. (i) The receptor mutants are expressed in *E. coli* and biotinylated *in vivo*. (ii) The receptors are solubilized and partially purified by immobilization on streptavidin-coated paramagnetic beads. (iii) The receptors are exposed to stability screening conditions that induce receptor unfolding (e.g., heat, detergent, buffer). (iv) The amount of residual folded receptor is determined by a LBA after exposing the receptor to the stability screening conditions. Comparison of the amount of correctly folded receptor before and after heat treatment yields a stability index for each mutant. The stability index is calculated by dividing the residual amount of receptor that has been exposed to the harsh conditions (e.g., high temperature or specific detergent) by the initial amount that is determined from an aliquot of nonexposed beads.

We applied this method to 96 randomly picked clones of a library of NTR1 mutants, which had been evolved for higher expression by FACS [5]. The mutants had gone through four rounds of random mutagenesis and after each round they had been sorted for highest expression by FACS for three to four rounds. The selected mutants showed on average a 5-fold higher functional expression level than wild-type NTR1 as a result of an average of nine amino acid mutations per receptor. For screening their stability, the mutants were expressed in 24-well plates, solubilized in the presence of a detergent mixture containing dodecyl maltoside (DDM), 3-(3-cholamidopropyl)-dimethylammoniopropane sulfonate (CHAPS), and cholesteryl hemisuccinate (CHS), and immobilized on streptavidin-coated magnetic beads in a 96-well plate. After washing the beads with the help of magnetic capturing, equal amounts of beads of each purified mutant were dispensed into two 96-well thermocycler plates. One plate was kept on ice. The second plate was placed in a thermocycler and exposed to 37 °C for 20 min to induce receptor inactivation. After cooling the heated plate on ice, a radioligand binding assay was performed on both plates to determine the amount of correctly folded receptor for each mutant before and after the heat inactivation step. After incubation with radioligand, unbound radioligand can be separated from bound ligand by magnetic capturing of the beads. A major advantage of this new screening method is that the LBA is performed directly in the wells of a thermocycler plate. For measuring the radioactivity the beads are simply transferred to a 96-well plate containing liquid scintillation cocktail.

Two screens were performed on the 96 mutants to identify the most stable mutants in either of the detergents, DDM or decyl maltoside (DM). Sixteen mutants showed higher thermostability than wild-type NTR1 in both detergents. Figure 10.4 shows that the difference in stability between the mutants and wild-type NTR1 was more pronounced in DM (Figure 10.4b) than in DDM (Figure 10.4a). The two clones that performed best in the screens (clones 70 and 73) were further characterized to get a better picture of the stability improvement. Melting curves in Figure 10.4(c) were recorded by exposing the mutants in DM, CHAPS, and CHS to different temperatures for 20 min. For both mutants the melting temperature T_m was increased by 6 °C compared to wild-type NTR1. The relatively moderate increase in stability most probably reflects the fact that the original pool

a)

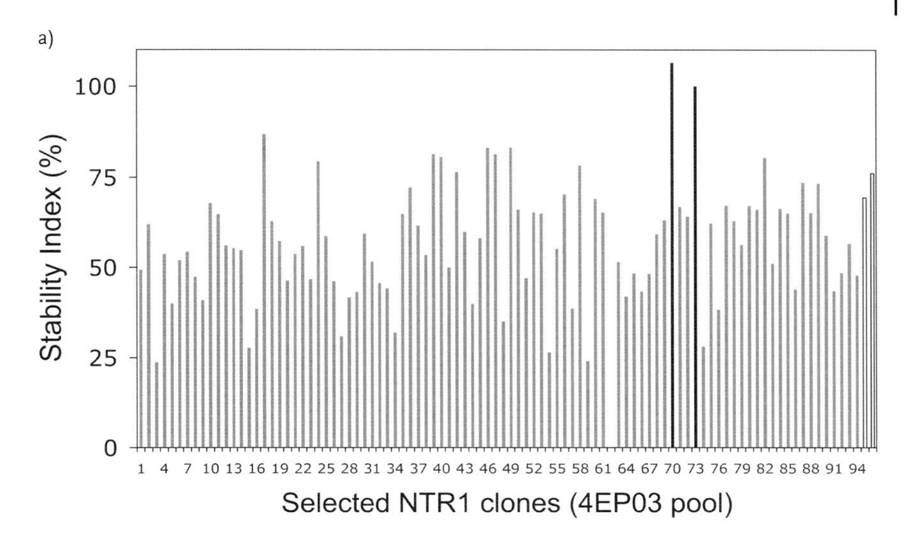

b)

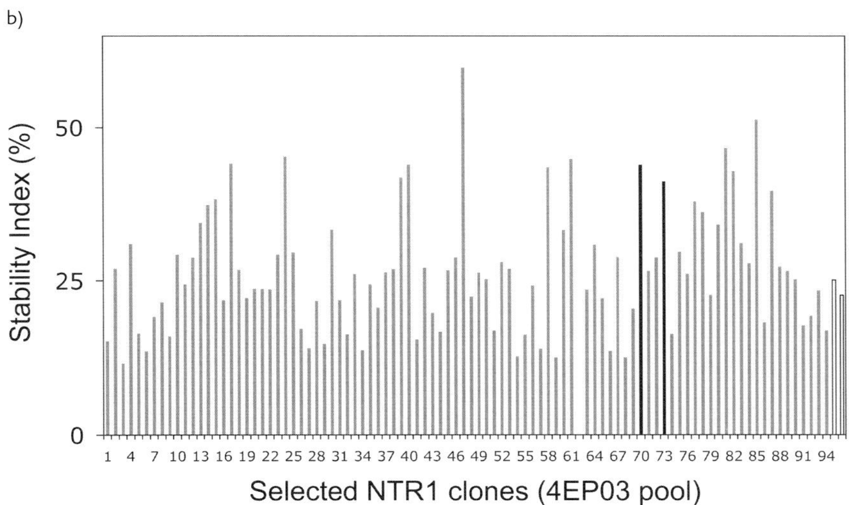

Figure 10.4 Thermal stability analysis of 96 clones isolated from selections for maximum receptor expression level (4EP03 pool, four rounds of epPCR, each followed by three to four rounds of sorting by FACS). (a and b) To identify the most thermostable clones, the stability index was measured at 37 °C in (a) DDM, CHAPS, and CHS or (b) DM, CHAPS, and CHS. The stability index indicates the fraction of receptors that retain the ability to bind ligand after exposure at 37 °C for 20 min (relative to receptors which are kept on ice). The open bars at positions 95 and 96 indicate duplicate measurements of the stability index of wild-type NTR1. The two black bars indicate the stability index of clone 70 and clone 73, which have been characterized in more detail. (c) The results from the screen were confirmed by reanalyzing the stability of clone 70 and clone 73 in more detail. Receptor aliquots (in DM, CHAPS, and CHS) were exposed to increasing temperatures and the fraction of intact receptors, which still binds radioligand, was measured. The resulting stability curves show that the evolved clone 70 (open triangles) and clone 73 (open circles) show a T_m (temperature at which 50% of the receptors retain ligand binding) that is 6 °C higher than for wild-type NTR1 (filled circles).

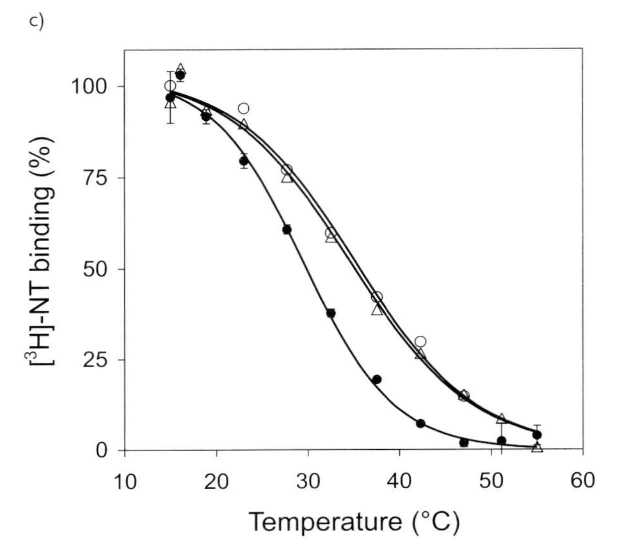

Figure 10.4 (*Continued*)

of selected mutants was strongly evolved for higher expression and not stability *per se*. In the meantime, however, from additional rounds of random mutagenesis and selection for functional expression, much higher stability increases have been found than described above or elsewhere [15] (Schlinkmann *et al.*, unpublished), suggesting that functional expression levels of GPCRs and their stability are coupled.

The principal advantage of this stability screening method based on immobilized receptors on paramagnetic beads is its high throughput. This allows for the parallelization of the rate-limiting steps in stability testing in a 96-well format, such as the LBA. Moreover, the immobilized receptors can easily be purified and exchanging detergents is straightforward, when mutants need to be screened in different detergents.

10.4 Conclusions

The presented methods for engineering the properties of IMPs all follow the same idea, namely that small changes to the amino acid sequence of a difficult-to-handle IMP can have major favorable effects. Functional expression as well as thermal stability can be increased by relatively few mutations, and can be screened and selected by comparatively simple methods. The ability to evolve functional expression level for IMPs that are very difficult to express removes one of the biggest roadblocks in IMP structural studies. Similarly, the possibility to generate IMP variants that are more stable in a variety of detergents will be important for the future structural biology of IMPs. Two results give reason for optimism: (i) stabilizing mutations do not seem to be rare, and (ii) functional expression and stability seem to be correlated to some extent. Nonetheless, the long-term goal from applying these methods is not only to use these methods for solving the practical problem of generating IMPs that can be studied structurally more efficiently, but also apply them to understand the rules that make membrane proteins well expressed and more stable.

Abbreviations

BODIPY-NT	BODIPY-neurotensin
CHAPS	3-(3-cholamidopropyl)-dimethylammoniopropane sulfonate
CHS	cholesteryl hemisuccinate
CoFi	colony filtration
DDM	dodecyl maltoside
DGK	diacylglycerol kinase
DM	decyl maltoside
epPCR	error-prone PCR
FACS	florescence activated cell sorting
GPCR	G-protein-coupled receptor
IMP	integral membrane protein
LBA	radioligand binding assay
MBP	maltose-binding protein
MFI	mean fluorescence intensity
TrxA	thioredoxin A

References

1 Stemmer, W.P. (1994) Rapid evolution of a protein *in vitro* by DNA shuffling. *Nature*, **370**, 389–391.

2 Zaccolo, M., Williams, D.M., Brown, D.M., and Gherardi, E. (1996) An approach to random mutagenesis of DNA using mixtures of triphosphate derivatives of nucleoside analogues. *J. Mol. Biol.*, **255**, 589–603.

3 Virnekäs, B., Ge, L., Plückthun, A., Schneider, K.C., Wellnhofer, G., and Moroney, S.E. (1994) Trinucleotide phosphoramidites: ideal reagents for the synthesis of mixed oligonucleotides for random mutagenesis. *Nucleic Acids Res.*, **22**, 5600–5607.

4 Molina, D.M., Cornvik, T., Eshaghi, S., Haeggstrom, J.Z., Nordlund, P., and Sabet, M.I. (2008) Engineering membrane protein overproduction in *Escherichia coli*. *Protein Sci.*, **17**, 673–680.

5 Sarkar, C.A., Dodevski, I., Kenig, M., Dudli, S., Mohr, A., Hermans, E., and Plückthun, A. (2008) Directed evolution of a G protein-coupled receptor for expression, stability, and binding selectivity. *Proc. Natl. Acad. Sci. USA*, **105**, 14808–14813.

6 Tucker, J. and Grisshammer, R. (1996) Purification of a rat neurotensin receptor expressed in *Escherichia coli*. *Biochem. J.*, **317**, 891–899.

7 Zhao, H., Giver, L., Shao, Z., Affholter, J.A., and Arnold, F.H. (1998) Molecular evolution by staggered extension process (StEP) *in vitro* recombination. *Nat. Biotechnol.*, **16**, 258–261.

8 Barroso, S., Richard, F., Nicolas-Etheve, D., Kitabgi, P., and Labbe-Jullie, C. (2002) Constitutive activation of the neurotensin receptor 1 by mutation of Phe^{358} in helix seven. *Br. J. Pharmacol.*, **135**, 997–1002.

9 Labbe-Jullie, C., Barroso, S., Nicolas-Eteve, D., Reversat, J.L., Botto, J.M., Mazella, J., Bernassau, J.M., and Kitabgi, P. (1998) Mutagenesis and modeling of the neurotensin receptor NTR1. Identification of residues that are critical for binding SR 48692, a nonpeptide neurotensin antagonist. *J. Biol. Chem.*, **273**, 16351–16357.

10 Middleton, R.J. and Kellam, B. (2005) Fluorophore-tagged GPCR ligands. *Curr. Opin. Chem. Biol.*, **9**, 517–525.

11 Daly, C.J. and McGrath, J.C. (2003) Fluorescent ligands, antibodies, and proteins for the study of receptors. *Pharmacol. Ther.*, **100**, 101–118.

12 Cornvik, T., Dahlroth, S.L., Magnusdottir, A., Herman, M.D., Knaust, R., Ekberg, M., and Nordlund, P. (2005) Colony filtration blot: a new screening method for soluble protein expression in *Escherichia coli*. *Nat. Methods*, **2**, 507–509.

13 Lau, F.W., Nauli, S., Zhou, Y., and Bowie, J.U. (1999) Changing single side-chains can greatly enhance the resistance of a membrane protein to irreversible inactivation. *J. Mol. Biol.*, **290**, 559–564.

14 Zhou, Y. and Bowie, J.U. (2000) Building a thermostable membrane protein. *J. Biol. Chem.*, **275**, 6975–6979.

15 Shibata, Y., White, J.F., Serrano-Vega, M.J., Magnani, F., Aloia, A.L., Grisshammer, R., and Tate, C.G. (2009) Thermostabilization of the neurotensin receptor NTS1. *J. Mol. Biol.*, **390**, 262–277.

16 Serrano-Vega, M.J., Magnani, F., Shibata, Y., and Tate, C.G. (2008) Conformational thermostabilization of the beta1-adrenergic receptor in a detergent-resistant form. *Proc. Natl. Acad. Sci. USA*, **105**, 877–882.

17 Magnani, F., Shibata, Y., Serrano-Vega, M.J., and Tate, C.G. (2008) Co-evolving stability and conformational homogeneity of the human adenosine A2a receptor. *Proc. Natl. Acad. Sci. USA*, **105**, 10744–10749.

18 Warne, T., Serrano-Vega, M.J., Baker, J.G., Moukhametzianov, R., Edwards, P.C., Henderson, R., Leslie, A.G., Tate, C.G., and Schertler, G.F. (2008) Structure of a beta1-adrenergic G-protein-coupled receptor. *Nature*, **454**, 486–491.

19 Dodevski, I. and Plückthun, A. (2011) Evolution of three human GPCRs for higher expression and stability. *J. Mol. Biol.*, in press.

11
Expression and Purification of G-Protein-Coupled Receptors for Nuclear Magnetic Resonance Structural Studies

Fabio Casagrande, Klaus Maier, Hans Kiefer, Stanley J. Opella, and Sang Ho Park

11.1
Introduction: G-Protein-Coupled Receptor Superfamily

G-protein coupled receptors (GPCRs), also known as seven-transmembrane receptors, represent the largest and most diverse protein family in the human genome [1–3]; because of their pervasive roles as drug receptors, they may also be the most important protein family for medicine. Embedded in the plasma membrane, they mediate cellular response by activating heterotrimeric guanine nucleotide-binding proteins (G-proteins) upon extracellular ligand stimulation. Defects in signal transduction result in human diseases like cancer, and neurodegenerative, autoimmune and inflammatory disorders [4, 5]. Currently, one-third of the drugs approved for human use target GPCRs [6, 7]. However, these drugs address only a few of the potential candidate receptors within the GPCR superfamily, such as the histamine receptor (Zantac, Tagamet, Benadryl, Claritin), dopamine receptor (Haloperidol), acetylcholine receptor (Scopolamine, Tropicamide), and adrenergic receptor (Sotalol, Timolol, Atenolol, Propranolol). Genome analysis has identified 100 (non-olfactory) newly recognized orphan GPCRs without established functions, which highlights the potential of the GPCR superfamily for the development of new therapeutic targets for drug discovery [8].

The discovery and development of drugs that act upon GPCRs is hindered by the lack of information about the three-dimensional structures of these proteins. Atomic-resolution structures of GPCRs would provide the essential details of how ligands interact with specific residues, improving the prospects for structure-based drug discovery. In addition, these structures would advance the understanding of how they function as well as how they interact with their respective ligands.

However, GPCRs are large, hydrophobic membrane proteins that function in the liquid crystalline phospholipid bilayer of the cell membrane, and the structure determination of membrane proteins is very challenging due to problems involving their instability, lack of solubility, and difficulty in crystallization. Although several recent crystal structures of GPCRs demonstrate that these obstacles can be overcome, they required extraordinary measures to modify the proteins to

Production of Membrane Proteins: Strategies for Expression and Isolation, First Edition.
Edited by Anne Skaja Robinson.

facilitate crystallization and the use of nonbilayer environments. This highlights the fundamental advantage of nuclear magnetic resonance (NMR) for structural studies of membrane proteins, which is that it enables both wild-type and mutated polypeptides to be analyzed in their native environment of the hydrated planar phospholipid bilayer under physiological conditions. As the proteins are not captured in three-dimensional crystals at very low temperatures, NMR offers opportunities to integrate descriptions of their local and global protein dynamics and ligand interactions with their primary, secondary, and tertiary structures. X-ray crystallography and NMR spectroscopy are complementary approaches. Both experimental methods require substantial quantities of highly purified protein for sample preparation trials and the samples themselves, and in the case of NMR the protein must be labeled with stable isotopes as well. Thus, there are great advantages in terms of flexibility of labeling and cost to express the protein in bacteria.

Consequently, our research and the topic of this chapter focus on bacterial expression of GPCRs. In particular, we use CXCR1 as an example of a rhodopsin-like GPCR that has an extensive background of biochemical and biophysical studies. We describe the cloning, expression, purification, functional reconstitution, and preparation of samples suitable for both solution NMR and solid-state NMR spectroscopy. Sodium dodecyl sulfate–polyacrylamide gel electrophoresis (SDS–PAGE) and functional binding assays demonstrate that the methods we utilize yield highly pure and active receptors in the multimilligram quantities required for NMR spectroscopy. Examples of both solution NMR and solid-state NMR spectra of ^{15}N-labeled proteins demonstrate that all steps from expression through sample preparation are sufficiently well developed for NMR structural studies to be initiated.

11.2
CXCR1

CXCR1 is a chemokine receptor, and is the example we use to illustrate the expression and labeling of a GPCR for NMR studies. Chemokines are small chemotactic cytokines that control a wide variety of biological and pathological processes such as allergic responses, infectious and autoimmune diseases, angiogenesis, inflammation, tumor growth, and hematopoietic development. Secreted chemokines interact with G-protein-coupled chemokine receptors expressed on the surface of their target cells. The first two chemokine receptors to be cloned were CXCR1 and CXCR2 [9, 10], named according to the CXC chemokine subfamily they bind. By interacting with the human chemokine ligand interleukin-8 (IL-8, CXCL8, NAP-1), CXCR1 and CXCR2 mediate cell migration and activation of polymorphonuclear neutrophils (PMN s) [11, 12]. IL-8-induced chemotaxis directs PMNs to sites of inflammation, where IL-8 is secreted by macrophages, lymphocytes, and epithelial and endothelial cells. Human CXCR1 and CXCR2 share 77% overall sequence identity, but differ in their N- and C-terminal domains (the ligand-binding regions).

Binding experiments in neutrophils showed that IL-8 affinity is decreased in CXCR1, whereas CXCR2 shows high affinity for IL-8 and other analogs such as macrophage inflammatory protein (MIP)-2, neutrophil-activating peptide (NAP)-2, CXCL1/GROα, and CXCL5/ENA-78 [13–16]. As a result, it is suggested that CXCR1 mediates CXCL8-induced chemotaxis at sites of inflammation (high IL-8 concentration), whereas CXCR2 participates in the initial phase of cell migration (low IL-8 concentration) [17]. By recruiting and activating PMN, IL-8 and related molecules have been implicated in a wide range of chronic inflammatory disorders. Apart from repertaxin, an allosteric inhibitor of CXCR1 and CXCR2 [18–20], no small molecular antagonist for chemokine receptors has been analyzed in advanced clinical trials. Therefore, any structural information disclosing the mode of action of ligand and receptor would be extremely valuable.

11.3 GPCR Structures

For almost a decade, the unique structure of the bovine photoreceptor rhodopsin in its dark, inactive state [21] provided the sole example of a structure of a GPCR determined at atomic resolution by X-ray crystallography. Efforts to model GPCRs on the structure of rhodopsin had limited success with regard to drug discovery [22]. On the one hand, rhodopsin is a highly specialized member within the GPCR family that responds to photons and not chemical ligands. On the other hand, highest homology among GPCRs is concentrated within the bundle of transmembrane helices, but very low in the residues that constitute the N- and C-termini and loop domains that interact with corresponding ligands or potential drugs.

However, several new GPCR structures have been determined in the last few years by X-ray crystallography: the human β_2-adrenergic receptor in the presence of the inverse agonist carazolol [23–25], the β_1-adrenergic receptor in complex with the high-affinity antagonist cyanopindolol [26], and the human A_{2A} adenosine receptor with a high-affinity subtype-selective antagonist [27]. The successful crystallization of these GPCRs was mainly enabled by a fusion protein strategy first described for the crystallization of the lactose permease [28–30]. The crystallization of the β_2-adrenergic and A_{2A} adenosine receptor were improved by replacing the residues of the disordered intracellular loop 3 by the sequence of the protein T4 lysozyme, which is known to readily crystallize. Additional sequence modifications, such as truncations of unstructured termini or mutations affecting thermal stability, ligand binding, and post-translational modifications, were essential in order to obtain high-diffracting three-dimensional crystals. Although it was necessary to modify the GPCRs for crystallization, comparisons among a total of four different GPCR structures provide many valuable biological insights and details about the common and variable features of their molecular architecture [31, 32]. During the same time period, insights into the activated state of rhodopsin were gained from the crystal structure of opsin in its ligand-free state [33] and in its G-protein-interacting conformation [34].

11.4
NMR Studies of GPCRs

In principle, NMR spectroscopy is a powerful approach for the study of all proteins, including membrane proteins. However, all studies of membrane proteins require careful consideration of the lipid environment needed to solubilize and stabilize them in their folded, active conformation. It is possible to study the structure and dynamics of membrane proteins solubilized in micelles or isotropic bicelles by solution NMR, although there is always concern about their properties outside of the native bilayer environment. Fortunately, solid-state NMR enables proteins in phospholipid bilayers to be studied. While solution NMR methods can be applied to small membrane proteins in micelles and isotropic bicelles, applications to membrane proteins with more than a few transmembrane helices are limited by the slow reorientation rates of the larger proteins in micelles, which result in broad linewidths, and correspondingly poor resolution and sensitivity. However, it is possible to observe signals from mobile residues in loop or terminal regions of these proteins in solution NMR spectra, as has been demonstrated for the detergent-solubilized vasopressin V2 receptor [35].

Numerous NMR studies on fragments of GPCRs, including N- and C-terminal regions, interhelical loops, and transmembrane domains have been summarized in a recent review [36]. The results provide information about the structure and function of the various fragments, the mapping of receptor–ligand interactions, and the conformational changes associated with the activation of the GPCR fragments. Dynamics and mobility studies on the C-terminus of full-length rhodopsin have been reported [37–39]. Recently, dimethyllysine NMR spectroscopy has demonstrated ligand-specific conformational changes of the full-length β_2-adrenergic receptor [40].

Solid-state NMR offers the possibility of studying all types of membrane proteins in planar phospholipid bilayers, since these methods do not rely on the overall rotational motion of the molecule to average the nuclear spin interactions and yield narrow resonance lines. Even proteins with only a single transmembrane helix are fully immobilized by their interactions in the phospholipid bilayers. Thus, in solid-state NMR radiofrequency irradiations replace the overall reorientation of the proteins as the line-narrowing mechanism, and once that is accomplished the linewidths of resonances from both small and large membrane proteins are equivalent. Of course, larger proteins have many more signals, and both more sophisticated isotopic labeling schemes and multidimensional spectroscopic methods are employed to resolve overlapping signals [41–44].

There are two approaches to protein structure determination by solid-state NMR spectroscopy: magic angle spinning (MAS) solid-state NMR and oriented sample (OS) solid-state NMR. MAS solid-state NMR techniques yield high-resolution isotropic spectra of unoriented membrane proteins, and have been applied to investigate the ligand-induced structural change of rhodopsin, the conformation of ligand bound to its GPCR, and receptor–ligand interactions [45–51]. OS solid-state NMR is particularly well suited for proteins in supramolecular assemblies, especially membrane proteins in planar phospholipid bilayers, whose structures are extended and dominated by regular secondary structure elements, and generally not well suited for isotropic approaches. In OS solid-state NMR, the frequencies observed in the spectra reflect the orientations of individual atomic sites with respect to a common axis of alignment defined by the direction of the applied magnetic field, and therefore the angular constraints are very strong and any experimental errors or uncertainties in spectroscopic parameters are not cumulative, contributing to the robustness of the approach. Three-dimensional structures and receptor–drug interactions of membrane proteins with one and two transmembrane helices have been determined using OS solid-state NMR [42, 44, 52, 53]. High-resolution spectra of the chemokine receptor CXCR1 in magnetically aligned bilayers demonstrated the feasibility of using OS solid-state NMR to determine the three-dimensional structure and address structure–activity relationships for the development of drugs [44].

11.5 Expression Systems

The recombinant production of membrane proteins is now sufficiently well developed that it can be routinely applied to nearly any GPCR. Getting to this point required extensive development in a number of laboratories over a long period of time, and we take advantage of the many ideas and results in the literature. In response to the difficulties encountered initially in expressing functional GPCRs, a broad selection of organisms have been utilized, including bacteria [54–58], yeast [59], insect cells [60], and mammalian cell cultures [61].

Bacterial expression, especially in *Escherichia coli*, offers distinct advantages with respect to simplicity and safety of laboratory procedures, ease and cost of isotopic labeling, and the availability of a variety of strains and the possibilities of well-established methods for genetic manipulation to optimize conditions for expression and purification. As usual for eukaryotic proteins expressed in bacteria, all post-translational modifications are absent. While the functionality of many GPCRs is unaffected by the absence of glycosylation, other GPCRs showed modified affinities towards ligands and G-proteins, which, however, is not indicative of misfolding. Therefore, the lack of post-translational modifications can be advantageous for structural studies where homogeneous samples are an important prerequisite, although caution in interpretation of functional studies is always warranted. A very common strategy to express GPCRs in *E. coli* is the use of fusion proteins. Functional GPCRs have been successfully expressed using the N-terminal fusion to the *E. coli* maltose-binding protein that targets the protein to the periplasmic membrane [62–67]. The disadvantage of this approach is toxic effects associated with membrane insertion of heterologously overexpressed proteins often limit the cell growth. To overcome this problem GPCRs can be expressed in fusion with ketosteroid isomerase or glutathione S-transferase (GST), which improves expression levels by targeting the expressed protein to inclusion bodies [55, 57]. Yields of 2–20 mg/l cell culture can be obtained with this approach. The principal disadvantage of this approach is that proteins expressed in bacterial inclusion bodies have to be refolded, which can now be accomplished routinely using a standard approach.

Although we utilize bacterial expression of GPCRs, for completeness we mention parallel developments of the expression systems in other organisms. At first glance yeast systems present an excellent combination of microbial growth with the availability of the eukaryotic cell compartmentation of the protein production and processing machinery. Thus, the incorporation of isotopes is feasible while most post-translational modifications are assisted. However, in contrast to bacteria, yeast strains such as *Saccharomyces cerevisiae, Saccharomyces pombe,* and the methylotrophic *Pichia pastoris* are much more delicate organisms, and are sensitive to minor changes in their environment. Therefore, cell growth as well as functional yields of produced GPCRs vary and have to be individually optimized by screening culture conditions. Although yeast is eukaryotic, the membrane composition differs from that of the mammalian cell membrane. As a result a decrease in functionality has been reported for some GPCRs [68]. An example of GPCR production in yeast was recently demonstrated by the expression and purification of high levels of the human adenosine A_{2A} receptor [69–71]. In addition, the number of expressed GPCRs in yeast is continuously increasing, indicating the feasibility of this method [72–74]. Insect and mammalian cell culture-based expression systems can be attractive alternatives because active and functional GPCRs are produced. On the one hand, these native-like GPCRs are excellent for functional studies; on the other hand, the potential heterogeneity due to variable post-translational modifications can make structural characterization by all methods more difficult. In addition, these systems are costly, difficult to handle, and often yield relatively small amounts of the GPCRs. Similar to the procedure in yeast, a time-consuming screening for a preferable insect or mammalian cell line as well as the refinement of expression parameters is necessary in order to obtain significant quantities of GPCRs for structural studies. Nevertheless, due to the increased popularity of insect and mammalian expression systems, their applicability is being improved by the commercial availability of a wide variety of vectors, media, and reagents. Several GPCRs have been reported to be expressed from insect and mammalian cell hosts in amounts appropriate for structural determination [75–78]. Finally, novel cell-free expression systems present another method of protein synthesis without the usage of living cells. While isotopic labeling is facilitated [79], the expression of membrane proteins is complex. Membrane mimicking liposomes or disks have to be added to stabilize freshly produced membrane proteins [80]. Unfortunately, all attempts to express GPCRs by cell-free translation resulted in precipitates that required the resolubilization with detergents [81, 82]. Apart from that, the most difficult step is the upscaling of cell-free systems that are usually performed in reaction volumes less than 1 ml. Future studies will prove

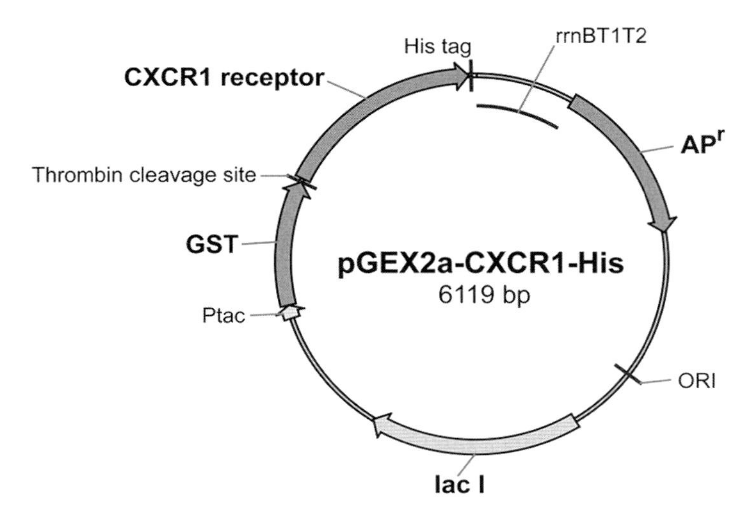

Figure 11.1 Map of expression vector pGEX2a-CXCR1-His. The CXCR1 gene is inserted immediately after the GST to express both proteins as CXCR1–GST fusion. The *lacI* repressor gene binds to the Ptac promoter and represses expression of the CXCR1–GST fusion until induction with IPTG. The thrombin cleavage site is located between the GST carrier protein and CXCR1, thus the GST moiety can be removed.

whether cell-free expression is applicable for the production of membrane proteins for structural studies.

Although expression in yeast, insect cells, and mammalian cell culture is becoming more practical, bacterial expression as described below has a big advantage in terms of isotopic labeling for NMR studies.

11.6 Cloning of CXCR1 into pGEX2a

DNA sequences coding for human full-length and truncated CXCR1 constructs were cloned into the pGEX2a vector. The polymerase chain reaction-amplified DNA fragments were gel-purified, digested by restriction enzymes (*Bam*HI, *Sam*I), and ligated to compatible restriction sites of pGEX2a. GST was fused to the N-terminus to enhance expression and to direct the expressed proteins to inclusion bodies. A thrombin cleavage site was introduced between the two regions encoding GST and CXCR1 to enable proteolytic removal of the GST. For purification by immobilized metal-chelating chromatography, a His_6 tag was appended to the C-terminus of the CXCR1 constructs. Expression is under control of the isopropyl-β-D-thiogalactopyranoside (IPTG)-inducible Ptac promoter (Figure 11.1). DH5 α-competent *E. coli* cells were used for subcloning and pGEX2a-CXCR1-His-transformed *E. coli* BL21 cells were used for expression.

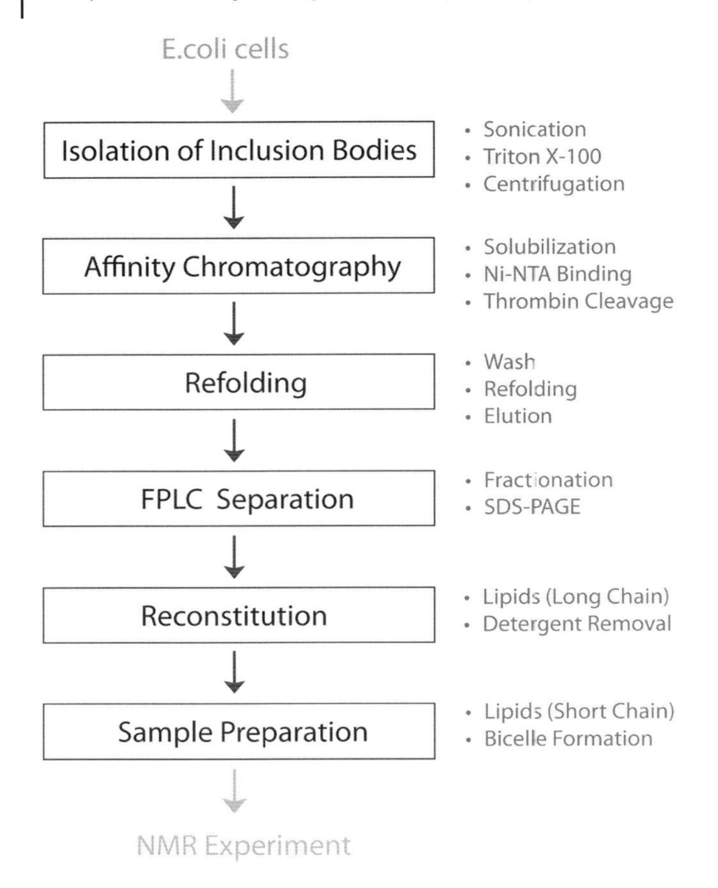

Figure 11.2 Flowchart describing the production of CXCR1 protein for NMR spectroscopy. CXCR1 is expressed in inclusion bodies that are isolated and solubilized in order to purify CXCR1. Refolding conditions are applied and functional CXCR1 is separated by FPLC. CXCR1 is reconstituted into proteoliposomes that present the starting material for NMR sample preparation.

11.7 Expression of CXCR1

The expression yields of CXCR1 are maximized using a GST-fusion approach, since large amounts of CXCR1–GST are accumulated in bacterial inclusion bodies without interfering with the normal membrane function of the cells. The flowchart in Figure 11.2 indicates the major steps in expression intended for sample preparation for NMR spectroscopy. Although individual NMR samples can be prepared from a few milligrams of protein, many milligrams are needed in the course of sample and experimental optimization. Also, expression yields need to be optimized to make the most efficient use of the various isotopically labeled growth media.

We express uniformly ^{15}N-labeled CXCR1 by culturing *E. coli* BL21 in M9 minimal media containing $^{15}NH_4Cl$ as the sole nitrogen source. Therefore, a 25-ml preculture of standard LB medium with 50 mg/l of carbenicillin was inoculated with single colonies of *E. coli* BL21 cells freshly transformed with pGEX2a-CXCR1-His and shaken overnight at 37 °C. Then 1 l of M9 minimal medium containing 1 g/l $^{15}NH_4Cl$ was inoculated with the total 25 ml of overnight preculture and shaken at 37 °C until the OD_{600} reached 0.5 (typically after around 4 h). At this time expression was induced with a final concentration of 100 μM IPTG. After 4 h of incubation, the cells were harvested by centrifugation. To express selectively ^{15}N labeled CXCR1, M9 media (unlabeled NH_4Cl) was supplemented with an amino acid mixture containing one single ^{15}N-labeled amino acid. *E. coli* cells growing in M9 media enriched with amino acids reached an OD_{600} of 0.5 more rapidly. In addition, the growth was harvested 2 h after IPTG induction to avoid scrambling of the ^{15}N-labeled amino acids.

SDS is used to solubilize the CXCR1–GST fusion protein efficiently from the inclusion bodies. The detergent-solubilized CXCR1–GST fusion protein is no longer aggregated and can be purified. As inclusion bodies contain mainly the overexpressed fusion proteins, a single-step purification using immobilized metal-affinity chromatography is often sufficient to yield highly purified protein. Whereas the purification and GST removal is straightforward, the goal is to refold the solubilized CXCR1 into its native structure, regaining full biological activity, and this must be done carefully. Protein folding results from the replacement of the solubilizing detergent by a milder detergent/lipid mixture that stabilizes the native fold of the protein while still keeping it in solution and preventing aggregation. Identification of the optimal refolding parameters including detergents, salts, pH conditions, and additives differs among GPCR constructs. To achieve the highest protein homogeneity, unfolded CXCR1 is removed by fast protein liquid chromatography (FPLC). CXCR1 obtained by this procedure is solubilized in detergent micelles that mimic a phospholipid bilayer environment in some ways. However, a much more native-like environment for membrane proteins are liposomes composed of natural or synthetic phospholipids. Therefore, detergent-solubilized CXCR1 is reconstituted into proteoliposomes by detergent removal. The lipid compositions of the proteoliposomes are then adjusted to produce the samples used for the NMR experiments.

In addition to the full-length receptor, essentially the same approach is applied on CXCR1 truncations and fragments. The study of receptor fragments can provide valuable insights into protein dynamics, receptor–ligand interactions, and conformations of individual domains.

11.8 Purification

The harvested cell pellets were resuspended 20 ml lysis buffer (20 mM Tris-HCl pH 8.0, 300 mM NaCl, 10% glycerol) and lysed with 10 mg/ml of lysozyme by

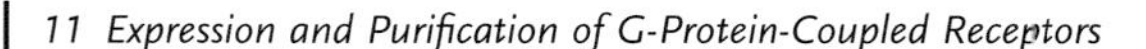

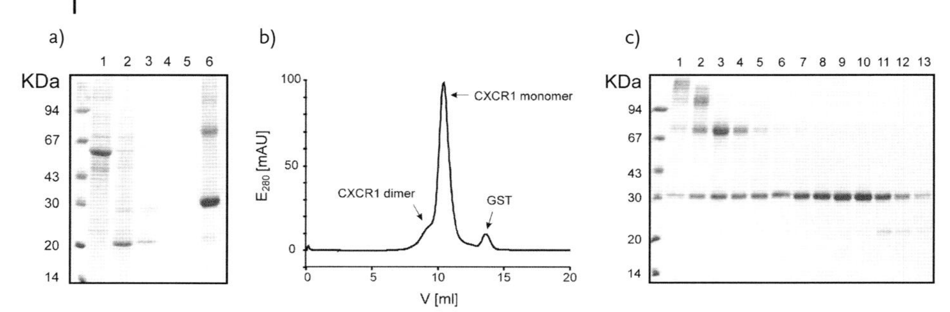

Figure 11.3 SDS–PAGE and FPLC analysis of His-tagged CXCR1 purification. (a) SDS–PAGE of different Ni-NTA affinity chromatography CXCR1 purification steps: inclusion bodies of CXCR1–GST fusion protein (lane 1), thrombin protease cleavage product (lane 2), Ni-NTA flow-through (lane 3) and Ni-NTA wash (lane 4 and 5), and CXCR1 elution (lane 6). (b) Gel-filtration chromatography of Ni-NTA-purified CXCR1. The profile indicates a prominent elution peak for monomeric CXCR1, and a minor peak for dimeric CXCR1 and contaminating GST. (c) SDS–PAGE of collected fractions of size-exclusion chromatography (lanes 1–13).

sonication on ice for 5 min (5 s on and 10 s off). Membrane fragments were solubilized by addition of 10 mg/ml of Triton X-100 and stirring for 10 min. Inclusion bodies were then collected by centrifugation at 20 000 g for 10 min at 16 °C and washed twice with lysis buffer containing 0.5% Triton X-100. The washed and centrifuged inclusion bodies were stored at −80 °C or solubilized directly in binding buffer (1% SDS, 1× phosphate-buffered saline, pH 7.5, 250 mM NaCl, and 0.15 tris(2-carboxyethyl)phosphine) by sonication on ice for 5 min (5 s on and 10 s off). The solution was centrifuged at 40 000 g for 30 min at 16 °C to remove insoluble particles, and the supernatant was loaded onto a chromatography column containing Ni-NTA agarose (Qiagen; 20 ml for 10-ml inclusion bodies suspension) equilibrated with binding buffer. The column was then washed with five column volumes of binding buffer. Subsequently, the column buffer is exchanged by washing with 10-bed volumes of thrombin cleavage buffer (0.1% hexadecyl-phosphocholine (FOS-16; Anatrace), 20 mM Tris-HCl, pH 8.0, and 250 mM NaCl). Then, 2000 U of thrombin are dissolved in 20 ml thrombin cleavage buffer onto the column that is sealed and gently agitated at room temperature overnight. To cleave off the GST fusion protein the exchange to a milder detergent is necessary to keep thrombin in its functional state. See Figure 11.3.

11.9
Refolding and Reconstitution

To induce refolding of CXCR1, detergent is exchanged to 0.5% dodecylphosphocholine (DPC, FOS-12; Anatrace) in 20 mM HEPES buffer, pH 7.2,

150 mM NaCl. Refolded CXCR1 was eluted from the column with the same buffer containing 400 mM imidazole. The eluate contained between 15–20 mg of recombinant CXCR1 in approximately 20 ml of buffer.

The resulting solution contained predominantly monomer, but the presence of significant amounts of dimer and higher oligomers of CXCR1 as could be visualized by analytical size-exclusion chromatography and SDS–PAGE. Monomers were isolated from the solution by preparative size-exclusion chromatography on a Superdex 200 column (Amersham). This chromatography step resulted in 50 ml of solution containing 10 mg of monomeric CXCR1.

For the reconstitution of purified CXCR1 into proteoliposomes, the monomeric CXCR1 solution and a mixture of 1,2-dimyristoyl-*sn*-glycero-3-phosphocholine (DMPC) and DPC lipids were mixed at a protein to lipid ratio of 1 : 10 (w:w). The mixed micelle solution of DMPC and DPC was prepared by dissolving DMPC and DPC (1 : 1, w/w) in chloroform in a round flask and evaporation of solvent in a rotary evaporator followed by high vacuum overnight. The DMPC/DPC film formed by this procedure was then dissolved in 20 mM HEPES buffer, pH 7.2, 150 mM NaCl at a final concentration of 5 mg/ml. After overnight incubation, the DPC was removed by adsorption to polystyrene beads (Calbiosorb).

11.10
Binding and Activity Measurements

Ligand binding is assayed by incubating proteoliposomes (0.2 μg refolded CXCR1 per sample) with 40 pM [^{125}I]IL-8 (2000 Ci/mmol; Amersham) and various concentrations of unlabeled IL-8 for 2 h at 4 °C in a 96-well glass fiber filter plate. After incubation, buffer and unbound ligand is removed by applying a vacuum. CXCR1 together with the bound ligand are retained in the well, dried, and suspended in scintillation cocktail. Radioactivity is measured in a 96-well scintillation counter. The data were fit to a one-site binding model using the Origin software and the K_d as well as the B_{max} was extracted from the fit parameters. The B_{max} was translated into concentration of binding sites (mol/l) and divided by the protein concentration (mol/l) previously determined by measuring the A_{280}. The resulting ratio is the relative specific activity expressed as percentage of the theoretical maximum. In the measurement in Figure 11.4(a), the specific activity was 83%. The function of any GPCR is to not only bind ligand, but also to change its conformation upon ligand binding and transmit a signal to the cognate G-protein. Refolded CXCR1 was reconstituted with $G_{i/o}$-protein trimer purified from porcine brain. This preparation was used to measure GTPγS binding as a function of agonist (IL-8) concentration. As shown in Figure 11.4(b), G-protein is activated in a ligand-dependent manner with an $EC_{50} \sim 1$ nM, which is very similar to the value reported in the literature [9]. The data in Figure 11.4 demonstrate that the refolded CXCR1 is fully functional and indistinguishable from the native receptor.

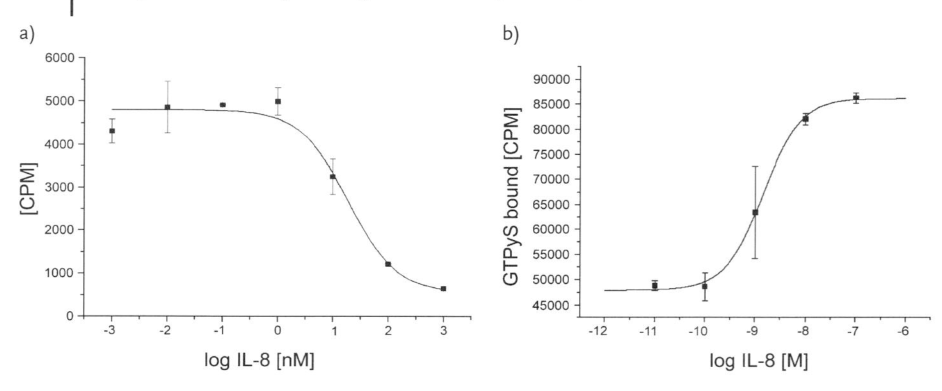

Figure 11.4 Binding and activity of refolded CXCR1. (a) Competition assay with radiolabeled (2000 Ci/mmol; Amersham) and unlabeled IL-8 shows binding to CXCR1 reconstituted into proteoliposomes (0.2 μg refolded CXCR1 per sample). (b) G-protein is activated in a ligand-dependent manner with $EC_{50} \sim 1$ nM. The data demonstrates that the refolded CXCR1 is fully functional and indistinguishable from the native receptor.

11.10.1
NMR Samples

Bicelles consist of long-chain phospholipids that form planar bilayers and short-chain lipids that cap the rims of the bilayer disks or form circular defects in the extended bilayer. The parameter q is the molar ratio of long-chain to short-chain lipids. Small isotropic bicelles ($q < 1.0$) reorient rapidly in solution and can be used for solution NMR studies, while larger bicelles ($q > 2.5$) align with the bilayer normal perpendicular to the direction of the applied magnetic field and can be used for solid-state NMR studies. The bicelle sample of CXCR1 immobilized in lipid bilayers provides an ideal NMR sample for structure determination in its active, native conformation by solid-state NMR spectroscopy.

For solution NMR, CXCR1 bicelle samples were prepared by adding 400 mM 1,2-dihexanoyl-*sn*-glycero-3-phosphocholine (DHPC; Avanti Polar Lipids) to the purified protein reconstituted into DMPC/1-palmitoyl-2-oleoyl-*sn*-glycero-3-phosphocholine (POPC) (long-chain lipid, 8 : 2 w/w) liposomes. Therefore, proteoliposomes consisting of 1 mg protein reconstituted in 10 mg of DMPC/POPC liposomes in 20 mM HEPES buffer, pH 7.2, 50 mM NaCl were collected by centrifuged at 300 000*g* for 2 h at 15 °C. The resulting supernatant was discarded and the hydrated proteoliposome pellet was resuspended with 400 μl H_2O containing the adequate amount of DHPC (short-chain lipid) in order to produce bicelle with $q = 0.1$. At that q value the solution will become immediately clear and isotropic bicelles are formed. An aliquot of 50 μl of D_2O (10%) is added to the bicelle solution and the solution NMR sample tube is filled.

The magnetically alignable bilayer samples for OS solid-state NMR spectroscopy are prepared similarly, using higher protein concentrations (4 mg protein and

40 mg of DMPC) and an increased q value of 3.2 with a lipid concentration of around 28% (w/v) in a 200-μl volume. The resulting solution is vortexed thoroughly and then allowed to equilibrate at room temperature. Upon bicelle formation, the CXCR1 proteoliposome becomes a clear, nonviscous solution at 4 °C and forms liquid crystal at 20–35 °C. A small, flat-bottomed NMR tube with 5 mm outer diameter (New Era Enterprises) was filled with 160 μl of the bicelle solution, using a precooled pipette at 4 °C. The NMR tube sealed with a tight-fitting rubber cap was pierced with a thin syringe to remove excess air from the sample.

11.11 NMR Spectra

Two-dimensional heteronuclear single-quantum correlation (HSQC) spectra of isotopically labeled CXCR1 samples are highly sensitive monitors of the expression, purification, refolding, and sample preparation process. For the uniformly and selectively ^{15}N-labeled CXCR1 samples in isotropic bicelles ($q = 0.1$) only the amide resonances from the mobile N- and C-terminal regions were observed in the ^{1}H/^{15}N-HSQC solution NMR spectra (Figure 11.5a and b). Approximately 40 amide peaks out of the expected 350 and two peaks (residues 1 and 9) out of the 12 methionine residues were observed (Figure 11.5b). Resonances from residues in the transmembrane helices and loops of CXCR1 are too broad to be observed in these solution NMR spectra. The HSQC NMR spectrum of the both N- and C-terminal truncated CXCR1 construct did not yield any corresponding signals compared to that of the wild-type CXCR1, suggesting that CXCR1, except for both terminal regions, is structured and undergoes slow reorientation in solution under these conditions.

CXCR1 undergoes rapid rotational diffusion ($D_{II} \geq 10^5$ s^{-1}) about the bilayer normal in phospholipid bicelles, yielding a motionally averaged single-line resonance spectrum from the bicelle sample aligned magnetically with the bilayer normal perpendicular to the applied magnetic field (Figure 11.5c and d). Significantly, there is no evidenc e of residual powder pattern intensity in the spectra, which demonstrates the near perfect alignment of the 350-residue GPCR in the bilayer. Since the radiofrequency irradiation and sample alignment with rapid rotational diffusion in solid-state NMR substitute for overall molecular reorientation as the line-narrowing mechanism, most of the amide signals of CXCR1 except for the mobile N- and C-terminal regions were observed in the solid-state NMR spectra. Studies of other membrane proteins demonstrate that residues in the mobile N- and C-terminal regions are not observed in the solid-state NMR spectra. The solid-state NMR spectrum of both N- and C-terminal truncated CXCR1 construct is identical to that of the wild-type construct, which is consistent with the signals of mobile terminal regions not being observed in the spectrum of the full-length protein.

We have shown that most of the signals of [^{15}N]Ile labeled CXCR1 were resolved in two-dimensional separated local field solid-state NMR spectrum [44]. Tailored isotopic labeling and a variety of three-dimensional solid-state NMR methods are

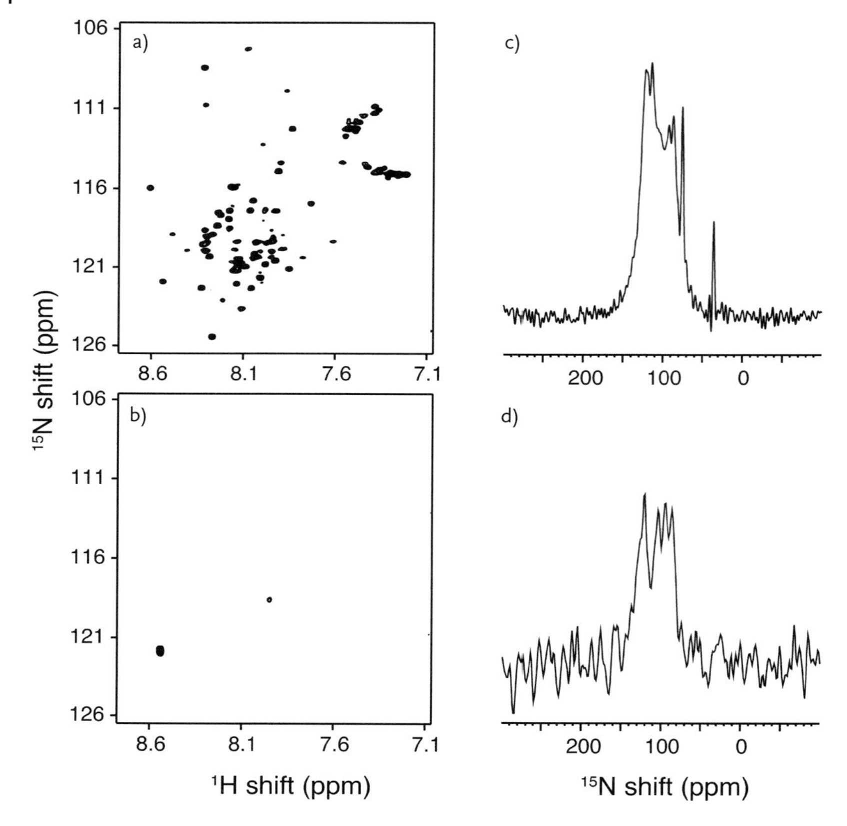

Figure 11.5 NMR spectra of CXCR1 reconstituted in bicelles. (a and b) Solution-state NMR $^{1}H/^{15}N$ correlation HSQC spectra of CXCR1 in isotropic bicelles ($q = 0.1$). (c and d) Solid-state ^{15}N chemical shift spectra of CXCR1 in magnetically aligned bicelles ($q = 3.2$). (a and c) Uniformly ^{15}N-labeled CXCR1. (b and d) Selectively [^{15}N]Met-labeled CXCR1.

able to provide high-resolution spectra of isotopically labeled GPCRs prepared as described in this chapter.

Acknowledgments

We thank Lauren Albrecht, Joel Bradley, Leah Cho, Mignon Chu, and Anna DeAngelis for their assistance. This research was supported by Bioengineering Research Partnership grant RO1 EB005161 from the National Institute of Biomedical Imaging and Bioengineering, and was performed at the Biomedical Technology Resource for NMR Molecular Imaging of Proteins at the University of California, San Diego, which is supported by grant P41 EB002031.

Abbreviations

DHPC	1,2-dihexanoyl-*sn*-glycero-3-phosphocholine
DMPC	1,2-dimyristoyl-*sn*-glycero-3-phosphocholine
DPC	dodecyl-phosphocholine
FPLC	fast protein liquid chromatography
GPCR	G-protein coupled receptor
GST	glutathione S-transferase
HSQC	heteronuclear single-quantum correlation
IL-8, CXCL8, NAP-1	interleukin-8
IPTG	isopropyl-β-D-thiogalactopyranoside
MAS	magic angle spinning
MIP	macrophage inflammatory protein
NAP	neutrophil-activating peptide
NMR	nuclear magnetic resonance
OS	oriented sample
SDS	sodium dodecyl sulfate
PAGE	polyacrylamide gel electrophoresis
PMN	polymorphonuclear neutrophil
POPC	1-palmitoyl-2-oleoyl-*sn*-glycero-3-phosphocholine

References

1 Foord, S.M. (2002) Receptor classification: post genome. *Curr. Opin. Pharmacol.*, **2**, 561–566.

2 Takeda, S., Kadowaki, S., Haga, T., Takaesu, H., and Mitaku, S. (2002) Identification of G protein-coupled receptor genes from the human genome sequence. *FEBS Lett.*, **520**, 97–101.

3 Kroeze, W.K., Sheffler, D.J., and Roth, B.L. (2003) G-protein-coupled receptors at a glance. *J. Cell. Sci.*, **116**, 4867–4869.

4 Schoneberg, T., Schulz, A., Biebermann, H., Hermsdorf, T., Rompler, H., and Sangkuhl, K. (2004) Mutant G-protein-coupled receptors as a cause of human diseases. *Pharmacol. Ther.*, **104**, 173–206.

5 Spiegel, A.M. and Weinstein, L.S. (2004) Inherited diseases involving G proteins and G protein-coupled receptors. *Annu. Rev. Med.*, **55**, 27–39.

6 Hopkins, A.L. and Groom, C.R. (2002) The druggable genome. *Nat. Rev. Drug Discov.*, **1**, 727–730.

7 Overington, J.P., Al-Lazikani, B., and Hopkins, A.L. (2006) How many drug targets are there? *Nat. Rev. Drug Discov.*, **5**, 993–996.

8 Jacoby, E., Bouhelal, R., Gerspacher, M., and Seuwen, K. (2006) The 7 TM G-protein-coupled receptor target family. *Chem. Med. Chem.*, **1**, 761–782.

9 Holmes, W.E., Lee, J., Kuang, W.J., Rice, G.C., and Wood, W.I. (1991) Structure and functional expression of a human interleukin-8 receptor. *Science*, **253**, 1278–1280.

10 Murphy, P.M. and Tiffany, H.L. (1991) Cloning of complementary DNA encoding a functional human interleukin-8 receptor. *Science*, **253**, 1280–1283.

11 Loetscher, P., Seitz, M., Clark-Lewis, I., Baggiolini, M., and Moser, B. (1994) Both interleukin-8 receptors independently mediate chemotaxis. Jurkat cells transfected with IL-8R1 or IL-8R2 migrate in response to IL-8, GRO alpha and NAP-2. *FEBS Lett.*, **341**, 187–192.

12 Fu, W., Zhang, Y., Zhang, J., and Chen, W.F. (2005) Cloning and characterization of mouse homolog of the CXC chemokine receptor CXCR1. *Cytokine*, **31**, 9–17.
13 Ahuja, S.K., Lee, J.C., and Murphy, P.M. (1996) CXC chemokines bind to unique sets of selectivity determinants that can function independently and are broadly distributed on multiple domains of human interleukin-8 receptor B. Determinants of high affinity binding and receptor activation are distinct. *J. Biol. Chem.*, **271**, 225–232.
14 Richardson, R.M., Pridgen, B.C., Haribabu, B., Ali, H., and Snyderman, R. (1998) Differential cross-regulation of the human chemokine receptors CXCR1 and CXCR2. Evidence for time-dependent signal generation. *J. Biol. Chem.*, **273**, 23830–23836.
15 Wuyts, A., Proost, P., Lenaerts, J.P., Ben-Baruch, A., Van Damme, J., and Wang, J.M. (1998) Differential usage of the CXC chemokine receptors 1 and 2 by interleukin-8, granulocyte chemotactic protein-2 and epithelial-cell-derived neutrophil attractant-78. *Eur. J. Biochem.*, **255**, 67–73.
16 McColl, S.R. and Clark-Lewis, I. (1999) Inhibition of murine neutrophil recruitment *in vivo* by CXC chemokine receptor antagonists. *J. Immunol.*, **163**, 2829–2835.
17 Wuyts, A., Van Osselaer, N., Haelens, A., Samson, I., Herdewijn, P., Ben-Baruch, A., Oppenheim, J.J., Proost, P., and Van Damme, J. (1997) Characterization of synthetic human granulocyte chemotactic protein 2: usage of chemokine receptors CXCR1 and CXCR2 and *in vivo* inflammatory properties. *Biochemistry*, **36**, 2716–2723.
18 Bertini, R., Allegretti, M., Bizzarri, C., Moriconi, A., Locati, M., Zampella, G., Cervellera, M.N., Di Cioccio, V., Cesta, M.C., Galliera, E., *et al.* (2004) Noncompetitive allosteric inhibitors of the inflammatory chemokine receptors CXCR1 and CXCR2: prevention of reperfusion injury. *Proc. Natl. Acad. Sci. USA*, **101**, 11791–11796.
19 Allegretti, M., Bertini, R., Cesta, M.C., Bizzarri, C., Di Bitondo, R., Di Cioccio, V., Galliera, E., Berdini, V., Topai, A., Zampella, G., *et al.* (2005) 2-Arylpropionic CXC chemokine receptor 1 (CXCR1) ligands as novel noncompetitive CXCL8 inhibitors. *J. Med. Chem.*, **48**, 4312–4331.
20 Casilli, F., Bianchini, A., Gloaguen, I., Biordi, L., Alesse, E., Festuccia, C., Cavalieri, B., Strippoli, R., Cervellera, M.N., Di Bitondo, R., *et al.* (2005) Inhibition of interleukin-8 (CXCL8/IL-8) responses by repertaxin, a new inhibitor of the chemokine receptors CXCR1 and CXCR2. *Biochem. Pharmacol.*, **69**, 385–394.
21 Palczewski, K., Kumasaka, T., Hori, T., Behnke, C.A., Motoshima, H., Fox, B.A., Le Trong, I., Teller, D.C., Okada, T., Stenkamp, R.E., *et al.* (2000) Crystal structure of rhodopsin: a G protein-coupled receptor. *Science*, **289**, 739–745.
22 Costanzi, S. (2008) On the applicability of GPCR homology models to computer-aided drug discovery: a comparison between in silico and crystal structures of the beta2-adrenergic receptor. *J. Med. Chem.*, **51**, 2907–2914.
23 Cherezov, V., Rosenbaum, D.M., Hanson, M.A., Rasmussen, S.G., Thian, F.S., Kobilka, T.S., Choi, H.J., Kuhn, P., Weis, W.I., Kobilka, B.K., *et al.* (2007) High-resolution crystal structure of an engineered human beta2-adrenergic G protein-coupled receptor. *Science*, **318**, 1258–1265.
24 Rasmussen, S.G., Choi, H.J., Rosenbaum, D.M., Kobilka, T.S., Thian, F.S., Edwards, P.C., Burghammer, M., Ratnala, V.R., Sanishvili, R., Fischetti, R.F., *et al.* (2007) Crystal structure of the human beta2 adrenergic G-protein-coupled receptor. *Nature*, **450**, 383–387.
25 Rosenbaum, D.M., Cherezov, V., Hanson, M.A., Rasmussen, S.G., Thian, F.S., Kobilka, T.S., Choi, H.J., Yao, X.J., Weis, W.I., Stevens, R.C., *et al.* (2007) GPCR engineering yields high-resolution structural insights into beta2-adrenergic receptor function. *Science*, **318**, 1266–1273.
26 Warne, T., Serrano-Vega, M.J., Baker, J.G., Moukhametzianov, R., Edwards, P.C., Henderson, R., Leslie, A.G., Tate, C.G. and Schertler, G.F. (2008) Structure of a beta1-adrenergic G-protein-coupled receptor. *Nature*, **454**, 486–491.

27 Jaakola, V.P., Griffith, M.T., Hanson, M.A., Cherezov, V., Chien, E.Y., Lane, J.R., Ijzerman, A.P., and Stevens, R.C. (2008) The 2.6 angstrom crystal structure of a human A_{2A} adenosine receptor bound to an antagonist. *Science*, **322**, 1211–1217.

28 Kaback, H.R., Frillingos, S., Jung, H., Jung, K., Prive, G.G., Ujwal, M.L., Weitzman, C., Wu, J., and Zen, K. (1994) The lactose permease meets Frankenstein. *J. Exp. Biol.*, **196**, 183–195.

29 Prive, G.G., Verner, G.E., Weitzman, C., Zen, K.H., Eisenberg, D., and Kaback, H.R. (1994) Fusion proteins as tools for crystallization: the lactose permease from *Escherichia coli*. *Acta Crystallogr. D*, **50**, 375–379.

30 Prive, G.G. and Kaback, H.R. (1996) Engineering the lac permease for purification and crystallization. *J. Bioenerg. Biomembr.*, **28**, 29–34.

31 Hanson, M.A. and Stevens, R.C. (2009) Discovery of new GPCR biology: one receptor structure at a time. *Structure*, **17**, 8–14.

32 Topiol, S. and Sabio, M. (2009) X-ray structure breakthroughs in the GPCR transmembrane region. *Biochem. Pharmacol.*, **78**, 11–20.

33 Park, J.H., Scheerer, P., Hofmann, K.P., Choe, H.W., and Ernst, O.P. (2008) Crystal structure of the ligand-free G-protein-coupled receptor opsin. *Nature*, **454**, 183–187.

34 Scheerer, P., Park, J.H., Hildebrand, P.W., Kim, Y.J., Krauss, N., Choe, H.W., Hofmann, K.P., and Ernst, O.P. (2008) Crystal structure of opsin in its G-protein-interacting conformation. *Nature*, **455**, 497–502.

35 Tian, C., Breyer, R.M., Kim, H.J., Karra, M.D., Friedman, D.B., Karpay, A., and Sanders, C.R. (2005) Solution NMR spectroscopy of the human vasopressin V2 receptor, a G protein-coupled receptor. *J. Am. Chem. Soc.*, **127**, 8010–8011.

36 Tikhonova, I.G. and Costanzi, S. (2009) Unraveling the structure and function of G protein-coupled receptors through NMR spectroscopy. *Curr. Pharm. Des.*, **15**, 4003–4016.

37 Klein-Seetharaman, J., Reeves, P.J., Loewen, M.C., Getmanova, E.V., Chung, J., Schwalbe, H., Wright, P.E., and Khorana, H.G. (2002) Solution NMR spectroscopy of [alpha-^{15}N]lysine-labeled rhodopsin: the single peak observed in both conventional and TROSY-type HSQC spectra is ascribed to Lys-339 in the carboxyl-terminal peptide sequence. *Proc. Natl. Acad. Sci. USA*, **99**, 3452–3457.

38 Getmanova, E., Patel, A.B., Klein-Seetharaman, J., Loewen, M.C., Reeves, P.J., Friedman, N., Sheves, M., Smith, S.O., and Khorana, H.G. (2004) NMR spectroscopy of phosphorylated wild-type rhodopsin: mobility of the phosphorylated C-terminus of rhodopsin in the dark and upon light activation. *Biochemistry*, **43**, 1126–1133.

39 Klein-Seetharaman, J., Yanamala, N.V., Javeed, F., Reeves, P.J., Getmanova, E.V., Loewen, M.C., Schwalbe, H., and Khorana, H.G. (2004) Differential dynamics in the G protein-coupled receptor rhodopsin revealed by solution NMR. *Proc. Natl. Acad. Sci. USA*, **101**, 3409–3413.

40 Bokoch, M.P., Zou, Y., Rasmussen, S.G., Liu, C.W., Nygaard, R., Rosenbaum, D.M., Fung, J.J., Choi, H.J., Thian, F.S., Kobilka, T.S., Puglisi, J.D., Weis, W.I., Pardo, L., Prosser, R.S., Mueller, L., and Kobilka, B.K. (2010) Ligand-specific regulation of the extracellular surface of a G-protein-coupled receptor. *Nature*, **463**, 108–112.

41 Wilson, J.R., Leang, C., Morby, A.P., Hobman, J.L., and Brown, N.L. (2000) MerF is a mercury transport protein: different structures but a common mechanism for mercuric ion transporters? *FEBS Lett.*, **472**, 78–82.

42 Park, S.H., Mrse, A.A., Nevzorov, A.A., Mesleh, M.F., Oblatt-Montal, M., Montal, M., and Opella, S.J. (2003) Three-dimensional structure of the channel-forming trans-membrane domain of virus protein "u" (Vpu) from HIV-1. *J. Mol. Biol.*, **333**, 409–424.

43 Howell, S.C., Mesleh, M.F., and Opella, S.J. (2005) NMR structure determination of a membrane protein with two transmembrane helices in micelles: merF of the bacterial mercury detoxification system. *Biochemistry*, **44**, 5196–5206.

44 Park, S.H., De Angelis, A.A., Nevzorov, A.A., Wu, C.H., and Opella, S.J. (2006) Three-dimensional structure of the transmembrane domain of Vpu from HIV-1 in aligned phospholipid bicelles. *Biophys. J.*, **91**, 3032–3042.
45 Grobner, G., Burnett, I.J., Glaubitz, C., Choi, G., Mason, A.J., and Watts, A. (2000) Observations of light-induced structural changes of retinal within rhodopsin. *Nature*, **405**, 810–813.
46 Luca, S., White, J.F., Sohal, A.K., Filippov, D.V., van Boom, J.H., Grisshammer, R., and Baldus, M. (2003) The conformation of neurotensin bound to its G protein-coupled receptor. *Proc. Natl. Acad. Sci. USA*, **100**, 10706–10711.
47 Ratnala, V.R., Kiihne, S.R., Buda, F., Leurs, R., de Groot, H.J., and DeGrip, W.J. (2007) Solid-state NMR evidence for a protonation switch in the binding pocket of the H1 receptor upon binding of the agonist histamine. *J. Am. Chem. Soc.*, **129**, 867–872.
48 Ahuja, S., Crocker, E., Eilers, M., Hornak, V., Hirshfeld, A., Ziliox, M., Syrett, N., Reeves, P.J., Khorana, H.G., Sheves, M., and Smith, S.O. (2009) Location of the retinal chromophore in the activated state of rhodopsin*. *J. Biol. Chem.*, **284**, 10190–10201.
49 Tiburu, E.K., Bowman, A.L., Struppe, J.O., Janero, D.R., Avraham, H.K., and Makriyannis, A. (2009) Solid-state NMR and molecular dynamics characterization of cannabinoid receptor-1 (CB1) helix 7 conformational plasticity in model membranes. *Biochim. Biophys. Acta.*, **1788**, 1159–1167.
50 Harding, P.J., Attrill, H., Ross, S., Koeppe, J.R., Kapanidis, A.N., and Watts, A. (2007) Neurotensin receptor type 1: Escherichia coli expression, purification, characterization and biophysical study. *Biochem. Soc. Trans.*, **35**, 760–763.
51 Smith, S.O. (2010) Structure and activation of the visual pigment rhodopsin. *Annu Rev. Biophys.*, **39**, 309–328.
52 Park, S.H., and Opella, S.J. (2007) Conformational changes induced by a single amino acid substitution in the trans-membrane domain of Vpu: implications for HIV-1 susceptibility to channel blocking drugs. *Protein. Sci.*, **16**, 2205–2215.
53 De Angelis, A.A., Howell, S.C., Nevzorov, A.A., and Opella, S.J. (2006) Structure Determination of a Membrane Protein with Two Trans-membrane Helices in Aligned Phospholipid Bicelles by Solid-State NMR Spectroscopy. *J. Am. Chem. Soc.*, **128**, 12256–12267.
54 Grisshammer, R., Averbeck, P., and Sohal, A.K. (1999) Improved purification of a rat neurotensin receptor expressed in *Escherichia coli. Biochem. Soc. Trans.*, **27**, 899–903.
55 Kiefer, H., Maier, K., and Vogel, R. (1999) Refolding of G-protein-coupled receptors from inclusion bodies produced in *Escherichia coli. Biochem. Soc. Trans.*, **27**, 908–912.
56 Baneres, J.L., Martin, A., Hullot, P., Girard, J.P., Rossi, J.C., and Parello, J. (2003) Structure-based analysis of GPCR function: conformational adaptation of both agonist and receptor upon leukotriene B_4 binding to recombinant BLT1. *J. Mol. Biol.*, **329**, 801–814.
57 Kiefer, H. (2003) *In vitro* folding of alpha-helical membrane proteins. *Biochim. Biophys. Acta*, **1610**, 57–62.
58 White, J.F., Trinh, L.B., Shiloach, J., and Grisshammer, R. (2004) Automated large-scale purification of a G protein-coupled receptor for neurotensin. *FEBS Lett.*, **564**, 289–293.
59 Schiller, H., Molsberger, E., Janssen, P., Michel, H., and Reilander, H. (2001) Solubilization and purification of the human ETB endothelin receptor produced by high-level fermentation in *Pichia pastoris. Receptors Channels*, **7**, 453–469.
60 Warne, T., Chirnside, J., and Schertler, G.F. (2003) Expression and purification of truncated, non-glycosylated turkey beta-adrenergic receptors for crystallization. *Biochim. Biophys. Acta*, **1610**, 133–140.
61 Reeves, P.J., Thurmond, R.L., and Khorana, H.G. (1996) Structure and function in rhodopsin: high level expression of a synthetic bovine opsin gene and its mutants in stable mammalian cell lines. *Proc. Natl. Acad. Sci. USA*, **93**, 11487–11492.

62 Bertin, B., Freissmuth, M., Breyer, R.M., Schutz, W., Strosberg, A.D., and Marullo, S. (1992) Functional expression of the human serotonin 5-HT1A receptor in *Escherichia coli*. Ligand binding properties and interaction with recombinant G protein alpha-subunits. *J. Biol. Chem.*, **267**, 8200–8206.

63 Grisshammer, R., Little, J., and Aharony, D. (1994) Expression of rat NK-2 (neurokinin A) receptor in *E. coli*. *Receptors Channels*, **2**, 295–302.

64 Hulme, E.C. and Curtis, C.A. (1998) Purification of recombinant M1 muscarinic acetylcholine receptor. *Biochem. Soc. Trans.*, **26**, S361.

65 Furukawa, H. and Haga, T. (2000) Expression of functional M2 muscarinic acetylcholine receptor in *Escherichia coli*. *J. Biochem.*, **127**, 151–161.

66 Hampe, W., Voss, R.H., Haase, W., Boege, F., Michel, H., and Reilander, H. (2000) Engineering of a proteolytically stable human beta 2-adrenergic receptor/maltose-binding protein fusion and production of the chimeric protein in *Escherichia coli* and baculovirus-infected insect cells. *J. Biotechnol.*, **77**, 219–234.

67 Weiss, H.M. and Grisshammer, R. (2002) Purification and characterization of the human adenosine A_{2a} receptor functionally expressed in *Escherichia coli*. *Eur. J. Biochem.*, **269**, 82–92.

68 Opekarova, M. and Tanner, W. (2003) Specific lipid requirements of membrane proteins – a putative bottleneck in heterologous expression. *Biochim. Biophys. Acta*, **1610**, 11–22.

69 Niebauer, R.T. and Robinson, A.S. (2006) Exceptional total and functional yields of the human adenosine (A2a) receptor expressed in the yeast *Saccharomyces cerevisiae*. *Protein Expr. Purif.*, **46**, 204–211.

70 Wedekind, A., O'Malley, M.A., Niebauer, R.T., and Robinson, A.S. (2006) Optimization of the human adenosine A2a receptor yields in *Saccharomyces cerevisiae*. *Biotechnol. Prog.*, **22**, 1249–1255.

71 O'Malley, M.A., Lazarova, T., Britton, Z.T., and Robinson, A.S. (2007) High-level expression in *Saccharomyces cerevisiae* enables isolation and spectroscopic characterization of functional human adenosine A2a receptor. *J. Struct. Biol.*, **159**, 166–178.

72 Reilander, H. and Weiss, H.M. (1998) Production of G-protein-coupled receptors in yeast. *Curr. Opin. Biotechnol.*, **9**, 510–517.

73 Sarramegna, V., Talmont, F., Demange, P., and Milon, A. (2003) Heterologous expression of G-protein-coupled receptors: comparison of expression systems from the standpoint of large-scale production and purification. *Cell. Mol. Life Sci.*, **60**, 1529–1546.

74 Sarramegna, V., Muller, I., Mousseau, G., Froment, C., Monsarrat, B., Milon, A., and Talmont, F. (2005) Solubilization, purification, and mass spectrometry analysis of the human mu-opioid receptor expressed in *Pichia pastoris*. *Protein Expr. Purif.*, **43**, 85–93.

75 Mazina, K.E., Strader, C.D., Tota, M.R., Daniel, S., and Fong, T.M. (1996) Purification and reconstitution of a recombinant human neurokinin-1 receptor. *J. Recept. Signal Transduct. Res.*, **16**, 191–207.

76 Massotte, D., Pereira, C.A., Pouliquen, Y., and Pattus, F. (1999) Parameters influencing human mu opioid receptor over-expression in baculovirus-infected insect cells. *J. Biotechnol.*, **69**, 39–45.

77 Akermoun, M., Koglin, M., Zvalova-Iooss, D., Folschweiller, N., Dowell, S.J., and Gearing, K.L. (2005) Characterization of 16 human G protein-coupled receptors expressed in baculovirus-infected insect cells. *Protein Expr. Purif.*, **44**, 65–74.

78 Hassaine, G., Wagner, R., Kempf, J., Cherouati, N., Hassaine, N., Prual, C., Andre, N., Reinhart, C., Pattus, F., and Lundstrom, K. (2006) Semliki Forest virus vectors for overexpression of 101 G protein-coupled receptors in mammalian host cells. *Protein Expr. Purif.*, **45**, 343–351.

79 Kigawa, T., Yabuki, T., Yoshida, Y., Tsutsui, M., Ito, Y., Shibata, T., and Yokoyama, S. (1999) Cell-free production and stable-isotope labeling of milligram quantities of proteins. *FEBS Lett.*, **442**, 15–19.

80 Shimizu, Y., Kuruma, Y., Ying, B.W., Umekage, S., and Ueda, T. (2006) Cell-free translation systems for protein engineering. *FEBS J.*, **273**, 4133–4140.

81 Ishihara, G., Goto, M., Saeki, M., Ito, K., Hori, T., Kigawa, T., Shirouzu, M., and Yokoyama, S. (2005) Expression of G protein coupled receptors in a cell-free translational system using detergents and thioredoxin-fusion vectors. *Protein Expr. Purif.*, **41**, 27–37.

82 Klammt, C., Schwarz, D., Fendler, K., Haase, W., Dotsch, V., and Bernhard, F. (2005) Evaluation of detergents for the soluble expression of alpha-helical and beta-barrel-type integral membrane proteins by a preparative scale individual cell-free expression system. *FEBS J.*, **272**, 6024–6038.

12
Solubilization, Purification, and Characterization of Integral Membrane Proteins

Víctor Lórenz-Fonfría, Alex Perálvarez-Marín, Esteve Padrós, and Tzvetana Lazarova

12.1
Introduction

To study the structure and function of integral membrane proteins (IMPs), it is generally necessary to remove them from their natural environment. Due to their hydrophobic nature, the use of detergents that mimic lipid properties is indispensable, allowing their extraction from the lipid bilayer. After this step, the solubilized protein is purified and usually reconstituted into liposomes for functional/structural studies (see Chapter 13). Obviously, these procedures should be as gentle and mild as possible to maintain the native conformation of the protein and its activity, but as effective as possible to maximize the recovery of the protein of interest and to minimize the presence of contaminant proteins. Direct reconstitution methods using cell-free expression systems (cell-free expression in the presence of liposomes or nanodisks allows direct insertion of membrane proteins into liposome or diskoidal phospholipid bilayers encircled by lipoproteins) are treated elsewhere (see Chapter 8 for G-protein-coupled receptors [GPCRs]). The aim of this chapter is to cover the solubilization and purification of IMPs, and their characterization in the detergent-solubilized form (Figure 12.1). The extraction and purification of peripheral membrane proteins is described in Chapter 7.

In Section 12.2, the solubilization course of IMPs is discussed. An efficient solubilization depends on the nature of the protein, the lipid composition of the membrane, the detergent, and the buffer. In this process, the lipid/protein/detergent ratio is of the utmost importance. Table 12.A1 included in the Appendix illustrates the advantages and the disadvantages of the most common detergents in biochemistry research. The necessity of adding essential lipids to preserve the protein integrity or using a milder solubilization protocol that leaves some essential lipids bound to the protein is also described briefly.

Section 12.3 covers the purification process of the solubilized IMP. This is not so different from the purification of soluble proteins and many similar protocols can be used. Owing to the unique physicochemical characteristics of each IMP, the choice of the detergent and the purification strategy vary from protein to protein. Obviously, all purification steps should be done in the presence of a

Production of Membrane Proteins: Strategies for Expression and Isolation, First Edition.
Edited by Anne Skaja Robinson.

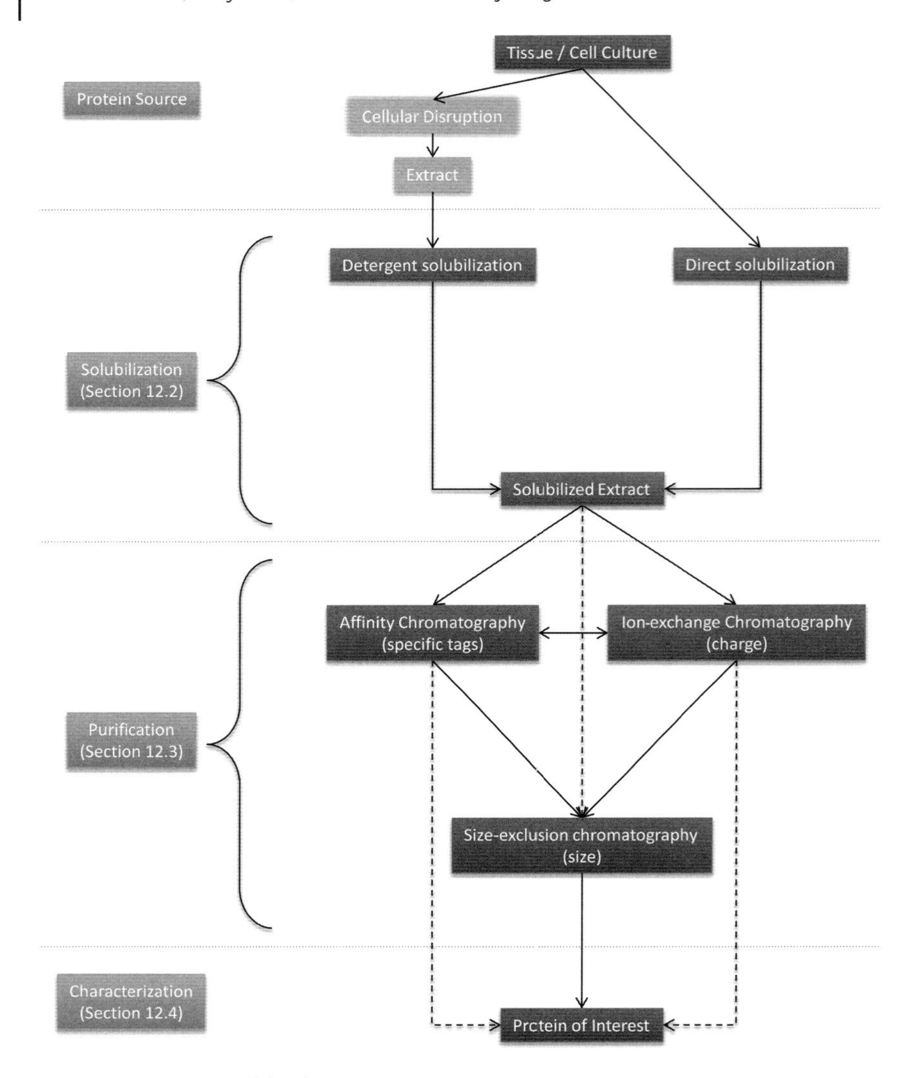

Figure 12.1 Structure of the chapter.

minimum detergent concentration to avoid protein precipitation. Frequently, the detergent used for IMP solubilization results in long-term protein destabilization and it should be exchanged for a milder detergent during the subsequent purification steps. Emphasis is given to the critical aspects of purification and to some recent progress in these strategies to produce pure and functional IMPs for subsequent structural and functional studies.

Once the membrane protein has been successfully purified, it should be functionally and structurally characterized. This is covered in Section 12.4, which discusses some of the main methods in use. Determination of the homogeneity and the oligomeric state of the solubilized protein is of help to assess the protein integrity, whereas spectroscopic methods can give valuable information about possible denaturation or aggregation problems that may have occurred during the solubilization/purification steps. In appropriate cases, the analysis of ligand binding can give more direct information about the functional integrity of the solubilized IMPs.

12.2 Solubilization of IMPs

To solubilize IMPs, it is usually necessary to disrupt or lyse the cells either by mechanical (e.g., mechanical homogenizers; nitrogen burst methods; ultrasound; glass or ceramic beads; French press; microfluidizers) or nonmechanical means (e.g., osmotic shock; repeated freezing and thawing) in the presence of detergents [1]. In some cases, the protein can be directly extracted from the membrane by incubating the cellular suspension with detergents or enzymes mixed with chaotropic agents [2]. Protease inhibitors are used extensively to avoid degradation of the protein of interest. It is advisable to pay attention to the buffer used in the solubilization, to avoid incompatibility with the detergent [3].

In some cases, high-level expression of the protein in heterologous host cells leads to the formation of highly aggregated proteins, called inclusion bodies. Proteins in inclusion bodies usually have the correct primary sequences, identical to those of their native and authentic counterparts, but are aggregated and inactive. The procedure for extraction of insoluble inclusion body protein from *Escherichia coli* is described in Chapter 1 and 8.

12.2.1 Physicochemical Characteristics of Detergents

The detergents, known also as surfactants, are amphiphilic compounds composed of hydrophobic (tails) and hydrophilic (heads) groups combined in one molecule. A polar or charged group constitutes the hydrophilic head-group of the detergent and an extended hydrocarbon chain, the hydrophobic tail, respectively. Due to their amphipathic nature detergents present a higher degree of hydrophilicity than lipids, making them appropriate agents for the solubilization of IMPs out of their native membrane environment [2, 4–6]. At a sufficiently low concentration detergents exist as monomer molecules in water. When the concentration of detergent reaches a certain threshold the monomers self-organize into noncovalent aggregates, called micelles [4, 7, 8] (see Figure 12.2). The lowest concentration above which the monomers come together to form micelles is known as the detergent critical micelle concentration (CMC). Above the CMC, detergent monomers and micelles coexist in equilibrium, and increasing the detergent concentration results in an increase of the amount of detergent in micelles, while the concentration of the detergent monomers remains constant [4]. The onset of the formation of

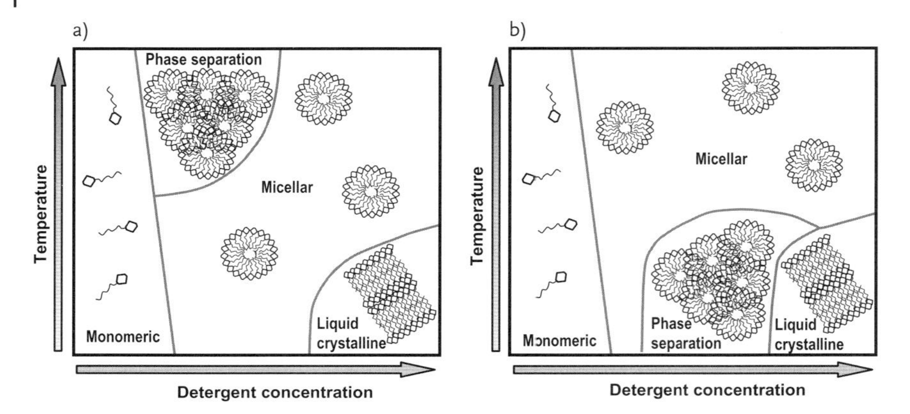

Figure 12.2 Simplified phase diagrams displaying the phase behavior of two types of detergents. The borderline between micelle solution and phase separation in the diagram is called the consulate boundary. (a) Phase diagram of detergent with a lower consulate boundary (usually most of the nonionic detergents fall in this group). (b) Phase diagram of detergents with an upper consulate boundary (frequently observed for zwitterionic and for glycosidic detergents). (Adapted from Linke [4].)

micelles, and so the CMC, can be properly evaluated by carrying out light scattering, surface tension, or isothermal calorimetry experiments [9]. The size of a micelle is described by its average molecular weight and aggregation number (the average number of detergent monomers per micelle), which depends on the solvent environment [10]. The aggregation number of the detergent is a useful measure in foreseeing the size of the micelles formed when solubilizing IMPs by a detergent. Note, however, that when the size of an IMP approaches or exceeds the size of the detergent micelle, the protein, not the detergent, dictates the size of the protein–detergent mixed micelles [6].

Detergents exhibit a very complex behavior in aqueous solutions. Depending on temperature and detergent concentration they can form several types of well-defined aggregates [4, 8] (Figure 12.2). Knowledge about temperature and concentration ranges in which a given detergent will form micelles is a key issue for the successful solubilization of IMPs. Apart from existing as monomers and micelles, detergents can also form well-defined nonmicellar aggregates immiscible with the water phase. This process is called phase separation and the temperature at which it occurs is known as the cloud point [4, 7, 8]. The cloud point is specific for each type of detergent. At the cloud point, the formerly clear homogeneous detergent solution separates into two immiscible solutions – one detergent-poor and other detergent-rich. Depending on the density of the buffer, the detergent-rich phase can be found above or below the aqueous phase [4, 7, 8]. At very high detergent concentrations and low temperature, conditions normally not relevant for biochemical studies, the detergent may also form well-ordered liquid crystalline phases immiscible with water and solid in appearance

(Figure 12.2). The temperature above which the detergent solubility is greater than the CMC is known as the Krafft point. This is the minimum temperature at which micelles can form.

The phase behavior (i.e., the type of surfactant aggregates and CMC) is mostly controlled by the physicochemical properties of detergents. The ionic strength and pH of the solution, as well as the presence of different additives to the buffer system, can affect significantly these parameters [4, 7]. Detergent-based phase separation is a simple way to concentrate or purify the protein (see Section 12.3.3.2). Most of the crystallization methods also rely on the phase behavior of detergents [11]. Detailed protocols for detergent-based phase separation, detergent exchange, and detergent removal can be found elsewhere [7, 8].

12.2.2 Classification of Detergents

According to their structure, detergents can be classified in three main groups: nonionic, zwitterionic, and ionic [5]. Table 12.A1 lists some important physicochemical properties of the most commonly used detergents for IMP solubilization and purification. It includes also some useful information related to the compatibility of detergents with some biochemical and biophysical assays, as well as examples of IMPs that have been successfully solubilized using particular detergents. The structures of some commonly used detergents in IMP solubilization can be found in the Appendix (Figures 12.A1–12.A3).

12.2.2.1 Nonionic Detergents

Nonionic detergents (Figure 12.A1) have uncharged hydrophilic head groups, either polyoxyethylene or glycosidic groups. They are generally considered as mild and relatively "nondenaturing" detergents. Their most recognized advantage is that they tend to solubilize proteins in functional state, due to their ability to break lipid–lipid and lipid–protein interactions, but not protein–protein interactions [6]. Common examples of nonionic detergents are Triton X-100, octyl-β-D-glucopyranoside (OG), decyl-β-D-maltopyranoside (DM), dodecyl-β-D-maltopyranoside (DDM), and digitonin.

12.2.2.2 Ionic Detergents

Ionic detergents (Figure 12.A2) are composed of cationic or anionic head-groups. The hydrophobic tail can be made of either a hydrocarbon straight chain, like in sodium dodecyl sulfate (SDS) and cetyltrimethylammonium bromide (CTAB), or a rigid steroidal structure, as in sodium deoxycholate (see bile acid salts in the following section).

Anionic Detergents The most representative member is SDS, mainly used for cell lysis and SDS–polyacrylamide gel electrophoresis (SDS–PAGE), because of its very harsh character. Bile acid salts (e.g., sodium salts of cholic acid and deoxycholic acid) are anionic detergents with a rigid steroidal hydrophobic group known also to work like nonionic surfactants in solubilization at concentrations near the CMC. They are frequently used for preparation of liposomes or two-dimensional crystals.

Cationic Detergents Cationic detergents are also incompatible with SDS–PAGE because they can modify the charge of the SDS micelles. Although they usually denature proteins, the presence of small amounts of cationic detergents can improve the phase separation of nonionic detergents [8].

12.2.2.3 **Zwitterionic Detergents**

Zwitterionic detergents (Figure 12.A3) are composed of head-groups bearing anionic and cationic net charges. They combine the properties of ionic and nonionic detergents. In general, they are more deactivating than nonionic detergents, but some like 3-[(3-cholamidopropyl)dimethylammonio]-1-propanesulfonate (CHAPS), 3-[(3-cholamidopropyl)dimethylammonio]-2-hydroxy-1-propanesulfonate (CHAPSO), or fos-choline have been recognized for their ability to solubilize membrane proteins with retained function [12].

12.2.2.4 **Recently Developed Detergents**

Cymal(N) and zwiterionic cyclo-fos and fos-choline series are mainly used for the solubilization of some recombinant proteins and for structural studies [42–45].

12.2.3 New Solubilizing Agents

One significant drawback of the use of detergents for IMP solubilization is the potential instability of the protein enwrapped in the protein–detergent complex (PDC; see below), which may significantly impair the function and native structure of the protein [6, 46]. In an attempt to overcome this problem newly solubilizing compounds have been designed in recent years, namely lipopeptide detergents, hemifluorinated surfactants, and tripod amphiphiles. Lipopeptide detergents form small micelles and have shown to preserve IMPs in an active state for a longer period than traditional detergents [47–49]. Hemifluorinated surfactants are not surfactants in the classical sense and cannot be used for IMP solubilization, but they can be used to replace the detergent from a solubilized protein to optimize its monodispersity [50, 51]. Tripod amphiphiles consist of a new series of detergents, which have shown success in substituting for traditional detergents, keeping some IMPs in an active, soluble state for weeks [52, 53]. The structure, properties, and uses of amphiphilic polymers ("amphipols") and fluorinated surfactants have been reviewed [50].

12.2.4 Solubilization Process

An IMP can be considered "solubilized" from the membrane if, after centrifugation of the detergent-treated membranes at around 105 000g for 1 h at 4 °C, the

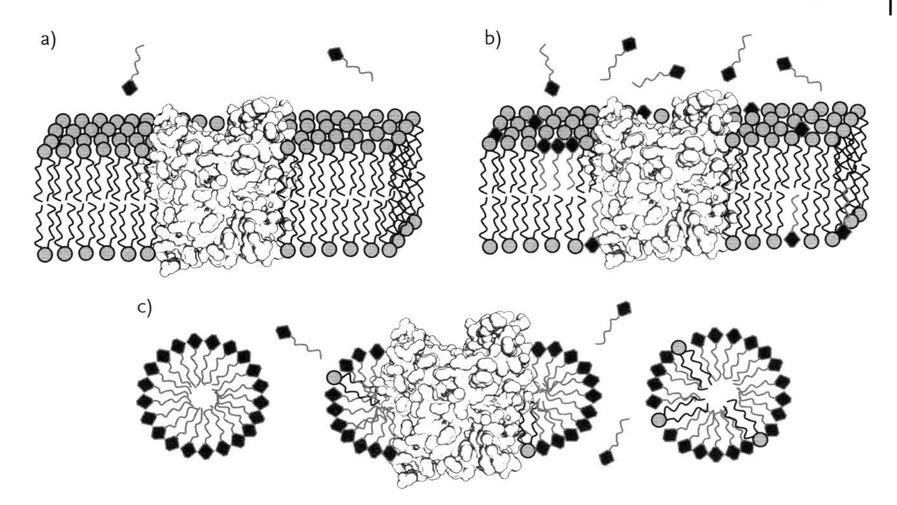

Figure 12.3 Schematic representation of the different steps during the solubilization of IMPs. (a) At low concentrations of detergent the detergent monomers minimally perturb the membrane, (b) followed by penetration into the membrane bilayer. (c) At higher detergent concentrations, the membrane starts to disintegrate and in the final step of the solubilization process the solution contains two types of mixed micelles: lipid–detergent and protein–lipid–detergent micelles together with detergent micelles and some free detergent monomers. Solubilized IMP in a complex with bound detergent is called a PDC [47].The amount of bound detergent in the PDC depends on the detergent and the protein, with a detergent contribution from 30 to 70% of the mass of the PDC.

protein appears in the supernatant fraction [2]. The solubilization of IMPs by detergents can be schematically represented by three different steps (Figure 12.3). First, at low concentrations, the detergents solely perturb the membrane structurally by partitioning into the lipid bilayer, but not yet solubilizing it [54, 55]. At higher detergent concentrations, the membrane starts to disintegrate, losing a substantial part of the lipids. This process gives rise to lipid–protein–detergent mixed micelles. In these complex micelles, the hydrophobic parts of the IMPs plus some lipids are enwrapped by detergent molecules, forming protective coat. It can be said that at this stage the IMPs are "solubilized". A detergent/protein ratio of around 1–2 (w/w) is generally believed to be sufficient to solubilize IMPs to form lipid–protein–detergent micelles [2]. A further increase of the detergent concentration causes a progressive delipidation of the lipid–protein–detergent mixed micelles. Finally, in the last step of the solubilization process, the solution contains two types of mixed micelles: lipid–detergent and protein–lipid–detergent micelles. On average, a detergent/protein ratio of around 10 (w/w) or higher will lead to complete delipidation [56].

12.2.5
The Means of a Successful Solubilization of IMPs

There is common agreement that the establishment of appropriate conditions for a successful solubilization of IMPs is mainly empirical [4, 6]. However, for a successful solubilization some important factors and rules are worthy of serious consideration. In this section we will try to account for some of the most critical of them. The choice of detergent is the first decisive step. Although several detergents have been shown to fulfill reasonably the requirement of maintaining IMP functionality, some may appear problematic or conflict with the methodology used subsequently for structural or functional characterization of solubilized IMPs (see Table 12. A1). In such cases it is recommendable to shift to a more appropriate detergent or to employ a detergent that can be readily exchanged for one that is compatible with the downstream experimental work. A comprehensive list of detergents used for the solubilization and purification of several GPCRs with the retention of their functionality can be found in a recent excellent review by Sarramegna *et al.* [12]. If protein yield is a limiting factor, a good strategy is to screen for the most effective detergent by simultaneously carrying out functional tests [44]. Protocols for small-scale experiments and useful guidelines are described in [57].

The question about how the behavior of the detergent will change in the presence of the protein is also of essential concern. Indeed, PDCs are not expected to behave like their pure isolated components nor necessarily to have properties intermediary between them. Hence, in PDC the micelle size, the CMC, and the phase behavior of the detergent will not be the same as in pure detergent micelles. Exploring physicochemical properties of the detergent in the PDC is another factor to address for a successful solubilization. The effectiveness of a detergent for a particular protein has to be *a priori* addressed and experimentally validated. As we discussed above, the CMC is an important parameter for the solubilization process, because at this concentration the detergent starts to associate with the hydrophobic surfaces of membrane proteins and form water-soluble PDCs [4, 6]. A detailed protocol for the determination of detergent CMC values has been published [58]. The behavior of the detergent (CMC, cloud point, and aggregation number) during the solubilization process can be significantly influenced by many other factors apart from the initial protein and detergent concentrations, such as buffer additives, pH, or temperature.

The complex behavior of the solubilization process of IMPs has given rise to the concept of critical solubilization concentration (CSC), which takes into account the ionic strength, pH, and temperature, and the possible contributions of all components in the system, such as lipids and proteins [10, 59]. A qualitative determination of the CSC can be made by adding different amounts of the detergent to a series of membrane suspensions and monitoring the solutions with time. The CSC is highly dependent on the exact starting conditions, especially the lipid concentration. As discussed above (Figure 12.3), depending on the detergent/lipid/protein ratio, different stages of solubilization could be achieved. The use of a high concentration of detergent can affect the activity of the protein (see below), while too low detergent/lipid ratio may result in poor solubilization or heterogeneous samples [6, 60]. The control of the lipid/detergent ratio is particularly important during purification, where excess detergent may cause loss of activity. Unfortunately, there is no a strict recipe on how to reach the optimal detergent/lipid ratio for the solubilization of IMPs and in practice this needs to be determined empirically, which often becomes a long trial-and-error process. The most effective way to identify the CSC is to monitor the solubilization over a wide range of detergent/lipid ratios or, alternatively, by varying detergent/protein ratios since concentrations of protein and lipid are related in a given membrane. The solubilization can be controlled simply by eye or more precisely by performing light-scattering measurements. A detailed protocol for measuring the lipid/protein ratio in a membrane can be found in the literature [58].

12.2.6
"All" or "Not All" Lipids and If "Purer Is Better"

Usually a high amount of detergent is needed for solubilization, not only to accommodate all protein in the PDC, but also to remove the lipids, which by mediating nonspecific protein contacts may cause aggregation of the PDC. However, excessive detergent concentrations may eliminate some essential bound lipids from the protein that may be critical for the structure, stability, or function of the protein [61, 62]. There is growing evidence that some lipids that are tightly associated with IMPs are essential for their activity and stability [63]. Several works support this view, reporting that even partial delipidation during solubilization may lead to the loss of enzymatic activities of membrane-bound enzymes [64, 65] and the function of IMPs [66]. The structural flexibility and integrity of some GPCRs have also been shown to depend on associated endogenous lipids [62]. Nevertheless, some IMPs have been successfully crystallized while being nearly completely delipidated [67, 68]. For other IMPs, however, the presence of some specific lipids appears to be compulsory to obtain crystals [69, 70]. Ordered lipids are observed in several high-resolution IMPs crystal structures and, importantly, some structures have shown significant differences when crystals are grown in the presence of lipids [71]. Thus, we come to a new paradigm if purer IMPs is better for us. Complete delipidation may not be a very appropriate approach when designing purification procedures with the aim of structure or functional studies.

12.2.7
Stability of the Protein–Detergent Solutions

The mechanism(s) by which protein stability is affected upon detergent solubilization is rather complex and it is not easily defined. A detailed analysis of "how do proteins die" in detergent solutions has been given by Bowie [46]. In the following we therefore will briefly account for some possible reasons for deactivation of IMPs. (i) As indicated above, during solubilization and purification lipids that may be important for protein stability and activity may be inadvertently extracted. (ii) The inability of the detergent to mask fully the protein hydrophobic regions can lead to nonspecific interactions, allowing non-native conformations. This usually results in destabilization, aggregation, and precipitation of the protein, in some cases being an irreversible process [46]. (iii) Detergent-induced deoligomerization of some IMPs may cause destabilization of these proteins [72]. For example, cross-linking of the functional oligomer of heterotetrameric basophil IgE receptor has been reported to stabilize the enzyme against detergent denaturation [73].

To enhance protein stability, several procedures appear helpful. In some cases, adding back specific lipids or using more than one detergent in supplementing buffers during the solubilization process leads to the maintenance of active protein conformation. For example, addition of phospholipids during Ni-affinity purification yields larger amounts of ligand-bound neurotensin receptor NTS1 active protein [74]. Addition of relatively small amounts of specific lipids to DDM micelles during solubilization has been shown to prolong significantly the lifetime of the β_2-adrenergic receptor in solution [75] or to improve dopamine D_1 receptor stability and to recover activity to 41–48% [76]. Using more than one detergent (mixed lipid–detergent system) sometimes is also constructive to attain a better environment for the stability and function of mixed PDC systems [77]. Likewise, the addition of a lipid-like substance such as cholesterol hemisuccinate (a soluble analog of cholesterol) during the solubilization and purification has been shown to stabilize the adenosine A_{2A} receptor [78] and the β_2-adrenergic receptor [79]. The role of cholesterol, an essential component of eukaryotic membranes, in the function and organization of several membrane proteins and receptors is well-recognized [79–81].

For some proteins addition of substrates during solubilization is another effective way for stabilizing solubilized IMPs in their native conformation, enhancing the yield of the recovered protein [82]. For example, addition of the agonist *N*-[^{3}H] methylscopolamine in conjunction with detergents in a purification of m3 muscarinic receptor was found to recover the receptor function considerably [83]. Finally, recently developed alternative compounds to traditional detergents have offered new ways for the stabilization of IMPs in the PDC [51].

12.3
IMP Purification

12.3.1
Strategy Definition

Depending on the type of studies for which the protein is needed, several factors should be taken into account during IMPs purification:

- **Type of IMP.** IMPs comprise two kinds of prototypical protein folds – α-helical (all-α) and β-barrel (all-β). These two kinds of protein folds are represented in membranes with different lipid composition [84], such as plasma membranes, bacterial outer membranes, and cellular compartments, making necessary the definition of slightly different solubilization techniques. For solubilization and purification of β-barrel proteins, see [85].

- **Protein source.** The first source to obtain a protein is its native host. Isolation from cells and tissues has been carried out for several IMPs from bacteria, animals, and plants [86–93]. The main drawback with this approach is the low yield and the high amount of contaminant proteins. Subcellular fractionation is normally carried out to achieve fraction enrichment and to overcome this limitation [90, 94]. The second and most used source nowadays is recombinant technologies, presenting several advantages, which are discussed extensively elsewhere [95–97].
- **Protein purity, yield, and activity.** The ideal IMP purification would yield high amounts of pure and functional protein, but often these three goals cannot be fulfilled simultaneously. Therefore, the final aim of the study (with its sample requirements) should largely determine the strategy for the purification process. For instance, whereas for protein detection and preliminary biochemical characterization the picogram to microgram scale suffices, the milligram scale (and high purity) is a requirement for most biophysical and structural biology approaches (NMR, X-ray crystallography, spectroscopy). High-throughput (HTP) methods allow screening studies to determine the yield and purity of the protein of interest at each of the purification steps [33, 98].

12.3.1.1 HTP Methods

HTP methods are being developed due to the lack of X-ray structural information on IMPs, in order to optimize the time from developing recombinant expression of the protein to the crystallization trials [34, 99, 100]. As stated in McLusky *et al.* [100], the idea is to find optimal expression and purification methods for the protein of interest. The ideal method should include, among other aspects:

i) Short time frame.

ii) Use of small amounts of protein (96-well plates).

iii) Standardized protocol that allows changes of parameters, starting from the vector transformation, recombinant protein expression, and protein solubilization and purification.

iv) Work with different proteins in parallel: recombinant His tag, Green Fluorescent Protein (GFP) tag, FLAG tag, and so on.

v) HTP methods are mainly oriented to express and purify proteins for crystallization attempts, thus focusing on parameters that have yielded quality crystals of membrane proteins. It should be highlighted that HTP methods have been developed for crystallization trials, but could be exported to other biophysical methods.

12.3.2 Purification Process

Most of the purification steps start with a solubilized IMP extract and involve different chromatographic approaches. A deep study of chromatography is beyond the scope of the present chapter and the reader is redirected to excellent reviews [3, 101, 102]. The most relevant features to take into account are discussed below.

12.3.2.1 Hydrophobicity

Being solubilized in detergent, hydrophobicity is rarely used as a differential feature in IMP purification. In general, hydrophobicity-based separation methods such as reversed-phase and hydrophobic protein interaction chromatographies are not indicated for IMPs [2], but they can be used for peripheral membrane proteins. However, the hydrophobic/hydrophilic amino acid proportion is extremely relevant and should be taken into account to avoid aggregation problems in the case of decreases in detergent concentration [103]. If aggregation is reversible, and the structure and function of proteins are not severely affected, aggregation and fractional precipitation may lead to an advantage in concentrating the protein and removing other contaminants [101].

12.3.2.2 Charge

As in the case of hydrophobicity, the proportions of charged amino acids determine specific features for the protein. The pH at which the net charge of the protein is zero (isoelectric point, p*I*) is a very useful characteristic for the purification of the protein of interest.

Ion-Exchange Chromatography (IEX) IEX is a common and "low-cost" chromatography technique with high capacity, capable of providing high protein concentration. Knowing the p*I* of the protein of interest is extremely valuable in order to establish the best starting conditions [104]. This method is based on the binding of the protein to the matrix at low salt conditions. To elute the protein, increasing salt conditions are used (continuous gradient or step gradient). As the salt concentration increases, a competitive process between the salt ions and the bound proteins allows the protein elution from the ion-exchange matrix. Since protein charge is the distinctive feature of this method, the selection of the chromatography buffer ion (usually Na^+ or Cl^-) and ion-exchange matrix (anion or cation exchanger) is very important. The optimal pH range for the buffer is also critical and should not be overlooked. IEX matrices and protocols are widely used, and are commercially available from different vendors in prepacked and loose-bead supports [104].

Chromatofocusing (CF) and Isoelectric Focusing (IEF) CF is a high-resolution technique to separate proteins based on their p*I*. It may provide high protein purity in intermediate or final steps (polishing) of the purification process. The CF column sorts proteins in a descending p*I* order. The CF column or gel consists of a buffer containing several buffering species available in a range from pH 3 to 11 (polybuffer exchangers based, for example, on cross-linked Sepharose 6B) [105]. These buffering species will carry out ion exchange with the protein of interest, focusing it in a narrow band at its p*I*. The critical step for CF is the proper selection of the CF gel for the pH range of the protein of interest; thus, knowledge on the p*I* of the protein is essential. However, if the p*I* is not known, IEF is a very useful tool.

In recent years, IEF has become a technique used on a regular basis as the first resolving dimension in two-dimensional gel electrophoresis [105]. IEF is a high-resolution technique that permits isolating a protein from a heterogeneous mixture by means of the distinctive feature of its p*I* (resolution of 0.01 p*I* units between two proteins). The basis of this method consists of separating the proteins in a stable pH gradient because of the different charged amino acids of proteins (both in gel phase or liquid phase [105, 106]). When voltage is applied, the positively charged proteins migrate to the negative side of the gradient, whereas negatively charged proteins migrate to the positive side. The protein stops migrating at the pH at which its net charge is zero (p*I*). Thus, a protein becomes sharply focused at its p*I*. For both gel- and liquid-phase IEF, the pH gradient consists of a mixture of carrier ampholytes with specific p*I* values [105–107].

12.3.2.3 Size

Size is a feature that may speed up the purification process when the protein is at one of the extremes of the size range. It is also very useful to isolate membrane protein homo- or heterocomplexes (functional unit) from membrane protein monomers (basic unit).

Size-Exclusion Chromatography (SEC) SEC (also previously known as gel filtration), as its name indicates, is a technique to separate proteins by their size. This is not an adsorption technique, since it is based in the partitioning of the total volume into two liquid volumes determined by the porous bead size (100–300 μm) composing the solid-phase matrix. The elution time is dependent on the protein molecular weight and hydrodynamic radius. Large proteins are excluded from the porous bead volume, being rapidly eluted in what is known as the void volume. Small proteins are capable of accessing the porous bead volume and are retained for longer times, being eluted more slowly. A very important aspect for SEC is the calibration of the column using the appropriate molecular weight standards [108, 109], but special attention should be paid in the case of IMPs since the solubilization and micellization effects may vary the hydrodynamic radius of the protein, affecting the elution time accordingly [60, 108, 110]. Several of these matrices and standards are available commercially. Standard matrices for SEC such as Sephadex, Superose, Superdex, or Sephacryl are all suitable for detergent-solubilized IMPs [111], but should be pre-equilibrated with a buffer containing detergent above its CMC (and possibly some lipid) to avoid protein instability caused by lost of detergent (and lipid) from the protein–detergent micelles during the column run [111, 112]. Aggregated IMPs may bind tightly to the matrix and clog columns;

thus, samples should be clarified by centrifugation and any pellet discarded before loading it onto the SEC column [112, 113]. It is also important to select a column with the most appropriate resolution range, taking into account the possible oligomeric state of the protein and the amount of absorbed detergent (typically 1.5–3.5 times the protein mass) [111, 112].

Since SEC is not a high-resolution technique, it is commonly used in the late steps of the purification process, when the protein is already pure, for instance to separate different aggregated states of the protein of interest (monomers from oligomers or vice versa). In the case of IMPs it may represent an advantage to use it in the early steps of the purification process [103] in order to increase the yield and the purity of the initial steps of the purification. However, to do so, sample concentration should be carried out before loading onto the SEC column.

Membrane Filtration Another technique that uses size as a distinctive feature of proteins is membrane filtration. This has been a standard procedure in purification processes for desalting, concentration, fractionation, and so on [114], and it is also being rediscovered as a protein purification HTP method [115].

12.3.2.4 Affinity

The main difference between affinity and adsorption chromatography is that the interaction of the protein with the affinity matrix is based on specific protein–matrix interactions. Thus, the affinity feature refers to the medium to high selectivity of binding by protein–protein (or peptide) or protein–ligand interactions. The principles of this method are not solely based on the specificity of the binding, but also on its reversibility. The specificity of this technique yields very pure protein and simultaneously enables concentration, thus reducing the number of purification steps for certain proteins [116]. This approach is also suitable to clean protein preparations, to trap undesired proteins, or to differentiate isoforms of proteins. The use of recombinant DNA techniques has proven very useful to produce fusion proteins that can be trapped by several commercially available affinity matrices and customizing affinity columns is a common practice to reach high levels of specificity [117–119].

Recombinant Proteins with Engineered Tags In order to obtain tagged proteins, the use of recombinant DNA techniques is essential. The protein source or expression system may be mammalian cells, *E. coli*, yeast, insect cells, and so on. *E. coli* is the system of choice for the design and engineering of the fusion proteins (beyond the scope of this chapter and reviewed elsewhere [81, 107, 120]). The basic idea is to tag the protein at its N- and/or C-termini in order to facilitate its ultimate purification after high-yield expression. It is important to consider that the activity or structure of the protein might be modified by the insertion of the tag(s); thus, it is wise to include specific cleavage sites to remove the tags after the purification has been carried out [121, 122]. Table 12.1 presents some examples of affinity tags used for membrane protein purification.

Table 12.1 Overview of tags used in IMP purification (adapted from [12, 13]).

Tag	Features	Antibody for detection	References
His	2–10 histidines (normally 6); at N- or C-terminal position; matrix containing divalent cations; elution by imidazole or low pH; tag or elution effects on protein properties; very common purification tag; high purity; unspecific antibody detection	yes; low specificity	[14–21]
FLAG	tag of 8 residues (DYKDDDK); at N- or C-terminal position; matrix containing anti-FLAG monoclonal antibody (immunoaffinity); elution by FLAG peptide, EDTA, or low pH; cleavage of tag by enterokinase	yes; M1 antibody only binds when tag at N-terminal position	[21–27]
KT3	tag of 11 residues (KPPTPPPEPET); at N- or C-terminal position; immunoaffinity; elution by low pH	yes	[28]
Mannose-binding protein (MBP)	Fusion protein of 396 residues; at N- or C-terminal position; matrix based on cross-linked amylose; elution by maltose; very large tag	yes; very specific	[29, 30]
Strep	tag of 8–9 residues (WSHPQFEK or AWAHPQPGG); at N- or C-terminal position; matrix consists of a modified streptavidin; elution by biotin or desthiobiotin	yes	[31, 32]
GFP/enhanced GFP	fusion protein of 220 residues; at N-terminal, C-terminal, or internal position; only suitable for detection in the absence of antibody; not for purification, used for screening of the solubilization and purification processes [33, 34]; very large tag and prone to dimerize (GFP)	yes; very specific; intrinsic fluorescence allows detection without antibody	[31, 35, 36]
Hemagglutinin (HA)	tag of 31 residues; at N-terminal, C-terminal, or internal position; immunoaffinity; elution by HA peptide or low pH	yes; specific	[37, 38]
Myc	tag of 11 residues (CEQKLISEEDL); at N-terminal, C-terminal, or internal position; immunoaffinity; elution at low pH	yes; low specificity	[27, 30, 35, 39–41]

The His tag is an outstanding example of an affinity tag. Immobilized metal ions (Co^{2+}, Cu^{2+}, Ni^{2+}, etc.) have been used for the isolation and purification of proteins containing sequences rich in histidine or tryptophan. The introduction of His affinity tagging in the form of fusion proteins provided a purification system that readily coordinates His with transition metal ions (Ni^{2+} being the most used of all) (see Table 12.1 for examples). Imidazole is the molecule of choice for protein elution. Reduction of the immobilized metal ion should be avoided; thus, special caution should be taken with the use of dithiothreitol and other reducing agents [13].

High-Specificity

- **Immunoaffinity and Customized Affinity Chromatography** The most specialized and high-specificity form of affinity chromatography is based on antibody–antigen or protein–protein interaction [123]. First, an antibody or antibody fragment (or a protein or protein fragment) is immobilized onto a matrix. These matrices have to be activated in order to perform the immobilization, since the matrix has to react with primary amines, thiol groups, or other reactive groups in the antibody or ligand protein, maximizing the antibody–antigen or protein–protein binding capacity. Suitable matrices that are well established are cyanogen bromide, trisacryl, and other agarose-bound resins. The aim is to raise antibodies against specific epitopes or to identify peptidic binding sequences in the partner protein. In the case of the antibodies, obtaining high amounts of pure antibody can be economically costly, so having the hybridoma technology to supply high amounts of monoclonal antibodies would be an advantage [124, 125]. In the case of known, strong, and specific protein–protein or ligand–protein interactions, the peptide or molecule can be purchased or chemically synthesized and directly immobilized onto the matrix [126]. Glutamate receptors are examples of proteins that have been purified using both customized ligand-based and immunoaffinity chromatographies [127, 128].

  In immunoaffinity, the elution of the protein of interest is achieved by mild acidic or alkali conditions [123]. Sometimes the conditions are more extreme because of the specificity of the antibody–antigen interaction. A way to overcome these extreme conditions is the usage of the peptide used to generate the antibody to competitively elute the protein from the antibody matrix [129].

Mid-Specificity

- **Lectin Affinity** Lectins are carbohydrate-binding proteins that can be used to purify integral membrane glycoproteins. Since glycosylation is a common modification of membrane proteins, this affinity method becomes very useful for IMP purification. We have stated lectin affinity as a medium-specificity affinity chromatography because a mixture of glycoproteins may be present when extracting proteins from the membrane. However, if used with an already purified protein, it can be extremely specific in order to distinguish the glycosylated and nonglycosylated forms of a membrane protein [130]. Once the glycoprotein is bound to the solid-support-immobilized lectin, the elution is carried out by competition with specific sugars. One of the limitations for lectin affinity is the choice of detergent, since some ionic detergents may inactivate lectins [2]. The most used lectins are concanavalin A and wheat germ agglutinin [131].

- **Dye-Ligand Affinity Chromatography** This protein purification procedure takes into account the high affinity of dyes for the binding sites of many enzymes and other proteins [132]. Most of these dyes come from the textile industry and they contain chloro groups that facilitate the immobilization in the support matrix, such as agarose and nylon membranes. The elution of bound proteins is achieved in a competitive way with the substrate or cofactor for the specific protein [132]. Apart from the specificity (medium if compared to immunoaffinity), dye affinity chromatography provides an additional advantage: whereas proteases present in crude membrane extracts may degrade antibodies or proteins bound to the matrix, affecting in some case the efficiency of binding, these proteases and enzymes do not degrade the bound dye, because of the differences in chemistry. This method becomes highly specific if used in the polishing steps, and it has been successfully used for the purification of IMPs, such as the case of Ca^{2+}-ATPase using the dye Reactive Red 120 [133–135].

12.3.3 New Approaches and Advances in Purification

12.3.3.1 Magnetic Beads

Taking advantage of the magnetic properties of certain matrices, it is possible to specifically fractionate the targeted molecules against a crude biological extract by just applying a magnetic field after the binding to the magnetic beads. This makes the use of magnetic beads an alternative method to ultracentrifugation. In this way membrane protein separation is faster, and easy to scale-up and to setup in an automatic way [136]. A particular example is the one-step isolation of membrane proteins achieved by immobilizing the lectin concanavalin A (lectin affinity) to magnetic beads [137]. Some of the hallmarks of this technique are:

i) Immobilization of biotinylated concanavalin A onto magnetic streptavidin beads. To separate the beads from the solution, or to perform the washing steps, a magnet is held close to the tube where the purification is carried out.

ii) The concanavalin A magnetic beads were used against several protein sources, such as liver extract, PC-3 cells, and HeLa cells, in order to enrich plasma membrane proteins with lectin affinity, such as CEACAM-1 (3-fold enrichment) and cadherin-type proteins (7-fold enrichment).

12.3.3.2 Phase Separation Methods

Phase separation is an alternative to chromatographic separation methods (see Section 12.2.1). It is an efficient and cheap way to purify and concentrate detergent-solubilized IMPs. It can be used as a first purification step or at a later stage of purification. When a polymer is added to a detergent-solubilized IMP (or the temperature or ionic strength is increased), the cloud point can be attained and two immiscible phases appear, in which the detergent-rich phase contains the IMP. After collection, the excess of detergent should be removed. A possible disadvantage of using phase separation is the high concentration of the detergent used, which can affect protein stability and/or interfere with biochemical assays and binding processes Sections 12.2.5–12.2.6 and Table 12.A1. Another approach uses metal-chelating polymers to bind a poly-His-tagged IMP. In this case, the protein of interest partitions in the polymeric phase allowing its separation from untagged proteins [138].

12.4 Characterization of Solubilized IMPs

Generally, activity is considered as the best indicator of protein integrity. However, for detergent-solubilized IMPs the physiological activity is habitually not measurable, lacking a compartmentalized membrane as cells (vesicles or liposomes) have. Thus, characterization of the homogeneity and oligomeric state of solubilized IMPs is a general alternative to indirectly assess protein structural integrity. Spectroscopic methods can be also used to obtain insight into the structural integrity of solubilized IMPs, allowing identification of some signatures of denaturation or aggregation. Testing for ligand binding can be even a better approximation to assess protein functional integrity, especially for membrane receptors and secondary transporters.

12.4.1 Sample Homogeneity and Protein Oligomeric State

12.4.1.1 SEC

SEC is useful not only as a final step for purification of proteins (see Section 12.3.2.3 for the basis of SEC), but it is also a popular tool in assessing their homogeneity and oligomeric state. It is a relatively fast method (30–60 min), which makes it advantageous for unstable IMPs that tend to denature in detergent micelles. SEC is often coupled in-line with an UV/Vis detector to probe the eluted protein at 280 nm, but other in-line or out-line detection systems are possible [139, 140].

The shape of the SEC chromatogram is habitually used to evaluate the monodispersity of the protein–detergent micelles. A sharp and narrow Gaussian peak is regarded as an excellent indication of sample homogeneity, and considered a prerequisite for successful crystallization of IMPs [113]. To reduce instrumental contributions to the peak width, and so to better assess sample monodispersity, it is advisable to inject the sample in a small volume (1–2%) relative to the column volume [111].

The Stoke's radius of soluble proteins can be inferred from the elution time, and thus their approximate size and their oligomeric state. However, one should note that for detergent-solubilized IMPs the inferred molecular size forcedly represents that of the PDC. For quantitative analysis, the protein/detergent (and protein/lipid) weight ratio in the micelles needs to be independently evaluated and accounted for [141]. In this regard, both spectroscopic [142, 143] and radioactive [110] methods are available. Experiments performed with a series of detergents giving rise to micelles with different protein/detergent weight ratios, and so of different sizes, have been described to provide to better estimates of the protein oligomeric state in the micelles [141].

12.4.1.2 **Static Light Scattering (SLS)**

LS is a method to determine the average mass of macromolecules in solution. It can be used to determine the CMC of a detergent, as well as the mass of a protein in a protein–detergent micelle independently of the protein/micelle size or of the amount of detergent (or lipid) per protein [144]. For this last application LS is often coupled to SEC (LS-SEC), allowing the resolution of the masses of a mixture of particles of different size. This makes LS-SEC one of the most appropriate methods for determining the oligomeric state or subunit stoichiometry of detergent solubilized IMPs [112, 145].

The light scattered by a particle at a given angle is proportional to its mass per particle and its concentration (mass/volume). In the so-called "two-detector method" the change in scattered light (ΔLS) and the refractive index change (ΔRI) are measured from the eluting SEC column, allowing us to estimate the weight of detergent micelles and soluble proteins [146]. For solubilized IMPs a "three-detector method" is required [144, 146], which combines simultaneous measurements of ΔLS, ΔRI, and $\Delta A_{280\,\mathrm{nm}}$. Then, provided that the present detergents (or lipids) do not absorb light at 280 nm (see Table 12.A1), a direct estimate of the mass of the protein, M, can be obtained as: $M = K \times (\Delta \mathrm{LS} \times \Delta A_{280\,\mathrm{nm}})/(\varepsilon_{280\,\mathrm{nm}} \times \Delta \mathrm{RI})$, where $\varepsilon_{280\,\mathrm{nm}}$ is the protein extinction coefficient in terms of weight concentration and K is the instrumental calibration constant [146].

12.4.1.3 **Analytical Ultracentrifugation (AUC)**

AUC is usually used in two different but complementary modes: sedimentation velocity (SV) and sedimentation equilibrium (SE). In both cases real-time acquisition of the solute concentration profiles along the radial axis is performed, recording the UV/Vis absorbance (at 280 nm for proteins) supplemented by interference detection, which reports the total solute concentration [147, 148]. As AUC relies on the principal property of mass and fundamental laws of gravitation it requires no standards for calibration [148]. However, while both SEC and LS-SEC can be run in around 30–60 min, AUC requires much longer experimental times, which limits its applicability to IMPs stable in detergent micelles for at least a full day.

SV is a hydrodynamic method. Appropriate analysis of the time-dependent concentration profiles provides a distribution of sedimentation coefficients [147], and from them, the masses of the particles present in a solution can be deduced with some assumptions regarding their shape and density. When applied to solubilized IMPs, SV is a powerful method to assess sample monodispersity [149], although the estimated hydrodynamic properties are those of the PDC. An example of a detergent-solubilized IMPs that has been studied by AUC using modern approaches is the mitochondrial ADP/ATP carrier [149].

SE is a thermodynamic method, characterized by the time-invariant exponential concentration gradient that develops as the flux of sedimenting and diffusing molecules balance, which depends only on the buoyant mass of the particle. Detergents in the PDC can become gravitationally transparent in SE when the solvent density is appropriately adjusted with D^2H_2O or $H_2{}^{18}O$, allowing for the estimation of the mass of the IMP in the micelles and thus its oligomeric state [150, 151]. However, density matching is not possible for detergents denser than 1.1 mg/ml, excluding some frequently used detergents such as DDM and OG [152].

12.4.1.4 Blue-Native Electrophoresis (BN-PAGE)

In BN-PAGE protein electrophoretic migration stops when trapped in a gel region of appropriated pore size, allowing for size determination using a protein standard of known size. The best advantage of BN-PAGE with respect to previous methods is the small amount of protein required and the possibility to analyze many samples simultaneously. Determination of the size and oligomeric state for detergent-solubilized IMPs is only possible if the amount of Coomassie Brilliant Blue bound is accounted for [153]. A way to overcome this limitation is the use of immunodetection techniques.

12.4.2 Structural Characterization

12.4.2.1 Circular Dichroism (CD)

CD is based on the differential absorbance of chromophores in front of left and right circularly polarized light. The amide group of the peptide bond dominates the CD spectra of proteins in the far UV (below 250 nm), containing information about the protein secondary structure [154]. Many methods have been developed to extract this information, mostly considering that the CD spectrum of a protein can be represented as the combination of a library of characteristic CD spectra of secondary structural elements or proteins of known three-dimensional structure [155]. Most of these methods have been included in a web server (dichroweb.cryst.bbk.ac.uk) [156]. Although tailored for soluble proteins, these spectral libraries can be used with lower but still reasonable success to estimate the secondary structure of IMPs in detergents [157, 158]. Even without a quantitative analysis, CD spectroscopy can be still useful as a sample quality control, reporting protein secondary structural changes with time, or with the detergent used. Since the characteristic CD spectrum of β-sheets is less intense and more variable than for α-helices [155],

CD might not be the method of choice to detect and quantify the formation of intermolecular β-sheets – the signature of protein aggregation.

CD requires some caution to avoid/minimize the use of buffers (e.g., Tris, 2-(*N*-morpholino)ethanesulfonic acid (MES), HEPES, sulfate, acetate, etc.), solvents (e.g., chloride salts) and detergents (see Table A1 in the Appendix) with substantial UV absorbance [154]. To minimize solvent absorption short sample path lengths (1–0.01 mm) are common, compensated by proportionally higher protein concentrations (0.1–10 mg/ml).

12.4.2.2 **IR Spectroscopy**

IR spectra arise from the absorption of light with a frequency resonant with the vibration of polar chemical bonds (primary stretching and bending motions). For proteins, structural information can be deduced from the so-called amide bands – several normal vibrations localized in the peptide bond [159]. From these, the amide I band frequency (around 1700–1620 cm^{-1}), mostly representing the C=O stretching of the peptide bond, is a sensitive reporter of protein structure. This sensitivity has been shown in polypeptides and proteins with a dominant secondary structure, factor analysis from proteins with known three-dimensional structure, and theoretical analysis [159]. However, retrieving this information for IMPs using spectral libraries optimized for soluble proteins remains less successful than for CD [160], most likely because the amide I band is sensitive to other structural aspects besides secondary structure (e.g., length and hydration of α-helices, number and length of strands in β-sheets, etc.) [159]. A common approach for the structural analysis of the amide I band is its decomposition in component bands (aided by mathematical band-narrowing and curve-fitting methods, see Figure 12.4) followed by the assignment of the resolved components bands to specific secondary structures based on their maximum wavenumber [161]. Interestingly, intermolecular β-sheets give characteristic and clearly observable bands in the amide I, making IR spectroscopy a suitable method to detect aggregation of IMPs [162].

In comparison with CD, IR spectroscopy can be performed with fewer restrictions in the usable buffers and detergents, and sample turbidity is not of concern. One drawback of IR spectroscopy is the strong absorbance of water at around 1645 cm^{-1}, which overlaps the protein amide I band. In transmittance, this demands very short path lengths (around 6 μm) and precise sample temperature control for the accurate digital subtraction of the water absorbance using a reference buffer spectrum, and in turn high protein concentrations (10 mg/ml or above) [163]. The need of high protein concentrations is relaxed in D_2O (0.1–1 mg/ml or above) [163]. In attenuated total reflection (ATR) the effectively path length is short and highly reproducible, allowing to work at lower proteins concentrations in H_2O (1 mg/ml or above) [164].

12.4.2.3 **NMR Spectroscopy**

One of the most celebrated applications of NMR spectroscopy is the resolution of three-dimensional structures of proteins in solution. Although IMPs solubilized

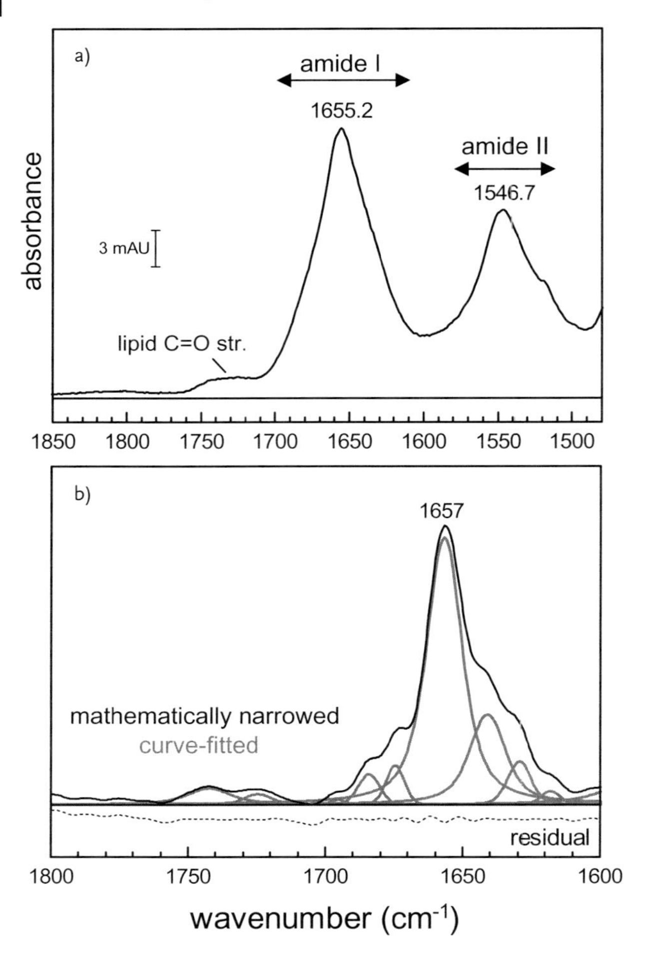

Figure 12.4 Example of structural analysis of a detergent-solubilized IMP by IR spectroscopy. (a) Absorption IR spectrum of the mitochondrial ADP/ATP carrier (AAC) solubilized with DDM, after subtracting the large absorbance contribution of the buffer. The amide I and II wavenumber regions are indicated by arrows. These band maxima correspond to typical values for α-helical IMP. A band assigned to lipid suggested that 7 ± 1 molecules of phospholipid are retained per solubilized AAC monomer. (b) Mathematical band-narrowing and curve-fitting of the amide I band. Band-narrowing was performed with Fourier self-deconvolution (FSD) [Kauppinen, J.K., Moffatt, D.J., Mantsch, H.H., and Cameron, D.G. (1981) Fourier selfdeconvolution: a method for resolving intrinsically overlapped bands. *Appl. Spectrosc.*, **35**, 271–276.], using a Lorentzian band of 18 cm^{-1} width and a narrowing factor of 2.0. Curve-fitting was performed for more accurate results using Voigtian bands, which partially take into account the band-shape modifications than accompany band-narrowing by FSD [Lórenz-Fonfría, V.A., Villaverde, J., and Padrós, E. (2002) Fourier deconvolution in nonself-deconvolving conditions. Effective narrowing, signal-to-noise degradation, and curve fitting. *Appl. Spectrosc.*, **56**, 232–242.]. The curve-fitting residual is shown multiplied by 5 and shifted down for a clear display. From the area of the central band at 1657 cm^{-1} it is estimated that the AAC contains a minimum of 65% of α-helical structures. (Adapted from Lórenz *et al.* [Lórenz, V.A., Villaverde, J., Trézéguet, V., Lauquin, G.J., Brandolin, G., and Padrós, E. (2001) The secondary structure of the inhibited mitochondrial ADP/ATP transporter from yeast analyzed by FTIR spectroscopy. *Biochemistry*, **40**, 8821–8833.].)

in detergent micelles are too large for conventional solution NMR studies, the resolution of their structure to the atomic level has started to become possible using transverse relaxation-optimized spectroscopy (TROSY) in combination with isotopic-labeling strategies [165]. In spite of some recent notorious advances [98], the atomic resolution of IMPs in micelles by solution NMR, specially for α-helical IMPs, is still very challenging due to the difficulty to obtain sufficient distance constrains. Alternatively, NMR can also provide insight into the secondary structure of IMPs, founded in an empirical correlation between protein backbone conformation and the secondary Hα, Cα, and Cβ chemical shifts (the difference between observed shifts and random coil chemical shifts). A smoothed plot of the chemical shifts as a function of the sequence can be used to identify regions of secondary structure in detergent solubilized IMPs [166].

12.4.3 Measurement and Characterization of Ligand Binding

Ligand binding is generally studied by titrating a solution containing a protein with the stepwise addition of the ligand and measuring a signal proportional to the formation of the protein–ligand complex [167]. Radioactive methods have traditionally dominated this area and are still in use [168]. However, for measurement of ligand binding to detergent-solubilized IMPs other methods described below are more convenient. Note that besides titration, ligand binding to IMPs can be conveniently confirmed by an increase on the protein stability to thermal or chemical denaturation by an added ligand [169, 170]. Not discussed here is surface plasmon resonance (SPR), a very powerful method to detect and characterize protein–protein and protein–ligand interactions [171]. SPR has not been applied to detect ligand binding to detergent solubilized IMPs to our knowledge, but only to IMPs once reconstituted into lipid membranes [172].

12.4.3.1 Isothermal Titration Calorimetry (ITC)

ITC measures the heat released or absorbed for each stepwise addition of concentrated solution of a ligand molecule (around 10 µl/injection) to a typically around 1.5-ml solution containing the macromolecule under study [173, 174], with a volume reduced to 200 µl in some modern calorimeters. After integration of the spiky responses of the calorimeter, the data consists of a binding saturation curve as a function of the added ligand concentration. We should note that in spite of their sensitivity, modern calorimeters still require a concentration of 10–100 µM of protein for accurate binding measurements (e.g., 0.5–5 mg/ml for a 50-kDa protein). ITC has been applied to several detergent-solubilized IMPs [175, 176].

For ligands with moderate binding constants ($10^4 < K_a < 10^8\,M^{-1}$), the binding affinity constant (K_a), the binding enthalpy, and the binding stoichiometry factor can be estimated applying an appropriate binding model to the ITC binding saturation curve [173]. Note that other techniques require additional experiments as a function of the temperature to estimate the enthalpy of binding. Competitive binding displacements protocols have been developed to study high- and low-affinity systems with ITC [174].

12.4.3.2 Spectroscopic Methods

Among the many spectroscopies available, fluorescence is probably the most versatile to detect and quantify protein–ligand interactions, requiring also low amounts of sample. Other spectroscopies are generally less sensitive, and also more demanding in terms of the amounts of sample and experimental complexity, but also generally more informative about the nature of the protein–ligand interaction.

Fluorescence There are three basic fluorescence observables (intensity, wavelength maximum (color), and anisotropy), all of them useful for characterizing ligand binding [177]. Fluorescence binding experiments are based on the observation of either intrinsic fluorescence (mainly tryptophan from proteins) or that of extrinsic probes. In some cases, the binding of the ligand to the protein modulates the intensity/color of the intrinsic tryptophan fluorescence of the protein. In this case, binding is characterized by titrating the protein with stepwise additions of the ligand.

If no change in intrinsic fluorescence occurs upon ligand binding, which can occur by the relative low sensitivity of tryptophan fluorescence or by an inadequate location of protein's tryptophan to sense ligand binding, one must resort to extrinsic fluorescence probes. If the experimental design is based on a fluorescent ligand, it is ideal to follow the binding by the changes in the fluorescence anisotropy of the ligand (reporting changes in its rotational mobility), which only depend on the bound/free ligand ratio [177, 178]. An alternative is to monitor the fluorescence intensity of a ligand analog in a competition assay. Here, the fluorescence intensity changes caused by the release of the fluorescent ligand by the stepwise addition of an unlabeled natural ligand are used to calculate the unlabeled ligand true affinity constant. However, care must be taken with the use of fluorescent ligands analogs. They are habitually hydrophobic molecules that can also interact nonspecifically with proteins and detergents, often being not fully displaceable by any amount of a natural ligand [179].

Other fluorescent methods to probe ligand binding are also available. Changes in the fluorophore size as it binds a protein can be monitored by fluorescence correlation spectroscopy (FCS). FCS measures the fluorescence intensity fluctuations in a tiny volume. The autocorrelation of this signal decays with a time constant proportional to the size (Stokes radius) of the diffusing fluorescent particle [178]. Förster resonance energy transfer (FRET) between a protein's tryptophan and a fluorescent ligand can be also used for binding titration experiments [180].

CD For CD to report ligand binding, either the ligand should be a chromophore binding in an asymmetric fashion (generating an *extrinsic* CD signal) or the binding should result in a protein conformational change affecting its *intrinsic* CD spectrum [181]. In either case, the change in ellipticity as a function of the substrate concentration provides the binding constant and, in the case of changes in the *intrinsic* CD, clues about the nature of the protein conformational change induced by ligand binding [155]. However, because ligand binding is not always associated with significant rearrangements in secondary structure, CD can give substantial false negatives [181].

IR Spectroscopy IR spectroscopy is a well-suited spectroscopy to reveal and characterize protein–ligand interactions of IMPs. Due to the large and widely distributed number of IR-active vibrations in a protein and a ligand [159], false-negatives (spectrally silent genuine protein–ligand interactions) are not to be expected. However, since ligand binding can be conveniently studied by ATR-IR spectroscopy on IMPs once reconstituted in lipid membranes [182], applications to detergent-solubilized IMPs are not common.

NMR In spite of its low sensitivity and relatively large protein concentration requirements (around 10 mg/ml), NMR has opened a way into studying protein–ligand interactions, as reviewed elsewhere [183]. Several NMR observables, such as ^{1}H relaxation rates or saturation transfer, depend on the overall motion of a compound and so they can be used to follow the interaction of a small molecule (e.g., ligand) with a larger molecule (e.g., protein) [183, 184]. Information at the atomic level about where and how a ligand interacts with a protein can be obtained by measuring and assigning chemical shifts that occur upon binding [183]. For the study of large macromolecules, such as IMPs in micelles, these studies are possible by isotope-labeling combined with TROSY [165].

Appendix

The advantages and the disadvantages of the most common detergents in biochemistry research are listed in Table 12.A1. The structures of some commonly used detergents in IMP solubilization are given in Figures 12.A1–12.A3.

Table 12.A1 Advantages and disadvantages of the most common detergents used in biochemistry research.

Detergent	CMC	Aggregation number	Cloud point/ exploring for phase separation [8]	Compatibility/incompatibility with biochemical/biophysical assays	References (membrane proteins solubilized with retained functionality)
Nonionic detergents	relatively unaffected by ionic strength, increased by temperature[a)]				
Digitonin	NA	NA	temperature decreases with PEG 6000	disadvantages: not suitable for dialysis, crystallography, or spectroscopic studies; large size micelles [187, 7]	[186, 12]
Triton X-100	0.17–0.3	100–150	64–65 °C (can be decreased to room temperature [8, 7])	disadvantages: difficult to remove by dialysis; interferes with UV/Vis and CD measurements [187, 7]; gives false-positives in Bradford assay [2]	[7, 12, 188, 189]
Triton X-114	0.2–0.35	NA	20–25 °C (can be reduced to 4 °C [7, 190, 191])	advantages: low cloud point ideal for phase separation disadvantages: interferes with UV/Vis and CD measurements [7]	[190]
C12E8	0.07–0.11	123	74–79 °C (can be reduced to 43 °C)		[10, 212, 213]

DDM	0.15	98	<0 °C	advantages: one of best for maintaining protein structure disadvantages: forms large micelles (60 kDa); not suitable for structural studies; gives moderate resolution in crystallization [10]	[192, 193]
DM	1.6–0.15	69	<0 °C (addition of PEG 40000 increases cloud point)	advantages: easily removed by dialysis	[12, 194–196]
OG	20–25	84	<0 °C (addition of PEG 40000 and lipids increases cloud point, glycerol reduces)	advantages: easily removed by dialysis [7, 2] disadvantages: interferes with concanavalin A chromatography [2]; often leads to deactivation of protein [6, 197]	[10]
BIG CHAPS	2.4–3.4	10		alternative to CHAPS for two-dimensional gels disadvantages: interferes with protein assays [187]	[198, 199]
Ionic detergents	depends on ionic strength, relatively unaffected by temperature[a] [6]			disadvantages: incompatible with SDS–PAGE; affinity chromatography (His tag) not recommended [7]; ion-exchange chromatography or electrophoresis should be avoided [187, 2]	

(*Continued*)

Table 12.A1 (*Continued*)

Detergent	CMC	Aggregation number	Cloud point/ exploring for phase separation [8]	Compatibility/incompatibility with biochemical/biophysical assays	References (membrane proteins solubilized with retained functionality)
SDS	7–10	62	>100 °C	advantages: SDS solubilization facilitates folding studies [197] disadvantages: detergent binding may result in anomalous SDS–PAGE migration of membrane proteins [204]	[200–203]
Sodium cholate	9–15	2–3 (increases with ionic strength [207])	NA	advantages: easily removed by dialysis [2] disadvantages: forms precipitate in the Bradford assay; forms insoluble complexes with divalent metals [187]	[12, 205, 206]
Sodium deoxycholate	2–6	3–12 (increases with ionic strength [207])	NA	advantages: easily removed by dialysis; disadvantages: forms precipitate in the Bradford assay; prevents nonspecific binding in affinity chromatography [187]	[6]
CTAB	0.9	61–169	NA	advantages: useful for the separation of highly basic proteins and hydrophobic proteins; improves the identification of low-content proteins in CTAB/SDS mixture two-dimensional electrophoresis [7]	[208]

				do not bind to ion-exchange resins	
Zwitterionic					
CHAPS	6–10 (depends on ionic strength [59])	4–14	>100 °C	advantages: easily removed by dialysis [2]; good for spectroscopic studies; in combination with nonionic detergents used in two-dimensional gel electrophoresis [7] disadvantages: removes cholesterol from the membrane [80]	[205, 209, 210]
CHAPSO	8	11	90	advantages: easily removed by dialysis	[211]
Lauryldimethylamine-*N*-oxide (LDAO)	1.7–2.2	74	phase separation at pH range 6.0–7.5; addition of small amounts of cationic detergents influences cloud point [7]	advantages: small micelle size; high transparency at low wavelength disadvantages: aggressive denaturizing ability	[14]

NA, data not available.
Data are compiled from [8, 7, 4, 185] and a) www.emdbiosciences.com/calbiochem.

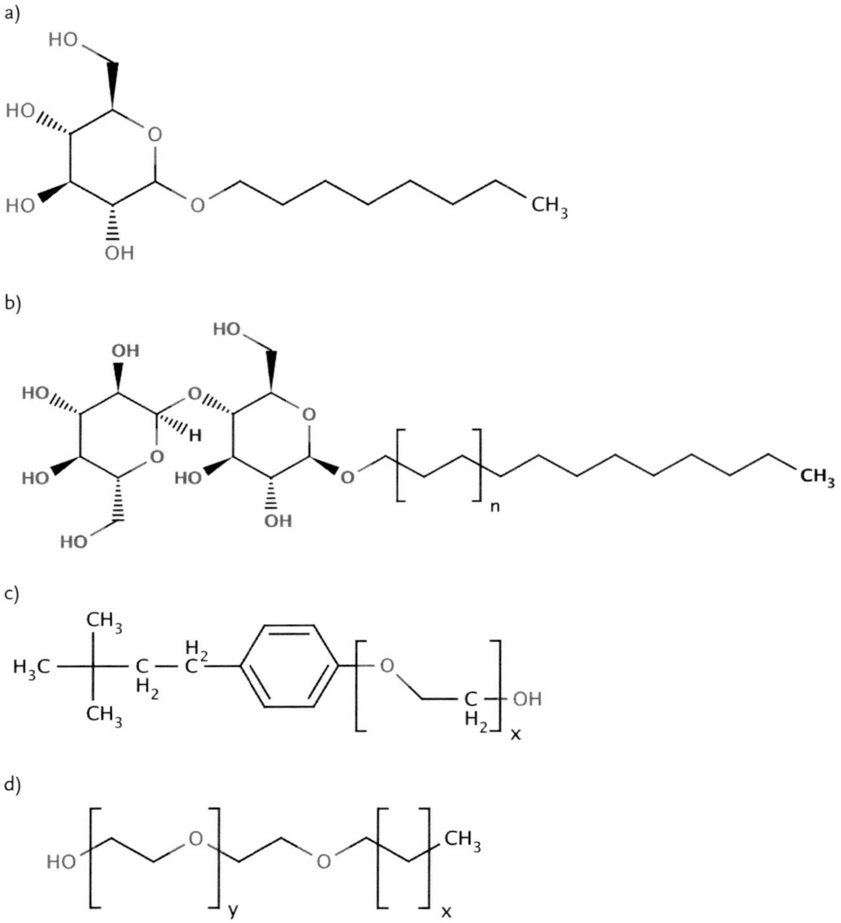

Figure 12.A1 Structure of commonly used nonionic detergents in membrane protein solubilization. (a) OG. (b) DM ($n = 1$); DDM ($n = 2$). (c) Triton X-100 ($x = 10$); Triton X-114 ($x = 14$). (d) Brij35 ($x = 10$, $y = 22$); nonaethylene glycol monododecyl ether C12E9 ($x = 10$, $y = 8$)

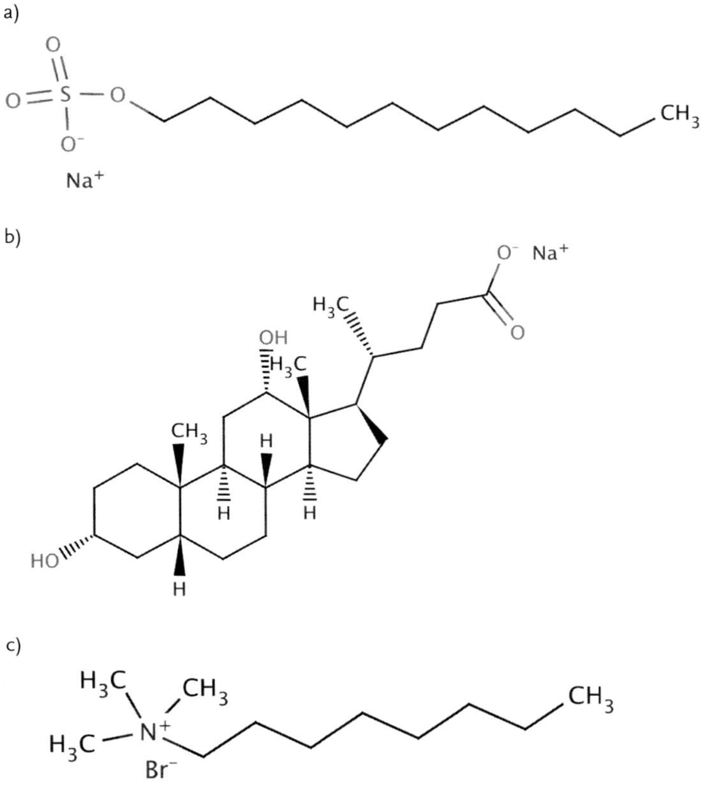

a)

O
O=S—O
O⁻
Na+
CH3

b)

O⁻ Na+
H3C
OH
H3C
CH3
H
O
H
H
HO
H

c)

H3C
CH3
H3C—N+
Br⁻
CH3

Figure 12.A2 Structure of commonly used ionic detergents in membrane protein solubilization. Anionic: SDS (a) and sodium deoxycholate (b). Cationic: CTAB (c).

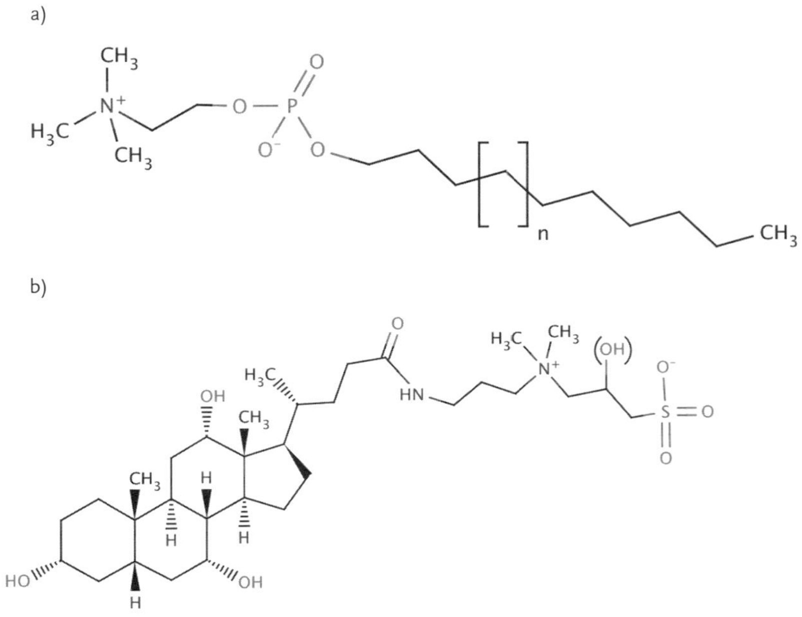

Figure 12.A3 Structure of commonly used zwitterionic detergents in membrane protein solubilization. (a) *N*-decylphosphocholine (fos-choline-10) ($n = 1$); *N*-dodecylphosphocholine (fos-choline-12) ($n = 2$). (b) CHAPSO/CHAPS.

Acknowledgments

This work was supported by Universitat Autònoma de Barcelona Postdoctoral Fellowship 40607 and Marie Curie Reintegration grant PIRG03-6A-2008-231063 (to V.L.-F.), Marie Curie International Outgoing Fellowship PIOF-GA-2009-237120 (to A.P.-M.), and by Ministerio de Ciencia e Innovación grants BFU2006-04656/BMC and BFU200908758/BMC. V.L.-F. and A.P.-M. contributed equally to this work.

Abbreviations

AAC	mitochondrial ADP/ATP carrier
ATR	attenuated total reflection
AUC	Analytical Ultracentrifugation
BN	Blue-Native
CD	Circular Dichroism
CF	Chromatofocusing
CHAPS	3-[(3-cholamidopropyl)dimethylammonio]-1-propanesulfonate
CHAPSO	3-[(3-cholamidopropyl) dimethylammonio]-2-hydroxy-1-propanesulfonate

CMC	critical micelle concentration
CSC	critical solubilization concentration
CTAB	cetyltrimethylammonium bromide
DDM	dodecyl-β-D-maltopyranoside
DM	decyl-β-D-maltopyranoside
FCS	fluorescence correlation spectroscopy
FRET	Förster resonance energy transfer
FSD	Fourier self-deconvolution
GFP	Green Fluorescent Protein
GPCR	G-protein-coupled receptor
HA	Hemagglutinin
HTP	High-throughput
IEF	Isoelectric Focusing
IEX	Ion-Exchange Chromatography
IMP	integral membrane protein
ITC	Isothermal Titration Calorimetry
LDAO	Lauryldimethylamine-*N*-oxide
LS	Light Scattering
MBP	Mannose-binding protein
MES	2-(*N*-morpholino)ethanesulfonic acid
OG	octyl-β-D-glucopyranoside
PAGE	polyacrylamide gel electrophoresis
PDC	protein–detergent complex
SDS	sodium dodecyl sulfate
SE	sedimentation equilibrium
SEC	Size-Exclusion Chromatography
SPR	surface plasmon resonance
SV	sedimentation velocity
TROSY	transverse relaxation-optimized spectroscopy

References

1 Ohlendieck, K. (2004) Extraction of membrane proteins. *Methods Mol. Biol.*, **244**, 283–293.

2 Lin, S.H. and Guidotti, G. (2009) Purification of membrane proteins. *Methods Enzymol.*, **463**, 619–629.

3 Rosenberg, I.M. (2005) *Protein Analysis and Purification: Benchtop Techniques*, 2nd edn, Birkhäuser, Boston, MA.

4 Linke, D. (2009) Detergents: an overview. *Methods Enzymol.*, **463**, 603–617.

5 Neugebauer, J.M. (1990) Detergents: an overview. *Methods Enzymol.*, **182**, 239–253.

6 le Maire, M., Champeil, P., and Moller, J.V. (2000) Interaction of membrane proteins and lipids with solubilizing detergents. *Biochim. Biophys. Acta*, **1508**, 86–111.

7 Arnold, T. and Linke, D. (2008) The use of detergents to purify membrane proteins. *Curr. Protoc. Protein Sci.*, **4**, 8.1–8.30.

8 Arnold, T. and Linke, D. (2007) Phase separation in the isolation and purification of membrane proteins. *Biotechniques*, **43**, 427–430, 432, 434 passim.

9 Chatterjee, A., Moulik, S.P., Sanyal, S.K., Mishra, B.K., and Puri, P.M. (2001) Thermodynamics of micelle formation

of ionic surfactants: a critical assessment for sodium dodecyl sulfate, cetyl pyridinium chloride and dioctyl sulfosuccinate (Na salt) by microcalorimetric, conductometric, and tensiometric measurements. *J. Phys. Chem. B*, **105**, 12823–12831.

10 Prive, G.G. (2007) Detergents for the stabilization and crystallization of membrane proteins. *Methods*, **41**, 388–397.

11 Koszelak-Rosenblum, M., Krol, A., Mozumdar, N., Wunsch, K., Ferin, A., Cook, E., Veatch, C.K., Nagel, R., Luft, J.R., Detitta, G.T., and Malkowski, M.G. (2009) Determination and application of empirically derived detergent phase boundaries to effectively crystallize membrane proteins. *Protein Sci.*, **18**, 1828–1839.

12 Sarramegna, V., Muller, I., Milon, A., and Talmont, F. (2006) Recombinant G protein-coupled receptors from expression to renaturation: a challenge towards structure. *Cell. Mol. Life Sci.*, **63**, 1149–1164.

13 Kimple, M.E. and Sondek, J. (2004) Overview of affinity tags for protein purification. *Curr. Protoc. Protein Sci.*, **9**, 9.9.

14 Baneres, J.L., Martin, A., Hullot, P., Girard, J.P., Rossi, J.C., and Parello, J. (2003) Structure-based analysis of GPCR function: conformational adaptation of both agonist and receptor upon leukotriene B_4 binding to recombinant BLT1. *J. Mol. Biol.*, **329**, 801–814.

15 Baneres, J.L., Mesnier, D., Martin, A., Joubert, L., Dumuis, A., and Bockaert, J. (2005) Molecular characterization of a purified 5-HT_4 receptor: a structural basis for drug efficacy. *J. Biol. Chem.*, **280**, 20253–20260.

16 de Jong, L.A., Grunewald, S., Franke, J.P., Uges, D.R., and Bischoff, R. (2004) Purification and characterization of the recombinant human dopamine D_{2S} receptor from *Pichia pastoris*. *Protein Expr. Purif.*, **33**, 176–184.

17 Kiefer, H., Krieger, J., Olszewski, J.D., Von Heijne, G., Prestwich, G.D., and Breer, H. (1996) Expression of an olfactory receptor in *Escherichia coli*: purification, reconstitution, and ligand binding. *Biochemistry*, **35**, 16077–16084.

18 Mazina, K.E., Strader, C.D., Tota, M.R., Daniel, S., and Fong, T.M. (1996) Purification and reconstitution of a recombinant human neurokinin-1 receptor. *J. Recept. Signal Transduct. Res.*, **16**, 191–207.

19 Nekrasova, E., Sosinskaya, A., Natochin, M., Lancet, D., and Gat, U. (1996) Overexpression, solubilization and purification of rat and human olfactory receptors. *Eur. J. Biochem.*, **238**, 28–37.

20 Ratnala, V.R., Swarts, H.G., VanOostrum, J., Leurs, R., DeGroot, H.J., Bakker, R.A., and DeGrip, W.J. (2004) Large-scale overproduction, functional purification and ligand affinities of the His-tagged human histamine H_1 receptor. *Eur. J. Biochem.*, **271**, 2636–2646.

21 Schiller, H., Molsberger, E., Janssen, P., Michel, H., and Reilander, H. (2001) Solubilization and purification of the human ETB endothelin receptor produced by high-level fermentation in *Pichia pastoris*. *Receptors Channels*, **7**, 453–469.

22 Andersen, B. and Stevens, R.C. (1998) The human D_{1A} dopamine receptor: heterologous expression in *Saccharomyces cerevisiae* and purification of the functional receptor. *Protein Expr. Purif.*, **13**, 111–119.

23 Christoffers, K.H., Li, H., and Howells, R.D. (2005) Purification and mass spectrometric analysis of the delta opioid receptor. *Brain Res. Mol. Brain Res.*, **136**, 54–64.

24 Christoffers, K.H., Li, H., Keenan, S.M., and Howells, R.D. (2003) Purification and mass spectrometric analysis of the mu opioid receptor. *Brain Res. Mol. Brain Res.*, **118**, 119–131.

25 David, N.E., Gee, M., Andersen, B., Naider, F., Thorner, J., and Stevens, R.C. (1997) Expression and purification of the *Saccharomyces cerevisiae* alpha-factor receptor (Ste2p), a 7-transmembrane-segment G protein-coupled receptor. *J. Biol. Chem.*, **272**, 15553–15561.

26 Grisshammer, R. and Tucker, J. (1997) Quantitative evaluation of neurotensin receptor purification by immobilized

metal affinity chromatography. *Protein Expr. Purif.*, **11**, 53–60.

27 Park, P.S. and Wells, J.W. (2003) Monomers and oligomers of the M2 muscarinic cholinergic receptor purified from Sf9 cells. *Biochemistry*, **42**, 12960–12971.

28 Kwatra, M.M., Schreurs, J., Schwinn, D.A., Innis, M.A., Caron, M.G., and Lefkowitz, R.J. (1995) Immunoaffinity purification of epitope-tagged human beta 2-adrenergic receptor to homogeneity. *Protein Expr. Purif.*, **6**, 717–721.

29 Hampe, W., Voss, R.H., Haase, W., Boege, F., Michel, H., and Reilander, H. (2000) Engineering of a proteolytically stable human beta 2-adrenergic receptor/maltose-binding protein fusion and production of the chimeric protein in *Escherichia coli* and baculovirus-infected insect cells. *J. Biotechnol.*, 219–234.

30 Yeliseev, A.A., Wong, K.K., Soubias, O., and Gawrisch, K. (2005) Expression of human peripheral cannabinoid receptor for structural studies. *Protein Sci.*, **14**, 2638–2653.

31 Shi, C., Shin, Y.O., Hanson, J., Cass, B., Loewen, M.C., and Durocher, Y. (2005) Purification and characterization of a recombinant G-protein-coupled receptor, *Saccharomyces cerevisiae* Ste2p, transiently expressed in HEK293 EBNA1 cells. *Biochemistry*, **44**, 15705–15714.

32 Tucker, J. and Grisshammer, R. (1996) Purification of a rat neurotensin receptor expressed in *Escherichia coli*. *Biochem. J.*, **317**, 891–899.

33 Hammon, J., Palanivelu, D.V., Chen, J., Patel, C., and Minor, D.L., Jr (2009) A green fluorescent protein screen for identification of well-expressed membrane proteins from a cohort of extremophilic organisms. *Protein Sci.*, **18**, 121–133.

34 Eshaghi, S., Hedren, M., Nasser, M.I., Hammarberg, T., Thornell, A., and Nordlund, P. (2005) An efficient strategy for high-throughput expression screening of recombinant integral membrane proteins. *Protein Sci.*, **14**, 676–683.

35 Sarramegna, V., Muller, I., Mousseau, G., Froment, C., Monsarrat, B., Milon, A., and Talmont, F. (2005) Solubilization, purification, and mass spectrometry analysis of the human mu-opioid receptor expressed in *Pichia pastoris*. *Protein Expr. Purif.*, **43**, 85–93.

36 Zhang, L., Salom, D., He, J., Okun, A., Ballesteros, J., Palczewski, K., and Li, N. (2005) Expression of functional G protein-coupled receptors in photoreceptors of transgenic *Xenopus laevis*. *Biochemistry*, **44**, 14509–14518.

37 Asmann, Y.W., Dong, M., and Miller, L.J. (2004) Functional characterization and purification of the secretin receptor expressed in baculovirus-infected insect cells. *Regul. Pept.*, **123**, 217–223.

38 Blackburn, P.E., Simpson, C.V., Nibbs, R.J., O'Hara, M., Booth, R., Poulos, J., Isaacs, N.W., and Graham, G.J. (2004) Purification and biochemical characterization of the D6 chemokine receptor. *Biochem. J.*, **379**, 263–272.

39 Alves, I.D., Cowell, S.M., Salamon, Z., Devanathan, S., Tollin, G., and Hruby, V.J. (2004) Different structural states of the proteolipid membrane are produced by ligand binding to the human delta-opioid receptor as shown by plasmon wave guide resonance spectroscopy. *Mol. Pharmacol.*, **65**, 1248–1257.

40 Feng, W., Benz, F.W., Cai, J., Pierce, W.M., and Kang, Y.J. (2006) Metallothionein disulfides are present in metallothionein-overexpressing transgenic mouse heart and increase under conditions of oxidative stress. *J. Biol. Chem.*, **281**, 681–687.

41 Kim, T.K., Zhang, R., Feng, W., Cai, J., Pierce, W., and Song, Z.H. (2005) Expression and characterization of human CB1 cannabinoid receptor in methylotrophic yeast *Pichia pastoris*. *Protein Expr. Purif.*, **40**, 60–70.

42 Mirzabekov, T., Bannert, N., Farzan, M., Hofmann, W., Kolchinsky, P., Wu, L., Wyatt, R., and Sodroski, J. (1999) Enhanced expression, native purification, and characterization of CCR5, a principal HIV-1 coreceptor. *J. Biol. Chem.*, **274**, 28745–28750.

43 Zvonok, N., Xu, W., Williams, J., Janero, D., Krishnan, S., and Makriyannis, A.

(2010) Mass spectrometry-based GPCR proteomics: comprehensive characterization of the human cannabinoid 1 receptor. *J. Proteome Res.*, **9**, 1746–1753.

44 Ren, H., Yu, D., Ge, B., Cook, B., Xu, Z., and Zhang, S. (2009) High-level production, solubilization and purification of synthetic human GPCR chemokine receptors CCR5, CCR3, CXCR4 and CX3CR1. *PLoS ONE*, **4**, e4509.

45 Zhang, Q., Horst, R., Geralt, M., Ma, X., Hong, W.X., Finn, M.G., Stevens, R.C., and Wuthrich, K. (2008) Microscale NMR screening of new detergents for membrane protein structural biology. *J. Am. Chem. Soc.*, **130**, 7357–7363.

46 Bowie, J.U. (2001) Stabilizing membrane proteins. *Curr. Opin. Struct. Biol.*, **11**, 397–402.

47 Prive, G.G. (2009) Lipopeptide detergents for membrane protein studies. *Curr. Opin. Struct. Biol.*, **19**, 379–385.

48 McGregor, C.L., Chen, L., Pomroy, N.C., Hwang, P., Go, S., Chakrabartty, A., and Prive, G.G. (2003) Lipopeptide detergents designed for the structural study of membrane proteins. *Nat. Biotechnol.*, **21**, 171–176.

49 Zhao, X., Nagai, Y., Reeves, P.J., Kiley, P., Khorana, H.G., and Zhang, S. (2006) Designer short peptide surfactants stabilize G protein-coupled receptor bovine rhodopsin. *Proc. Natl. Acad. Sci. USA*, **103**, 17707–17712.

50 Breyton, C., Pucci, B., and Popot, J.L. (2010) Amphipols and fluorinated surfactants: two alternatives to detergents for studying membrane proteins *in vitro*. *Methods Mol. Biol.*, **601**, 219–245.

51 Popot, J.L., Berry, E.A., Charvolin, D., Creuzenet, C., Ebel, C., Engelman, D.M., Flotenmeyer, M., Giusti, F., Gohon, Y., Hong, Q., Lakey, J.H., Leonard, K., Shuman, H.A., Timmins, P., Warschawski, D.E., Zito, F., Zoonens, M., Pucci, B., and Tribet, C. (2003) Amphipols: polymeric surfactants for membrane biology research. *Cell. Mol. Life Sci.*, **60**, 1559–1574.

52 McQuade, D.T., Quinn, M.A., Yu, S.M., Polans, A.S., Krebs, M.P., and Gellman, S.H. (2000) Rigid amphiphiles for membrane protein manipulation. *Angew. Chem. Int. Ed.*, **39**, 758–761.

53 Yu, S.M., McQuade, D.T., Quinn, M.A., Hackenberger, C.P., Krebs, M.P., Polans, A.S., and Gellman, S.H. (2000) An improved tripod amphiphile for membrane protein solubilization. *Protein Sci.*, **9**, 2518–2527.

54 Cladera, J., Rigaud, J.L., Villaverde, J., and Dunach, M. (1997) Liposome solubilization and membrane protein reconstitution using Chaps and Chapso. *Eur. J. Biochem.*, **243**, 798–804.

55 de la Maza, A. and Parra, J.L. (1997) Solubilizing effects caused by the nonionic surfactant dodecylmaltoside in phosphatidylcholine liposomes. *Biophys. J.*, **72**, 1668–1675.

56 Hjelmeland, L.M. (1990) Solubilization of native membrane proteins. *Methods Enzymol.*, **182**, 253–264.

57 Psakis, G., Nitschkowski, S., Holz, C., Kress, D., Maestre-Reyna, M., Polaczek, J., Illing, G., and Essen, L.O. (2007) Expression screening of integral membrane proteins from *Helicobacter pylori* 26695. *Protein Sci.*, **16**, 2667–2676.

58 Schimerlik, M.I. (2001) Overview of membrane protein solubilization. *Curr. Protoc. Neurosci.*, **5**, 5.9.

59 Kalipatnapu, S. and Chattopadhyay, A. (2005) Membrane protein solubilization: recent advances and challenges in solubilization of serotonin1A receptors. *IUBMB Life*, **57**, 505–512.

60 Chabre, M. and Maire, M. (2005) Monomeric G-protein-coupled receptor as a functional unit. *Biochemistry*, **44**, 9395–9403.

61 Lee, A.G. (2004) How lipids affect the activities of integral membrane proteins. *Biochim. Biophys. Acta*, **1666**, 62–87.

62 Opekarova, M. and Tanner, W. (2003) Specific lipid requirements of membrane proteins – a putative bottleneck in heterologous expression. *Biochim. Biophys. Acta*, **1610**, 11–22.

63 Watts, A. (1989) Membrane structure and dynamics. *Curr. Opin. Cell Biol.*, **1**, 691–700.

64 Esmann, M. (1984) The distribution of C12E8-solubilized oligomers of the (Na^{++}

K$^+$)-ATPase. *Biochim. Biophys. Acta*, **787**, 81–89.

65 Breyton, C., Tribet, C., Olive, J., Dubacq, J.P., and Popot, J.L. (1997) Dimer to monomer conversion of the cytochrome b_6f complex. Causes and consequences. *J. Biol. Chem.*, **272**, 21892–21900.

66 Lange, C., Nett, J.H., Trumpower, B.L., and Hunte, C. (2001) Specific roles of protein–phospholipid interactions in the yeast cytochrome bc_1 complex structure. *EMBO J.*, **20**, 6591–6600.

67 Cowan, S.W., Schirmer, T., Rummel, G., Steiert, M., Ghosh, R., Pauptit, R.A., Jansonius, J.N., and Rosenbusch, J.P. (1992) Crystal structures explain functional properties of two *E. coli* porins. *Nature*, **358**, 727–733.

68 Cyrklaff, M., Auer, M., Kuhlbrandt, W., and Scarborough, G.A. (1995) 2-D structure of the *Neurospora crassa* plasma membrane ATPase as determined by electron cryomicroscopy. *EMBO J.*, **14**, 1854–1857.

69 Guan, L., Smirnova, I.N., Verner, G., Nagamori, S., and Kaback, H.R. (2006) Manipulating phospholipids for crystallization of a membrane transport protein. *Proc. Natl. Acad. Sci. USA*, **103**, 1723–1726.

70 Toyoshima, C., Nakasako, M., Nomura, H., and Ogawa, H. (2000) Crystal structure of the calcium pump of sarcoplasmic reticulum at 2.6 A resolution. *Nature*, **405**, 647–655.

71 Luecke, H., Schobert, B., Richter, H.T., Cartailler, J.P., and Lanyi, J.K. (1999) Structure of bacteriorhodopsin at 1.55 A resolution. *J. Mol. Biol.*, **291**, 899–911.

72 Silvius, J.R. (1992) Solubilization and functional reconstitution of biomembrane components. *Annu. Rev. Biophys. Biomol. Struct.*, **21**, 323–348.

73 Rizzolo, L.J. (1981) Kinetics and protein subunit interactions of *Escherichia coli* phosphatidylserine decarboxylase in detergent solution. *Biochemistry*, **20**, 868–873.

74 Attrill, H., Harding, P.J., Smith, E., Ross, S., and Watts, A. (2009) Improved yield of a ligand-binding GPCR expressed in *E. coli* for structural studies. *Protein Expr. Purif.*, **64**, 32–38.

75 Yao, Z. and Kobilka, B. (2005) Using synthetic lipids to stabilize purified beta2 adrenoceptor in detergent micelles. *Anal. Biochem.*, **343**, 344–346.

76 Balen, P., Kimura, K., and Sidhu, A. (1994) Specific phospholipid requirements for the solubilization and reconstitution of D-1 dopamine receptors from striatal membranes. *Biochemistry*, **33**, 1539–1544.

77 Allen, S.J., Ribeiro, S., Horuk, R., and Handel, T.M. (2009) Expression, purification and *in vitro* functional reconstitution of the chemokine receptor CCR1. *Protein Expr. Purif.*, **66**, 73–81.

78 O'Malley, M.A., Lazarova, T., Britton, Z.T., and Robinson, A.S. (2007) High-level expression in *Saccharomyces cerevisiae* enables isolation and spectroscopic characterization of functional human adenosine A2a receptor. *J. Struct. Biol.*, **159**, 166–178.

79 Paila, Y.D. and Chattopadhyay, A. (2009) The function of G-protein coupled receptors and membrane cholesterol: specific or general interaction? *Glycoconj. J.*, **26**, 711–720.

80 Banerjee, P., Joo, J.B., Buse, J.T., and Dawson, G. (1995) Differential solubilization of lipids along with membrane proteins by different classes of detergents. *Chem. Phys. Lipids*, **77**, 65–78.

81 Chattopadhyay, A., Jafurulla, M., Kalipatnapu, S., Pucadyil, T.J., and Harikumar, K.G. (2005) Role of cholesterol in ligand binding and G-protein coupling of serotonin1A receptors solubilized from bovine hippocampus. *Biochem. Biophys. Res. Commun.*, **327**, 1036–1041.

82 Weiss, H.M. and Grisshammer, R. (2002) Purification and characterization of the human adenosine A_{2a} receptor functionally expressed in *Escherichia coli*. *Eur. J. Biochem.*, **269**, 82–92.

83 Vasudevan, S., Hulme, E.C., Bach, M., Haase, W., Pavia, J., and Reilander, H. (1995) Characterization of the rat m3 muscarinic acetylcholine receptor produced in insect cells infected with recombinant baculovirus. *Eur. J. Biochem.*, **227**, 466–475.

84 van Meer, G., Voelker, D.R., and Feigenson, G.W. (2008) Membrane lipids: where they are and how they behave. *Nat. Rev.*, **9**, 112–124.

85 Rummel, G. and Rosenbusch, J. (2003) Crystallization of bacterial outer membrane proteins from detergent solution: porin as a model, in *Methods and Results in Crystallization of Membrane Proteins* (ed. S. Iwata), International University Line, La Jolla, CA, pp. 101–129.

86 Altendorf, K., Lukas, M., Kohl, B., Muller, C.R., and Sandermann, H., Jr (1977) Isolation and purification of bacterial membrane proteins by the use of organic solvents: the lactose permease and the carbodiimide-reactive protein of the adenosine triphosphatase complex of *Escherichia coli*. *J. Supramol. Struct.*, **6**, 229–238.

87 Goncharuk, S.A., Shulga, A.A., Ermolyuk, Y.S., Kuzmichev, P.K., Sobol, V.A., Bocharov, E.V., Chupin, V.V., Arseniev, A.S., and Kirpichnikov, M.P. (2009) Bacterial synthesis, purification, and solubilization of membrane protein KCNE3, a regulator of voltage-gated potassium channels. *Biochemistry*, **74**, 1344–1349.

88 Singh, P., Jansch, L., Braun, H.P., and Schmitz, U.K. (2000) Resolution of mitochondrial and chloroplast membrane protein complexes from green leaves of potato on blue-native polyacrylamide gels. *Indian J. Biochem. Biophys.*, **37**, 59–66.

89 Mitra, S.K., Walters, B.T., Clouse, S.D., and Goshe, M.B. (2009) An efficient organic solvent based extraction method for the proteomic analysis of Arabidopsis plasma membranes. *J. Proteome Res.*, **8**, 2752–2767.

90 Chang, Y.C., Wu, T.Y., Li, B.F., Gao, L.H., Liu, C.I., and Wu, C.L. (1996) Purification and biochemical characterization of alpha-amino-3-hydroxy-5 methyl-4-isoxazolepropionic acid/kainate-sensitive L-glutamate receptors of pig brain. *Biochem. J.*, **319**, 49–57.

91 Marmagne, A., Salvi, D., Rolland, N., Ephritikhine, G., Joyard, J., and Barbier-Brygoo, H. (2006) Purification and fractionation of membranes for proteomic analyses. *Methods Mol. Biol.*, **323**, 403–420.

92 Hirano, M., Rakwal, R., Shibato, J., Agrawal, G.K., Jwa, N.S., Iwahashi, H., and Masuo, Y. (2006) New protein extraction/solubilization protocol for gel-based proteomics of rat (female) whole brain and brain regions. *Mol. Cells*, **22**, 119–125.

93 Davidsson, P., Westman, A., Puchades, M., Nilsson, C.L., and Blennow, K. (1999) Characterization of proteins from human cerebrospinal fluid by a combination of preparative two-dimensional liquid-phase electrophoresis and matrix-assisted laser desorption/ionization time-of-flight mass spectrometry. *Anal. Chem.*, **71**, 642–647.

94 Schindler, J., Jung, S., Niedner-Schatteburg, G., Friauf, E., and Nothwang, H.G. (2006) Enrichment of integral membrane proteins from small amounts of brain tissue. *J. Neural. Transm.*, **113**, 995–1013.

95 Grisshammer, R. (2006) Understanding recombinant expression of membrane proteins. *Curr. Opin. Biotechnol.*, **17**, 337–340.

96 Miroux, B. and Walker, J.E. (1996) Over-production of proteins in *Escherichia coli*: mutant hosts that allow synthesis of some membrane proteins and globular proteins at high levels. *J. Mol. Biol.*, **260**, 289–298.

97 Wingfield, P. (2009) Production of recombinant proteins, *Curr. Protoc. Protein Sci.*, **5**, 5.0.

98 Van Horn, W.D., Kim, H.J., Ellis, C.D., Hadziselimovic, A., Sulistijo, E.S., Karra, M.D., Tian, C., Sönnichsen, F.D., and Sanders, C.R. (2009) Solution nuclear magnetic resonance structure of membrane-integral diacylglycerol kinase. *Science*, **324**, 1726–1729.

99 Eshaghi, S. (2009) High-throughput expression and detergent screening of integral membrane proteins. *Methods Mol. Biol.*, **498**, 265–271.

100 McLuskey, K., Gabrielsen, M., Kroner, F., Black, I., Cogdell, R.J., and Isaacs, N.W. (2008) A protocol for high throughput methods for the expression and purification of inner membrane proteins. *Mol. Membr. Biol.*, **25**, 599–608.

101 Cutler, P. (2004) *Protein Purification Protocols*, 2nd edn, Humana, Totowa, NJ.

102 Hostettmann, K. and Marston, A. (1998) *Preparative Chromatography Techniques: Applications in Natural Product Isolation*, 2nd edn, Springer, Berlin.

103 Findlay, J. (1990) Purification of membrane proteins, in *Protein Purification Applications* (eds E. Harris and S. Angal), IRL, Oxford, pp. 59–82.

104 Selkirk, C. (2004) Ion-exchange chromatography. *Methods Mol. Biol.*, **244**, 125–131.

105 Friedman, D.B., Hoving, S., and Westermeier, R. (2009) Isoelectric focusing and two-dimensional gel electrophoresis. *Methods Enzymol.*, **463**, 515–540.

106 Hey, J., Posch, A., Cohen, A., Liu, N., and Harbers, A. (2008) Fractionation of complex protein mixtures by liquid-phase isoelectric focusing. *Methods Mol. Biol.*, **424**, 225–239.

107 Coligan, J.E. (1995) *Current Protocols in Protein Science*, John Wiley & Sons, Inc., New York.

108 Le Maire, M., Aggerbeck, L.P., Monteilhet, C., Andersen, J.P., and Moller, J.V. (1986) The use of high-performance liquid chromatography for the determination of size and molecular weight of proteins: a caution and a list of membrane proteins suitable as standards. *Anal. Biochem.*, **154**, 525–535.

109 le Maire, M., Rivas, E., and Moller, J.V. (1980) Use of gel chromatography for determination of size and molecular weight of proteins: further caution. *Anal. Biochem.*, **106**, 12–21.

110 Moller, J.V. and le Maire, M. (1993) Detergent binding as a measure of hydrophobic surface area of integral membrane proteins. *J. Biol. Chem.*, **268**, 18659–18672.

111 Kunji, E.R., Harding, M., Butler, P.J., and Akamine, P. (2008) Determination of the molecular mass and dimensions of membrane proteins by size exclusion chromatography. *Methods*, **46**, 62–72.

112 Wei, Y., Li, H., and Fu, D. (2004) Oligomeric state of the *Escherichia coli* metal transporter YiiP. *J. Biol. Chem.*, **279**, 39251–39259.

113 Newby, Z.E., O'Connell, J.D., 3rd, Gruswitz, F., Hays, F.A., Harries, W.E., Harwood, I.M., Ho, J.D., Lee, J.K., Savage, D.F., Miercke, L.J., and Stroud, R.M. (2009) A general protocol for the crystallization of membrane proteins for X-ray structural investigation. *Nat. Protoc.*, **4**, 619–637.

114 Ghosh, R. (2003) *Protein Bioseparation Using Ultrafiltration: Theory, Applications and New Developments*, Imperial College Press, London.

115 Thommes, J. and Etzel, M. (2007) Alternatives to chromatographic separations. *Biotechnol. Prog.*, **23**, 42–45.

116 Ladisch, M.R. (ed.) (1990) *Protein Purification: From Molecular Mechanisms to Large-Scale Processes*, American Chemical Society, Washington, DC.

117 Jones, C., Patel, A., Griffin, S., Martin, J., Young, P., O'Donnell, K., Silverman, C., Porter, T., and Chaiken, I. (1995) Current trends in molecular recognition and bioseparation. *J. Chromatogr. A*, **707**, 3–22.

118 Andersson, C., Hansson, M., Power, U., Nygren, P., and Stahl, S. (2001) Mammalian cell production of a respiratory syncytial virus (RSV) candidate vaccine recovered using a product-specific affinity column. *Biotechnol. Appl. Biochem.*, **34**, 25–32.

119 Matsubayashi, Y., Ogawa, M., Morita, A., and Sakagami, Y. (2002) An LRR receptor kinase involved in perception of a peptide plant hormone, phytosulfokine. *Science*, **296**, 1470–1472.

120 Sambrook, J. and Russell, D.W. (2001) *Molecular Cloning: A Laboratory Manual*, 3th edn, Cold Spring Harbor Laboratory Press, Cold Spring Harbor, NY.

121 Arnau, J., Lauritzen, C., Petersen, G.E., and Pedersen, J. (2006) Current strategies for the use of affinity tags and tag removal for the purification of recombinant proteins. *Protein Expr. Purif.*, **48**, 1–13.

122 Arnau, J., Lauritzen, C., Petersen, G.E., and Pedersen, J. (2008) The use of TAGZyme for the efficient removal of N-terminal His-tags. *Methods Mol. Biol.*, **421**, 229–243.

123 Cutler, P. (2004) Immunoaffinity chromatography. *Methods Mol. Biol.*, **244**, 167–177.

124 Kohler, G. and Milstein, C. (1975) Continuous cultures of fused cells secreting antibody of predefined specificity. *Nature*, **256**, 495–497.

125 Desai, M.A. (1990) Immunoaffinity adsorption: process-scale isolation of therapeutic-grade biochemicals. *J. Chem. Technol. Biotechnol.*, **48**, 105–126.

126 Whitehurst, C.E., Nazef, N., Annis, D.A., Hou, Y., Murphy, D.M., Spacciapoli, P., Yao, Z., Ziebell, M.R., Cheng, C.C., Shipps, G.W., Jr, Felsch, J.S., Lau, D., and Nash, H.M. (2006) Discovery and characterization of orthosteric and allosteric muscarinic M2 acetylcholine receptor ligands by affinity selection-mass spectrometry. *J. Biomol. Screen.*, **11**, 194–207.

127 Chazot, P.L. and Stephenson, F.A. (1997) Molecular dissection of native mammalian forebrain NMDA receptors containing the NR1 C2 exon: direct demonstration of NMDA receptors comprising NR1, NR2A, and NR2B subunits within the same complex. *J. Neurochem.*, **69**, 2138–2144.

128 Kumar, K.N., Babcock, K.K., Johnson, P.S., Chen, X., Eggeman, K.T., and Michaelis, E.K. (1994) Purification and pharmacological and immunochemical characterization of synaptic membrane proteins with ligand binding properties of *N*-methyl-D-aspartate receptors. *J. Biol. Chem.*, **269**, 27384–27393.

129 Nakagawa, T., Cheng, Y., Ramm, E., Sheng, M., and Walz, T. (2005) Structure and different conformational states of native AMPA receptor complexes. *Nature*, **433**, 545–549.

130 Andre, M., Morelle, W., Planchon, S., Milhiet, P.E., Rubinstein, E., Mollicone, R., Chamot-Rooke, J., and Le Naour, F. (2007) Glycosylation status of the membrane protein CD9P-1. *Proteomics*, **7**, 3880–3895.

131 West, I. and Goldring, O. (2004) Lectin affinity chromatography. *Methods Mol. Biol.*, **244**, 159–166.

132 Garg, N., Galaev, I.Y., and Mattiasson, B. (1996) Dye-affinity techniques for bioprocessing: recent developments. *J. Mol. Recognit.*, **9**, 259–274.

133 McGettrick, A.F. and Worrall, D.M. (2004) Dye-ligand affinity chromatography. *Methods Mol. Biol.*, **244**, 151–157.

134 Lenoir, G., Menguy, T., Corre, F., Montigny, C., Pedersen, P.A., Thines, D., le Maire, M., and Falson, P. (2002) Overproduction in yeast and rapid and efficient purification of the rabbit SERCA1a Ca^{2+}-ATPase. *Biochim. Biophys. Acta*, **1560**, 67–83.

135 Yao, Q., Chen, L.T., and Bigelow, D.J. (1998) Affinity purification of the Ca-ATPase from cardiac sarcoplasmic reticulum membranes. *Protein Expr. Purif.*, **13**, 191–197.

136 Franzreb, M., Siemann-Herzberg, M., Hobley, T.J., and Thomas, O.R. (2006) Protein purification using magnetic adsorbent particles. *Appl. Microbiol. Biotechnol.*, **70**, 505–516.

137 Lee, Y.C., Block, G., Chen, H., Folch-Puy, E., Foronjy, R., Jalili, R., Jendresen, C.B., Kimura, M., Kraft, E., Lindemose, S., Lu, J., McLain, T., Nutt, L., Ramon-Garcia, S., Smith, J., Spivak, A., Wang, M.L., Zanic, M., and Lin, S.H. (2008) One-step isolation of plasma membrane proteins using magnetic beads with immobilized concanavalin A. *Protein Expr. Purif.*, **62**, 223–229.

138 Sivars, U., Abramson, J., Iwata, S., and Tjerneld, F. (2000) Affinity partitioning of a poly(histidine)-tagged integral membrane protein, cytochrome bo_3 ubiquinol oxidase, in a detergent–polymer aqueous two-phase system containing metal chelating polymer. *J. Chromatogr.*, **743**, 307–316.

139 Striegel, A.M. (2005) Multiple detection in size-exclusion chromatography of macromolecules. *Anal. Chem.*, **77**, 104A–113A.

140 Jasti, J., Furukawa, H., Gonzales, E.B., and Gouaux, E. (2007) Structure of acid-sensing ion channel 1 at 1.9Å resolution and low pH. *Nature*, **449**, 316–323.

141 Bamber, L., Harding, M., Butler, P.J., and Kunji, E.R. (2006) Yeast

mitochondrial ADP/ATP carriers are monomeric in detergents. *Proc. Natl. Acad. Sci. USA*, **103**, 16224–16229.

142 daCosta, C.J. and Baenziger, J.E. (2003) A rapid method for assessing lipid:protein and detergent:protein ratios in membrane-protein crystallization. *Acta Crystallogr. D*, **59**, 77–83.

143 Urbani, A. and Warne, T. (2005) A colorimetric determination for glycosidic and bile salt-based detergents: applications in membrane protein research. *Anal. Biochem.*, **336**, 117–124.

144 Slotboom, D.J., Duurkens, R.H., Olieman, K., and Erkens, G.B. (2008) Static light scattering to characterize membrane proteins in detergent solution. *Methods*, **46**, 73–82.

145 Albright, R.A., Ibar, J.L., Kim, C.U., Gruner, S.M., and Morais-Cabral, J.H. (2006) The RCK domain of the KtrAB K^+ transporter: multiple conformations of an octameric ring. *Cell*, **126**, 1147–1159.

146 Wen, J., Arakawa, T., and Philo, J.S. (1996) Size-exclusion chromatography with on-line light-scattering, absorbance, and refractive index detectors for studying proteins and their interactions. *Anal. Biochem.*, **240**, 155–166.

147 Lebowitz, J., Lewis, M.S., and Schuck, P. (2002) Modern analytical ultracentrifugation in protein science: a tutorial review. *Protein Sci.*, **11**, 2067–2079.

148 Laue, T.M. and Stafford, W.F., 3rd (1999) Modern applications of analytical ultracentrifugation. *Annu. Rev. Biophys. Biomol. Struct.*, **28**, 75–100.

149 Nury, H., Manon, F., Arnou, B., le Maire, M., Pebay-Peyroula, E., and Ebel, C. (2008) Mitochondrial bovine ADP/ATP carrier in detergent is predominantly monomeric but also forms multimeric species. *Biochemistry*, **47**, 12319–12331.

150 Reynolds, J.A. and Tanford, C. (1976) Determination of molecular weight of the protein moiety in protein–detergent complexes without direct knowledge of detergent binding. *Proc. Natl. Acad. Sci. USA*, **73**, 4467–4470.

151 Center, R.J., Schuck, P., Leapman, R.D., Arthur, L.O., Earl, P.L., Moss, B., and Lebowitz, J. (2001) Oligomeric structure of virion-associated and soluble forms of the simian immunodeficiency virus envelope protein in the prefusion activated conformation. *Proc. Natl. Acad. Sci. USA*, **98**, 14877–14882.

152 Fleming, K.G. (2008) Determination of membrane protein molecular weight using sedimentation equilibrium analytical ultracentrifugation. *Curr. Protoc. Protein Sci.*, **53**, 7.12.11–17.12.13.

153 Heuberger, E.H., Veenhoff, L.M., Duurkens, R.H., Friesen, R.H., and Poolman, B. (2002) Oligomeric state of membrane transport proteins analyzed with blue native electrophoresis and analytical ultracentrifugation. *J. Mol. Biol.*, **317**, 591–600.

154 Kelly, S.M., Jess, T.J., and Price, N.C. (2005) How to study proteins by circular dichroism. *Biochim. Biophys. Acta*, **1751**, 119–139.

155 Greenfield, N.J. (2004) Analysis of circular dichroism data. *Methods Enzymol.*, **383**, 282–317.

156 Whitmore, L. and Wallace, B.A. (2008) Protein secondary structure analyses from circular dichroism spectroscopy: methods and reference databases. *Biopolymers*, **89**, 392–400.

157 Wallace, B.A., Lees, J.G., Orry, A.J., Lobley, A., and Janes, R.W. (2003) Analyses of circular dichroism spectra of membrane proteins. *Protein Sci.*, **12**, 875–884.

158 Sreerama, N. and Woody, R.W. (2004) On the analysis of membrane protein circular dichroism spectra. *Protein Sci.*, **13**, 100–112.

159 Barth, A. and Zscherp, C. (2002) What vibrations tell us about proteins. *Q. Rev. Biophys.*, **35**, 369–430.

160 Haris, P.I. and Chapman, D. (1996) Fourier transform infrared spectroscopic studies of biomembrane systems, in *Infrared Spectroscopy of Biomolecules* (eds H.H. Mantsch and D. Chapman), Wiley-Liss, New York, pp. 239–278.

161 Byler, D.M. and Susi, H. (1986) Examination of the secondary structure of proteins by deconvolved FTIR spectra. *Biopolymers*, **25**, 469–487.

162 Lórenz-Fonfría, V.A., Villaverde, J., Trézéguet, V., Lauquin, G.J., Brandolin, G., and Padrós, E. (2003) Structural and functional implications of the instability of the ADP/ATP transporter purified from mitochondria as revealed by FTIR spectroscopy. *Biophys. J.*, **85**, 255–266.

163 Jackson, M. and Mantsch, H.H. (1995) The use and misuse of FTIR spectroscopy in the determination of protein structure. *Crit. Rev. Biochem. Mol. Biol.*, **30**, 95–120.

164 Oberg, K.A. and Fink, A.L. (1998) A new attenuated total reflectance Fourier transform infrared spectroscopy method for the study of proteins in solution. *Anal. Biochem.*, **256**, 92–106.

165 Fernández, C. and Wider, G. (2003) TROSY in NMR studies of the structure and function of large biological macromolecules. *Curr. Opin. Struct. Biol.*, **13**, 570–580.

166 Gautier, A., Kirkpatrick, J.P., and Nietlispach, D. (2008) Solution-state NMR spectroscopy of a seven-helix transmembrane protein receptor: backbone assignment, secondary structure, and dynamics. *Angew. Chem. Int. Ed.*, **47**, 7297–7300.

167 Wang, Z.X. and Jiang, R.F. (1996) A novel two-site binding equation presented in terms of the total ligand concentration. *FEBS Lett.*, **392**, 245–249.

168 Singh, S.K., Piscitelli, C.L., Yamashita, A., and Gouaux, E. (2008) A competitive inhibitor traps LeuT in an open-to-out conformation. *Science*, **322**, 1655–1661.

169 Weber, P.C. and Salemme, F.R. (2003) Applications of calorimetric methods to drug discovery and the study of protein interactions. *Curr. Opin. Struct. Biol.*, **13**, 115–121.

170 Villaverde, J., Cladera, J., Hartog, A., Berden, J., Padrós, E., and Duñach, M. (1998) Nucleotide and Mg^{2+} dependency of the thermal denaturation of mitochondrial F_1-ATPase. *Biophys. J.*, **75**, 1980–1988.

171 Cooper, M.A. (2002) Optical biosensors in drug discovery. *Nat. Rev. Drug Discov.*, **1**, 515–528.

172 Salamon, Z., Brown, M.F., and Tollin, G. (1999) Plasmon resonance spectroscopy: probing molecular interactions within membranes. *Trends Biochem. Sci.*, **24**, 213–219.

173 Freire, E., Mayorga, O.L., and Straume, M. (1990) Isothermal titration calorimetry. *Anal. Chem.*, **62**, 950A–959A.

174 Velázquez-Campoy, A., Ohtaka, H., Nezami, A., Muzammil, S., and Freire, E. (2004) Isothermal titration calorimetry. *Curr. Protoc. Cell Biol.*, **17**, 17.18.

175 Nie, Y., Smirnova, I., Kasho, V., and Kaback, H.R. (2006) Energetics of ligand induced conformational flexibility in the lactose permease of *Escherichia coli*. *J. Biol. Chem.*, **281**, 35779–35784.

176 Reyes, N., Ginter, C., and Boudker, O. (2009) Transport mechanism of a bacterial homologue of glutamate transporters. *Nature*, **462**, 880–885.

177 Royer, C.A. and Scarlata, S.F. (2008) Fluorescence approaches to quantifying biomolecular interactions. *Methods Enzymol.*, **450**, 79–106.

178 Hovius, R., Vallotton, P., Wohland, T., and Vogel, H. (2000) Fluorescence techniques: shedding light on ligand–receptor interactions. *Trends Pharmacol. Sci.*, **21**, 266–273.

179 Roehrl, M.H., Wang, J.Y., and Wagner, G. (2004) A general framework for development and data analysis of competitive high-throughput screens for small-molecule inhibitors of protein–protein interactions by fluorescence polarization. *Biochemistry*, **43**, 16056–16066.

180 Piston, D.W. and Kremers, G.J. (2007) Fluorescent protein FRET: the good, the bad and the ugly. *Trends Biochem. Sci.*, **32**, 407–414.

181 Wallace, B.A. and Janes, R.W. (2003) Circular dichroism and synchrotron radiation circular dichroism spectroscopy: tools for drug discovery. *Biochem. Soc. Trans.*, **31**, 631–633.

182 León, X., Lórenz-Fonfría, V.A., Lemonnier, R., Leblanc, G., and Padrós, E. (2005) Substrate-induced conformational changes of melibiose permease from *Escherichia coli* studied by infrared difference spectroscopy. *Biochemistry*, **44**, 3506–3514.

183 Pellecchia, M. (2005) Solution nuclear magnetic resonance spectroscopy techniques for probing intermolecular interactions. *Chem. Biol.*, **12**, 961–971.

184 Pellecchia, M., Bertini, I., Cowburn, D., Dalvit, C., Giralt, E., Jahnke, W.,

James, T.L., Homans, S.W., Kessler, H., Luchinat, C., Meyer, B., Oschkinat, H., Peng, J., Schwalbe, H., and Siegal, G. (2008) Perspectives on NMR in drug discovery: a technique comes of age. *Nat. Rev. Drug. Discov.*, **7**, 738–745.

185 GE Healthcare (2007) *Purifying Challenging Proteins: Principles and Methods*. GE Healthcare, Little Chalfont.

186 Braun, H.P., Sunderhaus, S., Boekema, E.J., and Kouril, R. (2009) Purification of the cytochrome *c* reductase/cytochrome *c* oxidase super complex of yeast mitochondria. *Methods Enzymol.*, **456**, 183–190.

187 Hjelmeland, L.M. and Chrambach, A. (1984) Solubilization of functional membrane proteins. *Methods Enzymol.*, **104**, 305–318.

188 Saint, N., Lacapere, J.J., Gu, L.Q., Ghazi, A., Martinac, B. and Rigaud, J.L. (1998) A hexameric transmembrane pore revealed by two-dimensional crystallization of the large mechanosensitive ion channel (MscL) of *Escherichia coli*. *J. Biol. Chem.*, **273**, 14667–14670.

189 Keinanen, K., Kohr, G., Seeburg, P.H., Laukkanen, M.L. and Oker-Blom, C. (1994) High-level expression of functional glutamate receptor channels in insect cells. *Biotechnology*, **12**, 802–806.

190 Everberg, H., Gustavsson, N., and F. Tjerneld, F. (2009) Extraction of yeast mitochondrial membrane proteins by solubilization and detergent/polymer aqueous two-phase partitioning. *Methods Mol. Biol.*, **528**, 73–81.

191 Zuobi-Hasona, K., Crowley, P.J., Hasona, A., Bleiweis, A.S., and Brady, L.J. (2005) Solubilization of cellular membrane proteins from *Streptococcus mutans* for two-dimensional gel electrophoresis. *Electrophoresis*, **26**, 1200–1205.

192 Staudinger, R. and Bandres, J.C. (2000) Solubilization of the chemokine receptor CXCR4. *Biochem. Biophys. Res. Commun.*, **274**, 153–156.

193 Hunte, C., Screpanti, E., Venturi, M., Rimon, A., Padan, E., and Michel, H. (2005) Structure of a Na^+/H^+ antiporter and insights into mechanism of action and regulation by pH. *Nature*, **435**, 1197–1202.

194 Kobilka, B.K. (1995) Amino and carboxyl terminal modifications to facilitate the production and purification of a G protein-coupled receptor. *Anal. Biochem.*, **231**, 269–171.

195 Lund, S., Orlowski, S., de Foresta, B., Champeil, P., le Maire, M., and Moller, J.V. (1989) Detergent structure and associated lipid as determinants in the stabilization of solubilized Ca^{2+}-ATPase from sarcoplasmic reticulum. *J. Biol. Chem.*, **264**, 4907–4915.

196 le Maire, M., Arnou, B., Olesen, C., Georgin, D., Ebel, C., and Moller, J.V. (2008) Gel chromatography and analytical ultracentrifugation to determine the extent of detergent binding and aggregation, and Stokes radius of membrane proteins using sarcoplasmic reticulum Ca^{2+}-ATPase as an example. *Nat. Protoc.*, **3**, 1782–1795.

197 Seddon, A.M., Curnow, P., and Booth, P.J. (2004) Membrane proteins, lipids and detergents: not just a soap opera. *Biochim. Biophys. Acta*, **1666**, 105–117.

198 Aigner, A., Jager, M., Pasternack, R., Weber, P., Wienke, D., and Wolf, S. (1996) Purification and characterization of cysteine-*S*-conjugate *N*-acetyltransferase from pig kidney. *Biochem. J.*, **317**, 213–218.

199 Ueno, S., Kaieda, N., and Koyama, N. (2000) Characterization of a P-type Na^+-ATPase of a facultatively anaerobic alkaliphile, *Exiguobacterium aurantiacum*. *J. Biol. Chem.*, **275**, 4537–4540.

200 Booth, P.J. and Paulsen, H. (1996) Assembly of light-harvesting chlorophyll *a/b* complex *in vitro*. Time-resolved fluorescence measurements. *Biochemistry*, **35**, 5103–5108.

201 Booth, P.J., Farooq, A., and Flitsch, S.L. (1996) Retinal binding during folding and assembly of the membrane protein bacteriorhodopsin. *Biochemistry*, **35**, 5902–5909.

202 Booth, P.J. (2000) Unravelling the folding of bacteriorhodopsin. *Biochim. Biophys. Acta*, **1460**, 4–14.

203 Kiefer, H. (2003) *In vitro* folding of alpha-helical membrane proteins. *Biochim. Biophys. Acta*, **1610**, 57–62.

204 Rath, A. Glibowicka, M., Nadeau, V.G., Chen, G., and Deber, C.M. (2009) Detergent binding explains anomalous SDS–PAGE migration of membrane proteins. *Proc. Natl. Acad. Sci. USA*, **106**, 1760–1765.

205 Asmar-Rovira, G.A., Asseo-Garcia, A.M., Quesada, O., Hanson, M.A., Cheng, A., Nogueras, C., Lasalde-Dominicci, J.A., and Stevens, R.C. (2008) Biophysical and ion channel functional characterization of the *Torpedo californica* nicotinic acetylcholine receptor in varying detergent–lipid environments. *J. Membr. Biol.*, **223**, 13–26.

206 Sidhu, A. (1990) A novel affinity purification of D-1 dopamine receptors from rat striatum. *J. Biol. Chem.*, **265**, 10065–10072.

207 Furth, A.J., Bolton, H., Potter, J. and Priddle, J.D. (1984) Separating detergent from proteins. *Methods Enzymol.*, **104**, 318–328.

208 Yamaguchi, Y., Miyagi, Y., and Baba, H. (2008) Two-dimensional electrophoresis with cationic detergents: a powerful tool for the proteomic analysis of myelin proteins. Part 2: analytical aspects. *J. Neurosci. Res.*, **86**, 766–775.

209 Snell, P.H., Phillips, E., Burgess, G.M., Snell, C.R., and Webb, M. (1990) Characterization of bradykinin receptors solubilized from rat uterus and NG108-15 cells. *Biochem. Pharmacol.*, **39**, 1921–1928.

210 Liu, R., Lu, P., Chu, J.W., and Sharom, F.J. (2009) Characterization of fluorescent sterol binding to purified human NPC1. *J. Biol. Chem.*, **284**, 1840–1852.

211 Babcock, G.J., Farzan, M., and Sodroski, J. (2003) Ligand-independent dimerization of CXCR4, a principal HIV-1 coreceptor. *J. Biol. Chem.*, **278**, 3378–3385.

212 Rigaud, J., Chami, M., Lambert, O., Levy, D., and Ranck, J. (2000) Use of detergents in two-dimensional crystallization of membrance proteins. *Biochim. Biophys. Acta*, **1508**, 112–128.

213 Romero, F. (2009) Solubilization and partial characterization of ouabain-insensitive Na^+-ATPase from basolateral plasma membranes of the intestinal epithelial cells. *Invest. Clin.*, **50**, 303–314.

13
Stabilizing Membrane Proteins in Detergent and Lipid Systems

Mark Lorch and Rebecca Batchelor

13.1
Introduction

Membrane proteins exist in extremely complex environments. Within the membrane the protein experiences a range of physical states dictated by the hydrophobic thickness of the membrane, differing charges and pressures, as well as specific interactions which all serve to modulate the activity and stability of the protein. These myriad of environments are well illustrated by White *et al.* neutron diffraction data on just a simple 1,2-dioleoyl-*sn*-glycero-3-phosphocholine (di(C18 : 1)PC) bilayer [1]. Even this simple model membrane belies the added layer of complexity given by the massive number of different lipids (not to mention other proteins) found in a biological membrane. Given this it is surprising that membrane proteins function at all when placed in simple mimetic environments such as detergent micelles. Therefore, finding an environment in which to study isolated membrane proteins necessitates a compromise between a complex system that accurately mirrors the biological membrane and the technical restrictions that require a simpler mimetic compatible with the analytical technique of choice. This chapter aims to review the methods used to best achieve this balance and maintain purified proteins in a native state for as long as possible.

13.2
Choice of Detergent: Solubilization versus Stability

With a few exceptions membrane proteins are purified from the expression source (be it an organism or cell-free expression (CFE)) with the assistance of detergents. The job of the detergent is 2-fold. They are initially required for the primary solubilization step of the expressed protein. Since the protein is likely to have been expressed in the membrane, as an aggregate (in the case of CFE), or in a few cases as inclusion bodies, the detergent must have strong solubilizing characteristics. A relatively harsh detergent may be needed for this step. Following this a detergent must then keep the protein in a native-like state during the purification and subsequent analysis and/or reconstitution steps. For these parts of the procedure it

Production of Membrane Proteins: Strategies for Expression and Isolation, First Edition.
Edited by Anne Skaja Robinson.

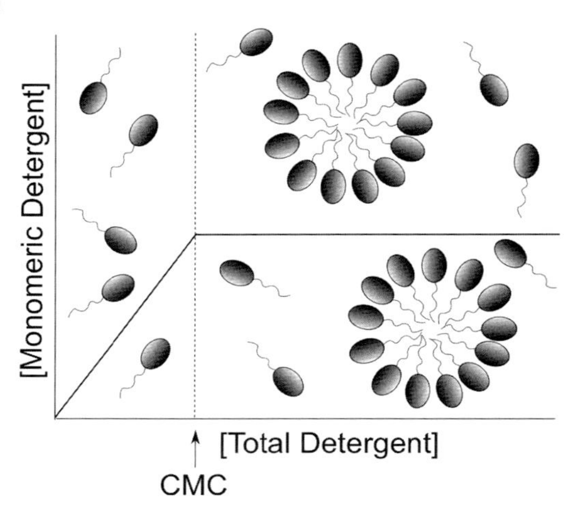

Figure 13.1 Ideal behavior of detergents with respect to concentration. Above the solubility limit of the monomer detergents form micelles. This point is known as the CMC. Increasing the concentration further results in more micelles while the monomer concentration remains static.

is desirable to use a milder detergent that is less likely to cause protein denaturation. Therefore, choosing the best detergent is a case of balancing the need to solubilize with the need to maintain stability.

13.2.1
Detergents: General Characteristics

To choose the most appropriate detergent one must first have a grasp of the general characteristics of detergents. Lipids and detergents are both amphipathic molecules that generally consist of a hydrophilic head-group linked to one or two hydrophobic chains. This duality means the monomeric molecules are not particularly soluble in water and so instead form structures that shield their nonpolar regions from the aqueous environment. Lipids may form bilayers (at least under the conditions found during most membrane protein experiments); detergents, however, form approximately spherical micelles.

Two important general characteristics of detergents are the aggregation number and the critical micelle concentration (CMC). The number of molecules in a micelle (the aggregation number) varies dramatically from detergent to detergent and is highly dependent on the conditions under which the measurements are performed. For example, the aggregation numbers for sodium cholate and sodium dodecyl sulfate (SDS) are 2–5 (http://www.anatrace.com/category.asp?cat=224&id=S1001) and 62–101 [2], respectively. The CMC is the concentration of detergent above which micelles spontaneously form. Increasing the concentration above the CMC results in an increase in the number of micelles while the concentration of monomeric detergent stays constant (Figure 13.1).

The CMC is heavily dependent on the physiochemical properties of the detergent in question as well as the pH, temperature, ionic strength, and presence of other surfactants and proteins. Increasing the length of the aliphatic chain, or in the case of ionic detergents, the presence of counterions results in a decrease in the CMC. Increasing temperature increases the solubility of monomeric detergent molecules and so increases the CMC. The CMC and the aggregation number become particularly important should it become necessary to remove the detergent from the system. This will be explained in more detail in Section 13.3.3 on reconstitution.

As a rule of thumb detergents with net charge are harsher than zwitterionic detergents, while nonionic detergents are mild. Furthermore, the shorter the aliphatic chain and smaller the head-group, the harsher the detergent. Examples of this behavior along with significant exceptions that prove the rule are given below.

13.2.1.1 Ionic Detergents

Ionic detergents have an anionic or cationic head-group and a hydrocarbon or steroidal backbone. They are generally harsh, best suited to applications where good solubilization is important or if the membrane protein is particularly hardy and amenable to refolding once the detergent is removed. Proteins are rarely active in these detergents so they are only used in an intermediate stage of a protein preparation or when denaturation is desired (such as protein folding experiments). Common examples include the anionic detergent SDS, which is used primarily for cell lysis and denaturation of proteins during electrophoresis. It is strongly denaturing, but nevertheless there are several examples where proteins have been successfully refolded after denaturation in SDS [3–5]. Likewise, cationic detergents are particularly harsh and are generally used for applications where protein denaturation is desirable.

Ionic detergents that resemble (or indeed are) naturally occurring steroids, on the other hand, are generally much less harsh than their artificial counterparts. Cholates, derived from cholesterol, are naturally secreted from the gall bladder and hence are sometimes know as bile acid detergents. The planar nature of these molecules, with a polar and nonpolar face, results in a high CMC and small micelles; characteristics that make them easy to remove by dialysis – a consideration when it comes to reconstituting the protein.

13.2.1.2 Zwitterionic Detergents

Zwitterionic detergents are generally milder than ionic detergents and so are more commonly used in the preparation of membrane protein samples. They are still too harsh to be suitable for activity measurements. However, they are commonly used during crystallography. The most notable examples in this class are:

- 3-((3-Cholamidopropyl)dimethylammonio)-1-propanesulphonate (CHAPS) and [3-((3-Cholamidopropyl)dimethylammonio)-2-hydroxy-1-propanesulfonate] (CHAPSO) are steroid-based detergents developed primarily for work with proteins. They are good solubilizers and rarely denature proteins.
- Lauryldimethylamine-*N*-oxide (LDAO) is quite harsh, but has been used on a few notable occasions with robust proteins for both X-ray and nuclear magnetic resonance (NMR) studies. However, its use is limited by its denaturing

properties; it has been estimated that 80% of membrane proteins will be denatured by it.

- Dodecylphosphocholine (DPC) is also known as fos-choline 12. It has a phosphocholine head-group directly bonded to an aliphatic chain and so closely resembles a phospholipid. It is especially mild and so lends itself to studies where the activity of the protein needs to be maintained. Furthermore, the small micelle size makes DPC an ideal choice for NMR studies. To date, the largest membrane protein structure determined by NMR was conducted in DPC [6].

13.2.1.3 **Nonionic Detergents**

Nonionic detergents are mild, nondenaturing, and are commonly used for all aspects of membrane protein research, including solubilization of expressed proteins, structural studies, reconstitution, refolding, and activity measurements.

n-Octyl-β-D-glucoside (β-OG) is relatively harsh for a nonionic detergent and so is often used for solubilizing proteins. It is also a detergent of choice in crystallographic trials, having proved extremely successful on numerous occasions [7,8].

n-Dodecyl-β-D-maltoside (DDM) and *n*-decyl-β-D-maltoside (DM) are particularly mild nondenaturing detergents, and so are firm favorites. They are both used extensively for reconstitution, activity, and solubilization. DM, with its shorter chain (10 versus 12 carbon atoms), is the harsher of the two and also has a higher CMC (1.8 [9] versus 0.17 mM [10]), making it a better choice for dialysis-based reconstitution methods (see below).

Triton X-100 and X-114 are nonhomogenous detergents with variable length polyethylene glycol polar regions and branched hydrophobic alkyl chains separated by an aromatic ring structure. This ring absorbs below 300 nm making them incompatible with UV measurements. They have extremely low CMCs and so are very difficult to remove by dialysis. However, they are nondenaturing, and seem to break lipid–lipid and lipid–protein interactions in preference to protein–protein interactions, making them ideal for solubilizing membranes while keeping the protein intact.

Polyoxyethylenes (e.g., $C_{12}E_8$) are a simple range of detergents with variable alkyl and oxyethylene chains. They follow the rule of thumb relating chain length and harshness. The shorter chains are generally used for solubilization and crystallographic trials [8], while the longer chains are used for reconstitution procedures.

A factor worth noting with the Tritons and polyoxyethylenes is that they can contain highly variable mixtures, and are particularly prone to containing harmful peroxide impurities. Therefore, only the purest versions should be used. In addition, they should be stored under inert gas and in the dark.

13.2.1.4 **Detergent-Like Phospholipids**

The single chain of lysolipids and very short chains of lipids such as 1,2-dihexanoyl-*sn*-glycero-3-phosphocholine (di(C6 : 0)PC) [11] means they form micelles instead of bilayers (Figure 13.2). The head-group defines their ionic properties and so they

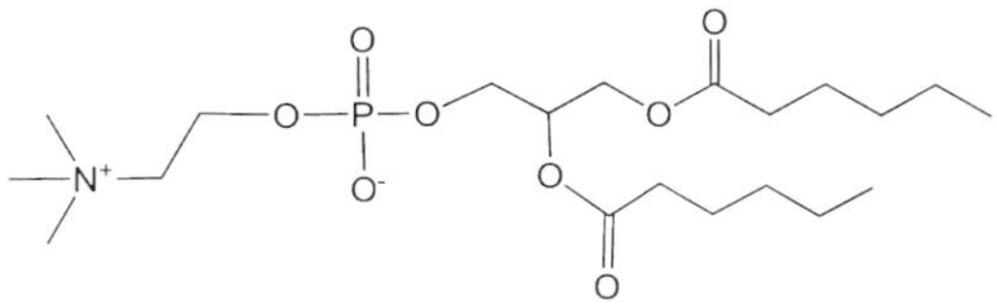

1,2-Dihexanoyl-sn-glycero-3-phosphocholine (di(C6:0)PC). CMC 15mM 0.68% w/v.

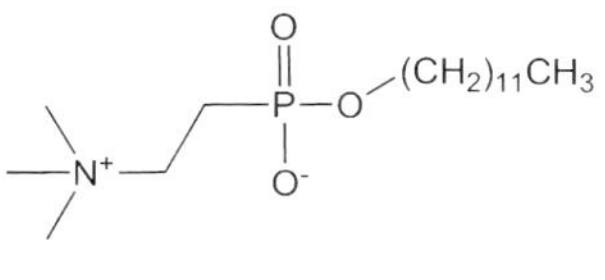

Dodecylphosphocholine (FOS-Choline 12, DPC). CMC 1.5 mM 0.047% w/v. Aggregation number 54

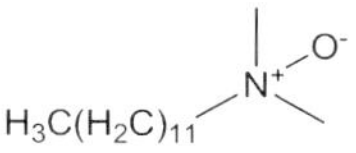

Lauryldimethylamine-N-oxide (LDAO). CMC 1-2 mM 0.023% w/v. Aggregation number 76

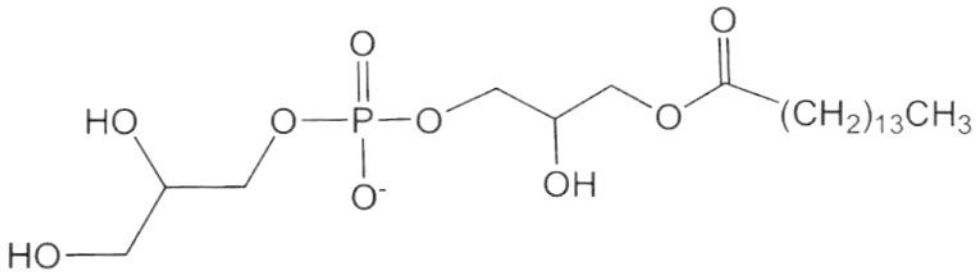

1-Myristoyl-2-hydroxy-sn-glycero-3-phosphoglycerol (C15:0PG). CMC 0.09 mM 0.0042% w/v.

Figure 13.2 Chemical structures of some detergents commonly used in mixed micelles and bicelles, for protein reconstitution into liposomes and crystallography. More examples are given in Chapter 12. CMCs and aggregation numbers are given for detergents in water [2].

can be anionic or zwitterionic. They are only mildly solubilizing and nondenaturing. This, coupled with their similarity to native phospholipids, has led many to try using them as convenient membrane mimetics. The most notable successes have been in the CFE systems, where 1-myristoyl-2-hydroxy-*sn*-glycero-3-phosphoglycerol (C14 : 0PG) is added to the expression mix. The mild nature of the lysolipid means the translation and transcription machinery is not hampered by the detergent, but a safe membrane mimetic environment is provided [12]. Di(C6 : 0)PC has also been successfully used as an environment into which membrane proteins have been refolded and is often used as the basis of bicelles (see Section 13.3.1.2).

13.2.2 Solubilization

A membrane protein purification procedure starts with a solubilization step. The mechanisms by which this process takes place have been covered in an excellent review by Le Marie [2]. Methods for purifying membrane proteins have been covered in detail in Chapter 12 of this book. However, there are some aspects of the solubilization procedure relating to maintaining protein stability that warrant further mention.

Solubilization of the membrane containing a protein will usually require concentrations in excess of the CMC. How much of an excess will depend on the lipid/detergent ratio (the denser the cell pellet the higher the concentration of detergent that will be needed to solubilize it) and the harshness of the detergent. For example, both β-OG and DDM have been used at about 1% w/v to solubilize membranes, which is a concentration that is only just above the CMC of β-OG, but two orders of magnitude above that of DDM. It is also important to consider that too much detergent can lead to the removal of essential tightly bound lipids, resulting in inactivation of the protein [13–15].

Taking a rational approach based on the characteristics of detergents will get you some way to selecting the most appropriate detergent for the job; however, one needs to be aware that sometimes changing detergents used during a procedure can have unpredictable consequences. For example, using DDM during the reconstitution of the lactose symporter from *Streptococcus thermophilus* resulted in the protein inserting in random orientations, then by switching to Triton X-100 proteoliposomes are produced with proteins biased towards one orientation [16].

13.3 Mitigating Protein Denaturation

Even the mildest detergents do not accurately mirror the membrane. Proteins that are successfully solubilized in detergents rarely maintain their activity for as long as those reconstituted into liposomes. The detergent environment cannot maintain the protein's native state, which leads to misfolding, aggregation, and more often than not permanent loss of activity. Clearly membrane proteins need lipids. One possible route for loss of protein stability is via removal of potentially critical tightly bound lipids during the solubilization and purification procedure. Secondly, some proteins may require specific interactions with lipids, without which they may appear stable and folded, but the lack of lipid interaction has subtle effects on the protein's structure. This has been clearly demonstrated by direct comparisons between structures, dynamics, and activities of the same protein in lipid and detergent systems [17,18]. However, probably of greatest importance are the host of nonspecific interactions that exist in the membrane but are lacking in micelles. The following section covers methods for reintroducing the native-like surfactant/protein interactions with a view to maintaining protein stability during the course of a procedure.

13.3.1 Mixed Detergent Systems

13.3.1.1 Micelles

Constraints imposed by many biophysical techniques means that often the only tractable approach is to use detergents as the basis of a membrane mimetic environment. This is most notably the case where the large size of a proteoliposome is an issue, as for liquid-state NMR or optical spectroscopy, and of course crystallography is normally conducted from a detergent-solubilized starting state. Thus, while it may not be the ideal system, it is important to consider methods for maintaining protein stability in detergents.

As discussed above, the single largest factor to consider is the choice of detergent. There is no universal detergent, and no hard and fast rule as to which detergents are best for a particular class of protein, so it becomes a case trial and error. This already arduous task is further compounded by suggestions that mixing micelles consisting of different detergents or detergent–lipid mixtures may provide increased stability. Mixed detergent micelles have been effective with TM0026, a small two-transmembrane spanning protein of unknown function from *Thermotoga maritima* [19]. Here, a more rational approach was taken by engineering mixed micelles with dimensions that matched those of the protein. The resulting DDM/fos-choline 10 micelles significantly improved the quality of NMR spectra over pure DDM, allowing several additional assignments to be made.

Probably a more productive approach is the addition of lipids to micelles, the rationale being that some native-like lipid–protein interactions, lost during solubilization, may be re-established. This approach has worked extremely well with some crystal trials where the presence of lipids has proved crucial (reviewed by Hunte *et al.* [20]). A case in point is cyanobacterial cytochrome b_6f complex where, along with the two native copurified lipids, addition of 10 di(C18 : 1)PC lipids per monomer greatly improved the quality of the crystals [21]. Furthermore, mixed detergent–lipid systems have been particularly effective with one of the most interesting and challenging classes of membrane protein – the G–protein-coupled receptor (GPCR). Rhodopsin is the most studied protein of the group, with both its functionality and stability being dependent on the presence of lipids. The presence of di(C14 : 0)PC in CHAPS-solubilized rhodopsin dramatically increases the protein's thermal stability [22]. Egg PC lipids added to DDM, CHAPS, or digitonin results in enhanced light-stimulated activity [23]. Di(C18 : 1)PC, 1,2-dioleoyl-*sn*-glycero-3-phosphoserine (di(C18 : 1)PS), 1,2-dioleoyl-*sn*-glycero-3-phosphoethanolamine (di(C18 : 1)PE), and asolectin in DDM micelles significantly affect the stability of the G-protein–rhodopsin complexes [24].

Similar phenomena have been observed with other GPCRs; the binding capacity of β_2-adrenergic receptor, when purified into DDM, drops by 80% within 4 h; however, addition of phospholipids (di(C18 : 1)PE, di(C18 : 1)PC, or di(C18 : 1)PS) to the micelles reduced the loss of activity to 50% over the same timescale. Even more significant was the addition of cholesterol hemisuccinate, this resulted in 60% activity after 7 days [25]. The most detailed investigation of this kind to date involves the chemokine receptors CXCR4 and CCR5, which were subjected to an extensive study involving 22 detergents systems and 12 lipid–detergent combinations [26]. The proteins' stabilities and capacity to bind conformational dependent antibodies and binding partners were followed using surface plasmon resonance. The results show that solubilization in the presence of all lipids tested increases the binding capacity of the proteins, with a combination of cholesteryl hemisuccinate/DDM/CHAPS along with di(C18 : 1)PC and di(C18 : 1)PS giving the best results. Together these studies demonstrate a clear dependence of GPCR stability on the presence of lipids.

Further examples of lipid requirements for maximum activity are seen across numerous classes of membrane protein. The phenomenon was reported as early as 1979 when it was observed that an Na,K-ATPase extracted from dogfish rectal glands, solubilized with deoxycholate, and delipidated by gel filtration was largely inactive; however, addition of cooked brains, di(C18 : 1)PC, or di(C18 : 1)PE reactivated the protein [13]. Other examples include the much-studied diacylglycerol kinase (DGK) from *Escherichia coli*. This trimeric three-transmembrane enzyme catalyses the formation of phosphatidic acid lipids from diacyglycerols and ATP [6, 27]. When solubilized it requires the presence of a further lipid (as well as its substrate) to function [28, 29]. Meanwhile, the *E. coli* heavy metal P-type ATPase ZntA is a little more fussy about its lipid requirements; addition of 1,2-dioleoyl-*sn*-glycero-3-phosphoglycerol (di(C18 : 1)PG) lipid to $C_{12}E_8$ solubilized protein results in a 4-fold increase in ATPase activity but di(C18 : 1)PS, 1,2-dioleoyl-*sn*-glycero-3-phosphosphate (di(C18 : 1)PA), and cardiolipin had no measurable effect [30].

Many other examples exist of activation and stabilization of proteins following relipidation, including an ion channel [31], light harvesting complex II [32], bacteriorhodopsin (bR) [33], and further ATPases [15, 34, 35]. What is striking is that often the proteins do not appear to be overly fussy about the type of lipid required to stimulate activity; sterols and phospholipids with a variety of chain length, head-group charge, and size all seem able to function as well as the other. Thus, the lipids are not acting as a cofactor, but instead provide some, as yet undetermined, general requirement that is unfulfilled by detergent alone. Since we are still discussing micellular systems it cannot be a function of a bilayer environment, the variety in head-groups rules out direct interactions here, and the detergents generally provide a hydrocarbon chain with the same characteristics as a lipid. The only thing left that is unique to a lipid (but not of course sterols) is the phosphogycerol group; we can therefore only speculate that many membrane proteins require an interaction with phosphate and/or glycerol groups to function optimally.

13.3.1.2 **Bicelles**

In mixed detergent–lipid micelle systems the lipid is the minor component. However, under specific conditions and where the lipid is in excess (lipid/detergent mole ratios of around 5 : 1 to 2 : 1) the two components can phase separate, resulting in a lipid bilayer disk, the edges of which are stabilized by detergents (Figure 13.3). This phenomenon was first described by Muller in 1981 [36] using egg yolk lectin and bile salts. Sanders then refined the system into a well defined binary mixture using CHAPSO and di(C14 : 0)PC lipids. At the same time it was demonstrated that the di(C14 : 0)PC bilayer component adopted a liquid disordered phase similar to pure lipid bilayers [37]. Sanders later developed a new binary mixture, replacing CHAPSO with di(C6 : 0)PC [38], resulting in a pure phospholipid system. He went on, in a seminal paper, to coin the term bicelle (from *bi*layer mixed mi*celle*) [39]. However, the crux of the study was the demonstration that bicelles could provide an environment in which membrane proteins could be function. Fifteen integral and peripheral membrane proteins and peptides were reconstituted in di(C14 : 0)PC/di(C6 : 0)PC and di(C14 : 0)PC/CHAPSO bicelles with a variety of peptides; the di(C14 : 0)PC mole ratios ranged from 1 : 8 for gramicidin to 1 : 360 for bR. NMR was used to confirm that the presence of protein did not disrupt the bicellular morphology (and hence negated the advantages of the

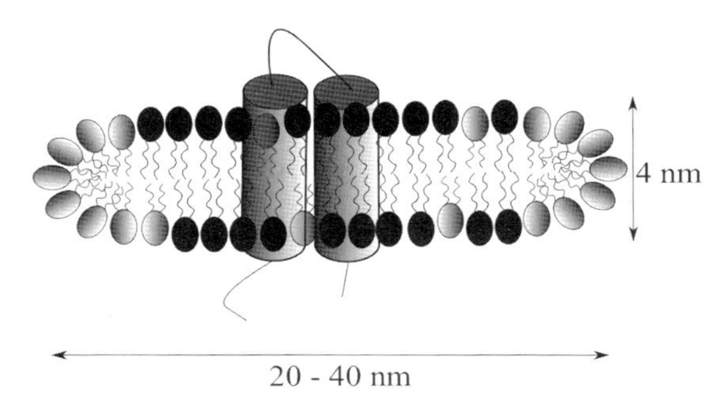

Figure 13.3 Cross-section model of membrane protein incorporated into a di(C6 : 0) PC/di(C14 : 0)PC bicelle. The longer-chain di(C14 : 0)PC (black head-group) lipids form a bilayer disk, the edges of which are stabilized by shorter-chain di(C6 : 0)PC (gray head-group). Small amounts of di(C6 : 0)PC partition into the bulk di (C14 : 0)PC bilayer domain. The dimensions of the bicelle can be controlled by varying the ratio of its components.

system). It was also demonstrated for the first time that a protein could be active in a bicelle. It was observed that the appropriateness of the system was largely protein specific; some proteins could be easily accommodated in bicelles with no sign of aggregation or disruption of bicelles, while others could not be reconstituted even at very low concentrations of protein. It was speculated that part of the limitations arose from the relatively short chain length of di(C14:0)PC, which could lead to hydrophobic mismatch and protein aggregation.

Sander's methods for forming bicelles has not been significantly altered or improved on. The longer chain lipid is dispersed in an aqueous environment and the detergent is then added to give a total surfactant concentration of 5–40% w/v. Alternatively the surfactants are codissolved in an organic solvent, then dried under vacuum before being taken up in a buffer (with 50–200 mM salt) at pH 5–8 and at a temperature above the gel melting point of the bilayer lipid; in the case of DMPC this is 23 °C. The resulting mixtures are often refrozen, thawed, and sonicated to ensure a homogenous solution [40].

The most common longer-chain lipid in bicelle recipes is di(C14 : 0)PC, which is shorter than many natural lipid membranes. Consequently, many proteins may suffer from hydrophobic mismatch that may disrupt structure and function (see Section 13.3.2.2). Furthermore, di(C14:0)PC has a relatively high gel-to-liquid disordered phase transition (23 °C), restricting the temperature range in which experiments can be conducted. Bicelle recipes have been tweaked by using shorter-chain lipids (1,2-dilauroyl-*sn*-glycero-3-phosphocholine (di(C12:0)PC)) [41] to reduce the melting temperature (to −1 °C), but this exacerbates the mismatch problem. Conversely, bicelles have also been made with longer-chain lipids (1,2-dipalmitoyl-*sn*-glycero-3-phosphocholine (di(C16 : 0)PC) [41] to better match the hydrophobic area of a protein, but this raises the transition temperature, restricting experiments to above 41 °C. 1-Palmitoyl-2-oleoyl-*sn*-glycero-3-phosphocholine (C16 : 0/C18 : 1PC) bicelles have also been developed [42, 43]. The transition temperature of this lipid is −1 °C, extending the temperature range over which thicker bicelles can be used. Recently the range of lipid compositions available in bicelle form has been expanding to include PG [44, 45], PS lipids [46], cardiolipin [47], cholesterol [48], ceramides [49], and artificial lipids [50, 51]. Thus, while the most commonly used bicelles are only of use with proteins that are happy in relatively thin bilayers and/or elevated temperatures, there are numerous systems that dramatically increase the scope of bicelles and will undoubtedly lead to their increased use as membrane mimetic environments.

Proteins and peptides can be reconstituted into bicelles by a number of methods. If the protein or peptides can tolerate being solubilized in organic solvent then they can be codissolved with the surfactant mix [39]. Alternatively, proteins dissolved in detergent can be simply added to the preformed bicelle [40, 41, 52, 53], being careful not to upset the detergent/lipid ratio required for bicelle formation. It is also possible to purify the protein directly into the whole bicelle mixture [22].

The potential advantages of bicelle systems are manifold (reviewed in Sanders and Prosser [54]). They provide small, optically transparent, lipid bilayer environments, combining many of the advantages of micelles and liposomes without many of their disadvantages. There have also been some notable successes in crystallizing membrane proteins from bicelles (reviewed by Johansson *et al.* [55]). The technique was developed by Bowie with the archetypal bR [52] and has since yielded a rare structure of a GPCR, β_2-adrenergic receptor [56]. The optical properties of bicelles have proved especially useful for spectroscopic studies of function (e.g., DGK [40]) and folding (e.g., rhodopsin [22, 57]) of membrane proteins. However, bicelles have really found their niche when it comes to NMR of both membrane and water-soluble proteins. The enormous interest in bicelles/membrane protein NMR is highlighted by the number of reviews of the topic in the last few years [58–60]. Bicelles lend themselves to NMR studies primarily because they are relatively small, predominantly lipid systems and so provide a bilayer environment that tumbles rapidly enough to acquire well-resolved spectra. An added advantage is the fact that bicelles can align themselves in a magnetic field. This introduces some orientational order into the system, allowing information concerning the relative orientations of the protein and membrane to be extracted.

13.3.2
Detergent-Free Bilayer Systems

It is becoming increasingly clear that a membrane exerts a considerable influence on the proteins embedded within it. While proteins in detergent-based systems may appear to function correctly it is quite possible that their structures and therefore activities are subtly altered, leading to potentially misleading results. Thus, it is undoubtedly preferable (analytical technique allowing) to incorporate the protein into a pure lipid bilayer system. Only in this way can proteins experience an environment that approaches a native bilayer and so exhibit native-like behavior.

13.3.2.1 Lipid Nanodisk

A promising recent development are nanodisks – consisting of a lipid bilayer disk, the edges of which are stabilized by a membrane scaffold protein (MSP) engi-

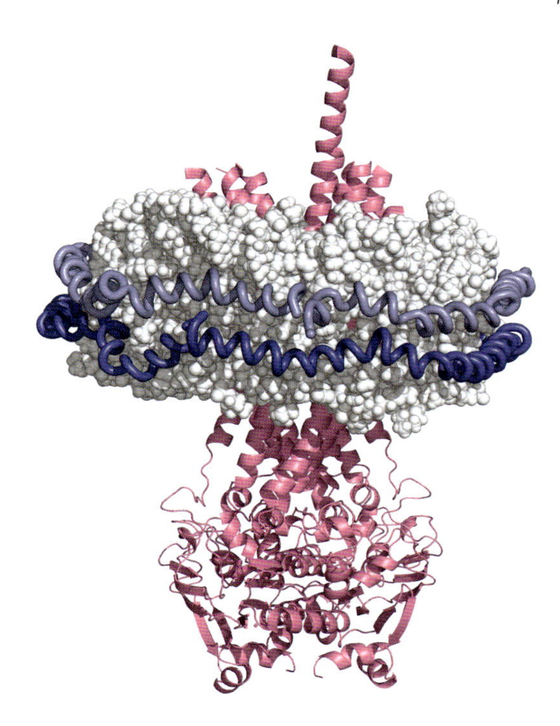

Figure 13.4 Model of P-glycoprotein incorporated into 9.8-nm nanodisks. A recent homology model of P-glycoprotein in the nucleotide-bound state based on a bacterial homolog [71] was overlaid with the model for small nanodisks, which is based on the molecular belt model of diskoidal high-density lipoprotein[72]. The graphic was generated using the PyMOL molecular graphics system (www.pymol.org). The image was provided by and reproduced with permission from William Atkins.

neered from an apolipoprotein [61]. The size of the disk can be carefully controlled using variable length MSPs leading to very well-defined monodispersed structures. For example, Sligar has shown that disks formed from C16:0/C18:1PC with a lipid/MSP mole ratio of 65:1 are 9.7 nm in diameter by 4.6 nm thick. Then by increasing the lipid/MSP ratio to 124:1 and using longer MSPs the disk diameter can be increased to 12.8 nm [62]. The disks are easily assembled by mixing MSP with cholate solubilized lipids and then removing the detergent by dialysis or detergent absorbing beads.

Nanodisks have proved to be excellent environments for membrane proteins and their use in the field has been the subject of two recent reviews [63, 64]. The old favorite bR was among the first proteins to be reconstituted into them [65]. Since then the list of membrane proteins that have been functionally incorporated and studied in nanodisks is expanding rapidly, and includes the GPCRs rhodopsin [66] and β_2-adrenergic receptor [67], a potassium channel KcsA [68], the large SecYEG channel [69], a voltage-gated ion channel [70], and an ABC transporter [64] (Figure 13.4). In all cases the reconstitution process is relatively simple; detergent-solubilized protein is added to the MSP, phospholipid, and cholate mix, and the detergents are then removed with polystyrene beads. One consideration is that the amount of lipid in the mix needs to be carefully controlled and reduced to take into account the area of the protein within the disk. Alternatively, the reconstitution step can be massively simplified by supplementing a CFE system with nanodisks [73]. In this way the nascent membrane protein is provided with a membrane environment the instant is comes off the ribosome. Furthermore, the detergent-free system means there is little danger of the membrane environment interfering with the translation/transcription machinery.

The potential applications for nanodisks are just being realized. Like bicelles, nanodisks have an obvious application in NMR, where the small, clearly defined structures allow for rapid tumbling and well-resolved spectra. The most promising study to date was conducted on the voltage-dependent anion channel VDAC-1 [70], where it was demonstrated that the spectra of the channel in nanodisks was essentially the same as in micelles. Other applications are less obvious but quite ingenious, such as hemagglutinin-containing nanodisks that provide protective immunity against influenza virus infection, demonstrating that the disks can maintain protein stability in an *in vivo* environment [74].

13.3.2.2 **Liposomes**

Of the possible environments that can be used to accommodate membrane proteins lipid bilayers most closely resemble the native environment. It is very clear that a lipid bilayer is the best choice if protein stability is the only factor. Furthermore, lipid bilayer systems afford the greatest degree of flexibility as their composition can be easily tailored to the tastes of a particular protein. Two main technical questions then arise from reconstituting membrane proteins into lipid bilayers:

i) How does one go about reconstituting a protein/lipid bilayer system?
ii) What lipid bilayer system should be used?

In the last 10 years the membrane protein field has largely settled on a (reasonably) reliable set of protocols to address the first question. However, the answer to the second is still largely a case of "hit and miss." Numerous studies have shown that a protein might prefer certain lipids with a given head-group, may bind a particular lipid, or happen to have a fondness for a tense bilayer. However, there is very little that can be done to predict how a new protein might react to a particular lipid environment. Thus, in the following section we will discuss reconstitution techniques and considerations when selecting a host bilayer. Reconstitution of membrane proteins has been the subject of many excellent reviews, so once again we do not intend to cover the nuts and bolts of the process, but instead provide an overview that points out the main considerations. For more detail we direct the reader to a must-read review by Jean-Louis Rigaud and Daniel Levy clearly entitled "Reconstitution of membrane proteins into liposomes" [75].

13.3.3 Detergent-Mediated Reconstitution of Proteoliposomes

In principle the process of mixing liposomes and membrane proteins to form proteoliposomes is fairly straightforward. In the vast majority of cases we start with detergent-solubilized protein that is then added to lipids. The detergent is removed, and the protein's hydrophobic surfaces are left naked and exposed to the aqueous environment. At this point it is thermodynamically unfavorable for the protein to stay in solution. What happens next primarily depends on whether the protein manages to insert itself into a liposome before it encounters another protein in the same state. In the first case a proteoliposome is formed, alternatively the proteins' hydrophobic surfaces interact resulting in aggregates. Controlling the formation of productive proteoliposomes then largely comes down to the "naked" protein/lipid ratio. The simplest way to control this is by keeping the concentration of protein low. However some techniques, most notably solid-state NMR and two-dimensional crystallography, require densely packed bilayers. In the case of solid-state NMR the problem is further compounded by the need for a homogenous sample, while for two-dimensional crystallography only relatively small patches of well-formed crystals are required. To form densely packed proteoliposomes extra care must be taken to develop a reconstitution process that is amiable to the protein. Furthermore, some strategies may be put in place to remove aggregated protein or empty liposomes. Ultracentrifugation and sucrose density gradients are particularly effective as empty liposomes will float to the top, aggregates will hit the bottom of the tube, and proteoliposome will be suspended at some point in between. A further consideration is the length of time the whole process takes; many proteins are only marginally stable in detergents, hence it is preferable to get them into liposomes as quickly as possible.

The starting state for any reconstitution is some form of mixture of protein, detergent and lipid (more detail later). The reconstitution process is initiated by reducing the detergent concentration to the point where there is insufficient detergent to mask the protein's hydrophobic surfaces. The vast majority of protein reconstitutions are conducted by the three methods given below. Other protocols have also been tried and tested (e.g., size-exclusion chromatography) but these are in a small minority [76].

13.3.3.1 Dilution Method

The simplest reconstitution method is to start with a suspension of preformed liposomes into which a small volume of detergent-solubilized protein is diluted. The dilution should be great enough for the resulting concentration of detergent to be well below the CMC. Below this point the detergent monomers disaggregate, leaving the protein exposed. This method is often very effective, but relies on a large excess of lipid; many thousands of lipids to one protein. At higher protein or lower lipid concentrations the chances of protein aggregation occurring increases resulting in a mixed population of proteliposomes and inactive protein. The protein aggregates (and empty liposomes) can be separated from proteoliposomes using ultracentrifugation and density gradients, but nevertheless the dilution method is limited to the formation of sparsely populated liposomes.

The dilution method has been used to great effect when studying the refolding processes of membrane proteins for optical spectroscopy and activity or binding studies when the concentration of protein is not a strongly limiting factor in the experimental process [22, 77–79]. If more densely packed proteoliposomes are required then alternative methods for detergent removal are called for.

13.3.3.2 Dialysis versus Hydrophobic Absorption

The low cost and simplicity of dialysis meant it was once the most widely used method for removing detergents. Lipid–protein–detergent mixtures are placed in a bag or device with an appropriate cutoff membrane (smaller than the size of the micelle, normally 14 kDa is fine) and the solution is left to dialyze against a large volume of buffer that is changed regularly. The free detergent monomers are the only species that can traverse the membrane, so over a period of time the detergent concentration inside the bag will drop to a point where it can no longer maintain the solubility of the lipids and proteins, which then come out of solution in the form of proteoliposomes.

The major drawback of dialysis is that it can be a very long process as the rate of removal of detergent is dependent on the concentration of free detergent monomers (given by the CMC). Detergents with high CMCs (and generally denaturing tendencies) such as β-OG and CHAPS are dialyzed away in a few days. However, using milder detergents with low CMCs, like DPC, DDM, and the Tritons, only extends the dialysis period; removing these detergents can take weeks, during which time the protein has ample opportunity to denature. Further disadvantages include the prospect of protein absorbing to the membrane and lack of control over the rate of dialysis, which will be dependent on numerous factors such as the stirring rate of the buffer, temperature, and the ever-changing relative concentrations on either side of the membrane.

In contrast, hydrophobic absorption offers much greater flexibility, control, and speed, and so is probably the best place to start when devising a new reconstitution procedure. It is now the most widely used reconstitution method having been successful with just about every class of membrane protein. Once again the guru of membrane protein reconstitution, Jean-Louis Rigaud, has provided several vital studies on the method [80, 81]. This means that the processes involved in hydrophobic absorption of detergents are very well characterized. The method is quite simple; once again lipid–detergent–protein mixtures are made, but this time polystyrene hydrophobic absorbing materials such as Bio-beads or Amberlite are used to remove the detergent from the bulk solution. This can be achieved by a batch method, whereby beads are added to the mixture or alternatively the solution can be passed through a column of beads. These beads preferentially absorb the detergent, leaving the protein and lipids to coalesce into proteoliposomes within a matter of minutes to hours. The binding capacity of the beads varies from detergent to detergent, but these have been carefully determined for the most commonly used systems and so it is quite easy to calculate appropriate amounts [75, 80, 81]. Furthermore, by varying the amount of beads in the mix it is possible to precisely control the rate of detergent removal. One disadvantage is the beads do have a small affinity for lipids so the final protein/lipid ratio may vary from the starting point. However this problem can be mitigated by preincubating the beads with lipids.

Combinations and variations on these three detergent removal methods are also prevalent. For example, Bio-beads can be placed in the dialysis or dilution buffer to facilitate complete removal of detergent or the dilution method can be per-

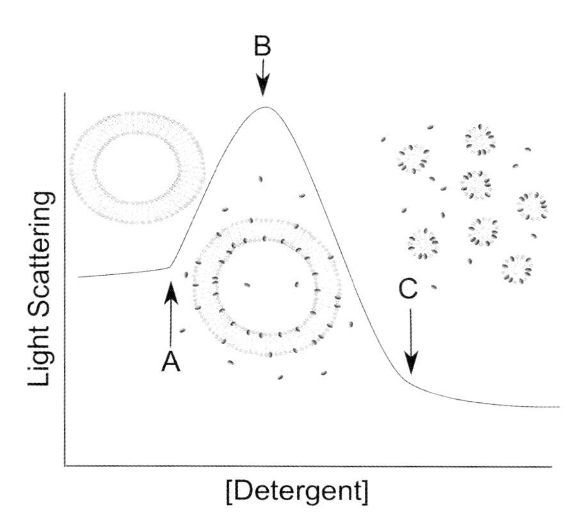

Figure 13.5 Schematic depicting the transition from liposomes to micelles with increasing detergent concentration. The turbidity of the suspension increases (from point A) as detergent molecules partition into the liposomes and causes them to swell. A maximum turbidity is reached after which the liposomes begin to break up (point B) into micelles. Finally, all the lipids have been completely solubilized and only micelles remain, leaving a clear, nonturbid solution (point C). (Adapted from [91].)

formed through multiple additions of protein to moderate the concentration of free protein and so limit aggregation.

13.3.3.3 **Detergent Saturation**

The starting state of the lipid–protein–detergent mixture can have a great effect on the efficiency of the reconstitution process. At one extreme detergent-solubilized protein is simply added to preformed liposomes (as in the dilution method mentioned above), while at the other the whole system is solubilized to form mixed detergent–lipid–protein micelles [12, 82–85]. However, the middle ground can prove the most effective [12, 86–90]. Here, liposomes are saturated with detergent so that they are on the cusp of being solubilized. At this point the bilayer is extremely fluid and as a consequence it is assumed that there is little barrier for the insertion of proteins. Excess detergent can then be removed using any of the methods given above.

The onset of saturation is accompanied by a dramatic increase in the turbidity of the suspension (Figure 13.5). Therefore, determining the saturation point is easily found by monitoring light scattering (optical density at 600 nm will suffice) and titrating in detergent. A maximum in the light scattering will be reached, followed by a drop as the phase transition from liposomes to micelles is crossed. The detergent concentration at the zenith represents the saturation point after which liposomes start to break up into micelles.

13.3.4 **Lipid Composition**

The composition of the host lipid bilayer can have a dramatic effect on the structure, function and stability of the host protein. The most striking examples are seen in the case of some bacterial secondary transporters. In *E. coli* lacking PE lipids several of these proteins do not fold to a functional state [92–94]. The lipid effects on lactose permease (LacY) folding has been studied in most detail, for which it has been shown that PE lipid is not required to maintain protein function once it has folded. Furthermore, the protein does not have a dependency for a PE head-group *per se*, as demonstrated by misfolding in the presence of lyso-PE [95]. Interestingly, PE lipids can be interchanged with polyunsaturated monoglucosyl-diacylglycerol [96], a lipid that imparts the same bilayer properties as the PE it replaces. It is still unclear what features of the PE lipids is required for proper LacY folding; charge, curvature stress, or specific interactions could all play a role. However, these studies clearly demonstrate that there is a complex interplay between proteins and their lipid environment.

Given the marked effect that lipids can have on proteins often the simplest approach when reconstituting proteoliposomes is to use a lipid mixture derived from the host (or related) organism. Consequently, bacterial membrane proteins are frequently reconstituted into *E. coli* lipid extracts [83, 87, 90, 97, 98], while egg yolk lipids are a firm favorite for reconstitution of eukaryotic proteins [82, 84, 85, 99, 100]. However it is not always practical or desirable to be limited to these complex and sometimes ill-defined mixtures. In that case careful consideration should be given to the nature of the lipids used in the reconstitution mix.

13.3.4.1 **Hydrophobic Mismatch**

The most obvious factor (and therefore the first to be studied in any detail) is the hydrophobic thickness of the membrane. This is given by the area bounded by the glycerol groups on the opposite sides of the bilayer. For optimum activity and stability the hydrophobic thickness should match that of the hydrophobic surface of an embedded membrane protein. A mismatch between the two will lead to compensating distortions in the bilayer and/or protein as the system adjusts to minimize the hydrophobic surfaces exposed to the aqueous environment (Figure 13.6). One way in which this is achieved is by the lipids immediately around the protein adapting their chain lengths to match the hydrophobic thickness of the protein. This is performed by changes in the lipid alkyl chain order and concomitant alteration in the phase of these lipids. For example, introducing a protein with a long hydrophobic length into a short-chain lipid bilayer can result in the annular lipids adopting a more gel-like phase characterized by extended, ordered chains. This phenomenon has been seen on multiple occasions using model peptides of varying lengths as well as polytopic membrane proteins (see review by Marsh and references therein [101]). However, the lipid alterations rarely compensate for the whole hydrophobic mismatch, the protein accommodates the rest.

The most extreme way in which the protein can react to hydrophobic mismatch is via aggregation to minimize exposed hydrophobic surfaces or even exclusion of the protein from the bilayer [102–106]. The risk of these gross incidents occurring increases with higher protein to lipid ratios. Both these events will clearly have highly undesirable effects on protein activity. However, the mechanisms that proteins use to accommodate hydrophobic mismatch can manifest in many more

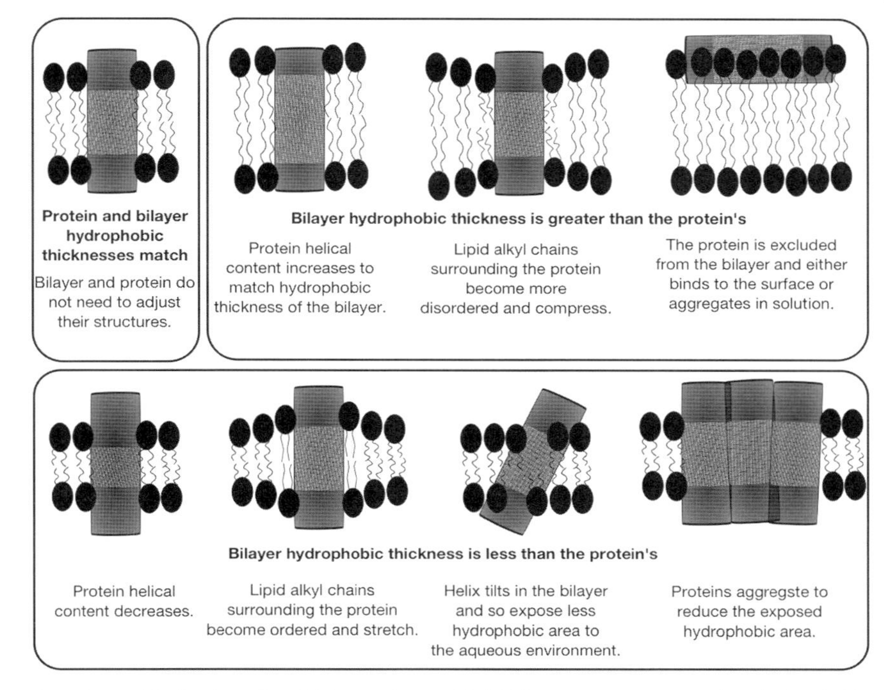

Figure 13.6 Mechanisms by which proteins and lipid bilayers may accommodate hydrophobic mismatch.

subtle ways. The transmembrane regions can tilt, kink [107], and even expand or contract. Altered tilts are most obvious with single span proteins and peptides [108–111]; however, this phenomenon is not generally enough to accommodate the mismatch and is usually accompanied by a change in bilayer properties. Changes in tilt angle have not as yet been observed for polytopic membrane proteins, instead these larger proteins may have quite an surprising degree of plasticity that allows them to alter their helical content to match the hydrophobic thickness of the bilayer [112]. A particularly notable case is rhodopsin where protein in C14 lipids has a helical content of 49%, which is increased to 57% in C18 lipids [113]. This change corresponds to a quite staggering increase in helix length of around 6Å or three to four amino acids per helix, enough to fully accommodate the thicker membrane! What is all the more surprising is that there is no noticeable change in protein activity or bilayer properties. Other proteins are not quite so tolerant; the activities of DGK and Ca^{2+}-ATPase are strongly modulated by the thickness of the bilayer [103, 114, 115]. Both proteins have maximum activity in monounsaturated C18 lipids while reconstituting into C14 or C24 lipids can reduce the activity by as much as 80%. These alterations in activity are not necessarily the result of a change in structure or stability, but instead may be caused by changes in equilibria between different conformational states or protein dynamics [116] or an inability of the protein to access its substrates.

When designing a reconstitution system the hydrophobic length of the bilayer is one aspect that can be approached in a rational way by simply comparing the thickness of the native membrane with that on the proposed system. For example, rat liver plasma membrane contains 49% C18 lipids and 26% C16 lipids plus a significant amount of cholesterol (around 20%) [117], which has an ordering effect that results in elongated lipid alkyl chains. Meanwhile, *E. coli* contains a much greater fraction of C16 lipids (49%), fewer C18 lipids (21%), and no sterols [118]. These contrasting compositions can translate into significantly different hydrophobic thicknesses; the hydrophobic length of a di(C16 : 1)PC bilayer is around 26 compared to around 30 for di(C18 : 1)PC [119]. Having said that, other factors may scupper the rational approach. For example, let us compare the preferences of Ca^{2+}-ATPase, derived from sacroplasmic reticulum with a plasma membrane Na,K-ATPase. The native membranes have similar lipid alkyl chain lengths and sterol contents, but in single component lipid systems the Ca^{2+}-ATPase is most active in C18 lipids [103], but the Na,K-ATPase displays maximum activity in C22 lipids. However adding cholesterol into the mix shifts the Na,K-ATPase's preference to C18 lipids, and greatly increases its activity over and above that seen for C18 or C22 lipids alone [120]. This implies either some specific dependency for cholesterol or that the sterol modulates the protein via some nonspecific alteration in the bilayer properties, possibly via changes in the bilayer's elastic stress.

13.3.4.2 Curvature Elastic Stress

Biological membranes are composed of a plethora of different lipids – a diversity that is required to precisely maintain the fluidity and elasticity of a membrane. Indeed, bacteria appear to go to great lengths to maintain these properties by changing the lipid compositions of their membranes in response to their environment [121]. They do so to maintain the structural integrity of the membrane, but it is increasingly clear that membrane proteins' structure, function, and folding can also be altered as the fluidity of the membrane changes.

The degree of fluidity of a bilayer arises from its curved elastic stress. This can be explained by considering the relative volumes of a lipid's head-group compared to its chain (Figure 13.7a). For example, a lyso-PC, with its single chain and relatively large head-group, is approximately cone shaped with the point at the end of the chain. Thus, it will pack together curving away from the water to form micelles. In contrast, di(C18 : 1)PC is approximately rod shaped and so packs nicely into flat bilayers. Meanwhile, PE and PA lipids have small head-groups compared to the chain region and unsaturated chains increase the chain volume compared to the head-group. Thus, unsaturated chains or small head-group lipids will pack with a curvature towards the water resulting in an inverse hexagonal phase. A lipid mixture containing PE and PC lipids is then in a quandary, the PC wants to form a bilayer, while the PE is a non-bilayer-forming lipid. The result is monolayers that have a desire to curve towards the water, but this is not possible, so instead they form flat bilayers that are under stress (Figure 13.7b). If more PE is added to the mix then eventually the PE will win out and the system will undergo a phase transition to the inverse hexagonal phase. Conversely, if lyso-lipids, detergents, or short-chain

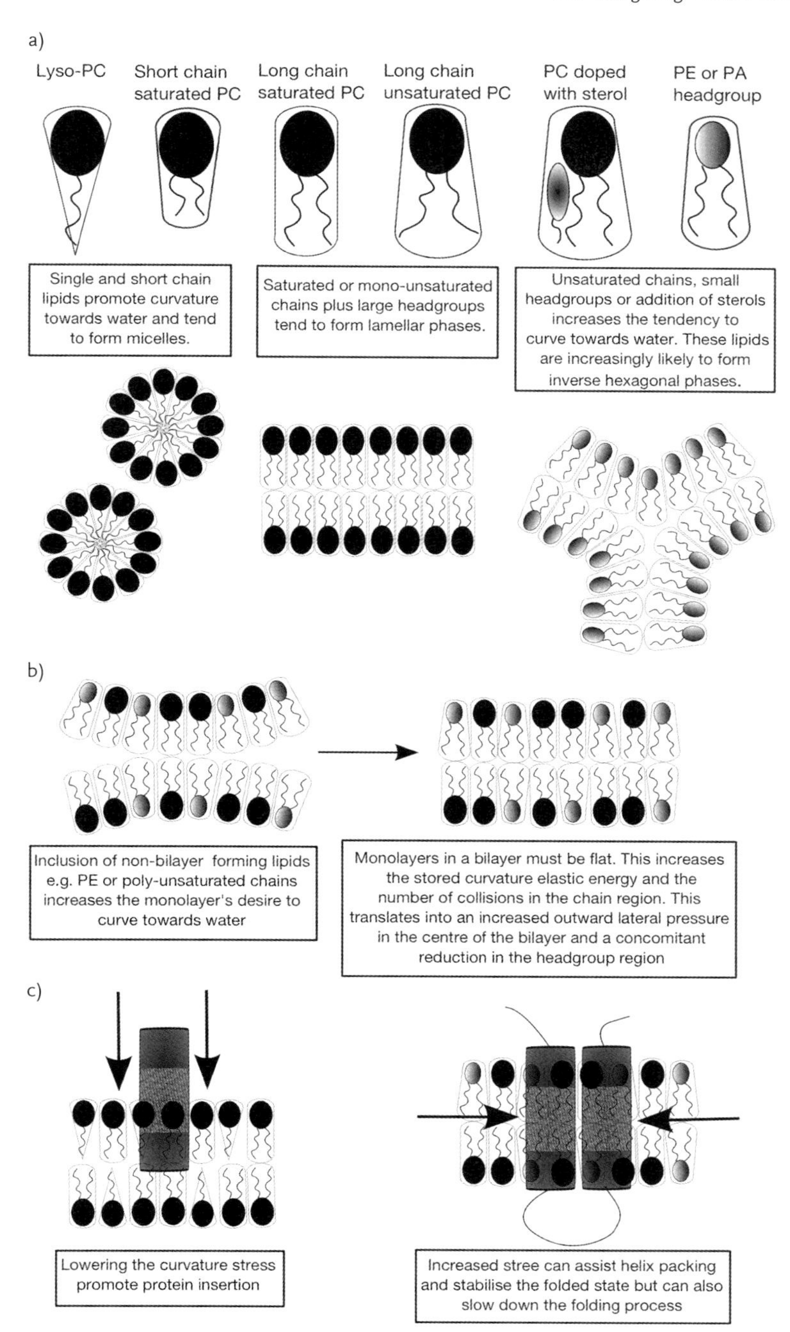

Figure 13.7 Schematic representation of (a) the shapes of lipids and the phases they prefer to adopt, (b) the manifestation of bilayer curvature stress, and (c) how bilayer stress can modulate membrane protein folding. (Adapted from [122].)

lipids are doped into PC bilayers the curvature stress is reduced and at a high enough lyso concentration the system will revert to a micellular phase (as in the detergent solubilized liposomes discussed earlier). Addition of other molecules will also cause changes in bilayer stress; sterols partition into the lower part of a monolayer and so increase stress; alcohols tend to locate in head-group regions and so have the converse effect. It is also worth noting that even seemingly innocuous materials such as buffers can have surprising effects on bilayer properties [123].

Changing the composition of a bilayer can lead to an embedded membrane protein experiencing very different forces, and this in turn can have dramatic effects on the stability, refolding, and reconstitution processes. One of the first demonstrations of this phenomenon was published in 1987 on reconstituted Ca^{2+}-ATPase. It was demonstrated that cholesterol markedly increased the protein's thermal stability while ethanol and anesthetics (both of which reduce bilayer stress) had the reverse effect. Similar observations were made with a Ca^{2+},Mg^{2+}- ATPase [124]. Despite these observations a link between curvature stress and protein folding/stability was not clearly established until much later when in 2002 Meijberg and Booth performed a detailed thermodynamic study on the effect of bilayer stress on the insertion of a synthetic peptide [125]. The peptide binds to bilayers, but only inserts at high pH. Thus, the insertion event can be triggered and followed in real-time. The system allowed the activation energy of insertion to be calculated, which increased with the fraction of di(C18 : 1)PE in the bilayer. The study clearly shows that insertion of a peptide becomes more difficult as the stress of the bilayer increases. Analogous experiments performed on polytopic membrane proteins reveal similar phenomena. The refolding efficiency of EmrE and bR into predominantly PC bilayers are hindered by di(C18 : 1)PE and promoted by lyso lipids [78, 126]. However, this trend is bucked by DGK. Here, the refolding efficiency of urea solubilized protein follows the opposite pattern, being promoted by di(C18 : 1)PE and hindered by lyso lipids [127]. The apparent contradiction is reconciled with reference to the two-stage model of membrane protein folding whereby formation of transmembrane helices (stage 1) is followed by helix packing and folding (stage 2) [128]. The first stage requires the protein to bind with and insert into the bilayer – a process that has been clearly shown to be hindered by increased curvature stress. The second stage requires the helices and possibly monomers to pack together and fold into the final state – in this case, increased pressure in the membrane will "push and hold" the helices together (Figure 13.7c). Thus, the limiting step in DGK folding may be the formation of the functional intertwined trimer [6] – a process which is assisted by bilayer stress. In contrast, EmrE and bR require less help to fold, and are instead limited by their capacity to insert into the membrane in the first place. This model is further born out by the observation that KcsA tetramers are destabilized by the addition of trifluoroethane (which reduces bilayer stress) and are stabilized in the presence of PE lipids [129].

13.3.4.3 Specific Lipid Effects

The examples given above demonstrate a correlation between bilayer stress and hydrophobic match with modulation of protein behavior. However, we must remember that correlation does not imply causality. A case in point is Band 3; the protein is stabilized by cholesterol and long alkyl chains, but destabilized by unsaturated chains. This demonstrates a much more complex situation than simply the protein's reaction to bulk bilayer properties [130]. Instead it is quite likely that a protein's direct interactions with lipids may modulate its behavior. This is very clearly the case for DGK, where the presence of di(C18 : 1)PG leads to higher refolding yields, which might suggest a requirement for an anionic lipid. However, substituting di(C18 : 1)PG for equally anionic di(C18 : 1)PS fails to elicit the same response [127]. There are numerous other examples of membrane proteins requiring lipids to function; cardiolipin is essential for activity of bovine cytochrome *c* oxidase [131] and assembly of mitochondrial carrier proteins [132]; a voltage-dependent K^+ channel from *Aeropyrum pernix* needs to interact with lipid phosphate group [133] as well as a PE or PG head-group [134]. Further compelling evidence comes from the majority of membrane protein crystal structures which contain bound lipids [20, 135].

13.4 Making or Selecting a Stable Protein

The bulk of this chapter has concentrated on manipulating a protein's environment in order to gain maximum stability and activity. Finally, we briefly turn our attention towards the protein itself. Nature does not always evolve proteins to achieve maximum stability. Consequently, there may be some room for improvement. In fact, it appears that stabilizing mutations of detergent-solubilized membrane proteins are surprisingly common. Random mutagenesis of β-OG-solubilized DGK identified 12 stabilizing point mutations and when four were combined the half-life of thermal inactivation went from 6 min at 55 °C for wild-type to 35 min at 80 °C [136]. Similarly bR [137], KcsA [138, 139] β_2-adrenergic receptor [140], and a neurotensin receptor [141] have all been significantly stabilized by mutation. What is striking is the frequency with which stabilizing mutations crop up – it would appear that better than 10% are more stable than the wild-type protein. This is probably a reflection of the fact that in all of the above examples the stability of the proteins has been tested in detergents and, of course, the proteins have not evolved to be stable in these simple membrane mimetics. Unfortunately, despite the high incidence of stabilizing mutants there appears to be little discernable rationale behind the substitutions' locations. Therefore, the best plan is to conduct a random screen.

Finally, before embarking on a protracted search for a stable mutant it is worth considering whether time would be better spent isolating a naturally thermostable candidate protein from a thermophilic organism. This strategy has proved fruitful with membrane protein crystallographers, yielding (at the time of writing) 16 structures from eight different thermophilic bacteria (http://blanco.biomol.uci.edu/Membrane_Proteins_xtal.html#Latest).

13.5
Conclusions

Reconstituting lipid protein systems and stabilizing membrane proteins is no longer the "black art" it once was. As shown here, there are many clearly defined and promising strategies one can take to optimize the conditions under which to keep the protein of choice. Despite this there are, unfortunately, very few, if any, hard and fast rules to follow, so a reasonable amount of trial and error is still par for the course.

Abbreviations

β-OG	*n*-Octyl-β-D-glucoside
bR	bacteriorhodopsin
C14 : 0PG	1-myristoyl-2-hydroxy-*sn*-glycero-3-phosphoglycerol
C16 : 0/C18 : 1PC	1-Palmitoyl-2-oleoyl-*sn*-glycero-3-phosphocholine
CFE	cell-free expression
CHAPS	3-((3-Cholamidopropyl) dimethylammonio)-1-propanesulphonate
CHAPSO	3-((3-Cholamidopropyl) dimethylammonio)-2-hydroxy-1-propanesulfonate
CMC	critical micelle concentration
DDM	*n*-Dodecyl-β-D-maltoside
DGK	diacylglycerol kinase
di(C12 : 0)PC	1,2-dilauroyl-*sn*-glycero-3-phosphocholine
di(C16 : 0)PC	1,2-dipalmitoyl-*sn*-glycero-3-phosphocholine
di(C18 : 1)PA	1,2-dioleoyl-*sn*-glycero-3-phosphosphate
di(C18 : 1)PC	1,2-dioleoyl-*sn*-glycero-3-phosphocholine
di(C18 : 1)PE	1,2-dioleoyl-*sn*-glycero-3-phosphoethanolamine
di(C18 : 1)PG	1,2-dioleoyl-*sn*-glycero-3-phosphoglycerol
di(C18 : 1)PS	1,2-dioleoyl-*sn*-glycero-3-phosphoserine
di(C6 : 0)PC	1,2-dihexanoyl-*sn*-glycero-3-phosphocholine
DM	*n*-decyl-β-D-maltoside
DPC	Dodecylphosphocholine
GPCR	G–protein-coupled receptor
LDAO	Lauryldimethylamine-*N*-oxide
MSP	membrane scaffold protein
NMR	nuclear magnetic resonance
PA	phosphatidate
PC	phosphocholine
PE	phosphoethanolamine
PG	phosphoglycerol
PS	phosphoserine
SDS	sodium dodecyl sulfate
VDAC	voltage-dependent anion channel

References

1 Wiener, M.C. and White, S.H. (1992) Structure of a fluid dioleoylphosphatidylcholine bilayer determined by joint refinement of X-ray and neutron-diffraction data 3. Complete structure. *Biophys. J.*, **61**, 434–447.

2 le Maire, M., Champeil, P., and Muller, J.V. (2000) Interaction of membrane proteins and lipids with solubilizing detergents. *Biochim. Biophys. Acta*, **1508**, 86–111.

3 Booth, P.J., Riley, M.L., Flitsch, S.L., Templer, R.H., Farooq, A., Curran, A.R., Chadborn, N., and Wright, P. (1997) Evidence that bilayer bending rigidity affects membrane protein folding. *Biochemistry*, **36**, 197–203.

4 Daghastanli, K.R.P., Ferreira, R.B., Thedei, G., Maggio, B., and Ciancaglini, P. (2004) Lipid composition-dependent incorporation of multiple membrane proteins into liposomes. *Colloids Surf. B*, **36**, 127–137.

5 Dornmair, K., Kiefer, H., and Jahnig, F. (1990) Refolding of an integral membrane-protein – OmpA of *Escherichia coli*. *J. Biol. Chem.*, **265**, 18907–18911.

6 Van Horn, W.D., Kim, H.J., Ellis, C.D., Hadziselimovic, A., Sulistijo, E.S., Karra, M.D., Tian, C.L., Sonnichsen, F.D., and Sanders, C.R. (2009) Solution nuclear magnetic resonance structure of membrane-integral diacylglycerol kinase. *Science*, **324**, 1726–1729.

7 Arnold, K. and Linke, D. (2008) The use of detergents to purify membrane proteins. *Curr. Protoc. Protein Sci.*, **53**, 4.8.1–4.8.30.

8 Prive, G.G. (2007) Detergents for the stabilization and crystallization of membrane proteins. *Methods*, **41**, 388–397.

9 Alpes, H., Apell, H.J., Knoll, G., Plattner, H., and Riek, R. (1988) Reconstitution of Na^+/K^+-ATPase into phosphatidylcholine vesicles by dialysis of nonionic alkyl maltoside detergents. *Biochim. Biophys. Acta*, **946**, 379–388.

10 Vanaken, T., Foxallvanaken, S., Castleman, S., and Fergusonmiller, S. (1986) Alkyl glycoside detergents – synthesis and applications to the study of membrane-proteins. *Methods Enzymol.*, **125**, 27–35.

11 Hauser, H. (2000) Short-chain phospholipids as detergents. *Biochim. Biophys. Acta*, **1508**, 164–181.

12 Klammt, C., Schwarz, D., Fendler, K., Haase, W., Dotsch, V., and Bernhard, F. (2005) Evaluation of detergents for the soluble expression of alpha-helical and beta-barrel-type integral membrane proteins by a preparative scale individual cell-free expression system. *FEBS J.*, **272**, 6024–6038.

13 Ottolenghi, P. (1979) Relipidation of delipidated Na,K-ATPase – analysis of complex-formation with dioleoylphosphatidylcholine and with dioleoylphosphatidylethanolamine. *Eur. J. Biochem.*, **99**, 113–131.

14 le Maire, M.L., Jorgensen, K.E., Roigaardpetersen, H., and Moller, J.V. (1976) Properties of deoxycholate solubilized sarcoplasmic-reticulum Ca^{2+}-ATPase. *Biochemistry*, **15**, 5805–5812.

15 Dufour, J.P. and Goffeau, A. (1980) Phospholipid reactivation of the purified plasma-membrane ATPase of yeast. *J. Biol. Chem.*, **255**, 591–598.

16 Knol, J., Sjollema, K., and Poolman, B. (1998) Detergent-mediated reconstitution of membrane proteins. *Biochemistry*, **37**, 16410–16415.

17 Cox, K. and Sansom, M.S.P. (2009) One membrane protein, two structures and six environments: a comparative molecular dynamics simulation study of the bacterial outer membrane protein PagP. *Mol. Membr. Biol.*, **26**, 205–214.

18 Klare, J.P., Bordignon, E., Doebber, M., Fitter, J., Kriegsmann, J., Chizhov, I., Steinhoff, H.J., and Engelhard, M. (2006) Effects of solubilization on the structure and function of the sensory rhodopsin II/transducer complex. *J. Mol. Biol.*, **356**, 1207–1221.

19 Columbus, L., Lipfert, J., Jambunathan, K., Fox, D.A., Sim, A.Y.L., Doniach, S., and Lesley, S.A. (2009) Mixing and

matching detergents for membrane protein NMR structure determination. *J.Am. Chem. Soc.*, **131**, 7320–7326.

20 Hunte, C. and Richers, S. (2008) Lipids and membrane protein structures. *Curr. Opin. Struct. Biol.*, **18**, 406–411.

21 Zhang, H., Kurisu, G., Smith, J.L., and Cramer, W.A. (2003) A defined protein–detergent–lipid complex for crystallization of integral membrane proteins: the cytochrome b_6f complex of oxygenic photosynthesis. *Proc. Natl. Acad. Sci. USA*, **100**, 5160–5163.

22 McKibbin, C., Farmer, N.A., Jeans, C., Reeves, P.J., Khorana, H.G., Wallace, B.A., Edwards, P.C., Villa, C., and Booth, P.J. (2007) Opsin stability and folding: modulation by phospholipid bicelles. *J. Mol. Biol.*, **374**, 1319–1332.

23 Bubis, J. (1998) Effect of detergents and lipids on transducin photoactivation by rhodopsin. *Biol. Res.*, **31**, 59–71.

24 Jastrzebska, B., Goc, A., Golczak, M., and Palczewski, K. (2009) Phospholipids are needed for the proper formation, stability, and function of the photoactivated rhodopsin–transducin complex. *Biochemistry*, **48**, 5159–5170.

25 Yao, Z.P. and Kobilka, B. (2005) Using synthetic lipids to stabilize purified beta$_2$ adrenoceptor in detergent micelles. *Anal. Biochem.*, **343**, 344–346.

26 Navratilova, I., Sodroski, J., and Myszka, D.G. (2005) Solubilization, stabilization, and purification of chemokine receptors using biosensor technology. *Anal. Biochem.*, **339**, 271–281.

27 Schneider, E.G. and Kennedy, E.P. (1976) Partial-purification and properties of diglyceride kinase from *Escherichia coli*. *Biochim. Biophys. Acta*, **441**, 201–212.

28 Walsh, J.P. and Bell, R.M. (1986) *sn*-1,2-Diacylglycerol kinase of *Escherichia coli*–structural and kinetic-analysis of the lipid cofactor dependence. *J. Biol. Chem.*, **261**, 5062–5069.

29 Bohnenberger, E. and Sandermann, H. (1983) Lipid dependence of diacylglycerol kinase from *Escherichia coli*. *Eur. J. Biochem.*, **132**, 645–650.

30 Zimmer, J. and Doyle, D.A. (2006) Phospholipid requirement and pH optimum for the *in vitro* enzymatic activity of the *E. coli* P-type ATPase *ZntA*. *Biochim. Biophys. Acta*, **1758**, 645–652.

31 Fan, Z. and Makielski, J.C. (1997) Anionic phospholipids activate ATP-sensitive potassium channels. *J. Biol. Chem.*, **272**, 5388–5395.

32 Reinsberg, D., Booth, P.J., Jegerschold, C., Khoo, B.J., and Paulsen, H. (2000) Folding, assembly, and stability of the major light-harvesting complex of higher plants, LHCII, in the presence of native lipids. *Biochemistry*, **39**, 14305–14313.

33 London, E. and Khorana, H.G. (1982) Denaturation and renaturation of bacteriorhodopsin in detergents and lipid–detergent mixtures. *J. Biol. Chem.*, **257**, 7003–7011.

34 Mandal, A.K., Cheung, W.D., and Arguello, J.M. (2002) Characterization of a thermophilic P-type Ag^+/Cu^+-ATPase from the extremophile *Archaeoglobus fulgidus*. *J. Biol. Chem.*, **277**, 7201–7208.

35 Deforesta, B., le Maire, M., Orlowski, S., Champeil, P., Lund, S., Moller, J.V., Michelangeli, F., and Lee, A.G. (1989) Membrane solubilization by detergent–use of brominated phospholipids to evaluate the detergent-induced changes in Ca^{2+}-ATPase lipid interaction. *Biochemistry*, **28**, 2558–2567.

36 Muller, K. (1981) Structural dimorphism of bile-salt lecithin mixed micelles–a possible regulatory mechanism for cholesterol solubility in bile–X-ray structure-analysis. *Biochemistry*, **20**, 404–414.

37 Sanders, C.R. and Prestegard, J.H. (1990) Magnetically orientable phospholipid-bilayers containing small amounts of a bile-salt analog, CHAPSO. *Biophys. J.*, **58**, 447–460.

38 Sanders, C.R. and Schwonek, J.P. (1992) Characterization of magnetically orientable bilayers in mixtures of dihexanoylphosphatidylcholine and dimyristoylphosphatidylcholine by solid-state NMR. *Biochemistry*, **31**, 8898–8905.

39 Sanders, C.R. and Landis, G.C. (1995) Reconstitution of membrane-proteins into lipid-rich bilayered mixed micelles for NMR-studies. *Biochemistry*, **34**, 4030–4040.

40 Czerski, L. and Sanders, C.R. (2000) Functionality of a membrane protein in bicelles. *Anal. Biochem.*, **284**, 327–333.
41 Lind, J., Nordin, J., and Maler, L. (2008) Lipid dynamics in fast-tumbling bicelles with varying bilayer thickness: effect of model transmembrane peptides. *Biochim. Biophys. Acta*, **1778**, 2526–2534.
42 Triba, M.N., Devaux, P.F., and Warschawski, D.E. (2006) Effects of lipid chain length and unsaturation on bicelles stability. A phosphorus NMR study. *Biophys. J.*, **91**, 1357–1367.
43 Lau, T.L., Partridge, A.W., Ginsberg, M.H., and Ulmer, T.S. (2008) Structure of the integrin beta 3 transmembrane segment in phospholipid bicelles and detergent micelles. *Biochemistry*, **47**, 4008–4016.
44 Soong, R., Majonis, D., and Macdonald, P.M. (2009) Size of bicelle defects probed via diffusion nuclear magnetic resonance of PEG. *Biophys. J.*, **97**, 796–805.
45 Wang, J.F., Schnell, J.R., and Chou, J.J. (2004) Amantadine partition and localization in phospholipid membrane: a solution NMR study. *Biochem. Biophys. Res. Commun.*, **324**, 212–217.
46 Struppe, J., Whiles, J.A., and Vold, R.R. (2000) Acidic phospholipid bicelles: a versatile model membrane system. *Biophys. J.*, **78**, 281–289.
47 Parker, M.A., King, V., and Howard, K.P. (2001) Nuclear magnetic resonance study of doxorubicin binding to cardiolipin containing magnetically oriented phospholipid bilayers. *Biochim. Biophys. Acta*, **1514**, 206–216.
48 Sasaki, H., Fukuzawa, S., Kikuchi, J., Yokoyama, S., Hirota, H., and Tachibana, K. (2003) Cholesterol doping induced enhanced stability of bicelles. *Langmuir*, **19**, 9841–9844.
49 Barbosa-Barros, L., de la Maza, A., Lopez-Iglesias, C., and Lopez, O. (2008) Ceramide effects in the bicelle structure. *Colloids Surf. A*, **317**, 576–584.
50 Soong, R. and Macdonald, P.M. (2005) Lateral diffusion of PEG-lipid in magnetically aligned bicelles measured using stimulated echo pulsed field gradient H-1 NMR. *Biophys. J.*, **88**, 255–268.
51 Katsaras, J., Harroun, T.A., Pencer, J., and Nieh, M.P. (2005) "Bicellar" lipid mixtures as used in biochemical and biophysical studies. *Naturwissenschaften*, **92**, 355–366.
52 Faham, S. and Bowie, J.U. (2002) Bicelle crystallization a new method for crystallizing membrane proteins yields a monomeric bacteriorhodopsin structure. *J. Mol. Biol.*, **316**, 1–6.
53 Triba, M.N., Zoonens, M., Popot, J.L., Devaux, P.F., and Warschawski, D.E. (2006) Reconstitution and alignment by a magnetic field of a beta-barrel membrane protein in bicelles. *Eur. Biophys. J. Biophys. Lett.*, **35**, 268–275.
54 Sanders, C.R. and Prosser, R.S. (1998) Bicelles: a model membrane system for all seasons? *Structure*, **6**, 1227–1234.
55 Johansson, L.C., Wohri, A.B., Katona, G., Engstrom, S., and Neutze, R. (2009) Membrane protein crystallization from lipidic phases. *Curr. Opin. Struct. Biol.*, **19**, 372–378.
56 Rasmussen, S.G.F., Choi, H.J., Rosenbaum, D.M., Kobilka, T.S., Thian, F.S., Edwards, P.C., Burghammer, M., Ratnala, V.R.P., Sanishvili, R., Fischetti, R.F., Schertler, G.F.X., Weis, W.I., and Kobilka, B.K. (2007) Crystal structure of the human $beta_2$ adrenergic G-protein-coupled receptor. *Nature*, **450**, 383-U4.
57 McKibbin, C., Farmer, N.A., Edwards, P.C., Villa, C., and Booth, P.J. (2009) Urea unfolding of opsin in phospholipid bicelles. *Photochem. Photobiol.*, **85**, 494–500.
58 Marcotte, I. and Auger, M. (2005) Bicelles as model membranes for solid- and solution-state NMR studies of membrane peptides and proteins. *Concepts Magn. Reson. A*, **24A**, 17–37.
59 Prosser, R.S., Evanics, F., Kitevski, J.L., and Al-Abdul-Wahid, M.S. (2006) Current applications of bicelles in NMR studies of membrane-associated amphiphiles and proteins. *Biochemistry*, **45**, 8453–8465.
60 Poget, S.F. and Girvin, M.E. (2007) Solution NMR of membrane proteins in bilayer mimics: small is beautiful, but sometimes bigger is better. *Biochim. Biophys. Acta*, **1768**, 3098–3106.

61 Bayburt, T.H., Grinkova, Y.V., and Sligar, S.G. (2002) Self-assembly of discoidal phospholipid bilayer nanoparticles with membrane scaffold proteins. *Nano Lett.*, **2**, 853–856.
62 Denisov, I.G., Grinkova, Y.V., Lazarides, A.A., and Sligar, S.G. (2004) Directed self-assembly of monodisperse phospholipid bilayer nanodiscs with controlled size. *J. Am. Chem. Soc.*, **126**, 3477–3487.
63 Borch, J. and Hamann, T. (2009) The nanodisc: a novel tool for membrane protein studies. *Biol. Chem.*, **390**, 805–814.
64 Ritchie, T.K., Grinkova, Y.V., Bayburt, T.H., Denisov, I.G., Zolnerciks, J.K., Atkins, W.M., and Sligar, S.G. (2009) Reconstitution of membrane proteins in phospholipid bilayer nanodiscs. *Methods Enzymol. F*, **464**, 211–231.
65 Bayburt, T.H. and Sligar, S.G. (2003) Self-assembly of single integral membrane proteins into soluble nanoscale phospholipid bilayers. *Protein Sci.*, **12**, 2476–2481.
66 Bayburt, T.H., Leitz, A.J., Xie, G.F., Oprian, D.D., and Sligar, S.G. (2007) Transducin activation by nanoscale lipid bilayers containing one and two rhodopsins. *J. Biol. Chem.*, **282**, 14875–14881.
67 Leitz, A.J., Bayburt, T.H., Barnakov, A.N., Springer, B.A., and Sligar, S.G. (2006) Functional reconstitution of beta$_2$-adrenergic receptors utilizing self-assembling nanodisc technology. *Biotechniques*, **40**, 601.
68 Shenkarev, Z.O., Lyukmanova, E.N., Solozhenkin, O.I., Gagnidze, I.E., Nekrasova, O.V., Chupin, V.V., Tagaev, A.A., Yakimenko, Z.A., Ovchinnikova, T.V., Kirpichnikov, M.P., and Arseniev, A.S. (2009) Lipid–protein nanodiscs: possible application in high-resolution NMR investigations of membrane proteins and membrane-active peptides. *Biochemistry*, **74**, 756–765.
69 Alami, M., Dalal, K., Lelj-Garolla, B., Sligar, S.G., and Duong, F. (2007) Nanodiscs unravel the interaction between the SecYEG channel and its cytosolic partner SecA. *EMBO J.*, **26**, 1995–2004.
70 Raschle, T., Hiller, S., Yu, T.Y., Rice, A.J., Walz, T., and Wagner, G. (2009) Structural and functional characterization of the integral membrane protein VDAC-1 in lipid bilayer nanodiscs. *J. Am. Chem. Soc.*, **131**, 17777–17779.
71 O'mara, M.L. and Tieleman, D.P. (2007) P-glycoprotein models of the apo and ATP-bound states based on homology with Sav1866 and MalK. *FEBS Lett.*, **581**, 4217–4222.
72 Segrest, J.P., Jones, M.K., Klon, A.E., Sheldahl, C.J., Hellinger, M., De Loof, H., and Harvey, S.C. (1999) A detailed molecular belt model for apolipoprotein A-I in discoidal high density lipoprotein. *J. Biol. Chem.*, **274**, 31755–31758.
73 Cappuccio, J.A., Blanchette, C.D., Sulchek, T.A., Arroyo, E.S., Kralj, J.M., Hinz, A.K., Kuhn, E.A., Chromy, B.A., Segelke, B.W., Rothschild, K.J., Fletcher, J.E., Katzen, F., Peterson, T.C., Kudlicki, W.A., Bench, G., Hoeprich, P.D., and Coleman, M.A. (2008) Cell-free co-expression of functional membrane proteins and apolipoprotein, forming soluble nanolipoprotein particles. *Mol. Cell. Proteomics*, **7**, 2246–2253.
74 Bhattacharya, P., Grimme, S., Ganesh, B., Gopisetty, A., Sheng, J.R., Martinez, O., Jayarama, S., Artinger, M., Meriggioli, M., and Prabhakar, B.S. (2010) Nanodisc-incorporated hemagglutinin provides protective immunity against influenza virus infection. *J. Virol.*, **84**, 361–371.
75 Rigaud, J.L. and Levy, D. (2003) Reconstitution of membrane proteins into liposomes. *Liposomes B*, **372**, 65–86.
76 Allen, T.M., Romans, A.Y., Kercret, H., and Segrest, J.P. (1980) Detergent removal during membrane reconstitution. *Biochim. Biophys. Acta*, **601**, 328–342.
77 Lorch, M. and Booth, P.J. (2004) Insertion kinetics of a denatured alpha helical membrane protein into phospholipid bilayer vesicles. *J. Mol. Biol.*, **344**, 1109–1121.
78 Allen, S.J., Curran, A.R., Templer, R.H., Meijberg, W., and Booth, P.J. (2004) Controlling the folding efficiency of an

integral membrane protein. *J. Mol. Biol.*, **342**, 1293–1304.

79 Tamm, L.K., Hong, H., and Liang, B. (2004) Folding and assembly of beta-barrel membrane proteins. *Biochim. Biophys. Acta*, **1666**, 250–263.

80 Rigaud, J.L., Levy, D., Mosser, G., and Lambert, O. (1998) Detergent removal by non-polar polystyrene beads – applications to membrane protein reconstitution and two-dimensional crystallization. *Eur. Biophys. J. Biophys. Lett.*, **27**, 305–319.

81 Rigaud, J.L., Mosser, G., Lacapere, J.J., Olofsson, A., Levy, D., and Ranck, J.L. (1997) Bio-beads: an efficient strategy for two-dimensional crystallization of membrane proteins. *J. Struct. Biol.*, **118**, 226–235.

82 Zhou, X.M. and Graham, T.R. (2009) Reconstitution of phospholipid translocase activity with purified Drs2p, a type-IV P-type ATPase from budding yeast. *Proc. Nat. Acad. Sci. USA*, **106**, 16586–16591.

83 Klammt, C., Lohr, F., Schafer, B., Haase, W., Dotsch, V., Ruterjans, H., Glaubitz, C., and Bernhard, F. (2004) High level cell-free expression and specific labeling of integral membrane proteins. *Eur. J. Biochem.*, **271**, 568–580.

84 Oppedisano, F., Pochini, L., Galluccio, M., and Indiveri, C. (2007) The glutamine/amino acid transporter (ASCT2) reconstituted in liposomes: transport mechanism, regulation by ATP and characterization of the glutamine/glutamate antiport. *Biochim. Biophys. Acta*, **1768**, 291–298.

85 Oppedisano, F. and Indiveri, C. (2008) Reconstitution into liposomes of the B degrees-like glutamine-neutral amino acid transporter from renal cell plasma membrane. *Biochim. Biophys. Acta*, **1778**, 2258–2265.

86 Geertsma, E.R., Mahmood, N.A.B.N., Schuurman-Wolters, G.K., and Poolman, B. (2008) Membrane reconstitution of ABC transporters and assays of translocator function. *Nat. Protoc.*, **3**, 256–266.

87 Hall, J.A. and Pajor, A.M. (2007) Functional reconstitution of SdcS, a Na^+-coupled dicarboxylate carrier protein from *Staphylococcus aureus*. *J. Bacteriol.*, **189**, 880–885.

88 Tan, W., Gou, D.M., Tai, E., Zhao, Y.Z., and Chow, L.M.C. (2006) Functional reconstitution of purified chloroquine resistance membrane transporter expressed in yeast. *Arch. Biochem. Biophys.*, **452**, 119–128.

89 Santos, H.D., Lopes, M.L., Maggio, B., and Ciancaglini, P. (2005) Na,K-ATPase reconstituted in liposomes: effects of lipid composition on hydrolytic activity and enzyme orientation. *Colloids Surf. B*, **41**, 239–248.

90 Mason, A.J., Siarheyeva, A., Haase, W., Lorch, M., van Veen, H., and Glaubitz, C. (2004) Amino acid type selective isotope labelling of the multidrug ABC transporter LmrA for solid-state NMR studies. *FEBS Lett.*, **568**, 117–121.

91 Seddon, A.M., Curnow, P., and Booth, P.J. (2004) Membrane proteins, lipids and detergents: not just a soap opera. *Biochim. Biophys. Acta*, **1666**, 105–117.

92 Zhang, W., Campbell, H.A., King, S.C., and Dowhan, W. (2005) Phospholipids as determinants of membrane protein topology – phosphatidylethanolamine is required for the proper topological organization of the gamma-aminobutyric acid permease (GabP) of *Escherichia coli*. *J. Biol. Chem.*, **280**, 26032–26038.

93 Dowhan, W. and Bogdanov, M. (2009) Lipid-dependent membrane protein topogenesis. *Annu. Rev. Biochem.*, **78**, 515–540.

94 Bogdanov, M., Sun, J.Z., Kaback, H.R., and Dowhan, W. (1996) A phospholipid acts as a chaperone in assembly of a membrane transport protein. *J. Biol. Chem.*, **271**, 11615–11618.

95 Bogdanov, M., Umeda, M., and Dowhan, W. (1999) Phospholipid-assisted refolding of an integral membrane protein – minimum structural features for phosphatidylethanolamine to act as a molecular chaperone. *J. Biol. Chem.*, **274**, 12339–12345.

96 Xie, J., Bogdanov, M., Heacock, P., and Dowhan, W. (2006) Phosphatidylethanolamine and monoglucosyldiacylglycerol are

interchangeable in supporting topogenesis and function of the polytopic membrane protein lactose permease. *J. Biol. Chem.*, **281**, 19172–19178.

97 Neves, P., Lopes, S.C.D.N., Sousa, I., Garcia, S., Eaton, P., and Gameiro, P. (2009) Characterization of membrane protein reconstitution in LUVs of different lipid composition by fluorescence anisotropy. *J. Pharm. Biomed. Anal.*, **49**, 276–281.

98 Fleischer, R., Heermann, R., Jung, K., and Hunke, S. (2007) Purification, reconstitution, and characterization of the CpxRAP envelope stress system of *Escherichia coli*. *J. Biol. Chem.*, **282**, 8583–8593.

99 Tonazzi, A., Galluccio, M., Oppedisano, F., and Indiveri, C. (2006) Functional reconstitution into liposomes and characterization of the carnitine transporter from rat liver microsomes. *Biochim. Biophys. Acta*, **1758**, 124–131.

100 Yao, S.Y.M., George, R., and Young, J.D. (1993) Reconstitution studies of amino-acid-transport system-L in rat erythrocytes. *Biochem. J.*, **292**, 655–660.

101 Marsh, D. (2008) Protein modulation of lipids, and vice-versa, in membranes. *Biochim. Biophys. Acta*, **1778**, 1545–1575.

102 Ge, M.T. and Freed, J.H. (1999) Electron-spin resonance study of aggregation of gramicidin in dipalmitoylphosphatidylcholine bilayers and hydrophobic mismatch. *Biophys. J.*, **76**, 264–280.

103 Cornea, R.L. and Thomas, D.D. (1994) Effects of membrane thickness on the molecular-dynamics and enzymatic-activity of reconstituted Ca-ATPase. *Biochemistry*, **33**, 2912–2920.

104 Lewis, B.A. and Engelman, D.M. (1983) Lipid bilayer thickness varies linearly with acyl chain-length in fluid phosphatidylcholine vesicles. *J. Mol. Biol.*, **166**, 211–217.

105 Ryba, N.J.P. and Marsh, D. (1992) Protein rotational diffusion and lipid protein interactions in recombinants of bovine rhodopsin with saturated diacylphosphatidylcholines of different chain lengths studied by conventional and saturation-transfer electron-spin-resonance. *Biochemistry*, **31**, 7511–7518.

106 Kusumi, A. and Hyde, J.S. (1982) Spin-label saturation-transfer electron-spin resonance detection of transient association of rhodopsin in reconstituted membranes. *Biochemistry*, **21**, 5978–5983.

107 Tiburu, E.K., Bowman, A.L., Struppe, J.O., Janero, D.R., Avraham, H.K., and Makriyannis, A. (2009) Solid-state NMR and molecular dynamics characterization of cannabinoid receptor-1 (CB1) helix 7 conformational plasticity in model membranes. *Biochim. Biophys. Acta*, **1788**, 1159–1167.

108 Holt, A., Koehorst, R.B.M., Rutters-Meijneke, T., Gelb, M.H., Rijkers, D.T.S., Hemminga, M.A., and Killian, J.A. (2009) Tilt and rotation angles of a transmembrane model peptide as studied by fluorescence spectroscopy. *Biophys. J.*, **97**, 2258–2266.

109 Duong-Ly, K.C., Nanda, V., Degrad, W.F., and Howard, K.P. (2005) The conformation of the pore region of the M2 proton channel depends on lipid bilayer environment. *Protein Sci.*, **14**, 856–861.

110 Harzer, U. and Bechinger, B. (2000) Alignment of lysine-anchored membrane peptides under conditions of hydrophobic mismatch: a CD, N-15 and P-31 solid-state NMR spectroscopy investigation. *Biochemistry*, **39**, 13106–13114.

111 Ozdirekcan, S., Rijkers, D.T.S., Liskamp, R.M.J., and Killian, J.A. (2005) Influence of flanking residues on tilt and rotation angles of transmembrane peptides in lipid bilayers. A solid-state H-2 NMR study. *Biochemistry*, **44**, 1004–1012.

112 Yeagle, P.L., Bennett, M., Lemaitre, V., and Watts, A. (2007) Transmembrane helices of membrane proteins may flex to satisfy hydrophobic mismatch. *Biochim. Biophys. Acta*, **1768**, 530–537.

113 Soubias, O., Niu, S.L., Mitchell, D.C., and Gawrisch, K. (2008) Lipid-rhodopsin hydrophobic mismatch alters rhodopsin helical content. *J. Am. Chem. Soc.*, **130**, 12465–12471.

114 Lee, A.G. (2003) Lipid–protein interactions in biological membranes: a

structural perspective. *Biochim. Biophys. Acta*, **1612**, 1–40.

115 Pilot, J.D., East, J.M., and Lee, A.G. (2001) Effects of bilayer thickness on the activity of diacylglycerol kinase of *Escherichia coli*. *Biochemistry*, **40**, 8188–8195.

116 Xu, Q., Kim, M., Ho, K.W.D., Lachowicz, P., Fanucci, G.E., and Cafiso, D.S. (2008) Membrane hydrocarbon thickness modulates the dynamics of a membrane transport protein. *Biophys. J.*, **95**, 2849–2858.

117 Ray, T.P., Skipski, V.P., Barclay, M., Essner, E., and Archibald, F.M. (1969) Lipid composition of rat liver plasma membranes. *J. Biol. Chem.*, **244**, 5528–5536.

118 Mavis, R.D. and Vagelos, P.R. (1972) Effect of phospholipid fatty-acid composition on membranous enzymes in *Escherichia coli*. *J. Biol. Chem.*, **247**, 652.

119 Gallova, J., Uhrikova, D., Kucerka, N., Teixeira, J., and Balgavy, P. (2008) Hydrophobic thickness, lipid surface area and polar region hydration in monounsaturated diacylphosphatidylcholine bilayers: SANS study of effects of cholesterol and beta-sitosterol in unilamellar vesicles. *Biochim. Biophys. Acta*, **1778**, 2627–2632.

120 Cornelius, F. (2001) Modulation of Na,K-ATPase and Na-ATPase activity by phospholipids and cholesterol. I. Steady-state kinetics. *Biochemistry*, **40**, 8842–8851.

121 Morein, S., Andersson, A.S., Rilfors, L., and Lindblom, G. (1996) Wild-type *Escherichia coli* cells regulate the membrane lipid composition in a "window" between gel and non-lamellar structures. *J. Biol. Chem.*, **271**, 6801–6809.

122 Booth, P.J. (2005) Sane in the membrane: designing systems to modulate membrane proteins. *Curr. Opin. Struct. Biol.*, **15**, 435–440.

123 Peiro-Salvador, T., Ces, O., Templer, R.H., and Seddon, A.M. (2009) Buffers may adversely affect model lipid membranes: a cautionary tale. *Biochemistry*, **48**, 11149–11151.

124 Ortega, A., SantiagoGarcia, J., MasOliva, J., and Lepock, J.R. (1996) Cholesterol increases the thermal stability of the Ca^{2+}/Mg^{2+}-ATPase of cardiac microsomes. *Biochim. Biophys. Acta*, **1283**, 45–50.

125 Meijberg, W. and Booth, P.J. (2002) The activation energy for insertion of transmembrane alpha-helices is dependent on membrane composition. *J. Mol. Biol.*, **319**, 839–853.

126 Curnow, P., Lorch, M., Charalambous, K., and Booth, P.J. (2004) The reconstitution and activity of the small multidrug transporter EmrE is modulated by non-bilayer lipid composition. *J. Mol. Biol.*, **343**, 213–222.

127 Seddon, A.M., Lorch, M., Ces, O., Templer, R.H., Macrae, F., and Booth, P.J. (2008) Phosphatidylglycerol lipids enhance folding of an alpha helical membrane protein. *J. Mol. Biol.*, **380**, 548–556.

128 Popot, J.L. and Engelman, D.M. (1990) Membrane-protein folding and oligomerization – the 2-stage model. *Biochemistry*, **29**, 4031–4037.

129 van den Brink-van der Laan, E., Chupin, V., Killian, J.A., and de Kruijff, B. (2004) Stability of KcsA tetramer depends on membrane lateral pressure. *Biochemistry*, **43**, 4240–4250.

130 Maneri, L.R. and Low, P.S. (1988) Structural stability of the erythrocyte anion transporter, band-3, in different lipid environments – a differential scanning calorimetric study. *J. Biol. Chem.*, **263**, 16170–16178.

131 Fry, M. and Green, D.E. (1980) Cardiolipin requirement by cytochrome-oxidase and the catalytic role of phospholipid. *Biochem. Biophys. Res. Commun.*, **93**, 1238–1246.

132 Claypool, S.M. (2009) Cardiolipin, a critical determinant of mitochondrial carrier protein assembly and function. *Biochim. Biophys. Acta*, **1788**, 2059–2068.

133 Xu, Y.P., Ramu, Y., and Lu, Z. (2008) Removal of phospho-head groups of membrane lipids immobilizes voltage sensors of K^+ channels. *Nature*, **451**, 826-U8.

134 Schmidt, D., Jiang, Q.X., and MacKinnon, R. (2006) Phospholipids and the origin of cationic gating charges in voltage sensors. *Nature*, **444**, 775–779.

135 Palsdottir, H. and Hunte, C. (2004) Lipids in membrane protein structures. *Biochim. Biophys. Acta*, **1666**, 2–18.

136 Zhou, Y.F. and Bowie, J.U. (2000) Building a thermostable membrane protein. *J. Biol. Chem.*, **275**, 6975–6979.

137 Faham, S., Yang, D., Bare, E., Yohannan, S., Whitelegge, J.P., and Bowie, J.U. (2004) Side-chain contributions to membrane protein structure and stability. *J. Mol. Biol.*, **335**, 297–305.

138 Chill, J.H., Louis, J.M., Miller, C., and Bax, A. (2006) NMR study of the tetrameric KcsA potassium channel in detergent micelles. *Protein Sci.*, **15**, 684–698.

139 Cortes, D.M., Cuello, L.G., and Perozo, E. (2001) Molecular architecture of full-length KcsA – Role of cytoplasmic domains in ion permeation and activation gating. *J. Gen. Physiol.*, **117**, 165–180.

140 Serrano-Vega, M.J., Magnani, F., Shibata, Y., and Tate, C.G. (2008) Conformational thermostabilization of the beta 1-adrenergic receptor in a detergent-resistant form. *Proc. Nat. Acad. Sci. USA*, **105**, 877–882.

141 Shibata, Y., White, J.F., Serrano-Vega, M.J., Magnani, F., Aloia, A.L., Grisshammer, R., and Tate, C.G. (2009) Thermostabilization of the neurotensin receptor NTS1. *J. Mol. Biol.*, **390**, 262–277.

14
Rapid Optimization of Membrane Protein Production Using Green Fluorescent Protein-Fusions and Lemo21(DE3)

Susan Schlegel, Mirjam Klepsch, Dimitra Gialama, David Wickström, David Drew, and Jan-Willem de Gier

14.1
Introduction

Optimizing the conditions for overexpression and purification of membrane proteins is usually a laborious and time-consuming process. In this chapter we describe in a protocol format how we identify the optimal conditions for the production of membrane proteins for functional and structural studies using *Escherichia coli* as an overexpression host. We make use of membrane protein Green Fluorescent Protein (GFP)-fusions to screen for optimal expression, solubilization, and purification conditions [1–6]. The exceptionally stable GFP moiety, which is attached to the C-terminus of the membrane protein, can easily be monitored and visualized at any stage during the procedure. Membrane protein expression levels can be estimated through measuring fluorescence in whole cells with a detection limit as low as 10 μg GFP per liter of culture, and solubilization and purification efficiency can be determined by measuring GFP fluorescence in solution [4]. Importantly, GFP fluorescence can be detected by standard sodium dodecyl sulfate–polyacrylamide gel electrophoresis (SDS–PAGE) with a detection limit of less than 5 ng of GFP [4]. In-gel fluorescence allows rapid assessment of the integrity of membrane protein–GFP fusions and can also be used for quantification.

To drive expression of the membrane protein of interest we utilize the widely used bacteriophage T7-based pET/T7-RNA polymerase (T7-RNAP) expression platform, in which expression of the gene encoding the target protein is governed by the T7-RNAP. As an overexpression host we use the BL21(DE3)-derived strain Lemo21(DE3). BL21(DE3) and its derivatives have a chromosomal copy of the T7-RNAP gene under the control of the Isopropyl β-D-1-thiogalactopyranoside (IPTG)-inducible *lac*UV5 promoter [7]. Upon addition of IPTG, repression of the *lac*UV5 promoter is released leading to expression of the gene encoding the target protein. In BL21(DE3) T7-RNAP activity is very strong and fixed. In contrast, in Lemo21(DE3) the activity of the T7-RNAP can be precisely tuned by coexpression of its natural inhibitor T7 lysozyme from the pLemo plasmid [8]. The pLemo plasmid is derived from pACYC184 and the expression of the gene encoding T7 lysozyme is controlled

Production of Membrane Proteins: Strategies for Expression and Isolation, First Edition.
Edited by Anne Skaja Robinson.

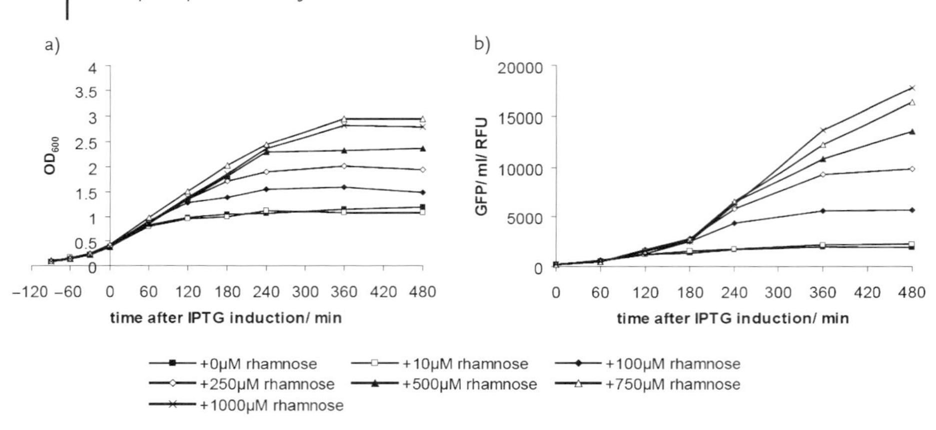

Figure 14.1 Characterization of Lemo21(DE3). In all experiments, Lemo21(DE3) overexpressed YidC-GFP in the presence of the indicated amounts of L-rhamnose. Growth (a) and expression (b) of YidC-GFP were monitored by measuring the OD_{600} and GFP fluorescence, respectively, every 1–2 h.

by the rhamnose promoter [8]. This promoter is extremely well titratable and covers a broad range of expression intensities [9]. The Lemo21(DE3) strain is tunable for membrane protein overexpression and conveniently allows optimizing the overexpression of any given membrane protein using only a single strain. The combination of the *lac*UV5 and the rhamnose promoter governing expression of T7-RNAP from the chromosome and T7 lysozyme from pLemo, respectively, guarantees the widest window of expression intensities possible. Therefore, in the Lemo21(DE3) strain the amount of membrane protein produced can be easily harmonized with the membrane protein biogenesis capacity of the cell [8, 10]. This harmonization will alleviate the toxic effects of membrane protein overexpression [8]. This will lead to the formation of more biomass that can overexpress membrane proteins resulting in increased yields (Figure 14.1). It should be noted that for a small number of overexpressed membrane proteins we have observed that in Lemo21(DE3) the membrane protein biogenesis capacity is sufficient without any inhibition of T7-RNAP activity by T7 lysozyme [8]. The generality and simplicity of this "all-in-one" solution for membrane protein expression in *Escherichia coli* guarantees the rapid identification of the conditions for the optimal overexpression yields of membrane proteins.

14.2 Main Protocol (Figure 14.2)

14.2.1 Determination of Membrane Protein Topology and Selection of Expression Vector

Using a membrane protein–GFP fusion to optimize expression/purification requires that the membrane protein of interest have a C_{in} topology. Only in the

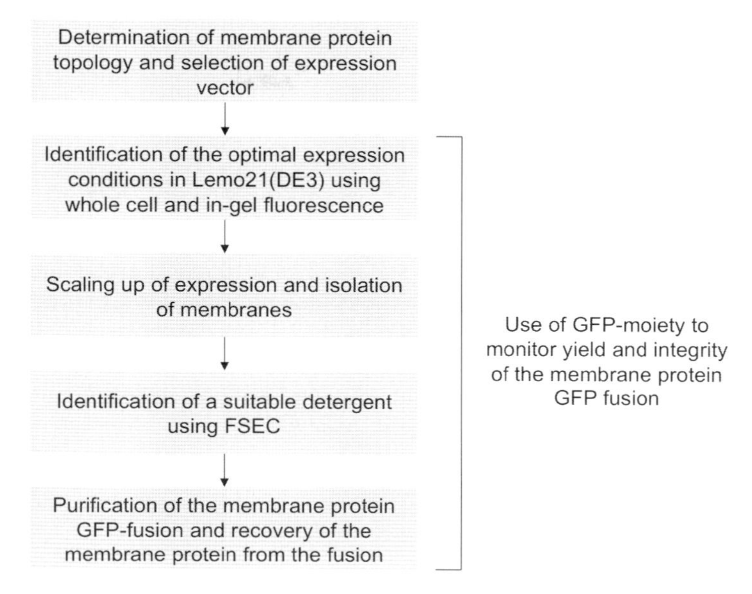

Figure 14.2 Flowchart illustrating the optimization of membrane protein production using GFP-fusions and Lemo21(DE3).

cytoplasm of *E. coli* does GFP fold correctly and become fluorescent [3, 11]. If the topology of the protein of interest is not known, use a topology predictor (e.g., MEMSAT3 [12], TOPCONS [13], or both). If the membrane protein has a C_{out} topology the membrane protein can be extended with one transmembrane segment, like the transmembrane segment of glycophorin A [14], at the C-terminus. Fusing GFP to the N-terminus of N_{in}/C_{out} membrane proteins is not recommended as the GFP moiety may interfere with the targeting of the protein to the membrane and its folding [15].

In our laboratories, we routinely use the pGFPd and e overexpression vectors [4]. These vectors are both derived from pET28a(+) and contain a tobacco etch virus (TEV) protease cleavage site between the multiple cloning site and the genetic information encoding the GFP-His_8 moiety. The pGFPd and e vectors confer resistance to kanamycin. Kanamycin resistance is preferred to ampicillin resistance for membrane protein overexpression. The antibiotic kanamycin binds to the 30S subunit of the prokaryotic ribosome. Its resistance gene codes for a cytoplasmic protein. The antibiotic ampicillin interferes with the biogenesis of the peptidoglycan layer in the periplasm and is neutralized by the protein encoded by the ampicillin resistance gene, β-lactamase. To reach the periplasm, β-lactamase must be translocated through the same protein-conducting channel in the cytoplasmic membrane that is also involved in mediating the biogenesis of membrane proteins into the cytoplasmic membrane. Therefore, the use of the ampicillin resistance marker during membrane protein overexpression would unnecessarily use up capacity of this protein-conducting channel.

For equipment needed and composition of reagents/buffers, we refer to the materials section at the end of the protocol (Section 14.3).

14.2.2 Identification of the Optimal Expression Conditions in Lemo21(DE3) Using Whole-Cell and In-Gel Fluorescence

The first step in optimizing membrane protein overexpression using Lemo21(DE3) is to identify the concentration of L-rhamnose that gives the highest expression yields (i.e., the highest amount of full-length membrane protein–GFP fusion/ml of culture). We standard perform the initial screen at 30 °C since we have noticed that using different temperatures with Lemo21(DE3) usually does not lead to improved overexpression yields. The following steps refer to the screening for one construct, but many constructs can be screened in parallel.

1) Transform the expression vector encoding the membrane protein–GFP fusion of interest into Lemo21(DE3). Always use fresh transformants (not older than 4–5 days) for overexpression experiments.

Note: The use of glycerol stocks as starting material can lead to severe reduction of expression yields and is not recommended.

2) Set up an overnight culture in a 15-ml Falcon tube containing 3 ml LB medium with 34 μg/ml chloramphenicol (for maintaining pLemo) and the corresponding antibiotic for the expression vector used (50 μg/ml kanamycin for pGFPd/e). Incubate in a shaking incubator at 30 °C, 220 rpm.

Note: If "leaky" expression of the target membrane protein is toxic, the addition of L-rhamnose to the plates used for the transformation and the overnight culture will reduce the toxicity of the "leaky" expression.

3) Prepare eight 50-ml Falcon tubes with 12 ml LB medium each, containing the appropriate antibiotics. Add L-rhamnose to seven of the Falcon tubes to a final concentration of 10, 50, 100, 250, 500, 1000, and 2000 μM.

Note: We use these concentrations of L-rhamnose by default for the overexpression screening but they may of course be adapted.

4) Inoculate each Falcon tube with a 50-fold dilution of the overnight culture. Incubate at 30 °C, 220 rpm and monitor the OD_{600} of the cultures.

5) At an OD_{600} of around 0.4–0.5 (this OD_{600} will be reached approximately 2–2.5 h after inoculation) induce expression of the membrane protein–GFP fusion by adding IPTG to a final concentration of 0.4 mM. At 4, 8 and 24 h after induction take 1 ml of culture for whole-cell fluorescence measurements using a plate reader (see Step 8). Simultaneously, take 100 μl for OD_{600} measurements and approximately 500 μl for measuring in-gel fluorescence (see Steps 9–14). The whole-cell and in-gel fluorescence measurements will

allow determining the optimal concentration of L-rhamnose for the overexpression of a membrane protein.

Note: As mentioned above, screening at different temperatures usually does not lead to improved overexpression yields in Lemo21(DE3). However, if severe degradation of the membrane protein–GFP fusion is observed, a switch to lower expression temperatures or a shortening of induction times may be considered. Expression at lower temperatures usually results in a different optimal L-rhamnose concentration.

6) Transfer 1 ml of culture volume to a 1.5-ml Eppendorf tube and collect the cells by centrifugation for 2 min at 15 700 g. Carefully remove the supernatant.

7) Resuspend the pellet in 100 μl ice-cold PBS and leave it on ice for at least 30 min. This will allow the GFP moiety to fold. Alternatively, wash the cell pellet once in between using 1 ml ice-cold PBS and repeat the centrifugation step.

8) Transfer the 100-μl suspension into to a black Nunc 96-well optical bottom plate and measure GFP fluorescence (emission: 512 nm, excitation: 485 nm) in a microtiter plate spectrofluorometer. For maximal sensitivity select the option "bottom read."

Note: The use of a black Nunc 96-well optical bottom plate is recommended to reduce background signal due to stray light. To estimate membrane protein overexpression levels from whole-cell fluorescence, see "Expression and isolation of GFP-His$_8$" (Section 14.4). Measuring whole-cell fluorescence does not allow discriminating between the full-length fusion protein and degradation products. GFP is an exceptionally stable molecule and remains fluorescent even if the membrane protein of interest has been degraded. The level of background fluorescence is usually quite low, but can lead to overestimation of membrane protein expression yields especially if protein expression levels are very low. Measure whole-cell fluorescence of cells harboring an empty expression vector to account for background fluorescence. If expression yields are lower than 200 μg/l, the signal-to-noise ratio may be improved by increasing the amount of cells analyzed; use 5 ml of culture for fluorescence measurements.

9) To measure in-gel fluorescence harvest the cells from approximately 500 μl of culture volume by centrifugation for 2 min at 15 700 g. Carefully remove the supernatant.

10) Measure the OD_{600} of the culture using an appropriate dilution.

Note: OD_{600} values vary between different spectrophotometers. Make sure the measured values are within the linear range of the spectrophotometer.

11) Resuspend the cell pellets in PBS to an equal OD_{600} (we dilute to a final concentration of 0.2 ODU/10 μl PBS). Add an equal volume of SB buffer to each suspension (final concentration 0.1 ODU/10 μl solution). Ensure homogeneity of the cell suspension. If available different concentrations of purified GFP

of a known concentration (see "Expression and isolation of GFP-His_8" (Section 14.4)) may be included in the analysis from here on. This will allow for an accurate estimation of overexpression yields, and help to discriminate between the full-length membrane protein–GFP fusion and degradation products.

Note: Instead of adjusting the cell suspensions to the same OD_{600} they may also be adjusted to the same fluorescence levels (useful if screening various constructs with different expression levels as determined by whole-cell fluorescence). That way, weak bands can be detected easily without interference from neighboring, stronger bands.

12) Incubate all samples (cell suspensions and purified GFP) for 5–10 min at 37 °C.

Note: Incubation at temperatures higher than 37 °C is not recommended as this can lead to aggregation of membrane proteins and loss of GFP fluorescence. If frozen cells are used add $MgCl_2$ to a final concentration of 1 mM and DNase I/benzonase (1–5 units/10 µl) to the samples and incubate for 15 min on ice before adding SB buffer to yield a homogenous suspension.

13) Analyze a fraction of each sample corresponding to 0.1–0.2 ODU by means of standard SDS–PAGE including an appropriate molecular weight marker.

14) Rinse the gel with distilled water and detect in-gel fluorescence with a CCD camera system. Expose the gel to blue light at 460 nm and capture images with increasing exposure time until the desired band intensity is reached. Fluorescence intensities can be quantified using appropriate software and expression yields estimated by comparing intensities to a GFP reference sample of known concentration.

Note: GFP-His_8 has a molecular weight of approximately 28 kDa; however, GFP remains folded in SDS and the apparent molecular weight in SDS gels is lower (approximately 20 kDa). For a protocol for the overexpression and isolation of GFP-His_8, see "Expression and isolation of GFP-His_8" (Section 14.4).

15) Stain the gel for 2 h in Coomassie staining solution and destain in destaining solution.

14.2.3 Scaling Up of Expression and Isolation of Membranes

In this protocol we use 2.5-l baffled shaker flasks for scaling up the expression. However, it is also possible to use fermenters (we have successfully used 15-l fermenters for scaling up expression using Lemo21(DE3)).

16) Setup an overnight culture in a 200-ml shaker flask containing 20 ml LB medium with the appropriate antibiotics (see Step 2). Incubate at 30 °C, 220 rpm in a shaking incubator.

17) Inoculate 1 l of LB medium (with appropriate antibiotics and the optimal concentration of L-rhamnose as determined in Step 5) with the overnight culture in a 2.5 l baffled shaker flask and incubate at 30 °C, 220 rpm.

18) Induce protein expression as described before, at an OD_{600} of approximately 0.4–0.5 using 0.4 mM IPTG (final concentration) for the time determined to be optimal by the overexpression screen. Before harvesting the cells take 1 ml of culture for measuring whole-cell fluorescence.

Note: From here on, all steps should be carried out on ice or at 4 °C. Centrifugation steps are performed at 4 °C.

19) Harvest the cells by centrifugation for 20 min at 6200 g. Discard the supernatant and carefully resuspend the pellet in 50 ml ice-cold PBS.

20) Pellet the cells according to the previous step, discard the supernatant, and resuspend the pellet in 10 ml ice cold PBS. If needed, the pellet or the suspension can be frozen in liquid nitrogen and stored at –80 °C up to 6 months. Freezing and thawing may facilitate breaking the cells.

21) Add 1 mg/ml Pefabloc SC (or another protease inhibitor mix of your choice), 1 mM EDTA, and 0.5 mg/ml lysozyme (final concentration), and incubate on ice for 30–60 min. Subsequently, add 5 μg/ml DNase I and 2 mM $MgCl_2$, and incubate for another 10–15 min. Break the cells using a French press (18 000 psi for at least two cycles) or a method of your choice. Most cells are broken when the suspension has turned translucent.

Note: The lysozyme version, LysY, expressed by Lemo21(DE3) is not lytic [8]. *Adding lysozyme is not essential; however, it facilitates breaking the cells. Other methods than French pressing to break cells (e.g., Constant Systems cell disrupter or sonication) can be used.*

22) Clear the suspension of unbroken cells/debris by centrifugation at 24 000 g for 20 min. Transfer the supernatant (containing the membranes) to a clean tube and repeat the centrifugation step.

23) To collect the membranes, centrifuge the supernatant for 45 min at 150 000 g. Discard the supernatant and resuspend the membrane pellet in 10 ml ice-cold PBS using a disposable 10-ml syringe with a 21-gage needle.

24) Fill up the centrifugation tube with ice-cold PBS and harvest the membranes once more for 45 min at 150 000 g. This step will remove residual EDTA which otherwise would interfere with the immobilized metal-affinity chromatography (IMAC) later on.

25) Resuspend the membrane pellet in 5 ml ice-cold PBS essentially as described before. Determine protein concentration with a standard BCA assay.

Note: If desired, membrane suspensions may be frozen in liquid nitrogen and stored at –80 °C for up to 6 months. However, some membrane protein crystallographers avoid freezing and storing of membranes, and continue with purification immediately as repeated freezing/thawing may negatively affect the material.

14.2.4
Identification of a Suitable Detergent Using Fluorescence-Detection Size-Exclusion Chromatography

Next, a detergent has to be identified that can be used to extract the overexpressed membrane protein from the membrane optimally. Importantly, the membrane protein should remain stable in the detergent used. Each detergent has a specific critical micellar concentration (CMC) that depends on both the temperature and composition of the solubilization buffer used (e.g., salt content and pH). To efficiently solubilize a membrane protein work well above the CMC of a detergent. In Table 14.1 we have listed the detergents and their concentrations we routinely use for the solubilization of membranes.

26) Adjust the portion of the membrane suspension used for detergent screening to a concentration of 3.75 mg of protein/ml. This will yield a final protein concentration of 3 mg/ml in the next step. Transfer aliquots of 800 µl of membrane suspension into 1.5-ml Beckman pollyallomer microcentrifuge tubes.
27) Dissolve the detergents to be screened for in PBS to 5 × the final concentration.
28) Add 200 µl of each detergent to each of the tubes and incubate the samples for 1 h under mild agitation.

Table 14.1 Detergents.

Name	Type[a)]	Molecular weight	Aggregation number[b)]	CMC (mM)[c)]	CMC (% w/v)[c)]	Percentage for solubilization (%)[d)]
DDM	N	510.6	78–149	0.17	0.0087	1
UDM	N	496.6	74	0.59	0.029	1
DM	N	482.6	69	1.8	0.087	1
Cymal 7	N	522.5	–	0.19	0.0099	1
Cymal 6	N	508.5	63	0.56	0.028	2
nOG	N	292.4	78	18	0.53	2
LDAO	Z	229.4	76	1–2	0.023	1
CHAPS	Z	614.9	10	8	0.49	1
Triton X-100	N	647	75–165	0.23	0.015	1 (v/v)

a) N = nonionic, Z = zwitterionic.
b) Molecular weight of the micelle divided by the molecular weight of the detergent.
c) CMC depends on temperature and solution conditions. For solubilizing and purification of membrane proteins one has to always work above the CMC.
d) Percentage (w/v) of detergent as used in the detergent screen (Steps 26–32).
Suppliers like Anatrace sell different detergent grades; some crystallographers prefer the highest grade (e.g., Anagrade), but for most applications less-expensive alternatives (e.g., Sol-Grade) are sufficient.

29) Pellet nonsolubilized material by centrifugation for 45 min at 100 000 g.

30) Transfer the supernatant to a clean tube and measure GFP fluorescence in 100 μl of the supernatant (see Step 8). To determine the percentage of detergent solubilized GFP resuspend the nonsolubilized pelleted membranes in the same buffer volume and compare fluorescence to that of the detergent solubilized membranes.

31) Repeat Steps 26–30 to determine the optimal protein to detergent ratio with the best detergent (as established in Step 30). Keep the percentage of detergent constant but increase the amount of protein (starting from 3 mg/ml). The optimal protein to detergent ratio is the point at which linear increase in protein still yields a linear increase in GFP fluorescence.

Note: None of the detergents listed in Table 14.1 interferes with GFP fluorescence.

32) To assess the stability of a membrane protein–GFP fusion in a particular detergent its monodispersity in this detergent can be rapidly monitored by size-exclusion column chromatography (SEC) followed by fractionation into a 96-well plate for measuring the fluorescence in each well (Step 8) [5, 6].

Note: As originally outlined by Kawate and Gouaux, an in-line detector, such as a Prominence high-performance liquid chromatography (HPLC) device (Shimadzu) can be used for higher sensitivity [14]. *Notably, fluorescence-detection SEC ((FSEC) has also successfully been used to screen for optimal solubilization conditions using whole cells as starting material (e.g.,* [14, 16, 17]*). In addition, it is to be kept in mind that the addition of lipids to the buffers used for the purification of a membrane protein can considerably improve the quality of the isolated material (e.g.,* [18]*).*

14.2.5 Purification of the Membrane Protein GFP-Fusion and Recovery of the Membrane Protein from the Fusion

33) Solubilize the membranes, using the established optimal protein/detergent ratio by incubating membrane–detergent mixture for 1 h under mild agitation.

34) Pellet unsolubilized material by centrifugation for 45 min at 100 000 g and transfer the supernatant to a fresh tube. Take 100 μl of the supernatant to measure GFP fluorescence as described in Step 8 and estimate the percentage of solubilization (see Step 30). Keep another 100 μl for subsequent analysis of the purification as described below (Steps 44–45).

35) Pack a XK 16/20 column with Ni-NTA resin (binding capacity 50 mg of protein/ml resin). It is recommended to use an excess of resin; we use around 1 ml of Ni-NTA resin per milligram of membrane protein–GFP fusion to be purified, but this can easily be scaled down.

36) Equilibrate the Ni-NTA column with five column volumes of Buffer A.

37) Add imidazole to a final concentration of 10 mM to the solubilized membranes, and load them onto the Ni-NTA column at a flow rate of approximately 0.3–0.5 ml/min.

38) Wash the column with around 20 column volumes of 4% Buffer B (see Section 14.3) at a flow rate of 1 ml/min to remove contaminants

39) Wash with a gradient of 4–25% Buffer B over 20 column volumes at a flow rate of 1 ml/min and collect fractions.

Note: Once the wash and elution conditions have been established, step gradients instead of continuous gradients can be used: wash for 20 column volumes at 2% less than the highest percentage of Buffer B where protein was still bound to the column.

40) Elute the membrane protein–GFP fusion with 50% Buffer B at a flow rate of 1 ml/min and collect all fractions.

Note: Save 100-μl samples from flow-through, wash, and elution fractions (store at 4 °C). These will later be used to assess the quality and integrity of the purified material (see Steps 44 and 45).

41) Measure GFP fluorescence in the different fractions (flow-through, wash, elution).

Note: If the amount of overexpressed protein has exceeded the capacities of the purification system this will be evident as a fluorescence signal in the flow-through fraction. To increase the yield of the membrane protein–GFP fusion increase the bed volume or, if necessary, change to a volume with a larger diameter. In case of nonspecific binding of membrane protein to Ni-NTA resin, add 5–10 mM imidazole to buffer used in batch binding.

42) Determine the amount of membrane protein–GFP fusion in each fraction as described in “Expression and isolation of GFP-His_8” (Section 14.4).

Note: The use of the BCA assay to estimate the amount of fusion protein in the fractions is not recommended as the BCA assay measures total protein and the presence of contaminants may lead to misestimating expression yields. Additionally, the imidazole used for elution cross-reacts with the assay at concentrations above 50 mM (http://www.piercenet.com/files/1296dh4.pdf).

43) To recover the membrane protein from the membrane protein–GFP fusion incubate the membrane protein–GFP fusion with His-tagged TEV protease for 10 h or overnight at 4 °C (or another site-specific protease if a different cleavage site was chosen). Dialyze into an appropriate crystallization buffer.

Note: For commonly used detergents like DDM and Triton X-100 equimolar amounts of TEV protease typically suffice for a complete overnight digest. We generally dialyze into a crystallization buffer containing 20 mM Tris-HCl, pH 7.5, 150 mM NaCl and the concentration of detergent at 3 times the CMC.

44) Determine total protein concentrations in the different, fractions (including the ones saved from the IMAC purification) using the BCA assay.

45) Add an appropriate amount of protein in a 10-μl volume to 10 μl of SB buffer. Analyze the different fractions using SDS–PAGE followed by in-gel fluorescence and Coomassie staining/destaining. This will allow assessing the efficiency of the IMAC purification (binding/elution of the membrane protein fusion), the integrity of the fusion protein, and the efficiency of protease cleavage.

Note: The different fractions obtained after purification may of course also be analyzed accordingly before proceeding with protease cleavage.

46) If the protease digest was complete, pass the digest through a 0.22-μM filter to remove any precipitation and then load the digested material through a 5-ml His-Trap™ column either by the use of a peristaltic pump or an appropriate sized syringe and collect the flow-through. The recovered fraction should not be fluorescent!

Note: As some membrane proteins are sensitive to high concentrations of imidazole we prefer to use the more weakly binding Ni-NTA resin in the first IMAC step in 35. However, for removal of the tag and of contaminants such as AcrB [19]*) the His-Trap™ column is preferred as it has better binding capacity.*

47) Wash the column with 20 ml of crystallization buffer containing 30 mM imidiazole and collect the flow-through. Repeat this step with buffer containing 250 mM imidiazole. Concentrate the digest in Centricon concentrators to around 0.5 ml using an appropriate molecular weight cut-off and measure concentration of protein in each fraction.

Note: As untagged protein can still bind to the resin we include an additional wash step to ensure that all untagged protein is recovered.

48) Analyze the purified protein found typically in the first fraction by standard gel filtration using a Superdex 200 10/30. At this stage one symmetric peak similar to that observed by FSEC in Step 32 is expected.

49) If the protein is to be used for crystallization trials concentrate to around 5–10 mg/ml in Centricon concentrators using an molecular weight cut-off as large as possible to minimize over-concentration of free detergent. Excessive detergent can interfere with crystallization.

Note: This is more problematic with mild detergents that have a large micelle-size (e.g., DDM is 72 kDa) [20]*. Typically we start with 100-kDa cut-off concentrators and only change to 50 kDa if our target protein is present in our 100-kDa concentrator flow-through fraction.*

14.3 Materials

14.3.1 Reagents

- BCA protein assay kit (Pierce).
- Buffer A: PBS with 0.1% DDM (or other detergent at 5 × CMC (for CMCs, see Table 14.1)).
- Buffer B: 500 mM imidazole in Buffer A.
- Buffer C: 20 mM Tris-HCl (pH 7.5), 5 mM EDTA (pH 8), and 100 mM NaCl.
- 3((3-Cholamidopropyl)dimethylammonio)propanesulfonic acid (CHAPS) (Sigma), 10% (w/v).
- Chloramphenicol (Sigma), 34 mg/ml stock solution in ethanol.
- Coomassie staining solution: Coomassie Brilliant Blue R-250 (Fluka) 0.1%, 40% (v/v) methanol, 7% (v/v) acetic acid.
- Cymal 6 (Anatrace), 10% (w/v).
- Cymal 7 (Anatrace), 10% (w/v).
- Destaining solution: 30% (v/v) methanol, 10% (v/v) acetic acid.
- DNase I from bovine pancreas Type IV lyophilized powder (Sigma).
- Imidazole, for molecular biology, minimum 99% (Sigma).
- Isopropyl β-D-1-thiogalactopyranoside (IPTG) (Saveen), 1 M solution, filter sterilized.
- Kanamycin monosulfate (Sigma), 50 mg/ml stock solution, filter sterilized.
- Lemo21(DE3) can be obtained from Xbrane Bioscience AB (www.xbrane.com) or from New England Biolabs as competent cells (http://www.neb.com/nebecomm/products/productC2528.asp).
- L-Rhamnose (Sigma).
- Lysogeny broth (LB medium) (Difco) LB medium is usually referred to as Luria Bertani broth.
- $MgCl_2$, 1 M solution.
- *N,N*-Dimethyldodecylamine *N*-oxide (lauryldimethylamine-oxide (LDAO)) (Fluka), 10% (w/v).
- *n*-Decyl-β-D-maltopyranoside (DM) (Anatrace), 20% (w/v).
- *n*-Dodecyl-β-D-maltopyranoside (DDM) (Anatrace), 20% (w/v).

- Ni-NTA Superflow (Qiagen).
- *n*-Octyl-β-D-glucopyranoside (nOG) (Anatrace), 20% (w/v).
- *n*-Undecyl-β-D-maltopyranoside (UDM) (Anatrace), 20% (w/v).
- Phosphate-buffered saline (PBS): 1.44 g $Na_2HPO_4 \cdot 2H_2O$ (8.1 mM phosphate), 0.25 g KH_2PO_4 (1.9 mM phosphate), 8.00 g NaCl, 0.2 g KCl in 1000 ml H_2O; adjust pH to 7.4 using 1 M NaOH or 1 M HCl.
- Pefabloc SC (Biomol).
- Solubilization buffer (SB): 200 mM Tris-HCl, pH 8.8, 20% glycerol, 5 mM EDTA, pH 8.0, 0.02% Bromphenol Blue, make aliquots of 700 µl and keep at −20 °C. Before use, add 200 µl 20% SDS and 100 µl 0.5 M dithiothreitol.
- Triton X-100 (Sigma), 20% (v/v).

14.3.2 Equipment

- ÄKTAprime or higher Äkta system (GE Healthcare).
- Beckman pollyallomer 1.5-ml microcentrifuge tubes.
- Beckman TLA100 benchtop ultracentrifuge equipped with Beckman TLA100 rotor.
- Centricon centrifugal filter unit (Millipore); cut-off 30 000, 50 000, and 100 000 nominal molecular weight limit) depending on size of protein and detergent.
- LAS-1000 CCD camera system (Fujifilm).
- Nunc 96-well optical bottom plate, black (Nunc).
- Poly-Prep chromatography columns (Bio-Rad).
- Shaking incubator with temperature control (we use New Brunswick Innova 44 shakers).
- SpectraMaxGeminiEM microplate spectrofluorometer (Molecular Devices).
- Superdex 200 10/300 GL Tricorn gel filtration column (GE Healthcare).
- Thermomixer comfort (Eppendorf) equipped with thermoblocks for 1.5-ml microcentrifuge tubes.
- Tunair 2.5-l baffled shaker flasks.
- UV-1601 UV/Vis spectrophotometer (Shimadzu).
- XK 16/20 column (GE Healthcare) or larger column.

14.4 Expression and Isolation of GFP-His$_8$

Both whole-cell fluorescence and in-gel fluorescence can be used to estimate expression yields and require purified GFP-His$_8$ as a reference. For a detailed protocol, see [4]. For the sake of completeness, we outline a slightly modified version of the protocol below.

1) Transform Lemo21(DE3) with a plasmid encoding GFP fused to a His_8 purification tag. We standard use pET20bGFP-His_8 (Amp^R) [4].

2) When using Lemo21(DE3) to express GFP-His_8 add L-rhamnose to a final concentration of 750 μM, chloroamphenicol (34 μg/ml), and ampicillin (100 μg/ml) [8].

3) Express GFP-His_8 for approximately 8 h as described in Steps 16–18 and process the cells according to Steps 19–22. Proceed with the supernatant rather than the pellet from Step 22.

4) Purify GFP-His_8 according to the purification procedure outlined in Steps 31–41 in the main protocol.

Note: Wash the Ni-NTA column with 20 column volumes of 10% Buffer B and elute with 50% Buffer B.

5) Pool the major GFP-His_8-containing fractions (as determined by fluorescence) and dialyze overnight in Buffer C. As soluble GFP-His_8 is expressed to very high yields and serves as a reference only it is not essential to retain all of it, but the protein should contain as little contaminants as possible.

6) Determine the protein concentration using a BCA assay according to the instructions of the manufacturer and measure GFP fluorescence from 0.01 to 0.3 mg/ml GFP-His_8. Check the purity of GFP-His_8 by using standard SDS–PAGE followed by Coomassie staining/destaining.

7) Plot the GFP fluorescence versus the protein concentration and use the slope of the plot to convert the GFP fluorescence from any 100-μl sample to mg/ml of GFP-His_8.

8) Estimate expression yields by dividing the molecular weight of the expressed membrane protein–GFP fusion by 28 kDa (molecular weight of GFP-His_8) and multiply the obtained value with the amount of GFP-His_8 as determined in the previous step.

Note: GFP fluorescence is dampened by whole cells/membranes. Whole cells dampen the GFP fluorescence by a factor of approximately 1.5, membranes by a factor of 1.3 [4].

14.5 Conclusions

GFP-fusions and Lemo21(DE3) form the ideal combination for the rapid optimization of membrane protein production in *E. coli*. It should be noted that GFP-fusions can also be used for the optimization of soluble proteins in the cytoplasm of *E. coli* [21], and that Lemo21(DE3) can also be used for optimizing the expression of soluble proteins in both the cytoplasm and periplasm.

Acknowledgments

Research in the laboratory of J.W.de G. is supported by grants from the Swedish Research Council, the Carl Tryggers Stiftelse, the Marianne and Marcus Wallenberg Foundation, National Institutes of Health grant 5R01 GM081827-03, and the SSF-supported Center for Biomembrane Research. D.D. acknowledges the support of the Royal Society (UK) through a University Research Fellowship.

Abbreviations

CHAPS	3((3-Cholamidopropyl)dimethylammonio)propanesulfonic acid
CMC	critical micellar concentration
DDM	*n*-Dodecyl-β-D-maltopyranoside
DM	*n*-Decyl-β-D-maltopyranoside
FSEC	fluorescence-detection size-exclusion column chromatography
GFP	Green Fluorescent Protein
HPLC	high-performance liquid chromatography
IMAC	immobilized metal-affinity chromatography
IPTG	Isopropyl β-D-1-thiogalactopyranoside
LB	Lysogeny broth
LDAO	lauryldimethylamine-oxide
nOG	*n*-Octyl-β-D-glucopyranoside
PAGE	polyacrylamide gel electrophoresis
PBS	Phosphate-buffered saline
RNAP	RNA polymerase
SB	Solubilization buffer
SDS	sodium dodecyl sulfate
SEC	size-exclusion column chromatography
TEV	tobacco etch virus
UDM	*n*-Undecyl-β-D-maltopyranoside

References

1 Drew, D.E., von Heijne, G., Nordlund, P., and de Gier, J.W. (2001) *FEBS Lett.*, **507**, 220–224.

2 Drew, D., Slotboom, D.J., Friso, G., Reda, T., Genevaux, P., Rapp, M., Meindl-Beinker, N.M., Lambert, W., Lerch, M., Daley, D.O., Van Wijk, K.J., Hirst, J., Kunji, E., and de Gier, J.W. (2005) *Protein Sci.*, **14**, 2011–2017.

3 Drew, D., Sjostrand, D., Nilsson, J., Urbig, T., Chin, C.N., de Gier, J.W., and von Heijne, G. (2002) *Proc. Natl. Acad. Sci. USA*, **99**, 2690–2695.

4 Drew, D., Lerch, M., Kunji, E., Slotboom, D.J., and de Gier, J.W. (2006) *Nat. Methods*, **3**, 303–313.

5 Newstead, S., Kim, H., von Heijne, G., Iwata, S., and Drew, D. (2007) *Proc. Natl. Acad. Sci. USA*, **104**, 13936–13941.

6 Drew, D., Newstead, S., Sonoda, Y., Kim, H., von Heijne, G., and Iwata, S. (2008) *Nat. Protoc.*, **3**, 784–798.

7 Studier, F.W. and Moffatt, B.A. (1986) *J. Mol. Biol.*, **189**, 113–130.

8 Wagner, S., Klepsch, M.M., Schlegel, S., Appel, A., Draheim, R., Tarry, M., Hogbom, M., van Wijk, K.J., Slotboom, D.J., Persson, J.O., and de Gier, J.W. (2008) *Proc. Natl. Acad. Sci. USA*, **105**, 14371–14376.

9 Giacalone, M.J., Gentile, A.M., Lovitt, B.T., Berkley, N.L., Gunderson, C.W., and Surber, M.W. (2006) *Biotechniques*, **40**, 355–364.

10 Wagner, S., Baars, L., Ytterberg, A.J., Klussmeier, A., Wagner, C.S., Nord, O., Nygren, P.A., van Wijk, K.J., and de Gier, J.W. (2007) *Mol. Cell. Proteomics*, **6**, 1527–1550.

11 Feilmeier, B.J., Iseminger, G., Schroeder, D., Webber, H., and Phillips, G.J. (2000) *J. Bacteriol.*, **182**, 4068–4076.

12 Nugent, T. and Jones, D.T. (2009) *BMC Bioinformatics*, **10**, 159.

13 Bernsel, A., Viklund, H., Hennerdal, A., and Elofsson, A. (2009) *Nucleic Acids Res.*, **37**, W465–W468.

14 Hsieh, J.M., Besserer, G.M., Madej, M.G., Bui, H.Q., Kwon, S., and Abramson, J. (2010) *Protein Sci.*, **19**, 868–880.

15 Kawate, T. and Gouaux, E. (2006) *Structure*, **14**, 673–681.

16 Gonzales, E.B., Kawate, T., and Gouaux, E. (2009) *Nature*, **460**, 599–604.

17 Kawate, T., Michel, J.C., Birdsong, W.T., and Gouaux, E. (2009) *Nature*, **460**, 592–598.

18 Guan, L., Mirza, O., Verner, G., Iwata, S., and Kaback, H.R. (2007) *Proc. Natl. Acad. Sci. USA*, **104**, 15294–15298.

19 Drew, D., Klepsch, M.M., Newstead, S., Flaig, R., de Gier, J.W., Iwata, S., and Beis, K. (2008) *Mol. Membr. Biol.*, **25**, 677–682.

20 Strop, P. and Brunger, A.T. (2005) *Protein Sci.*, **14**, 2207–2211.

21 Waldo, G.S., Standish, B.M., Berendzen, J., and Terwilliger, T.C. (1999) *Nat. Biotechnol.*, **17**, 691–695.

Index

Page numbers in *italics* refer to entries in figures or tables.

Production of Membrane Proteins: Strategies for Expression and Isolation, First Edition.
Edited by Anne Skaja Robinson.

d

e

k

l